Cell Biology Monographs

Volume 8

Springer-Verlag
Wien New York

Cytomorphogenesis in Plants

Edited by O. Kiermayer

Springer-Verlag
Wien New York

Prof. Dr. Oswald Kiermayer
Botanical Institute, University of Salzburg, Austria

With 202 Figures

Library of Congress Cataloging in Publication Data.
Main entry under title: Cytomorphogenesis in plants. (Cell biology monographs; v. 8)
Bibliography: p. Includes index.
1. Plant cytomorphogenesis. I. Kiermayer, O. (Oswald), 1930. II. Series.
QK725.C96.581.87.81-14354

ISSN 0172-4665

ISBN 3-211-81613-5 Springer-Verlag Wien — New York
ISBN 0-387-81613-5 Springer-Verlag New York — Wien

Preface

In 1958 E. Bünning published a book in the former series "Protoplasmatologia" entitled "Polarität und inäquale Teilung des pflanzlichen Protoplasten" (polarity and unequal division of the plant protoplast) in which for the first time results of experimental plant cytomorphogenesis were reviewed. This book was based completely on light microscopic observations and rather simple experimental techniques. Since then our knowledge of basic cytomorphogenetic mechanisms has greatly increased, especially with the introduction of modern ultrastructural, biochemical and sophisticated experimental methods so that the field of cytomorphogenesis in our days should be considered a separate discipline within the general field of cell biology.

This book, "Cytomorphogenesis in Plants", represents a necessary attempt to bring together current knowledge in this field of research on a comparable basis. Unfortunately enormous gaps in our understanding of the underlying principles of cytomorphogenetic events still exist. Therefore it seemed reasonable to present a book composed of individual chapters, each written by experts for a defined experimental system. Each chapter represent a separate treatise with its own references, hence it was not possible to avoid some overlap both in the text and in the literature of the chapters without destroying the uniformity of the respective article.

The term "cytomorphogenesis" is used in this book in a broad sense. Not only is the formation of the outer cell shape included in this term but also all other formative processes, even those occurring on the macromolecular and ultrastructural level during the differentiation of a cell. Thus, for example, the formation of different macromolecular patterns on bacterial cell walls, the formation of ultrastructural patterns on the scales of flagellates and cell walls of diatoms and green algae, as well as microfibril formation and orientation are considered as cytomorphogenetic processes and are described in this book.

The book is subdivided into three sections. The first two sections are entitled "I. Cytomorphogenesis in Unicellular Plants and Cell Aggregates" and "II. Cytomorphogenesis in Multicellular Plants". These two sections contain articles about the cytomorphogenesis of various well-studied plant cell systems (flagellates, *Acetabularia*, *Micrasterias*, *Nitella*, *Azolla* and others) arranged according to their taxonomic position within the plant kingdom. A third section, entitled "III. General Aspects of Cytomorphogenesis in Plants", is dedicated to discussing the role of electric currents in cytomorphogenesis as well as the role of lipid self-assembly in subcellular morphogenesis.

The main emphasis in the book has been put on causal cytoplasmic aspects of morphogenesis of plant cells and the ultrastructural events occurring during cell differentiation, leading to specific forms and patterns, whereas ontogenetic and genetic aspects are treated only to a small extent.

Most chapters deal with the formation of the cell wall, which normally determines the exterior shape of the plant cell. However, included into the entire process of cytomorphogenesis of plant cells is not only the formation of the external cell envelope but also the internal shaping and arrangement of the organelles, especially the nucleus and the chloroplast, often leading to an elaborate intracellular architecture. To establish this architecture, a number of intracellular morphogenetic processes occur, such as migration and local anchoring of organelles. During the last 15 years it has become obvious that many formative processes are controlled by cytoskeletal elements such as microtubules and microfilaments and that the plasma membrane and membrane systems such as the Golgi-complex and the endoplasmic reticulum play a most important functional role. Therefore, nearly all chapters deal to a certain degree with these cytological elements.

Although quite a number of important formative processes taking place in various plant cell systems could be elucidated during the last years, many basic facts of form and pattern formation are still unknown. We hope that this book, which presents for the first time a survey of our present knowledge on plant cytomorphogenesis, will help to increase the interest of more scientists in this highly interesting and fascinating field of research. This book also makes clear the present methodical limitations restricting a deeper understanding and search for the basic principles of cytomorphogenesis. There is no doubt that more inventive ideas and the introduction of the most advanced cytological techniques, especially for the study of membranogenesis and cytoskeletal structures and functions, will be necessary for a further search into cytomorphogenesis, which still belongs to one of the most mysterious events in biology.

I am most grateful to the authors of the chapters for the realization of this book. Without their appreciative cooperation this book would not have been completed. I am also thankful to professors R. M. BROWN (Chapel Hill), P. GREEN (Stanford), P. K. HEPLER (Amherst), and J. D. PICKETT-HEAPS (Boulder) for various suggestions, and to my coworkers Dr. URSULA MEINDL and SUSANNE HAMPL for various aids during the preparation of the manuscript. I am also thankful to Springer-Verlag in Vienna, especially to Dr. W. SCHWABL and Dr. ERNA UNGERSBÄCK for their interest and support of this publication and for the high printing quality of the many illustrations.

Salzburg, August 1981   OSWALD KIERMAYER

I. Cytomorphogenesis in Unicellular Plants and Cell Aggregates

Morphopoietic and Functional Aspects of Regular Protein Membranes Present on Prokaryotic Cell Walls

U. B. SLEYTR

Centre for Ultrastructure Research, University of Agriculture, Wien, Austria

With 9 Figures

Contents

I. Introduction

Most bacteria and cyanobacteria (blue-green algae) possess a supramolecular layered cell wall structure outside the cytoplasmic membrane. The cell wall and the cytoplasmic membrane constitute the prokaryotic cell envelope complex, which regulates the molecular exchange between the cell and its environment. The cell wall also functions in most organisms as an osmotic protection barrier by providing a rigid wall structure.

Until now, the cell envelope ultrastructure in a great variety of organisms has been investigated by electron microscopy of thin sectioning, freeze-etching, shadowing, and negatively stained preparations (SLEYTR and GLAUERT 1981). Although these studies have revealed a considerable variation in both complexity and structure, most prokaryotic cell wall profiles can be classified into three main categories as illustrated in Fig. 1. When viewed in thin sections, the cell walls of the majority of gram-positive bacteria appear as 15 to 85 nm thick, rigid, rather amorphous layers (Fig. 1 a). Depending on the staining and fixation procedures, the dimensions and appearance of the cell wall can vary, and indications of layering have been reported (BEVERIDGE and WILLIAMS 1978, GLAUERT and THORNLEY 1969). Chemical analysis has

1*

shown that with the exception of the gram-positively staining methanogenic bacteria (BALCH et al. 1979), peptidoglycan is the major component of the cell wall. Teichoic acids, teichuronic acids, and proteins in varying amounts may be found interspersed as additional components.

In contrast to the gram-positive cell wall profile, the gram-negative cell walls generally have a more complex structure (Fig. 1 b). They consist essentially of a unit membrane-like layered structure 7 to 9 nm thick (outer membrane) containing lipopolysaccharide, protein, and phospholipid, and a structureless electron-opaque dense layer of variable thickness in which the peptidoglycan is located (COSTERTON et al. 1974).

Although not a universal feature of all gram-positive and gram-negative bacteria, regular arrays of proteins or glycoproteins have been detected as an additional surface layer (S-layer) outside the formal cell wall in a wide range of species (THORNE 1977, SLEYTR 1978). Some Archaebacteria (organisms from diverse and highly specialized niches), including thermoacidophilic, halophilic, and a few methanogenic bacteria that have no rigid murein or pseudomurein cell wall layer (KANDLER 1979, KÖNIG and KANDLER 1978, WOESE et al. 1978), possess subunit type cell walls composed exclusively of two- dimensional arrays of macromolecules (Fig. 1 c). Recently, it was shown that S-layers may also be present in Cyanobacteria (KESSEL 1978, LOUNATMAA et al. 1980), and in a few gram-negative bacteria, more than one S-layer could be identified external to the outer membrane (THORNE 1977).

Since most ultrastructural studies on bacterial cell walls have been done by thin sectioning—a method less suitable than freeze-etching and negative staining for demonstrating S-layers—the presence of regular arrays of macromolecules may be a much more common feature than is generally realized (SLEYTR 1978). Furthermore, it has been shown that S-layers are often lost from wild-type bacteria when grown in serial culture in the laboratory (BUCKMIRE 1971, STEWART and BEVERIDGE 1980). Ultrastructural studies on intact bacteria, particularly those studies using freeze-etching techniques, have shown that S-layers cover the whole cell leaving no gaps. As outermost cell wall layers, they represent the primary interface between the cell and its environment, and accumulated data indicate that these regular arrays of proteins may help to protect the cell in a hostile environment (SLEYTR 1978). Thus, many interesting questions of general significance in connection with the biological function and the morphogenesis of S-layers can be asked.

This chapter gives major consideration to the assembly process that guarantees the maintenance of the highly ordered continuous (closed) two-dimensional arrays of macromolecules on growing and dividing cell surfaces. In a more speculative section, possible functional aspects of these regular cell surface protein membranes will be discussed.

II. Ultrastructure of Regular Surface Layers

Regular cell wall surface layers were first detected by heavy metal shadowing (HOUWINK 1953). At present, the best method for demonstrating S-layers on intact cells is freeze-fracture replication of deep-etched cell suspensions.

Unless unstructured amorphous material is present on the cell surfaces, this low temperature technique will give the closest picture of the *in vivo* structure and orientation of S-layers (SLEYTR and GLAUERT 1975). Nevertheless, most high resolution studies have been carried out on isolated S-layers using negatively stained or freeze- or air-dried and shadowed preparations.

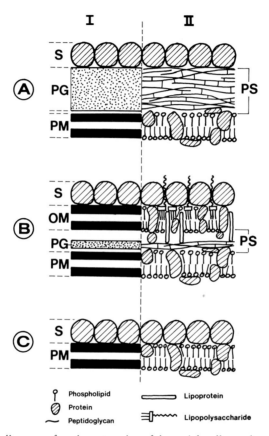

Fig. 1. Schematic diagram of main categories of bacterial cell envelopes including regular cell surface layers. I. Thin section profiles. II. Molecular architecture showing major components: *A* gram-positive cell envelope; *B* gram-negative cell envelope; *C* cell envelope structure of *Archaebacteria* that lack a rigid cell wall component: *OM* outer membrane; *PG* peptidoglycan containing layer; *PM* plasma membrane; *PS* periplasmatic space; *S* surface layer.

From these studies, most regular arrays are known to have hexagonal (Fig. 2 *a*) or tetragonal (Fig. 2 *b*) symmetry with a centre-to-centre spacing between adjacent morphological units within the range 5 to 20 nm. Some organisms show oblique translational S-layer lattices (Fig. 2 *c*) in which the subunits are arranged with different centre-to-centre spacings in rows lying at

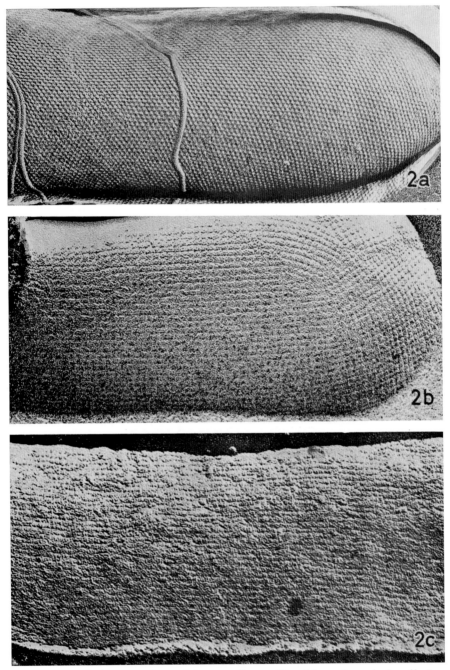

Fig. 2. Freeze-etched preparations of intact cells: *a Clostridium thermohydrosulfuricum* strain L 111/69 with a hexagonal array of subunits (×110,000); *b Bacillus sphaericus* strain S 66/67 with a tetragonally ordered S-layer (×140,000); *c B. stearothermophilus* strain L 32/65 showing an oblique translational surface lattice (×155,000).

an angle of less than 90° and more than 60° to each other. Negatively stained and shadowed S-layer preparations have also been analyzed by optical diffraction and filtration, electron diffraction, and digital computer Fourier transform and filtration methods, achieving a resolution of 1.5 to 2.5 nm. The averaged images show clear tetrameric morphological units in the tetragonal lattice and hexameric units in the hexagonal lattice (AEBI et al. 1973, CROWTHER and SLEYTR 1977, KÜBLER et al. 1980, STEWART and BEVERIDGE 1980, STEWART et al. 1980). Generally, in vitro S-layer self-assembly products (see section IV.) show an improved lattice long-range order when compared with the S-layer fragments of intact cells. Image processing procedures,

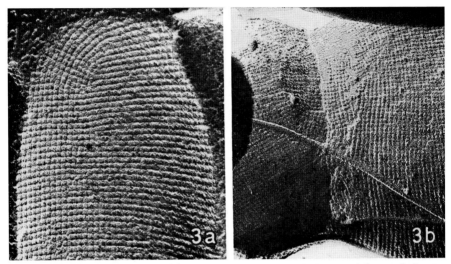

Fig. 3. Freeze-etched preparations of intact cells showing distortions and lattice faults at a cell pole (a) and a septation site (b): a Bacillus stearothermophilus strain 1 b₂/106 (×100,000); b Clostridium thermosaccharolyticum strain L 111-69 (×95,000).

including linear integration, optical filtering and digital filtering by computer, are best applied to these recrystallized assemblies, although methods have been developed that take account of and correct the spacial distortions prior to averaging (CROWTHER and SLEYTR 1977). The degree of information contained in regular arrays of identical subunits allows the application of low electron dose techniques that drastically reduce beam induced specimen damages (UNWIN and HENDERSON 1975, KELLENBERGER and KISTLER 1979). Finally, methods are available that allow computation of the topographical map of S-layer surfaces using digitized micrographs of metal shadowed preparation (SMITH and KISTLER 1977, SMITH and IVANOV 1980).

Freeze-etching of many gram-positive and gram-negative bacteria have shown that S-layer patterns are generally uniform over large areas of the cylindrical part of rod-shaped cells. Rectangular arrays are usually arranged with one axis nearly parallel to the long axis of the rod (Fig. 2 b), but a specific skew angle may be characteristic for certain strains (SLEYTR and

Glauert 1975, Sleytr 1978). More variations are observed in the alignment of the hexagonal arranged lattices, but there is a tendency for one axis to be approximately parallel to the long axis of the rod (Fig. 2 *a*) (Hollaus and Sleytr 1972, Sleytr and Glauert 1975).

In contrast to the cylindrical part, an accumulation of faults can be observed in the tetragonally or hexagonally ordered lattices occurring at the cell poles (Fig. 3 *a*) and septation sites (Fig. 3 *b*). In these areas, the orientation of the pattern frequently changes, giving the appearance of a mosaic composed of small crystallites. These morphological data obtained on cryofixed intact cells are in accordance with theoretical considerations which indicate that a cylinder can be covered with a hexagonally or tetragonally ordered array of macromolecules without any lattice faults. On the other hand, it is necessary to introduce lattice faults for covering the rounded surfaces at the cell poles and at the septation sites of a rod-shaped cell or the spherical surface of a coccus. Nevertheless, lattice faults in these areas have been shown to be much more numerous than the theoretical minimum required for a complete covering. By evaluating the local frequency of lattice faults at septation sites and cell poles, it was found that crystallites covering the cell poles are generally larger. This observation gave the first indication that S-layer crystallites may rearrange and fuse on the cell surface to form larger areas of regular patterns (Sleytr and Glauert 1975, Sleytr 1976).

III. Chemical Characterization and Subunit Bonding Properties

Various methods have been applied for the isolation and purification of the regular arranged surface layers (Thorne 1977, Sleytr 1978). Generally, the procedure for gram-positive cell walls consists of breaking the cell in a French Press or by ultrasonication. The crude cell wall preparations then obtained by differential centrifugation are subsequently treated with Triton X-100 to dissolve plasma membrane contaminants (Sleytr and Thorne 1976, Beveridge 1979).

Intact S-layer fragments are best obtained by digesting the supporting peptidoglycan containing layer with lysozyme (Sleytr 1976, Beveridge 1979). S-layer fragments have also been stripped off the supporting layer as sheets by incubation of either intact cells or cell wall fragments in low concentrations of urea (0.5 M) or guanidine hydrochloride (1 M) (Nermut and Murray 1967). Both a complete disintegration of S-layer fragments and a removal of S-layer subunits from isolated cell wall preparations is generally obtained by treatments with high concentrations of chaotropic agents such as urea (8 M) or guanidine hydrochloride (5 M). Most isolated S-layers will disintegrate into their subunits by lowering the pH ($<$ 3), but only in a few organisms can the S-layer be extracted from the supporting layer using acidic conditions (Plohberger and Sleytr 1980).

In gram-negative organisms, the surface layers have been detached from the outer membrane of isolated cell wall fragments by a variety of procedures (Thorne 1977). The method used included treatment with 1 M urea (Thornley *et al.* 1974 b), 0.5 M guanidine hydrochloride (Buckmire and

MURRAY 1970), metal chelating agents such as EDTA or EGTA (WATSON and REMSEN 1969), cation substitution (BEVERIDGE and MURRAY 1976 a, c) and acid treatment (BEVERIDGE and MURRAY 1976 b, c). Other gram-negative microorganisms lose patches of their regular surface structure during growth (BEVERIDGE and MURRAY 1975), and these can be isolated from the medium.

Chemical analyses of S-layer subunits isolated from gram-positive and gram-negative cell walls have shown that they are mainly composed of single homogeneous polypeptides with carbohydrates being occasionally present as a minor component. The molecular weights vary from 40,000 to 200,000 daltons. The amino acid composition of the subunits is predominantly acidic, as has been confirmed for several purified S-layers using isoelectric focusing (THORNE 1977, SLEYTR 1978, HASTIE and BRINTON 1979 a, SLEYTR and PLOHBERGER 1980).

Until now, most chemical characterizations have been done on the S-layers of single selected strains of any given species and occasionally their related mutants. Recent studies performed to elucidate the distribution and the degree of uniformity of S-layers within thermophilic spore-forming bacteria (*Bacillus stearothermophilus, B. coagulans, B. thermodenitrificans, B. sphaericus, Clostridium thermohydrosulfuricum, C. thermosaccharolyticum, C. tartarivorum, Desulfotomaculum nigrificans*) have shown that the individual strains of a species exhibit a remarkable heterogeneity in the morphology and chemistry of their regular arrays. Subunit species with molecular weights ranging from 100,000 to 180,000 daltons, with or without carbohydrate residues, have been observed (SLEYTR and PLOHBERGER 1980, PLOHBERGER and SLEYTR 1980). This heterogeneity is consistent with the amino acid composition of the S-layers determined for seven selected strains of *B. stearothermophilus* (PLOHBERGER *et al.* 1981). On the other hand, mutants of *B. sphaericus* with modified S-layer protomers showed little difference in the molar ratios of their constituent amino acids when compared with the protomers of the wild type strain (HOWARD and TIPPER 1973).

So far, only a few cell walls of *Archaebacteria* species, composed exclusively of regularly arranged subunits (Fig. 1 c), have been chemically characterized. SDS-polyacrylamide gel electrophoresis of isolated *Sulfolobus acidocaldarius* S-layer cell walls gave two glycoprotein bands with apparent molecular weights of 140,000 and 170,000 daltons (MICHEL *et al.* 1980). The authors believed that both bands resulted from only one protein species that can occur in two forms with different molecular weights: this would correlate with chemical studies on the obligate halophilic *Archaebacterium, Halobacterium salinarium*, which possesses a cell wall structure composed of a hexagonally arranged 200,000 dalton glycoprotein subunit (MESCHER and STROMINGER 1976).

Valuable information about the nature of the chemical bonds between S-layer monomers or between the monomers and the supporting layer can be obtained from consideration of the reagents that cause detachment and/or disintegration of the regular arrays. The available data show that S-layer subunits on both gram-positive and gram-negative envelopes are noncovalently linked to each other and to the underlying wall component. S-layer extraction and disintegration experiments further suggest that the bonds holding the lattice protomers together differ somewhat from those binding the S-layer lattice to the supporting envelope layer (SLEYTR 1975, 1978, THORNE 1977, HASTIE and BRINTON 1979 b).

Hydrogen bonding appears to be important for the stability of the S-layers of most gram-positive organisms since high concentrations of chaotropic agents cause a complete dissociation of the macromolecular layers into their component protomers. At low concentrations, these agents frequently loosen only the bonds between the S-layer subunits and the supporting peptidoglycan-containing layer, which allows the intact S-layer fragment to be detached (NERMUT and MURRAY 1967, SLEYTR 1978). Further indications of the differences between the intersubunit bonds of the S-layer and those bonds between the S-layer and the supporting layer are obtained from low pH treatments. When cell walls of *Bacilli* or *Clostridia* were treated with buffer at a pH below the isoelectric point of the S-layer subunit, the regular arrays were no longer visible in negatively stained or freeze-etched preparations, and no periodicity was detectable by optical diffraction analysis of electron micrographs. No protein was lost, however, and the regular S-layer pattern reappeared very clearly when the pH was adjusted to neutral. This showed that acid conditions that, in general, completely dissociate isolated S-layers into protomers can only cause an uncoiling of the S-layer monomers and do not result in their detachment from the layer that contains the peptidoglycan (SLEYTR 1975, SLEYTR and GLAUERT 1976).

Recently, MASUDA and KAWATA (1980) have attempted to determine which chemical component of the cell wall of *Lactobacilli* is responsible for the attachment of S-layer subunits. Homologous and heterologous reattachment experiments with mixtures of isolated subunits from *Lactobacillus brevis* and chemically treated cell wall fragments from the same organism and from *Lactobacillus fermentum, L. plantarum,* and *L. casei,* suggested that the neutral polysaccharide of the cell wall, and not the teichoic acid or the peptidoglycan, is responsible for the subunit binding.

Besides hydrogen bonds, ionic bonds involving divalent cations seem to be responsible for the stability of S-layers in many gram-negative species and for their attachment to the supporting outer membrane layer. In *Acinetobacter 199 A* (THORNLEY et al. 1974 b, THORNE et al. 1975), it was shown that Ca^{2+} is required for the formation of salt bridges between the carboxyl groups of the S-layer protomers and the outer membrane protein. Similarly, the binding of the S-layer protein of *Spirillum serpens* (BUCKMIRE and MURRAY 1973, 1976), *S. putridiconchylium* (BEVERIDGE and MURRAY 1976 b, c), and *S. anulus* strain "ordal" (BEVERIDGE and MURRAY 1976 a) to the outer membrane requires the presence of divalent cations. Particularly strong intermolecular bonds were reported for the hexagonally patterned monolayer of protein found in association with the outer membrane-like backing layer of *Micrococcus radiodurans* (KÜBLER et al. 1980, BAUMEISTER and KÜBLER 1978). This S-layer proved to be resistant to both nonionic and ionic detergents, to high concentrations of chaotropic and chelating agents, and to various organic solvents. To dissociate the regular array of protein, treatments with SDS above 90 °C or under acidic conditions (pH < 2) were necessary. SDS polyacrylamide gel electrophoresis of S-layer material led to their disintegration and yielded major bands corresponding to molecular weights of 100,000, 97,000, and 74,000 in a 1 : 3 : 3 stoichiometry (KÜBLER et al. 1980). Finally, special intersubunit bonding properties have been observed in the regulary arranged proteinaceous cell walls of certain *Archae-bacteria* species adapted to extreme environments. For example, the S-layer of *Sulfolobus acidocaldarius* (MICHEL et al. 1980) withstands environmental pH values as low as 0.9, and *Halobacteria* cell surfaces remain stable during contact with high molar salt concentrations (MESCHER and STROMINGER 1976).

IV. Self-Assembly and Recrystallization of Regular Surface Layers

Isolated S-layer protomers from numerous gram-positive and gram-negative bacteria have shown the ability to assemble into regular arrays with the same lattice dimensions as those observed in intact cells. If this assembly process takes place in the absence of a supporting layer, it is referred to as "self-assembly", whereas regular arrays formed on the surfaces of cell wall layers from which the subunits have been originally removed are described as "reattached" or "reconstituted" S-layers.

In most systems, *in vitro* self-assembly of S-layer subunits can be induced by removing the disrupting agents used for their isolation. S-layer subunits of gram-positive and gram-negative bacteria generally reaggregate either into flat sheets (Fig. 4 *a*), open ended cylinders (Figs. 4 *b–d*), or closed vesicles (Fig. 4 *e*) when low pH, guanidine hydrochloride, urea, or formamide extracts are dialyzed against water or buffer at the correct pH and ionic strength (SLEYTR 1976, 1978).

These self-assembly properties were first demonstrated for the isolated protomers of the tetragonally arranged S-layers present on the cell wall of *Bacillus brevis* P-1 (later reclassified as *Bacillus sphaericus* P-1 (HOWARD and TIPPER 1973). Low pH treatment caused the isolated regular arrays to dissociate, and, on neutralization, planar sheet-like aggregates were formed (BRINTON et al. 1969). Treament of intact S-layers with proteolytic enzymes (*e.g.,* pronase) reduced the molecular weight by 18,000 (AEBI 1973, HASTIE and BRINTON 1979 b). Native S-layer subunits modified in this way assembled into open-ended cylinders instead of into sheets but retained the biological specifity of the native subunits in that they had the same spectrum of phage neutralization and specific antiserum reactions (HENRY 1972). In other studies, it was shown that isolated S-layer subunits of a variety of *Bacilli* and *Clostridia* have the ability to reassemble into open-ended cylinders without previous proteolysis (SLEYTR 1976, SLEYTR and PLOHBERGER 1980). Self-assembly cylinders are mostly formed from subunits that have the ability to aggregate into rectangular or oblique lattices.

Although tetragonally ordered S-layers on rod-shaped cells may exhibit a strain-specific pitch angle, usually one row of the lattice will run parallel to the longitudinal axis of the cell. In contrast, most self-assembly cylinders have lattice orientations that differ from those observed on intact cells. In *Bacillus sphaericus*, the difference in lattice orientation between the *in vivo* and *in vitro* cylinders was shown to be approximately 39° (AEBI 1973, HASTIE and BRINTON 1979 a). Similarly, S-layer subunits from most strains of *Bacillus stearothermophilus* and *Clostridium thermosaccharolyticum* have the tendency to assemble into cylinders in which the rows lie at an angle of about 45°, although freeze-etching preparations of intact cells reveal tetragonal

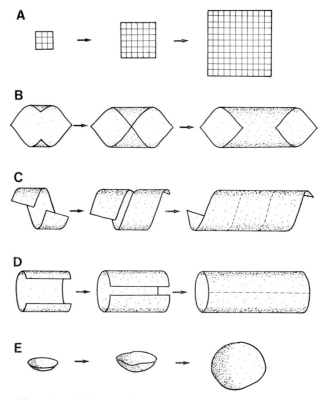

Fig. 4. Diagram illustrating different self-assembly routes of S-layer subunits leading to the formation of *A* sheets, *B, C, D* cylinders; and *E* spheres. Modified diagram from SLEYTR and PLOHBERGER 1980.

arrays with one axis parallel to the rod (HOLLAUS and SLEYTR 1972, SLEYTR and PLOHBERGER 1980, SLEYTR 1976). With the S-layer protomers of some thermophilic organisms, different assembly products were obtained when guanidine hydrochloride, urea, or low pH extracts were dialyzed at 4 or 60 °C. Furthermore, different assemblies could be created by changing the ionic strength or by dialyzing in the presence of bivalent cations. Since, with one strain of *Bacillus stearothermophilus*, conversion from flat aggregates to S-layer cylinders was reversible and was not accompanied by a change in the molecular weight of the lattice protomers, the possibility of a proteolytic cleavage, which accompanies the cylinder formation in *Bacillus sphaericus*

(AEBI et al. 1973, HASTIE and BRINTON 1979 a), can be excluded. These experimental data indicate that depending on the assembly conditions, chemically identical S-layer subunits may have the ability to aggregate into either sheets or cylinders (SLEYTR and PLOHBERGER 1980, PLOHBERGER and SLEYTR 1980).

Several experiments have shown that under proper conditions, S-layer fragments have a strong inclination to fuse and to recrystallize into arrangements having lower lattice-free energies. Self-assembly products such as cylinders

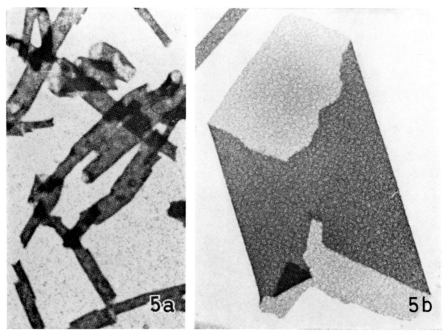

Fig. 5. a Negatively stained preparation of small S-layer fragments obtained by lysozyme digestion of cell wall fragments of Bacillus stearothermophilus strain NRS 2004 (×30,000); b cylindrical fusion product obtained upon incubation of fragments at 60 °C (×30,000).

or sheets, mechanically disintegrated by g-forces or by brief ultrasonication, may fuse at the onset of incubation. Similarly, when the peptidoglycan layer of small, irregularly shaped gram-positive cell fragments is digested with muramidases, the liberated S-layer fragments (Fig. 5 a) can recrystallize (Fig. 5 b), forming assembly products identical to those obtained by dialyzing guanidine hydrochloride, urea, or low pH extracts of the S-layer (SLEYTR and PLOHBERGER 1980, PLOHBERGER and SLEYTR 1980, SLEYTR et al. 1980).

As shown in Figs. 4 b–d, different assembly routes may lead to the formation of S-layer cylinders. Theoretically, one could assume that flat sheets are formed originally and, after reaching a critical size, have enough flexibility to curve over so that opposite edges may, by chance, meet and fuse. For square sheets with equal flexibility in all possible directions, the most likely points to meet would be the diagonally opposite corners. This could explain the frequently observed 45° pitch angle of the lattice rows in tetragonal arrays and the oblique ends of the cylinders that are preserved during the subsequent longitudinal cylinder growth. However, most experimental data suggest an alternative mechanism of cylinder formation. High resolution shadowing techniques in combination with image reconstruction procedures (SMITH and

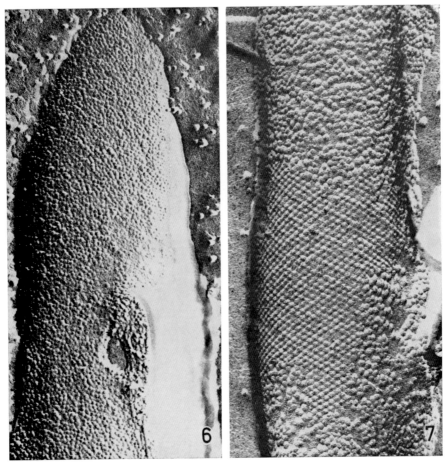

Fig. 6. Freeze-etched preparation of ferritin labeled cells of *B. sphaericus* strain S 66/67. Due to a net-negative surface charge, the regular lattice seen in Fig. 2 *b* has bound a mono-layer of polycationic ferritin (×110,000). From SLEYTR and FRIERS 1978.

Fig. 7. Freeze-etched preparation of *C. thermohydrosulfuricum* strain L 111-69 showing a partially reconstituted S-layer. Polycationic ferritin was only bound at areas where the peptidoglycan layer was exposed (×150,000). From SLEYTR and FRIERS 1978.

KISTLER 1977) and specific labelling experiments using morphologically distinct marker molecules (SLEYTR and PLOHBERGER 1980) have shown that S-layers forming cylinders curve in defined directions. Depending on the subunit species, cylinders may have either a curvature identical or opposite to that of the lattice occurring on intact cells. Thus, the cylinder width and the lattice orientation will depend on the morphology (mass distribution) and bonding properties of the protomers. Fig. 4 illustrates that cylinders with the same diameter can be formed by lattice fusion at diagonally opposed corners (Fig. 4 *b*), opposite parallel sides (Fig. 4 *d*), or in any possible overlapping fashion (Fig. 4 *c*).

Although most S-layer species form monolayers on intact cells, the *in vitro* assembly products may be composed of double sheets. Labelling experiments with polycationic ferritin have shown that the surfaces of S-layers may have different net charges (Figs. 6–8). In the presence of bivalent cations, the

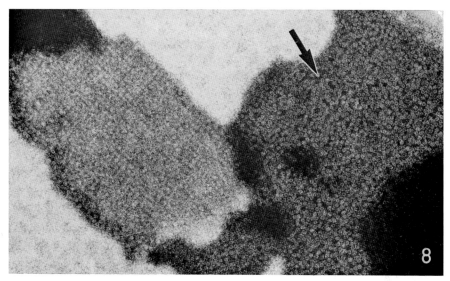

Fig. 8. Sheet-like assembly products of S-layer subunits from *B. stearothermophilus* strain NRS 2004. Polycationic ferritin only binds to one side of the S-layer (arrow). (×110,000).

negatively charged surfaces of two identical S-layer sheets can be linked by salt bridges, thereby creating a double layered assembly product, in which both lattices are superimposed in a well-defined orientation (SLEYTR and PLOH-BERGER 1980). In such double lattices, the polycation ferritin is only bound at the edges or at areas exposing the negatively charged surface of a single sheet.

V. Reattachment of Regular Arrays to Cell Walls

Isolated S-layer subunits or S-layer fragments from a variety of gram-positive and gram-negative bacteria have shown the ability to reattach to the surface of the cell wall from which they have been originally removed (THORNE 1977, SLEYTR 1975, 1976, 1978, SLEYTR and PLOHBERGER 1980, HASTIE and BRINTON 1979 b, MASUDA and KAWATA 1980). Reattachment of regular arrays on gram-positive cell walls can be induced when S-layer subunits dissociated in high molar urea or guanidine hydrochloride solutions are mixed with the peptidoglycan-containing wall fragments and subsequently dialyzed against distilled water or buffer to remove the chaotropic agent (SLEYTR 1978).

In contrast to the reattachment conditions for most gram-positive S-layer subunits, regular arrays that are frequently isolated from the outer membrane of gram-negative envelopes require the presence of bivalent cations and/or a defined ionic strength for a specific binding. This was demonstrated, for example, with S-layers derived from *Spirillum serpens* (BUCKMIRE and MURRAY 1976), *Spirillum putridiconchylium* (BEVERIDGE and MURRAY 1976 b, c), and *Acinetobacter* sp. (THORNE *et al.* 1975).

Isolated S-layer subunits from the gram-positive *Clostridium thermo-hydrosulfuricum* and *C. thermosaccharolyticum* cell walls have been shown to reattach to the peptidoglycan even at pH 2.5, although this pH value will disintegrate isolated S-layer assembly products (section IV). When reattachment experiments are performed at low pH, the regular patterns are not visible in freeze-etching or negatively stained preparations until the pH is subsequently readjusted to neutral. These experimental data suggest that the subunits at low pH reattach to the nucleation surface in a random uncoiled fashion and then reassemble into regular arrays during the change of pH to neutral (SLEYTR 1976). Intact S-layers from both *Clostridia* reattach to the wall fragments of naked sacculi when incubated as a mixture at neutral pH.

Generally, reattachment processes involving intact S-layer fragments or S-layer self-assembly products require that the specific binding sites on the regular arrays be exposed. This will be the case with monolayer sheets but not with cylindrical aggregates that have a reversed lattice curvature nor with double sheets combined at the sides originally facing the cell wall. Frequently, the S-layer reassembly process leads first to the formation of randomly orientated crystallites, indicating that the orientation of the lattice is not determined by any order of the underlying cell wall layer in the binding sites. S-layer fragments or subunits may reattach not only to cell walls from which they were removed (homologous reattachment) but also to those of other strains (heterologous reattachment).

When mixtures of two different S-layer protomers from two suitable organisms are used, both types of lattices can be formed on one cell. Using *C. thermohydrosulfuricum* and *C. thermosaccharolyticum* as a model system, it was shown that the strain specific hexagonal and tetragonal S-layer patterns are formed with equal frequency, and the murein layers do not favour the attachment of one type of subunit over the other. Provided that a surplus of subunits of both types of lattices is present, the initially formed hexagonally and tetragonally ordered S-layer "nucleation crystallites" will grow until they meet their synchronously growing neighbour. These heterologous re-attachment experiments provided further evidence that the reattachment process and the lattice orientation on intact cells are not influenced by any order within the binding sites of the peptidoglycan containing layer (SLEYTR 1975, SLEYTR and PLOHBERGER 1980).

Nevertheless, by testing 23 taxonomically well defined strains of *Bacillus stearothermophilus*, it was possible to ascertain that the ability for heterologous reattachment is not a common feature of all S-layers within a given species. Strains both suitable and non-suitable for S-layer exchange were found (PLOHBERGER et al. 1981, in preparation). Detailed studies on the specific interaction between the tetragonally arranged surface layer and the peptidoglycan sacculi were also reported for *Bacillus sphaericus* (HASTIE and BRINTON 1979 a, b).

S-layer subunits of a wild type strain (140,000 MW) and a variant strain (122,000 MW) bonded specifically to the peptidoglycan of either cell type but not to other tested strains of *B. sphaericus*, *Lactobacillus casei*, *B. cereus*, *B. subtilis*, *Corynebacterium hoagii*, and *Staphylococcus aureus*. Mild proteo-

Fig. 9. Completely reconstituted S-layer on the cell walls of *Bacillus stearothermophilus* strain NRS 2004 showing the same lattice orientation and long range order as observed on intact cells. (×145,000).

lysis of both the high and low molecular weight subunit type reduced the molecular weight by 18,000. Subunits from which this minor fragment had been cleaved retained the capacity to assemble *in vitro* but would no longer reattach to the peptidoglycan. It was concluded that the ability of S-layer subunits to aggregate into tetragonal arrays was separate from the ability to bind to cell walls, and the 18,000 dalton piece of both subunit species was considered to be the important part for their attachment to the cell wall.

Reconstitution experiments of the S-layer onto surfaces of cell wall fragments mostly lead to the formation of closely packed, randomly orientated crystallites (crazy paving appearance). S-layer subunits only reattached to naked cells that had maintained their cylindrical shape (Fig. 9), giving lattice orientations and long-range orders similar to those observed from freeze-etching studies made on intact cells (SLEYTR and PLOHBERGER 1980). These results provided evidence that the curvature at the cylindrical part of the cell induces the orientation of the array.

VI. Dynamic Assembly of Regular Surface Layers

Self-assembly and reattachment experiments, such as those described in the previous section, have clearly demonstrated that the information for the morphogenesis of highly ordered S-layers resides in the molecular structure of its protomers. Furthermore, it was shown that S-layer fragments have the ability to fuse and recrystallize into aggregates with a reduced number of lattice faults. For the maintenance of a highly ordered, two dimensional array of macromolecules on a growing cell surface, the bonds holding the subunits together have to differ from those binding the subunits to the under-lying layer of the cell wall. In other words, the adhesion forces between the regular array and the supporting layer must not restrain the macromolecular lattice from rearranging itself into an order having the least strain and minimum free energy between the directional bonds connecting the pro-tomers.

The most direct evidence for such bonding differences could be obtained from analyses of the S-layer structures in freeze-etched preparations of actively growing, rod-shaped cells (Sleytr and Glauert 1975). S-layers demonstrated by this technique appear uniform over a considerable region of the cylindrical part of the cell. They show a few faults at the cell poles and frequently, a mosaic of small crystallites at division sites. Since the cell poles are the result of a completed cell division, the observed decrease in the number of lattice faults must have occured by a rearrangement of the subunits on the supporting layer during cell separation.

Further evidence for bonding differences were provided by S-layer extraction and disintegration experiments (Section III). Since both homologous and, heterologous reattachment experiments have shown that the S-layer supporting cell envelope components act as a pattern-neutral (nonperiodic) binding surface, the orientation of the regular arrays of macromolecules on intact, rod-shaped cells can only be determined by the curvature of the cylindrical part of the cell. This means that during cell growth, newly incorporated lattice areas will have a tendency to align on the cylindrical surface in an orientation having the least strain between adjacent subunits. Consequently, only the spherical areas at cell poles and septation sites (or over the whole surface of a coccus) will allow a random orientation of S-layer crystallites. The number of crystallites, and consequently lattice faults, required for covering a rounded surface with a macromolecular array will strongly depend on both the flexibility of the intersubunit bonds and the elasticity of the constituent protomers.

Knowing the mechanisms that govern S-layer assembly and recrystallization, it becomes evident that the only requirement for maintaining highly ordered arrays of macromolecules with no gaps on a growing cell surface is a continuous synthesis of a surplus of subunits and their transfer to the site of lattice growth. Consequently, for the morphogenesis of the simplest type of S-layer, only the genetic information for a single polypeptide species will be necessary. It can be assumed that, as with many other secreted proteins, the S-layer subunits are synthesized on plasma membrane bound ribosomes and will cross the membrane as growing chains rather than after the completion of their synthesis and subsequent tertiary folding (Davis and Tai 1980). Thus, depending on the envelope structure, the incorporation sites for newly secreted S-layer protomers may be a considerable distance from the plasma membrane.

With the exception of some species of *Archaebacteria,* such as *Halobacteria* or *Sulfolobus acidocaldarius,* and some *Methanogenes* that have S-layers in contact with the plasma membrane, the released protomers may have to pass several intermediate layers before reaching the final assembly sites. Since it may not be valid to assume that the sites of protein secretion and final incorporation are accurately superimposed, a considerable lateral diffusion may also be required for the S-layer subunits to reach their pre-determined loci. In other words, once the secreted proteins have folded into their tertiary pattern in the aqueous environment at the membrane surface, their physical properties will determine their migration target. Labeling experiments with polycationic ferritin (Fig. 7) have shown that in gram-positive species, the peptidoglycan surface adjacent to the S-layer has a net negative charge (Sleytr and Friers 1978). Since polycationic ferritin is frequently only bound to one S-layer surface (which can be the outer or the inner surface) (Fig. 7), charge phenomena may be responsible for a proper alignment of the protomers once they have reached the surface of the supporting layers.

With several species of *Clostridia, Bacilli,* and strains of *Desulfotomaculum nigrificans,* it has been shown that on detachment of the plasma membrane during plasmolysis or cell autolysis, an additional layer of S-layer subunits assembles on extended areas of the inside of the murein layer (Sleytr and Glauert 1976, Sleytr 1978). Thus, at least for gram-positive species, a close contact between the plasma membrane and the peptidoglycan layer has been shown to be important for the penetration of S-layer protomers through the rigid wall layer; this being a prerequisite for reaching their migration targets (Sleytr 1978).

So far, very little is known about the interrelationship between S-layer synthesis and cell growth. One of the major questions to be asked is how the information about structural requirements is transferred from the regular array to the site of synthesis of the subunits. Such a feedback system could act at the level of transscription, translation, translocation, or incorporation in the regular array. One could even speculate that S-layer subunits that cannot be incorporated pile up in the cytoplasmic membrane or periplasmic space and, consequently, inhibit their own synthesis. Several studies on gram-positive and gram-negative species have confirmed that more subunits are synthesized than the minimum number required for a complete covering of the growing cell surface. Frequently, in freeze-etched preparations, or in thin sections of gram-positive organisms, a surplus of S-layer subunits or of small crystallites can be observed at septation sites. After the completion of cell septation, this surplus is no longer present at the new cell poles, having been discarded by the cell (Sleytr and Glauert 1976).

Quantitative analysis has revealed that 7% of the total S-layer protein synthesized by *B. sphaericus* can be found in the culture medium (Howard and Tipper 1973). Since such a small excess of S-layer material was detected, the authors assumed that the synthesis of subunits is strictly controlled. The most detailed biochemical analyses of the synthesis and turnover of S-layer proteins have been carried out on the regular array of macromolecules occuring on the gram-negative *Acinetobacter* sp. strain 199 A (Thorne *et al.* 1976 a). It was found that although the lipopolysaccharide and intrinsic protein of the outer membrane were stable, half the S-layer protein was lost to the growth medium. Whereas secretions of excess S-layer material in *Clostridia* and *Acinetobacter* sp. were observed on logarithmically growing cells, Beveridge and Murray (1975) have shown that in *Spirillum metamorphum,* S-layer fragments detach from the outer membrane in the late logarithmic and stationary phases of growth.

Few experiments have been done to identify the sites at which subunits are incorporated into the regular array on a growing cell. Areas showing an obvious surplus of S-layer material need not necessarily be the only sites of lattice growth. Furthermore, the recrystallization and rearrangement abilities of the regular arrays do not imply that the S-layer secretion sites accurately correspond to the growth areas of the underlying cell envelope layer. To identify the areas of S-layer growth, cells of *Bacilli* species having net-negatively charged surface lattices were isolated from logarithmic cultures and labeled with a dense monolayer of polycationic ferritin (Sleytr and Friers 1978). After five minutes of growth, small, randomly arranged polycationic ferritin aggregates were observed leading to larger but fewer aggregates after ten minutes incubation, which corresponded to approximately half the generation time. The patches of newly formed S-layer occuring between the aggregates frequently exhibited a less uniform orientation on the cylindrical part of the cell when compared to unlabeled controls. This phenomenon was explained by an inhibition of the dynamic rearrangement of newly incorporated subunits by the polycationic ferritin clusters. After about one generation time, most of the polycationic ferritin clusters were shed into the medium, and with subsequent cell growth, the lattice long-range order continuously improved. The formation of the randomly distributed polycationic ferritin clusters during cell growth was explained by a process involving a random growth of the surface lattice and a secretion of the potential surplus of protomers that were capable of multiple binding with the marker molecules. Recently, labeling experiments with ferritin conjugated S-layer specific antibodies have provided further evidence of random S-layer growth (Friers and Sleytr, unpublished).

In the theoretical papers of Harris and Scriven (1970) and Harris (1977, 1978), it was suggested that during lattice growth, new subunits are best added at sites of "migrating dislocations". Both dislocations and disclinations can be shown on the S-layers of intact cells using freeze-etching techniques, whereas neither polycationic ferritin nor specific antibody labeling experiments can provide the resolution required to confirm this attractive lattice growth theory. On the other hand, dislocations and disclinations may merely represent faults created by the stress that occurs when two dimensional protein crystals cover rounded surfaces or rearrange during cell growth (SLEYTR and GLAUERT 1975).

VII. Functional Aspects

Although a considerable amount of data is available about the chemistry and morphogenesis of S-layers, relatively little is known about their biological significance. For all of the gram-positive and gram-negative organisms studied, it has been shown that the entropy driven S-layer assembly mechanism leads to a closely packed two dimensional protein crystal that covers the whole surface of the cell. The few lattice faults observed in freeze-etching preparations of actively growing cells confirm that the regular arrays exist at a low energy state equilibrium. In such an assembly system, faulty subunits will not be incorporated and, consequently, the generation of leaks and discontinuities in the lattice will be low. Thus, the pore size and permeability properties of these protein or glycoprotein membranes will only be determined by the mass and charge distribution of its constituent subunits (SLEYTR 1975, CROWTHER and SLEYTR 1977, SLEYTR and PLOHBERGER 1980).

Because of their surface location, S-layers have most probably evolved as a result of cell-environmental interaction (THORNLEY et al. 1974 a, SLEYTR 1978). Theoretically, S-layers could act as a barrier or molecular sieve controlling the movement of both external and internal factors.

One of the best understood protective functions provided by S-layers is that of the hexagonal array of proteins found on the surface of *Spirillum serpens* (BUCKMIRE and MURRAY 1970, BUCKMIRE 1971). It was shown that strains of *S. serpens* that possess regular S-layers are resistant to attachment and invasion by the ubiquitous bacterial predator *Bdellovibrio bacteriovorus*. Possibly the regular arrays mask the receptor sites for the attachment of the parasite since a removal of the S-layer renders the cells susceptible to parasitism. Thus, under normal environmental conditions, a strong selective force may occur favour of strains that carry regular surface structures (BUCKMIRE 1971). However, this would not apply to gram-positive bacteria since these organisms are not invaded by *Bdellovibrio*.

So far, no permeability studies have been carried out on S-layers to support the concept of a "molecular sieve function". High resolution studies made with negatively stained S-layer preparations revealed specifically patterned areas of high stain density that presumably, correspond to lattice holes. The stain distribution patterns observed on, for example, the S-layers of *Clostridium thermosaccharolyticum* and *C. thermohydrosulfuricum* (CROWTHER and SLEYTR 1977), *Bacillus brevis* (AEBI et al. 1973), *Sporosarcina urea* (STEWART and BEVERIDGE 1980), *Spirillum serpens* (BUCKMIRE and MURRAY 1970, 1973), *Spirillum putridichonchylium* (BEVERIDGE and MURRAY 1976 b, c, STEWART et al. 1980), and *Micrococcus radiodurans* (KÜBLER and BAUMEISTER 1978) suggest the existence of 15 to 20 nm diameter holes. Such "pores" could act as a barrier for molecules with molecular weights larger than about 3,000 daltons but would not significantly inhibit the permeability of nutrients and metabolic degradation products. Frequently, gram-positive bacteria with S-layers have a much thinner peptidoglycan layer compared with those that do not possess regular surface structures (Sleytr and Glauert 1976). Consequently, it may be that for gram-positive bacteria, S-layers have a protective function for the underlying peptidoglycan layer itself (Crowther and Sleytr 1977).

In a natural (and consequently hostile) environment, lytic enzymes such as lysozyme may be a potential hazard to the cell. Evidence that S-layers provide protection from attack by muramidases has been suggested from several observations. For example, in *Bacillus polymyxa* (Nermut and Murray 1967), *Clostridium botulinum* (Takumi and Kawata 1974), *Lactobacillus fermentum* (Wallinder and Neujahr 1971), *Bacillus cereus* (Ellar and Lundgren 1967), and *Sporosarcina urea* (Stewart and Beveridge 1980), removal or partial degradation of the S-layer can make the peptidoglycan layer susceptible to digestion by lysozyme. A lattice pore size of 15 to 20 nm could also screen the cell from attack by other lytic enzymes. In addition, resistance of the S-layer protein to proteolytic enzymes such as papain, trypsin, pronase, and thermolysin has been reported, for *C. thermohydrosulfuricum* and *C. thermosaccharolyticum* (Sleytr 1975, Sleytr and Glauert 1976) amongst others. So far, the only enzyme activity that has been detected within S-layer subunits themselves was phospholipase A 2 activity in *Acinetobacter* 199 A surface protein. In the free form, the S-layer subunits have enzyme activity, but when bound to the cell surface, they do not (Thorne *et al.* 1976 b).

S-layers may also be involved in controlling the release of macromolecules from the periplasmic space. Wecke and coworkers (Wecke *et al.* 1974, Schallehn and Wecke 1974) found that toxigenic *Clostridia* that show an intracellular accumulation of toxins (such as *Clostridium botulinum*, *C. tetani*, and *C. novyi*) possess S-layers. Since mild disintegration of the regular array by sodium citrate buffer (pH 7.7) released the accumulated toxins, it was concluded that the S-layers act as a diffusion barrier to these toxic macromolecules.

Similarly, it is very likely that the S-layers found on gram-positive bacteria prevent cell wall-associated enzymes from being lost into the environment. Since gram-positive bacteria generally release a large proportion of their extracellular enzymes, it is advantageous for them to grow in conditions of high population densities and substrate concentration. Thus, a dilute environment could easily provide the necessary selection pressure for developing protein membranes that trap degradative enzymes in the periplasmic space in a similar manner to that of the outer membrane of gram-negative cell envelopes (Costerton *et al.* 1974). In this context, it was interesting to observe that all the strains of hyperthermophilic *Clostridia* and *Bacilli* isolated from the dilute environments of extraction plants in beet sugar factories possessed S-layers (Hollaus and Sleytr 1972, Plohberger and Sleytr 1980, Sleytr and Plohberger 1980).

In view of a possible barrier function against internal or external factors, the remarkable variety of both the lattice morphology and the molecular weight of subunits becomes understandable since lattices of different geometry can easily possess pores of identical size (CROWTHER and SLEYTR 1977, STEWART and BEVERIDGE 1980). Support for the concept of a selection pressure induced development of S-layers comes from observations that wild type strains frequently loose S-layers during prolonged cultivation under laboratory conditions (BEVERIDGE, personal communication; SLEYTR and PLOHBERGER, unpublished). In such an environment, mutants lacking the ability to synthesize S-layers will have a selective advantage and, eventually, may outgrow the wild type strain (STEWART and BEVERIDGE 1980).

Various other possible functions have been proposed for S-layers. It has been suggested that the regular arrays, due to their defined surface charge, may function in the adhesion of the bacterial cell to a suitable surface (SLEYTR and FRIERS 1978) or in conditioning or protecting the ionic and molecular environment of both the cell wall and the plasma membrane enzymes (COSTERTON et al. 1974, BEVERIDGE 1979).

Based on the observation that S-layers can assemble in vitro to form cylinders, BRINTON proposed a morphopoietic function for the regular arrays (BRINTON et al. 1969, HASTIE and BRINTON 1979 b). It was assumed that assembled S-layers may provide a mould or shell for binding and shaping newly synthesized peptidoglycan. In view of both the observed structural and chemical heterogeneity of S-layers and the existence of strains with and without regular arrays amongst the individual species studied, it is most unlikely that S-layers serve as an important parameter in the genetic expression that determines the cell shape. However, in several species of Archaebacteria that lack a rigid peptidoglycan layer, the subunit-type cell wall (Section I) associated with the plasma membrane may also have a morphopoietic function (MESCHER and STROMINGER 1976).

Since some Archaebacteria thrive under extreme environmental conditions (WOESE et al. 1978), it may also be that dynamically growing S-layer membranes could have fulfilled all the necessary barrier functions required by a self-reproducing system during the early stages of biological evolution.

Acknowledgements

The cultures of Bacilli and Clostridia were kindly provided by Dr. F. HOLLAUS, Oesterreichisches Zuckerforschungsinstitut, Fuchsenbigl. I wish to thank Dr. D. PITT for critical reading of the manuscript and Mrs. E. POHORALEK for her expert technical assistance. This work was supported by the Fonds zur Förderung der wissenschaftlichen Forschung, Projekt 3293, 3839.

References

AEBI, U., SMITH, P. R., DUBOCHET, J., HENRY, C., KELLENBERGER, E., 1973: A study of the structure of the T-layer of Bacillus brevis. J. Supramolec. Struct. 1, 498—522.

BALCH, W. E., FOX, G. E., MAGRUM, L. J., WOESE, C. R., WOLFE, R. S., 1979: Methanogens: re-evaluation of a unique biological group. Microbiol. Reviews 43, 260—296.

BAUMEISTER, W., KÜBLER, O., 1978: Topographic study of the cell surface of Micrococcus radiodurans. Proc. Nat. Acad. Sci. (U.S.A.) 75, 5525—5528.

BEVERIDGE, T. J., 1979: Surface arrays on the wall of Sporosarcina urea. J. Bacteriol. 139, 1039—1048.

BEVERIDGE, T. J., MURRAY, R. G. E., 1975: Surface arrays on the cell wall of *Spirillum metamorphum*. J. Bacteriol. **124**, 1529—1544.
— — 1976 a: Superficial cell wall layers on *Spirillum* "ordal" and their *in vitro* reassembly. Canad. J. Microbiol. **22**, 567—582.
— — 1976 b: Reassembly *in vitro* of the superficial cell wall components of *Spirillum putridiconchylium*. J. Ultrastruct. Res. **55**, 105—118.
— — 1976 c: Dependence of the superficial layers of *Spirillum putridiconchylium* on Ca^{2+} or Sr^{2+}. Canad. J. Microbiol. **22**, 1233—1244.
— WILLIAMS, F. M. R., 1978: The effect of chemical fixatives on cell walls of *Bacillus subtilis*. Canad. J. Microbiol. **24**, 1439—1451.
BRINTON, C. C., McNARY, J. C., CARNAHAN, J., 1969: Purification and *in vitro* assembly of a curved network of identical protein subunits from the outer surface of a *Bacillus*. Bacteriol. Proc., p. **48**.
BUCKMIRE, F. L. A., 1971: A protective role for a cell wall protein layer of *Spirillum serpens* against infection by *Bdellovibrio bacteriovorus*. Bacteriol. Proc., p. **43**.
— MURRAY, R. G. E., 1970: Studies on the cell wall of *Spirillum serpens*. 1. Isolation and partial purification of the outermost cell wall layer. Canad. J. Microbiol. **16**, 1011—1022.
— — 1973: Studies on the cell wall of *Spirillum serpens*. 2. Chemical characterization of the outer structured layer. Canad. J. Microbiol. **19**, 59—66.
— — 1976: Substructure and *in vitro* assembly of the outer, structured layer of *Spirillum serpens*. J. Bacteriol. **125**, 290—299.
COSTERTON, J. W., INGRAM, J. M., CHENG, K. J., 1974: Structure and function of the cell envelope of gram-negative bacteria. Bacteriol. Reviews **38**, 87—110.
CROWTHER, R. A., SLEYTR, U. B., 1977: An analysis of the fine structure of the surface layers from two strains of *Clostridia* including correction of distorted images. J. Ultrastruct. Res. **58**, 41—50.
DAVIS, B. D., TAI, P. C., 1980: The mechanism of protein secretion across membranes. Nature (London) **283**, 433—438.
ELLAR, D. J., LUNDGREN, D. G., 1967: Ordered substructure in the cell wall of *Bacillus cereus*. J. Bacteriol. **94**, 1778—1780.
GLAUERT, A. M., THORNLEY, M. J., 1969: The topography of the bacterial cell wall. Ann. Rev. Microbiol. **23**, 159—198.
HARRIS, W. F., 1977: Disclinations. Scientific American **237**, 130—145.
— 1978: Dislocations, disclinations and dispirations: distractions in very naughty crystals. S. Afr. J. Sci. **74**, 332—338.
— SCRIVEN, L. E., 1970: Function of dislocations in cell walls and membranes. Nature (London) **228**, 827.
HASTIE, A. T., BRINTON, C. C., 1979 a: Isolation, characterization, and *in vitro* assembly of the tetragonally arrayed layer of *Bacillus sphaericus*. J. Bacteriol. **138**, 999—1009.
— — 1979 b: Specific interaction of the tetragonally arrayed protein layer of *Bacillus sphaericus* with its peptidoglycan sacculus. J. Bacteriol. **138**, 1010—1021.
HENRY, C. M., 1972: Studies on the tetragonal self assembly protein layer of *Bacillus brevis* P 1. Ph.D. dissertat., Diss. Abstr. Int. B **33**, 73—2878, 4409-B.
HOLLAUS, F., SLEYTR, U. B., 1972: On the taxonomy and fine structure of some hyperthermophilic saccharolytic *Clostridia*. Arch. Microbiol. **86**, 129—146.
HOUWINK, A. L., 1953: A macromolecular mono-layer in the cell wall of *Spirillum* sp. Biochim. Biophys. Acta **10**, 360—366.
HORWARD, L., TIPPER, D. J., 1973: A polypeptide bacteriophage receptor: modified cell wall protein subunits in bacteriophage resistant mutants of *Bacillus sphaericus* strain P-1. J. Bacteriol. **113**, 1491—1504.
KANDLER, O., 1979: Zellwandstrukturen bei Methanbakterien. Zur Evolution der Prokaryonten. Naturwiss. **66**, 95—105.
— KÖNIG, H., 1978: Chemical composition of the peptidoglycan-free cell walls of methanogenic bacteria. Arch. Microbiol. **118**, 141—152.
KELLENBERGER, E., KISTLER, J., 1979: The physics of specimen preparation. In: Unconventional electron microscopy for molecular structure determination (HOPPE, W., MASON, R., eds.). Braunschweig-Wiesbaden: Friedrich Vieweg & Sohn.

KESSEL, M., 1978: A unique crystalline wall layer in the Cyanobacterium *Microcystis marginata*. J. Ultrastruct. Res. **62**, 203—212.

KÜBLER, O., BAUMEISTER, W., 1978: The structure of a periodic cell wall component (HPI-layer) of *Micrococcus radiodurans*. Cytobiologie **17**, 1—19.

— ENGEL, A., ZINGSHEIM, H. P., EMDE, B., HAHN, M., HEISSE, W., BAUMEISTER, W., 1980: Structure of the HPI-layer of *Micrococcus radiodurans*. In: Electron microscopy at molecular dimension (BAUMEISTER, W., VOGELL, W., eds.). Berlin-Heidelberg-New York: Springer.

LOUNATMAA, K., VAARA, T., ÖSTERLUND, K., VAARA, M., 1980: Ultrastructure of the cell wall of a *Synechocystis* strain. Can. J. Microbiol. **26**, 204—208.

MASUDA, K., KAWATA, T., 1980: Reassembly of the regularly arranged subunits in the cell wall of *Lactobacillus brevis* and their reattachment to cell walls. Microbiol. Immunol. **24**, 299—308.

MESCHER, M. F., STROMINGER, L., 1976: Structural (shape-maintaining) role of the cell surface glycoprotein of *Halobacterium salinarium*. Proc. Nat. Acad. Sci. (U.S.A.) **73**, 2687—2691.

MICHEL, H., NEUGEBAUER, D.-CH., OSTERHELT, D., 1980: The two-dimensional crystalline cell wall of *Sulfolobus acidocaldarius*: structure, solubilization, and reassembly. In: Electron microscopy at molecular dimensions (BAUMEISTER, W., VOGELL, W., eds.). Berlin-Heidelberg-New York: Springer.

NERMUT, M. V., MURRAY, R. G. E., 1967: Ultrastructure of the cell wall of *Bacillus polymyxa*. J. Bacteriol. **93**, 1949—1965.

PLOHBERGER, R., SLEYTR, U. B., 1980: Characterization of regular arrays of macromolecules on cell walls of thermophilic *Bacilli*. Proc. 7th Europ. Congr. Electr. Microsc., The Hague, **2**, 608—609.

— — MESSNER, P., 1981: Chemical characterization of surface arrays on cell walls of *Bacillus stearothermophilus* strains. (In preparation.)

SCHALLEHN, G., WECKE, J., 1974: Zur Feinstruktur der Zellwand von *Clostridium perfringens*, *C. septicum* and *C. novyi*. Zbl. Bakt. Mikrobiol. Hyg., Abt. 1, Orig. **A 228**, 63.

SLEYTR, U. B., 1975: Heterologous reattachment of regular arrays of glycoproteins on bacterial surfaces. Nature (London) **257**, 400—402.

— 1976: Self assembly of the hexagonally and tetragonally arranged subunits of bacterial surface layers and their reattachment to cell walls. J. Ultrastruct. Res. **55**, 360—377.

— 1978: Regular arrays of macromolecules on bacterial cell walls: structure, chemistry, assembly, and function. Int. Rev. Cytol. **53**, 1—64.

— FRIERS, G.-P., 1978: Surface charge and morphogenesis of regular arrays of proteins on bacterial cell walls. Proc. 9th Int. Congr. Electr. Microsc., Toronto, **2**, 346—347.

— GLAUERT, A. M., 1975: Analysis of regular arrays of subunits on bacterial surfaces: evidence for a dynamic process of assembly. J. Ultrastruct. Res. **50**, 103—116.

— — 1976: Ultrastructure of the cell walls of two closely related *Clostridia* that possess different regular arrays of surface subunits. J. Bacteriol. **126**, 869—882.

— — 1981: Bacterial cell walls and membranes. In: Electron microscopy of proteins (HARRIS, J. R., ed.). New York-San Francisco-London: Academic Press. In press.

— THORNE, K. J. I., 1976: Chemical characterization of the regularly arranged surface layers of *Clostridium thermosaccharolyticum* and *C. thermohydrosulfuricum*. J. Bacteriol. **126**, 377—383.

— PLOHBERGER, R., 1980: The dynamic process of assembly of two-dimensional arrays of macromolecules on bacterial cell walls. In: Electron microscopy at molecular dimensions (BAUMEISTER, W., VOGELL, W., eds.). Berlin-Heidelberg-New York: Springer.

— — EDER, J., 1980: Dynamically growing protein membranes as regular cell surface layers of *Bacillaceae*. Europ. J. Cell Biol. **22**, 216.

SMITH, P. R., KISTLER, J., 1977: Surface reliefs computed from micrographs of heavy metal shadowed specimens. J. Ultrastruct. Res. **61**, 124—133.

— IVANOV, I. E., 1980: Surface reliefs computed from micrographs of isolated heavy metal shadowed particles. J. Ultrastruct. Res. **71**, 25—36.

STEWART, M., BEVERIDGE, T. J., 1980: Structure of the regular surface layer of *Sporosarcina urea*. J. Bacteriol. **142**, 302—309.

Stewart, M., Beveridge, T. J., Murray, R. G. E., 1980: Structure of the regular surface layer of *Spirillum putridiconchylium*. J. Mol. Biol. **137**, 1—8.

Takumi, K., Kawata, T., 1974: Isolation of a common cell wall antigen from the proteolytic strains of *Clostridium botulinum*. Jap. J. Microbiol. **18**, 85—90.

Thorne, K. J. I., 1977: Regularly arranged protein on the surfaces of gram-negative bacteria. Biol. Rev. **52**, 219—234.

— Thornley, M. J., Naisbitt, P., Glauert, A. M., 1975: The nature of the attachment of a regularly arranged surface protein to the outer membrane of an *Acinetobacter* sp. Biochim. Biophys. Acta **389**, 97—116.

— Oliver, R. C., Glauert, A. M., 1976 a: Synthesis and turnover of the regularly arranged surface protein of *Acinetobacter* sp. relative to the other components of the cell envelope. J. Bacteriol. **127**, 440—450.

— — Heath, M. F., 1976 b: Phospholipase A_2 activity of the regularly arrange surface protein of *Acinetobacter* sp. 199 A. Biochem. Biophys. Acta **45**, 335—341.

Thornley, M. J., Glauert, A. M., Sleytr, U. B., 1974 a: Structure and assembly of bacterial surface layers composed of regular arrays of subunits. Phil. Trans. Roy. Soc. (London) **B 268**, 147—154.

— Thorne, K. J. I., Glauert, A. M., 1974 b: Detachment and chemical characterization of the regularly arranged subunits from the surface of an *Acinetobacter*. J. Bacteriol. **118**, 654—662.

Unwin, P. N. T., Henderson, R., 1975: Molecular structure determination by electron microscopy of unstained crystalline specimens. J. Mol. Biol. **94**, 425—440.

Wallinder, I. B., Neujahr, H. Y., 1971: Cell wall and peptidoglycan from *Lactobacillus fermentii*. J. Bacteriol. **105**, 918—926.

Watson, S. W., Remsen, C. C., 1969: Macromolecular subunits in the walls of marine nitrifying bacteria. Science **163**, 685—686.

Wecke, J., Reinicke, B., Schallehn, G., 1974: Remarkable differences in the ultrastructure of the cell wall of toxigenic *Clostridia*. 8th Int. Congr. Electr. Microsc., Canberra, **2**, 644—645.

Woese, C. R., Magrum, L. J., Fox, G. E., 1978: Archaebacteria. J. Mol. Evol. **11**, 245—252.

Scale Formation in Flagellates

D. K. ROMANOVICZ

Department of Biology, West Georgia College, Carrollton, Georgia, U.S.A.

With 14 Figures

Contents

I. Introduction

In the context of a cell wall, the term scale is used to describe a discrete, morphologically distinct wall subunit synthesized within the cell. Following exocytosis, the scales become arranged on the surface of the cell in one or more layers forming a cell wall. This simple functional definition belies the great diversity of scale morphology and composition. Scales range from 30 nm (MANTON and ETTL 1965) to 14 μm (MANTON et al. 1976 a) in diameter, some bearing spines up to 75 μm in length (MOESTRUP 1974 a, MANTON 1978 a). In some species, the scales are constructed of organic microfibrils (BROWN et al. 1970), and may bear crystals of calcium carbonate (MANTON and LEEDALE 1969). In others, the scales are composed of silica and may be three-dimensional, reticulate structures (PENNICK and CLARKE 1972) or flat plates with perforations (HARRIS and BRADLEY 1958). Still other species produce membranous scales with no discernible substructure (Belcher and Swale 1967, Darley et al. 1973). Despite this diversity in morphology and composition, for most organisms in which scale biosynthesis has been investigated, the scales have been shown to arise within the cisternae of the Golgi apparatus (Moestrup 1974 a, Olive 1975).

Although a scaly cell covering can be discerned with the light microscope, the study of scale morphology and biosynthesis began in earnest with the early electron microscopic investigations of *Chrysochromulina* by Parke et al. (1955, 1956, 1958, 1959) and of several Chrysophycean species by Fott (1955) and Harris and Bradley (1956, 1958). For the first ten years, the increase in knowledge of scale formation paralleled the rapid improvement of preparative techniques for electron microscopy. Thus the earliest work concentrated on scale morphology, as revealed by heavy metal shadowing or direct observation of mineralized scales. Thin section techniques based on osmium tetroxide fixation and methacrylate embeddment were at first insufficient to permit observation of the intracellular origin of scales (Manton and Leedale 1961 a), but the concept of intracellular origin was gaining acceptance (Manton and Leedale 1961 b) and was soon demonstrated (Manton and Leedale 1961 c, Manton and Parke 1962). This was followed closely by the observation that scales are synthesized within the Golgi apparatus (Manton and Leedale 1963 b, Parke and Rayns 1964). A similar sequence of events occurred in the study of the formation of coccoliths, organic scales with calcified rims; intellectual acceptance of the intracellular origin of coccoliths (Paasche 1962) preceeded the actual observation (Wilbur and Watabe 1963). With the adoption of glutaraldehyde as a primary fixative, details of the biosynthetic pathway began to emerge (Manton 1966 b).

Brown and co-workers (Brown 1969, Brown et al. 1969, 1970) presented an account of Golgi-mediated scale formation in the Prymnesiophycean (Haptophycean) alga, *Pleurochrysis scherffelii,* describing this event in terms of a cell surface-oriented membrane flow (Morré et al. 1971, Northcote 1971, Whaley et al. 1971). The proposed pathway was incomplete due to an inability to distinguish between the two microfibrillar scale subcomponents prior to assembly into a completed scale. With the application of a cytochemical procedure that stained only one of the microfibrillar subcomponents (Brown et al. 1973), a more complete pathway of scale biogenesis was presented (Romanovicz 1975, Brown and Romanovicz 1976). Brown also documented the polysaccharide nature of *Pleurochrysis* scales, in particular, an accumulation of evidence indicating the presence of cellulose (Brown et al. 1970). As there was considerable resistance to the idea of Golgi-derived cellulose (Colvin 1972), Herth et al. (1972) presented a careful study of alkali-purified scales, confirming the presence of a cellulosic structural polysaccharide that also contained a covalently-linked peptide moiety. However, the lack of crystallographic evidence caused doubt about the presence of cellulose (Preston 1974) until a cellulose I X-ray diffraction pattern was obtained from acid-purified scales (Romanovicz and Brown 1976). While Brown did not continue to offer *Pleurochrysis* as a model system for cellulose biosynthesis in all plants, it remains that cellulose is produced in the Golgi apparatus in this organism, and most likely in all other scales forming organisms that produce cellulosic scales.

During the quarter century since the first work on *Chrysochromulina*, many new species of scale-forming organisms have been described. Although the majority are contained within three algal classes *Chrysophyceae, Prasinophyceae,* and *Prymnesiophyceae* there are also several non-algal protists that make a cell wall of scales (GRELL and BENWITZ 1966, OLIVE 1975). Most significant is the discovery of scales on the male gamete in the *Charales* (PICKETT-HEAPS 1968, TURNER 1968) and on the zoospore of two Chlorophycean algae thought to be ancestors of bryophytes and higher plants (McBRIDE 1968, MOESTRUP 1974 b). In all, there are over one hundred species of scale-forming organisms currently known.

The unique nature of a scaly cell wall offers several advantages in an experimental organism. Whole scales or individual subcomponents can often be unequivocally recognized, whether within the cell or in cell-free wall preparations. Consequently, it is possible to determine the complete composition, as well as the precise intracellular site of synthesis, of a particular cell product, a task of greater difficulty in most other experimental systems. Some species alter the structure or composition of the scales at various stages in the life cycle, or alterations may be induced by treatment with chemical agents. Other species make several different types of scales simultaneously, arranging them on the cell surface in a consistent pattern. Such unique properties have already been exploited in the study of biological calcification (OUTKA and WILLIAMS 1971, DE JONG *et al.* 1976, PIENAAR 1976 b) and cellulose biosynthesis (BROWN and ROMANOVICZ 1976, ROMANOVICZ and BROWN 1976), while other uses await discovery.

II. Pleurochrysis: A Model System

A. Scale Structure and Composition

The non-motile vegetative cells of the marine Prymnesiophycean alga *Pleurochrysis scherffelii* produce unmineralized, microfibrillar scales, 1.5 to 2 µm in diameter. An amorphous, pectin-like coating causes the scales to adhere and form a compact cell wall (Fig. 1). When agar cultures of vegetative cells are flooded with seawater, the protoplasts divide and their division products escape into the medium as naked zoospores (BONEY 1967), leaving behind a cell-free scale preparation called scale fraction A. BROWN (1969) used negative staining of these isolated scales to demonstrate the presence of two distinct microfibrillar subcomponents (Fig. 2): a quadriradial network (radial microfibrils) to which is attached a flat, spirally wound band of microfibrils (spiral microfibrils). The length of the spiral microfibrils and the number of turns in the spiral vary proportionately with the diameter of the scale. Originally, the spiral microfibrils were termed "concentric", until the observation of a partially degraded scale demonstrated their spiral arrangement (BROWN *et al.* 1973, see Fig. 3). Although most investigators have adopted the notation "spiral microfibrils" (MANTON *et al.* 1977), some persist in using the term "concentric ridges". In addition to the morphologically-distinct amorphous, radial, and spiral subcomponents, two functional sub-components have been established in *Pleurochrysis*. One is a coating material on the spiral microfibrils that maintains the spiral arrangement (ROMANOVICZ and BROWN 1976). The other is an amorphous material at the scale rim that is recognizable only by its unique antigenic properties (KAVOOKJIAN and BROWN 1978). The function of this most recently discovered subcomponent is still a matter of speculation. However, rim modifications are found in many Prymnesiophytes, so this material at the rim of the *Pleurochrysis* scale may represent an unexpressed capability for modification.

The scale subcomponents can be separated for compositional analysis by serial extraction with increasing concentrations of trifluoroacetic acid (TFA), monitored by negative staining (ROMANOVICZ and BROWN 1976). Treatment with acidified water (pH 5) hydrolyzed the amorphous polysaccharide coat-ing, yielding a mixture of galactose, glucose, and L-fucose. Further extraction of the pellet with 0.5 N TFA hydrolyzed the radial microfibrils, which are composed of galactose, arabinose, L-fucose, and sulfate. These are the same monosaccharides reported by BROWN *et al.* (1970) for *Pleurochrysis* scales and by GREEN and JENNINGS (1967) for scales of the related species, *Chryso-chromulina chiton*, with the exception of L-fucose. These earlier investigations had indicated instead the presence of ribose, an atypical component of struc-tural polysaccharides whose presence was interpreted by some as indicative of cytoplasmic contamination. In a later study of *Chrysochromulina* scales, ALLEN and NORTHCOTE (1975) indicated the presence of both ribose and L-fucose, taking pains to dispute the notion of contamination by cellular components. ROMANOVICZ and BROWN (1976) used a chromatographic solvent system in which ribose and L-fucose had similar R_f values, but could be distinguished by the visualizer which yielded different colors for pentoses and hexoses. While it appeared to them that only L-fucose was present, it is apparent that further investigation of *Pleurochrysis* scales is needed to deter-mine whether one or both sugars are present. There have been reports of xylose in Prymnesiophycean scales (MARKER 1965, ALLEN and NORTHCOTE 1975), but there was no indication of this monosaccharide in *Pleurochrysis* scales. In addition to the sugars, both the amorphous and radial subcomponents con-tained small amounts of protein, a component of cell walls that has been receiving increased attention (LAMPORT 1965, 1970, 1973, ALBERSHEIM *et al.* 1973, KEEGSTRA *et al.* 1973, PARADIES *et al.* 1977).

Serial extraction with TFA also revealed the presence of a polysaccharide coating on the spiral microfibrils which maintains the spiral arrangement of the intact scale and is responsible for positive stainability in thin section. It is interesting to note that the coating material also contains protein, the most abundant amino acid of which is hydroxyproline (ROMANOVICZ and BROWN 1976). This fact, coupled with its role in maintaining scale structure, suggests an analogy with the peptide cell wall component "extensin" postulated by LAMPORT (1970, 1973).

The resultant pellet from hydrolysis of the coating material with 2 N TFA consisted of disorganized spiral microfibrils, approximately 3.7 × 4.9 nm in

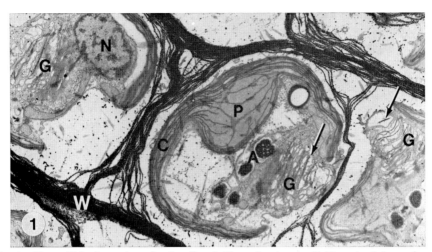

Fig. 1. Glutaraldehyde-osmium fixed cells of *Pleurochrysis scherffelii*, stained for polysaccharide localization with periodic acid-silver methenamine and counterstained with lead citrate. Parietal chloroplast (*C*) with pyrenoid (*P*), autophagic vacuoles (*A*), nucleus (*N*), and Golgi apparatus (*G*) are visible in the section. Silver deposition is evident on the cell wall (*W*) and intracisternal scales (arrows). From BROWN and ROMANOVICZ 1976. Courtesy of J. Wiley. ×8,500.

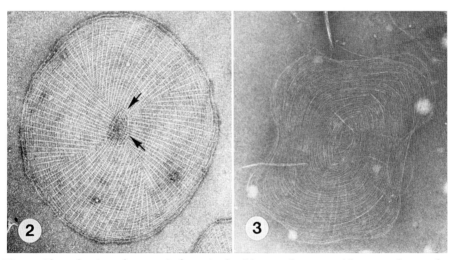

Fig. 2. *Pleurochrysis* scale, negatively stained with uranyl acetate. The amorphous polysaccharide has been hydrolyzed so that the quadriradial network and spiral microfibrils would be more evident. Note the "breakage loci" at the center of the scale (arrows). From BROWN and ROMANOVICZ 1976. Courtesy of J. Wiley. ×35,600.

Fig. 3. *Pleurochrysis* scale, after extraction with 1 N HCl, negatively stained with phosphotungstic acid. The spiralling band pattern is apparent in the upper right quadrant. From ROMANOVICZ and BROWN 1976. Courtesy of J. Wiley. ×49,500.

cross-section (ROMANOVICZ and BROWN 1976). As these dimensions are identical to those of spiral microfibrils in intact scales, there was apparently no degradation from the extraction. Upon drying, the acid-extracted scale material yielded a cellulose I X-ray diffraction pattern (Fig. 4), establishing beyond doubt the cellulosic nature of the spiral microfibrils. Calculation of crystallite size from line-broadening in the diffraction pattern gave a value of 3.8 nm, supporting the dimensions observed with negative staining (ROMANOVICZ and BROWN 1976). The requirement of acid extraction and drying before a cellulose I diffraction pattern is obtained indicates that scale cellulose has a low crystallinity with respect to cellulose from other

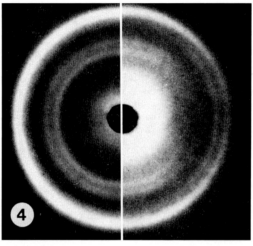

Fig. 4. X-ray diffraction pattern of dried, acid-extracted *Pleurochrysis* scales (left) compared with a pattern from cotton fibers (right). The 2.6 Å reflection is not visible due to its low intensity on the recording film. From ROMANOVICZ and BROWN 1976. Courtesy of J. Wiley.

sources that yield such a diffraction pattern while wet and in the native state (HEYN 1965, CAUFIELD 1971). This finding also suggests one reason for the inability of GREEN and JENNINGS (1967) to obtain a cellulose I diffraction pattern from *Chrysochromulina* scales, as they used dried scale material without prior acid extraction.

Examination of infrared absorption spectra (Fig. 5) provides support for several aspects of scale composition. The overall similarity between the spectra for cotton cellulose and scale material is immediately apparent. The broad band in the 3,000–3,600 cm^{-1} region is indicative of hydrogen bonding (TSUBOI 1957, LIANG and MARCHESSAULT 1959, McCALL et al. 1971), and the absence of distinct absorption peaks in this region supports an interpretation of low crystallinity as compared to cellulose from *Acetobacter* or *Valonia*, which yield as many as six peaks in this region (MARRINAN and MANN 1956, LIANG and MARCHESSAULT 1959, McCALL et al. 1971). The peaks in the 1,600–1,800 cm^{-1} region indicate the presence of protein (MITCHELL and

SCURFIELD 1970, PARADIES *et al.* 1977), supporting the finding of protein in the scale subcomponents, especially the covalently-bound protein moiety accounting for more than 30% by weight of scale cellulose (HERTH *et al.* 1972, ROMANOVICZ and BROWN 1976). The presence of sulfated polysaccharides in the non-cellulosic subcomponents is supported by the absorption at 1,240 cm^{-1} and in the 820–850 cm^{-1} region (LLOYD *et al.* 1961) in native scale material and the loss of this absorption upon acid extraction.

The correlation of structure and composition was made possible by the distinctive architecture of the scales. The demonstration that the spiral micro-

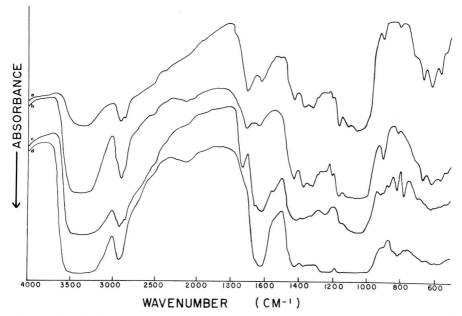

Fig. 5. Infrared absorption spectra of acid-extracted cotton (*a*), acid-extracted *Pleurochrysis* scales (*b*), native cotton (*c*), and native *Pleurochrysis* scales (*d*). The overall similarity is evident, see text for further interpretation. From ROMANOVICZ and BROWN 1976. Courtesy of J. Wiley.

fibrils are composed of cellulose suggested another reason why GREEN and JENNINGS (1967) were unable to establish the presence of cellulose in scales of *Chrysochromulina:* in their micrographs the spiral microfibrils appear few in number, consisting of a narrow band at the rim of the scale. Moreover, the discrete nature of the scales permitted measurements of cellulosic microfibrils that could not be disputed on the basis of disruptive isolation techniques (COLVIN 1972, PRESTON 1974) or interference resulting from overlapping images (PRESTON 1971).

B. Scale Biogenesis

The elaboration and secretion of a cellular product via the Golgi apparatus is conceived as a dynamic process of membrane flow (MORRÉ *et al.* 1971, WHALEY *et al.* 1971). Regions of the nuclear envelope give rise to endoplasmic reticulum, which blebs off small vesicles that coalesce to produce the forming face of the Golgi apparatus. Synthesis of the product is completed across the Golgi stack, after which cisternae from the secreting face fuse with the plasma membrane, releasing their contents into the surrounding extracellular space. The polarity of the Golgi apparatus is maintained in part by a process of membrane differentiation, observable as a progressive change in cisternal membrane characteristics from the forming face to the secreting face (GROVE et al. 1968). Although this entire process has not been demonstrated in every secretory cell (see WHALEY 1975), the polarity of the Golgi apparatus is unmistakeable in scale-forming organisms as a result of the gradual synthesis of the scale. In Pleurochrysis, membrane flow in the Golgi apparatus is accompanied by synthesis and assembly of the three scale subcomponents, establishing and maintaining the characteristic dorso-ventrality of the scales in the cell wall (spirals distal). Considering that each cisterna synthesizes all three subcomponents, the cisternal differentiation is actually temporal with respect to each cisterna. However, when viewed in static electron micrographs, the Golgi apparatus appears to be spatially differentiated into several regions, each of which is involved in a different synthetic event. With this distinction in mind, scale formation can be discussed in terms of spatially differentiated regions of the Golgi apparatus, beginning with the proximal, or forming face, and proceeding to the distal, or secreting face.

At the forming face of the Golgi apparatus, the cisternae are quite flattened except for characteristic dilations originally noted by MANTON (1966 a, 1967 a, see FRANKE 1970). Since these dilations frequently contained an electron-dense product, and always preceeded the appearance of any recognizable microfibrils, BROWN and colleagues (1970) called the dilations polymerization centers. Immediately distal to the polymerization centers, the first microfibrils appear as two parallel rows, one row adjacent to each of the cisternal membranes. With the observation that periodic acid-silver methenamine stained only the radial microfibrils, the staining of the polymerization centers and the double row of microfibrils by this method (Fig. 6) clearly identified these structures as events in the biosynthesis of the radial microfibrils (BROWN et al. 1973). Further support for this interpretation was provided by the localization within this Golgi region of aryl sulfatase (Fig. 7) and acid phosphatase (BROWN and ROMANOVICZ 1976), two enzymes involved in the elaboration of sulfated products (BALASUBRAMANIAN and BACHHAWAT 1970, HANKER et al. 1975). The polymerization center, or radial precursor pool, apparently contains a polysaccharide product which is fed into an adjacent region of the cisterna to be crystallized into the radial microfibrils. The close association of the microfibrils with the cisternal membranes suggests a templating mechanism or membrane-bound enzyme system as has been demonstrated in other microfibril-synthesizing systems (MÜHLETHALER 1969, BROWN et al. 1976, MUELLER and BROWN 1980, see p. 160 f.).

The process by which the double row configuration becomes the characteristic quadriradial arrangement is still a matter of speculation. Based on the observation of bent or funnel-shaped radial microfibril networks and Z-stage Golgi cisternae in which the double row is turned 90° from the typical

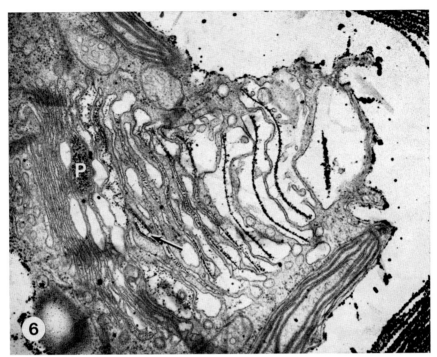

Fig. 6. Section of a glutaraldehyde-osmium fixed *Pleurochrysis* cell, stained with periodic acid-silver methenamine and counterstained with lead citrate. The secreting face of the Golgi apparatus is to the right. A polymerization center (*P*) and double row of radial microfibrils (arrow) are visible near the forming face. All scales distal to the double row are unfolded. From Brown and Romanovicz 1976. Courtesy of J. Wiley. ×29,500.

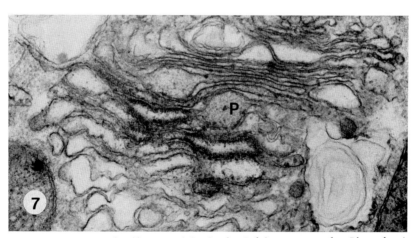

Fig. 7. Localization of aryl sulfatase activity in the Golgi apparatus of a *Pleurochrysis* cell. The proximity of a polymerization center (*P*) suggests that the localization is on a double row of radial microfibrils. The secreting face of the Golgi apparatus is downward in this figure. From Brown and Romanovicz 1976. Courtesy of J. Wiley. ×45,000.

orientation (Fig. 8), an unfolding of the radial microfibrils from the double row to the quadriradial arrangement was postulated (BROWN *et al.* 1973). Support for this proposal comes from the novel work of SLABY and SLABY (1979, personal communication) who envision the Z-stage cisterna as a point where cytoplasmic flow shifts from an inward spiral to an outward spiral, causing the separation of the two rows since they are attached to opposite cisternal membranes. With the quadriradial network a consistent feature of Prymnesiophycean scales, it would be expected that all species would share a common mechanism of quadriradial formation. MANTON (1972 b) took note of bent intracisternal scales in *Chrysochromulina*, and in a later investigation

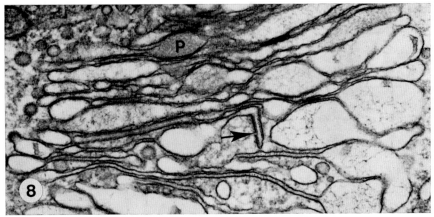

Fig. 8. A section of a *Pleurochrysis* cell, illustrating a Z-stage Golgi cisterna with folded radials (arrow), distal to a polymerization center (*P*). The secreting face of the Golgi apparatus is downward in this figure. From BROWN and ROMANOVICZ 1976. Courtesy of J. Wiley. × 44,000.

of this same organism, ALLEN and NORTHCOTE (1975) concurred with the unfolding hypothesis based on their own observations. On the other hand, PIENAAR (1969 b, 1971, 1976 b) has consistently demonstrated a process of quadriradial formation in *Hymenomonas* (*Cricosphaera*) *carterae* that does not involve unfolding. This is particularly surprising in that *Pleurochrysis* and another organism in the genus *Hymenomonas* may simply be different phases in the life cycle of a single species (LEADBEATER 1971, see section III A), so that one would expect an identical process of quadriradial formation. In addition, a double row of microfibrils is evident in a micrograph from the study of another species of *Hymenomonas* (PIENAAR 1976 a). So while it may be unwarranted to advance the unfolding process as the means of quadriradial formation in all scale-forming prymnesiophytes, this process remains the best interpretation of the observations in *Pleurochrysis*.

Addition of cellulosic spiral microfibrils to the completed quadriradial network occurs centripetally, requiring several successive cisternal differentiations. The cisternal membranes are closely appressed to the scale at the site of spiral attachment, with the cisterna becoming dilated after attachment is complete (Fig. 9). This separation of the completed scale from the cisternal

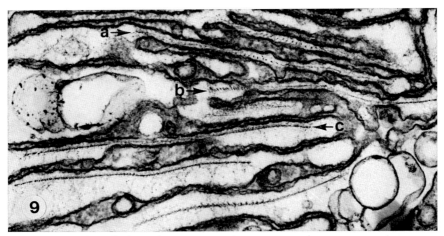

Fig. 9. A section through the Golgi apparatus of a *Pleurochrysis* cell, illustrating the centripetal addition of the spiral microfibrils (*a* to *c*). Note the proximity of the cisternal membranes to the scale at the point where the microfibrils are being added. The secreting face of the Golgi is downward in this figure. From BROWN and ROMANOVICZ 1976. Courtesy of J. Wiley. ×54,000.

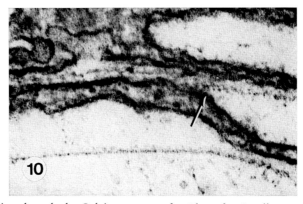

Fig. 10. A section through the Golgi apparatus of a *Pleurochrysis* cell, secreting face downward. Three spherical granules (arrow) are visible on the distal cisternal membrane at the site of spiral microfibril addition. From BROWN and ROMANOVICZ 1976. Courtesy of J. Wiley. ×108,000.

membranes had also been observed by MANTON (1972 a). As to the actual synthesis of the spiral microfibrils, there are two alternatives: 1. *in situ* synthesis directly at the point of attachment to the quadriradial network, and 2. synthesis at a single locus after which the microfibrils are fed into the lumen of the cisterna and deposited onto the radial microfibrils. The first mechanism is consistent with current theories of cellulose synthesis, which involve mobile enzyme complexes within the plasma membrane (PRESTON 1964, 1974, COLVIN 1972, ROBINSON and PRESTON 1972, ROBINSON et al. 1972, BROWN and MONTIZENOS 1976, GIDDINGS et al. 1980, MUELLER and BROWN 1980, see p. 160 f.). While these complexes are rarely demonstrated in thin section, several similar granules have been observed at the point of contact between the distal cisternal membrane and the innermost spiral microfibrils (Fig. 10). However, the central region of the scale contains sharply angled "breakage loci" (see Fig. 1), presumably the result of bending the spiral microfibrils beyond the limit of their flexibility. The "breakage loci" have been taken to imply that orientation of the spiral microfibrils occurs after their synthesis, leading to the second hypothesis. This interpretation is supported by the consistent presence of a central tubule on the distal side of the cisterna. While spiral microfibrils have never been observed in the central feeding tubule, it should be remembered that the stainability of the spiral microfibrils results not from the cellulose but from the coating material which maintains the spiral pattern. So it is conceivable that the spiral microfibrils are synthesized within the central tubule, while the coating material is added at the time of attachment to the quadriradial network. With either hypothesis the restricted space of the cisternal lumen may function as a passive orienting mechanism, constraining the microfibrils into a spiral pattern. While it may not be possible to choose between the alternatives, there is no doubt that cellulose biosynthesis in *Pleurochrysis* occurs within the Golgi cisternae. While this finding is contrary to the cell surface theory of cellulose synthesis, it has been demonstrated that the Golgi apparatus in corn root caps can synthesize microfibrillar β 1–4 glucans (WRIGHT and NORTHCOTE 1976), and KIERMAYER and colleagues (KIERMAYER and DOBBERSTEIN 1973, KIERMAYER and SLEYTR 1979) have observed that microfibrillar-synthesizing complexes on the plasma membrane of *Micrasterias* are derived from the Golgi apparatus (p. 158 f.). It is likely then, that the difference between cellulose synthesis in the Golgi apparatus and cellulose synthesis at the plasma membrane is nothing more than a difference in timing of enzyme activation. While Golgi synthesis is adequate for discrete wall subunits like scales, the relatively small space of the Golgi cisterna could not produce the extensive, coherent microfibrillar organization of typical plant cell walls.

Synthesis of the amorphous polysaccharide coating also occurs centripetally, either concomitantly with or immediately after spiral microfibril synthesis. In thin sections through the Golgi apparatus, stained with periodic acid-silver methenamine, addition of the amorphous coating can be observed as a dense staining of the scale that progresses centripetally through several successive cisternae (Fig. 11). An identical staining pattern is obtained by incubating cells for the localization of alkaline phosphatase activity (Fig. 12), frequently associated with Golgi-mediated synthesis both *in vivo* (GOLDFISHER et al. 1964, DAUWALDER et al. 1969) and *in vitro* (RAY et al. 1969). Proof that these cytochemical reactions are localizing amorphous polysaccharide synthesis rather than spiral microfibril synthesis comes from the observation that both reactions are

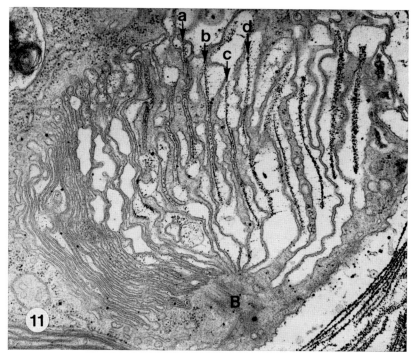

Fig. 11. A section through the Golgi apparatus of a *Pleurochrysis* cell, with the secreting face to the right. The density of periodic acid-silver methenamine staining increases centripetally as the amorphous polysaccharide coating is added to the scales (*a* to *d*). Note the orientation of the Golgi stack with respect to the basal body (*B*). From BROWN and ROMANOVICZ 1976. Courtesy of J. Wiley. ×28,000.

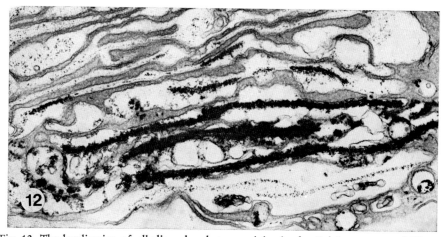

Fig. 12. The localization of alkaline phosphatase activity in the Golgi apparatus of *Pleurochrysis*, the secreting face downward. The staining increases centripetally in successive cisternae and then abruptly ceases at the most distal cisterna. From BROWN and ROMANOVICZ 1976. Courtesy of J. Wiley. ×36,000.

negative in *Pleurochrysis* zoospores, which synthesize spiral microfibrils but not the amorphous coating (BROWN and ROMANOVICZ 1976). It is possible that there is sufficient quantity of this hydrated amorphous polysaccharide to account for the dilation of the Golgi cisterna observed around the completed scale. Upon dehydration during preparation for electron microscopy, this material could collapse to yield the small, densely-staining strands seen in the dilated cisternae.

Once synthesis is complete, the scale is secreted by exocytosis, a process in which the distal-most Golgi cisterna fuses with the plasma membrane, releasing the scale into the extracellular space (MANTON 1967 b, BROWN et al. 1970). A role for microtubules in exocytosis is suggested by the apparent ability of colchicine to prevent scale secretion but not synthesis (BROWN et al. 1973, BROWN and ROMANOVICZ 1976). The Golgi apparatus in *Pleurochrysis* is flanked by an array of microtubules (BROWN and FRANKE 1971), and also exhibits a distinctive orientation with respect to the basal bodies (see Fig. 11) which may be characteristic of Prymnesiophycean Golgi apparatus (MANTON 1964, 1967 a, b, 1968 a).

Under optimal conditions, one scale is released every two minutes (BROWN 1969), implying fusion of Golgi cisternae with the plasma membrane at the same rate. For the cell to avoid a rapid increase in volume, most of the cisternal membrane must be recycled. BROWN and ROMANOVICZ (1976) suggested involvement of the extensive subsurface cisterna in this recycling process, based upon the localization of acid phosphatase activity in the distal Golgi cisterna and the subsurface cisterna, resembling the GERL system proposed by NOVIKOFF and colleagues (1971, NOVIKOFF and NOVIKOFF 1977). One hypothesis involves invaginations of the plasma membrane into the subsurface cisterna, suggesting an endocytotic mechanism (BROWN and ROMANOVICZ 1976). These invaginations have been observed in other scale-forming organisms (KLAVENESS 1973) and ingestion has been documented in *Chrysochromulina* (MANTON 1972 b). However, BROWN (personal communication) was unable to demonstrate the uptake of cationized ferritin by *Pleurochrysis*. Although the mechanism has not been elucidated, it seems certain that membrane recycling occurs in *Pleurochrysis*.

The data and interpretations presented in this section are summarized in a diagrammatic cross-sectional model of the Golgi apparatus in *Pleurochrysis* (Fig. 13). The validity of these interpretations is supported not only by morphologic observations, but also by cytochemical enzyme localization which illustrates the sharp distinction between functional regions of the Golgi apparatus as abrupt changes in enzyme activity. Cellulose biosynthesis is depicted as occurring in the distal tubule, although *in situ* synthesis is an equally viable explanation and is perhaps more consistent with data from other systems. Membrane recycling has not been included because the proposed mechanisms are still speculative. The reader is reminded once again that such static representations fail to convey the dynamic nature of the process in which a single cisterna arises at the forming face, conducts the synthesis and assembly of the entire scale, fuses with the plasma membrane, and is finally reduced to its constituent components for recycling.

III. Survey of Scale-Forming Organisms

There are more than 100 species of scale-forming organisms that have been observed with electron microscopy. These organisms are listed by genus in Table 1, along with major references and some not cited in the text. While the majority are in the class *Prymnesiophyceae*, there are also significant

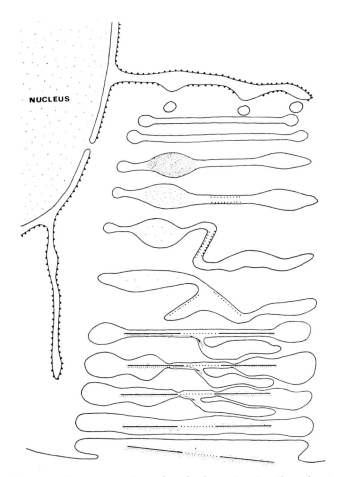

Fig. 13. A diagrammatic representation of scale formation in *Pleurochrysis*. The Golgi cisternae are depicted in median cross section, with the secreting face downward. From BROWN and ROMANOVICZ 1976. Courtesy of J. Wiley. (See also p. 164.)

numbers of species among the *Chrysophyceae* and *Prasinophyceae*. In addition, scale formation has been observed in *Dinophyceae*, *Charales*, two members of the *Chlorophyceae*, and several non-algal protists. Many of these organisms have been described from collection and are not in culture, so that laboratory studies have been limited to a small percentage of the total. For general treatments illustrating comparative scale morphology, the reader is directed to the work of HARRIS and BRADLEY (1958) on the *Chrysophyceae*,

Table 1. *Electron microscopic studies of scale-forming organisms*

Prymnesiophyceae

Apistonema	LEADBEATER 1970
Balaniger	THOMSEN and OATES 1978
Calciarcus	MANTON, SUTHERLAND, OATES 1977
Calyptrosphaera	KLAVENESS 1973
Chrysocampanula	FOURNIER 1971
Chrysochromulina	ALLEN and NORTHCOTE 1975
	LEADBEATER 1972 a
	LEADBEATER and MANTON 1969 b, 1971
	MANTON 1967 b, 1972 a, 1978 a
	MANTON, OATES, COURSE 1981
	MANTON, OATES, SUTHERLAND 1981
	MANTON and PARKE 1962
	PARKE, LUND, MANTON 1962
	THOMSEN 1977
Chrysotila	GREEN and PARKE 1975
Coccolithus (Emiliana)	KLAVENESS 1972 a and b
	KLAVENESS and PAASCHE 1971
Corymbellus	GREEN 1976
Crystallolithus	MANTON and LEEDALE 1963 b
	PARKE and ADAMS 1960
Hymenomonas (Cricosphaera)	MANTON and LEEDALE 1969
	MILLS 1975
	OUTKA and WILLIAMS 1971
	PIENAAR 1976 a and b
Imantonia	REYNOLDS 1974
Isochrysis	BILLARD and GAYRAL 1972
	GREEN and PIENAAR 1977
Navisolenia	LEADBEATER and MORTON 1973
Pappomonas	MANTON and OATES 1975
	MANTON and SUTHERLAND 1975
Papposphaera	MANTON, SUTHERLAND, McCULLY 1976 a
	TANGEN 1972
Phaeocystis	PARKE, GREEN, MANTON 1971
Platychrysis	CHRÉTIENNOT 1973
Pleurochrysis	BROWN and ROMANOVICZ 1976
	LEADBEATER 1971
	ROMANOVICZ and BROWN 1976
Prymnesium	MANTON 1966 a
	MANTON and LEEDALE 1963 a
Ruttnera	GREEN and PARKE 1974
Turrisphaera	MANTON, SUTHERLAND, OATES 1976 b
Wigwamma	MANTON, SUTHERLAND, OATES 1977
	THOMSEN 1980

Chrysophyceae

Chromulina	BELCHER and SWALE 1967
	MANTON and PARKE 1960
Chrysosphaerella	BRADLEY 1964
Mallomonas	BELCHER 1969
	HARRIS 1967
	ZIMMERMAN 1977

Table 1 (*continued*)

Mallomonopsis	HARRIS 1966
Paraphysomonas	LUCAS 1967, 1968
	TAKAHASHI 1976
	THOMSEN 1975
Phaester	BELCHER and SWALE 1971
Sphaleromantis	MANTON and HARRIS 1966
	PIENAAR 1976 c
Syncrypta	CLARKE and PENNICK 1975
Synura	HIBBERD 1973, 1978
	SCHNEPF and DEICHGRÄBER 1969
Prasinophyceae	
Halosphaera	BOALCH and MOMMAERTS 1969
	PARKE and HARTOG-ADAMS 1965
Heteromastix	MANTON, RAYNS, ETTL, PARKE 1965
Mesostigma	MANTON and ETTL 1965
Micromonas	MANTON and PARKE 1960
	MOESTRUP and ETTL 1979
Nephroselmis	PARKE and RAYNS 1964
Platymonas	LEADBEATER 1972 b
	MCLACHLAN and PARKE 1967
Prasinocladus	HORI and CHIHARA 1974
Pseudoscourfieldia	MANTON 1975
Pyramimonas	MANTON 1966 b, 1968 b
	MOESTRUP and WALNE 1979
	NORRIS and PEARSON 1975
	SWALE and BELCHER 1968
Pterosperma	PARKE, BOALCH, JOWETT, HARBOUR 1978
Dinophyceae	
Oxyrrhis	CLARKE and PENNICK 1972, 1976
Heterocapsa	PENNICK and CLARKE 1977
Chlorophyceae	
Chaetosphaeridium	MOESTRUP 1974 b
Choleochaete	MCBRIDE 1968
Charales	
Chara	MOESTRUP 1970
	PICKETT-HEAPS 1968
Nitella	TURNER 1968
Non-algal Protists	
Althornia	JONES and ALDERMAN 1970
Ceratiomyxella	FURTADO and OLIVE 1971, 1972
Labyrinthuloides	PERKINS 1973
Paramoeba	GRELL and BENWITZ 1966
Schizochytrium	DARLEY, PORTER, FULLER 1973
Sorodiplophrys	DYKSTRA 1976
Thraustochytrium	DARLEY, PORTER, FULLER 1973
	CHAMBERLAIN 1980

reports on nanoplankton collection by LEADBEATER (1972 a, b, 1974) that encompass several algal classes, and the work of OLIVE (1975) in which he discusses scale formation in the non-algal protists.

A. Prymnesiophyceae

Members of the class *Prymnesiophyceae* (HIBBERD 1976) characteristically possess two smooth flagella, a haptonema which may be greatly reduced or even lacking in some species (GREEN and PIENAAR 1977), and a cell covering of organic microfibrillar scales. The scaly covering may also extend to the haptonema (MANTON and PETERFI 1969, PIENAAR 1976 b), with haptonema scales typically smaller than body scales. The number of scale layers is quite variable, the thick wall of *Pleurochrysis* representing an upper limit. In species which produce more than one scale type, the different scales are frequently organized into separate layers within the wall (PARKE *et al.* 1959, LEADBEATER and MANTON 1969 a, THOMSON 1977). There have been suggestions of the scales being attached to a membrane external to the plasma membrane in some species (PARKE and MANTON 1962, FOURNIER 1971), and one demonstration of a "skin" external to the wall itself (KLAVENESS 1973), but these appear to be isolated cases not typical of the class as a whole. Members of the order *Pavlovales* do not produce scales (FOURNIER 1969, GREEN and PIENAAR 1977) and lack the Golgi dilations characteristic of scale-forming Prymnesiophytes (GREEN and HIBBERD 1977). However, several species of *Pavlova* exhibit small dense bodies on one flagellum (GREEN and MANTON 1970, GREEN 1973, 1975, 1980) which reportedly arise in the Golgi apparatus (VAN DER VEER 1976). These dense bodies could be interpreted as reduced scales, although they apparently lack the microfibrillar substructure typical of the class.

Prymnesiophycean scales are typically elliptical to circular, from 0.5 to over 10 µm in diameter, and consist of a quadriradial network of microfibrils on the proximal surface and a spiral arrangement of microfibrils on the distal surface. Although this scale pattern is consistent throughout the class, there is considerable variation in the numbers of microfibrils comprising the scales, even within the same genus. For example, the scales of *Chrysochromulina hirta* contain closely packed spiral microfibrils (MANTON 1978 a) resembling the arrangement in *Pleurochrysis*, while the scales of one strain of *C. chiton* appear to have no spiral microfibrils at all (LEADBEATER 1972 b). Radial microfibrils exhibit a similar variation in number, with *Imantonia* scales (REYNOLDS 1974) possessing the fewest radials of any Prymnesiophyte. Deviation from the basic pattern include the scales of *Chrysochromulina tenuispina* which are composed of two layers each of radials and spirals (MANTON 1978 b), the scales of *Navisolenia aprilei* which are rhombic in shape (LEADBEATER and MORTON 1973), and the scales of *Chrysochromulina parkeae* which are elongated into 20–30 µm spines (GREEN and LEADBEATER 1972). Despite the variation in shape, scales of the latter two species still possess recognizable proximal and distal microfibrils. ALLEN and NORTHCOTE (1975) used the term "woven" to describe the scales of *Chrysochromulina*, but their micrograph is not convincing. Such a structure would be quite different from other Prymnesiophycean scales and would require a different biosynthetic mechanism (see section II. B.). The one characteristic shared by all Prymnesiophytes that produce scales is use of the Golgi apparatus to synthesize and secrete the scales, as illustrated with *Pleurochrysis* in the previous section.

The largest genus in this class is *Chrysochromulina*, containing more than two dozen species that produce scales. Modifications of the basic plate scale shape include the trumpet scales of *C. erincina* (LEADBEATER 1972 b), the

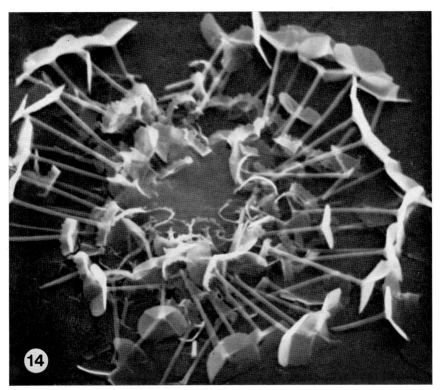

Fig. 14. Scanning electron micrograph of the coccoliths from *Papposphaera lepida*. The four petal-like lobes, supporting shaft, and basal "crown" are crystallites of calcium carbonate elaborated upon an organic base plate scale. From TANGEN 1972. Courtesy of Universitets-forlaget. ×8,200.

spines of *C. parkeae* (GREEN and LEADBEATER 1972) and the cylindrical elaborations found on the outer scales of *C. microcylindra* and *C. megacylindra* (LEADBEATER 1972 a). The process of scale formation has been investigated in *C. chiton* (MANTON 1967 a, b), *C. mactra* (MANTON 1972 a), and *C. microcylindra* (LEADBEATER 1972 a). These studies illustrate a Golgi-mediated synthesis similar to that in *Pleurochrysis*, particularly the proximity of the cisternal membrane to the developing scale and the separation of the two when synthesis is complete.

The coccolithophorids are unique members of the *Prymnesiophyceae* that modify the rim of the base plate scale with the deposition of discrete calcium

carbonate crystallites (MANTON and LEEDALE 1969, MANTON and PETERFI
1969, MANTON et al. 1977). Coccoliths typically form the outer layer of the
cell covering, with layers of non-calcified scales underneath. An exception
to this pattern is found in Coccolithus huxleyi, a species in which base plate
scales are absent from the coccoliths and there are no underlying layers of
organic scales (KLAVENESS 1972 a). In contrast to the Pavlovales, which also
do not produce scales, the Golgi apparatus of C. huxleyi does exhibit the
characteristic dilations. The morphology of coccoliths varies from the simple
ring of Hymenomonas (Cricosphaera) carterae (FRANKE and BROWN 1971) to
the elaborate lobed coccoliths of Papposphaera (TANGEN 1972, Fig. 14), and
is a reflection of the structure of the crystallites themselves (MANTON et al.
1977). The distinctive crystallite morphology and the persistance of calcium
carbonate in marine sediments has permitted the identification of several
hundred extant and fossil coccolithophorids (LOEBLICH and TAPPAN 1971).

An interest in biological calcification has led to investigation of the
physiological and ultrastructural aspects of coccolith formation, principally
in Coccolithus (Emiliana) huxleyi and Hymenomonas (Cricosphaera) carterae.
An early review by PAASCHE (1968) discussed the effects of culture conditions
on coccolith production and lamented the relative paucity of laboratory
studies on coccolithophorids. BLANKELY (1971) devised a defined growth
medium and investigated the utilization of organic carbon sources. In an
ultrastructural study, PIENAAR (1969 a, b) demonstrated the process of intracellular coccolith formation, but accepted an earlier misinterpretation of
ISENBERG et al. (1966) that identified the lysosome-like autophagic vacuole
as a precursor to the coccolith. This misconception was corrected by OUTKA
and WILLIAMS (1971) who demonstrated that coccolith synthesis occurs
entirely within the Golgi cisternae, and by PIENAAR (1971) himself who used
cytochemistry to support the lysosomal nature of the supposed coccolith
precursor. Calcification of the base plate scale occurs in a T-shaped cisterna,
in which the coccolith occupies the cross of the T, oriented along the edge of
the Golgi stack, perpendicular to the other cisternae (MANTON and LEEDALE
1969, PIENAAR 1969 b, OUTKA and WILLIAMS 1971). The actual precipitation
of calcium carbonate is preceeded by the appearance of an organic matrix
with a shape identical to that of the completed coccolith (OUTKA and
WILLIAMS 1971, DE JONG et al. 1976). Beyond the obvious implication that
this organic matrix is somehow involved in calcification, there is little knowledge of the actual mechanism.

Correlation of existing data on Prymnesiophytes is often hampered by the
complicated and sometimes poorly understood life histories of these organisms.
Manipulation of culture conditions can induce the non-motile cells of Pleurochrysis scherffelii to form a motile, coccolith-bearing phase (BONEY and
BURROWS 1966, LEADBEATER 1971, BROWN and ROMANOVICZ 1976), identified
as Hymenomonas (Cricosphaera) pringsheimii (PARKE and DIXON 1976). In
the study of the alternation between the non-motile scale-forming Apistonema
carterae and the motile coccolithophorid Hymenomonas (Cricosphaera)
carterae, RAYNS (1962) concluded that the coccolith-bearing Hymenomonas
phase is diploid and the scale-bearing Apistonema phase is haploid (see also

LEADBEATER 1970). However, OUTKA (personal communication) has recently obtained evidence that the *Hymenomonas* phase is haploid. KLAVENESS and PAASCHE (1971) and KLAVENESS (1972 b) demonstrated three cell types in the life history of *Coccolithus (Emiliana) huxleyi:* non-motile coccolithophorids (C cells), motile scale-formers (S cells), and naked non-motile cells (N cells). All three cell types are reportedly capable of vegetative reproduction. In one case there are even different types of coccoliths produced in the life history: the coccoliths of the non-motile *Coccolithus pelagicus* contain crystallites that are restricted to the rim of the base plate scale, while coccoliths of the motile phase (previously named *Crystallolithus hyalinus*) have crystallites over the entire distal surface (PARKE and ADAMS 1960, MANTON and LEEDALE 1963 b). It is likely that similar relationships will be discovered between other existing species. When these life histories are fully understood and reproducible, such organisms may prove useful in studying the genetic control of scale synthesis.

B. Chrysophyceae

Organisms placed in the class *Chrysophyceae* typically possess two unequal flagella, the longer bearing two rows of stiff hairs. Scales are not common in this class (HIBBERD 1976) but are found in nine genera representing approximately 20 species. The scaly covering on these organisms usually consists of a single scale type on the cell body, though at least one species of *Mallomonas* bears scales of three distinct shapes, segregated about the anterior, middle, and posterior of the cell (HARRIS 1967, 1970 a). Another member of this genus, *M. multiunca*, has only one type of body scale, but in addition bears scales on the flagella (ZIMMERMAN 1977). Flagellar scales are not typical of Chrysophycean scale-formers, but they are also found in *Synura sphagnicola* (HIBBERD 1978) and two species of *Sphaleromantis* (MANTON and HARRIS 1966, PIENAAR 1976 c).

Chrysophycean body scales can be divided into five distinct types. The most common are flat, non-fibrillar plates, often perforated or bearing spines, and reportedly containing silica (HARRIS 1966, HIBBERD 1978). Plate scales have been observed in species in the genera *Chrysosphaerella* (HARRIS and BRADLEY 1958), *Mallomonas* (HARRIS 1967, 1970 a, b), *Mallomonopsis* (HARRIS 1966), *Paraphysomonas* (LUCAS 1967, LEADBEATER 1974), and *Synura* (HARRIS and BRADLEY 1958, HIBBERD 1978). Several other species in the genus *Paraphysomonas* bear a second type of scale, a meshwork or reticulate structure constructed of discrete rods on the order of 30–50 nm thick (LUCAS 1968, PENNICK and CLARKE 1972, 1973, REES *et al.* 1974). No information on the origin or composition of the meshwork scales is available. Two chrysophytes, *Chromulina placentula* (BELCHER and SWALE 1967) and *Phaester pascheri* (BELCHER and SWALE 1971) possess thin membranous scales with no discernible substructure. A fourth type of scale, shaped like a flower pot and often developing branching processes, has been observed on the body and flagella of *Sphaleromantis tetragona* (MANTON and HARRIS 1966). Similar cylindrical scales, all bearing distal projections of consistent number and length, cover the body and flagella of *Sphaleromantis marina* (PIENAAR

1976 c). Finally, a fibrillar body scale has been demonstrated in *Syncrypta glomerifera* (Clarke and Pennick 1975). This organism appears to be a chrysophyte, but the random jumble of fibrils that comprises the scale is found in no other member of this class.

There have been only four investigations of scale formation in the *Chrysophyceae*, and these have yielded contradictory results. Manton and Leedale (1961 c) observed intracellular scales in *Paraphysomonas vestita*, but could not convincingly establish their origin. In *Sphaleromantis tetragona* (Manton and Harris 1966) and *Synura sphagnicola* (Hibberd 1978) the scales are clearly produced by the Golgi apparatus. However, Schnepf and Deichgräber (1969) ruled out a Golgi origin for the scales of *Synura petersenii*, suggesting instead that scale formation occurred in vesicles derived from the endoplasmic reticulum. It is clear that further investigation is required into the process of scale formation in this class of algae.

C. Prasinophyceae

The class *Prasinophyceae* contains green algae that possess one to eight flagella, and layers of scales on the flagella and cell body. An individual cell may have as many as four layers of body scales, with each layer composed of a morphologically distinct scale type. Moestrup and Walne (1979) demonstrated six different scale types in *Pyramimonas tetrarhynchus,* three types on the body and three on the flagella. The authors estimated that a 15 × 22 µm cell would bear more than 300,000 scales. Cells of the genus *Halosphaera* exhibit three layers of body scales and two layers of flagellar scales (Boalch and Mommaerts 1969). In contrast, on *Nephroselmis gilva* there is only a single layer of body scales and a single layer of flagellar scales (Parke and Rayns 1964, Leadbeater 1972 b). Cells in the genera *Platymonas* (Manton and Parke 1965, McLachlan and Parke 1967) and *Prasinocladus* (Parke and Manton 1965) bear scales on the flagella, but the cell body is covered with a theca instead of scales. However, it has been demonstrated in *Platymonas* that the theca is formed by the fusion of stellate plates made in the Golgi apparatus, and thus is analogous to a layer of scales (Manton and Parke 1965, Gooday 1971).

Prasinophycean scales are of three distinct types. The scales that form the innermost layer are polygonal, frequently diamond shaped, 30 to 50 nm in diameter (Manton and Ettl 1965, Swale 1973, Parke et al. 1978). The sculptured surface of these small scales gives them a lacy or spider's web appearance in shadowcast preparations. Scales comprising the outer layers, particularly on the cell body, are of an open meshwork construction, resembling the reticulate scale type found in the *Chrysophyceae*. These meshwork scales are larger than the plate scales, having dimensions in the 0.25 to 0.5 µm range (Manton and Ettl 1965, Parke et al. 1978). In some genera this scale type may have a stellate appearance, with as many as 20 arms radiating from a common center (Moestrup and Ettl 1979). The third type of scale is unique to this class: a long, hollow scale found only on the flagella. As these give the appearance of flagellar hairs, they are called hair scales (Moestrup and Thomsen 1974) but are quite distinct from the mastigonemes of heterokont flagella. Other than the obvious fact that *Prasinophycean* scales are unmineralized, there is no information on their composition. The theca of *Platymonas tetrahele* has been shown to be polysaccharide, composed of galactose, arabinose, and galacturonic acid (LEWIN 1958, GOODAY 1971).

Investigations of scale formation in the *Prasinophyceae* have consistently demonstrated a Golgi origin (PARKE and RAYNS 1964, MANTON *et al.* 1965), with all scale types being produced at the same time (MANTON and ETTL 1965, MOESTRUP and THOMSEN 1974). In *Pyramimonas tetrarhynchus*, four of the six scale types have been observed during synthesis within a single Golgi cisterna, the smaller scales at the periphery and the larger ones in more central regions (MOESTRUP and WALNE 1979). In *Platymonas*, however, the flagellar scales and the particles that form the theca are reportedly not synthesized simultaneously (PARKE and MANTON 1967). In their detailed account of scale synthesis in *P. tetrarhynchus*, MOESTRUP and WALNE (1979) emphasized the close relationship of the cisternal membrane to the developing scale, which has previously been discussed as a key feature of Prymnesio-phycean scale formation. A major difference between scale formation in Prasinophytes and that described for *Pleurochrysis* lies in the process of exo-cytosis. In the *Prasinophyceae*, the scales collect in a reservoir within the cytoplasm (MANTON 1966 b), and become attached to the reservoir membrane in the ordered pattern characteristic of the scales on the cell surface (MOESTRUP and THOMSEN 1974). The reservoir membrane apparently becomes new flagellar and plasma membrane with scales already in place, a process analogous to that described by BOUCK (1971) for the formation of mastigo-nemes in heterokont algae. This also implies the absence of any membrane recycling, the cell apparently increasing in size until mitosis (MOESTRUP and THOMSEN 1974).

Although the life histories of *Prasinophyceae* are apparently much less complicated than those of Prymnesiophytes, there have been reports of distinct phases within some species. MANTON *et al.* (1963) suggested that *Pyramimonas* may be a phase in the life cycle of *Halosphaera*. PARKE *et al.* (1978) have recently demonstrated a motile and a non-motile phase in the genus *Ptero-sperma*, with scales occurring only on the motile phase.

D. Other Plants

While the presence of scales on Prymnesiophytes, Chrysophytes, and Prasinophytes has been known for twenty years, the demonstration of scales on other plant cells is the result of more recent investigation. CLARKE and PENNICK (1972) observed flagellar scales in the dinophycean species *Oxyrrhis marina* and were later able to demonstrate the presence of body scales as well (CLARKE and PENNICK 1976). Both scale types are flat, framework plates of similar appearance, with the 0.25 μm body scales perhaps somewhat larger than the flagellar scales. The same authors also observed framework scales outside the theca in the dinoflagellate *Heterocapsa triquetra* (PENNICK and CLARKE 1977). Their studies give no information of the composition or origin of these scales. Potentially more significant are the observations of small flagellar scales on the male gamete of *Chara* (PICKETT-HEAPS 1968, MOESTRUP

1970) and *Nitella* (TURNER 1968), and on the zoospores of the Chlorophycean algae *Choleochaete* (MCBRIDE 1968) and *Chaetosphaeridium* (MOESTRUP 1974 b). The similarity between these scales and those found in the *Prasinophyceae*, along with ultrastructural similarities between *Chloeochaete/Chaetosphaeridium* and certain higher plants, led MOESTRUP (1974 b, MOESTRUP and ETTL 1979) to propose a phyletic scheme in which the scale-bearing organisms are the ancestors of higher plants. Finally, it should be noted that Golgi synthesis has been demonstrated for the scales of *Nitella* and *Chaetosphaeridium* (TURNER 1968, MOESTRUP 1974 b).

E. Non-Algal Protists

This section has been reserved for those organisms whose only apparent affinity with any previously discussed is the presence of scales on the cell surface. FURTADO and OLIVE (1971, 1972) observed membranous scales with inflexed rims on the surface of the mycetozoan *Ceratiomyxella tahitiensis*. The scales were also found within vesicles in the cytoplasm, and a Golgi origin was suggested but not clearly established. Layers of membranous scales cover the cell surface of the thraustochytrids, a group of related organisms nominally classified as fungi (see OLIVE 1975). The scales of *Schizochytrium* and *Thraustochytrium* are approximately 1 μm in diameter and are composed of protein and polysaccharide, with L-galactose as the principle monosaccharide (DARLEY et al. 1973). The scales of *Sorodiplophrys* are larger, 2 to 4 μm in diameter, and the principle monosaccharide component appears to be arabinose (DYKSTRA 1976). In both studies it was demonstrated that the scales are synthesized in the Golgi apparatus. The cells of the choanoflagellate *Choanoeca* are surrounded by a theca that appears to be formed from small granules produced in the Golgi apparatus (LEADBEATER 1977), similar to the thecal formation in *Platymonas*. Other members of this group assemble a lorica extracellularly from costal strips produced within the cell (LEADBEATER 1979 a and b). The choanoflagellates were originally classified in the *Chrysophyceae*, but probably do not belong in the plant kingdom at all (HIBBERD 1976). Finally, there is the work of GRELL and BENWITZ (1966), who observed box-like scales on the surface of *Paramoeba eilhardi*.

F. Phyletic Significance of Scales

Scales occur on such diverse organisms and are so varied in morphology, that it seems likely they are an ancient character. This idea has, of course, been put forth by numerous other investigators. MANTON and PARKE (1965), in discussing the *Prasinophyceae*, suggested that the presence of scales is more primitive than the presence of chlorophyll b. In papers on the *Prymnesiophyceae*, MANTON (1967 a, 1972 a) has considered the scales and the distinctive Golgi apparatus in which they are synthesized to be primitive within that class also. However, the situation in the *Chrysophyceae* is curious. The presence of scales is uncommon in this class and the scales themselves are quite diverse. The membranous scales of *Phaester* and *Chromulina* resemble those of the thraustochytrids, the meshwork scales of some *Paraphysomonas* species are similar in appearance to the outer body scales of the *Prasino-*

phyceae, and flagellar scales are present in *Sphaleromantis* and several of the *Synuraceae.* Also, there seems to be no common pathway of scale formation, not even within the single genus *Synura.* Despite all this, HIBBERD (1976) interprets the cellular organization of these genera to be modifications of the *Ochromonas* pattern typical of the *Chrysophyceae,* supporting their inclusion in that class.

OLIVE (1975) used the presence of scales in *Ceratiomyxella* to support the idea that the mycetozoans had a phytoflagellate ancestor. The scales of the thraustochytrids could no doubt be interpreted in the same manner, especially in light of the membranous scales of some *Chrysophyceae.* Based on the presence of discrete scales, and their composition, DARLEY *et al.* (1973) suggested that the thraustochytrids do not belong in the *Oomycetes,* where they had previously been grouped. It is interesting, however, to consider that certain biochemical characteristics have suggested that the *Oomycetes,* distinct from other fungi in many ways, have affinities with the green plant line.

Perhaps the most far-reaching phyletic interpretation with regard to scales is that of MOESTRUP (1974 b, MOESTRUP and THOMSEN 1974, MOESTRUP and ETTL 1979). He suggests that the *Prasinophyceae* are ancestral to the *Chlorophyceae,* even though the "Vierergruppe" flagellar root system of the scale-forming *Chlorophyceae* is distinct from the cruciate root system found in the *Prasinophyceae* and non-scaly members of the *Chlorophyceae.* However, in possession of the "Vierergruppe" and other cytological characteristics, the scale-forming *Chlorophyceae* and the *Charales* bear a striking similarity to higher plants. Consequently, MOESTRUP (1974 b) has proposed that the scale forming green algae are possible ancestors of higher plants.

IV. Concluding Remarks

The process of scale formation is surprisingly similar in all organisms in which it has been studied. A highly polarized Golgi stack, close association of the cisternal membranes with the developing scale, and a proximal/distal scale orientation that is maintained throughout synthesis are consistent features of this process. The only exception is *Synura petersenii,* in which the vesicles responsible for scale formation do not appear to be derived from the Golgi apparatus. Even in this organism, scale synthesis is accomplished by the endomembrane system, and appears to involve a similar process of membrane control over final scale morphology (SCHNEPF and DEICHGRÄBER 1969). There remain, however, many unanswered questions about scale formation.

The precise and specific pattern of scale arrangement exhibited by some scale-bearing organisms prompts one to consider how this arrangement is achieved. MANTON (1966 b), realizing that an arrangement mechanism would be distinct from the synthetic process and would necessarily operate after the scales had been released into the extracellular space, suggested the existence of "receiving sites" in the membrane surface. This is a specific example of the problem of "recognition" that arises in many cell processes. In *Pleurochrysis* there does not appear to be a need for an arrangement mechanism, as there is

only one scale type and no specific pattern in the scale layers. The coccolithophorids bear coccoliths only as an outer layer, with unmineralized scales
beneath this layer. While it seems reasonable that such an arrangement could
arise from a temporal separation of coccolith and plate scale synthesis, both
scale types are observed in the Golgi stack at the same time. It has been
demonstrated that the coccolith layer can be removed by lowering the pH of
the culture medium and will be replaced after transfer to the original medium
(see PAASCHE 1968). Perhaps closer scrutiny of this process would yield
observations relevant to the arrangement question. The most elaborate scale
arrangements occur on the *Prasinophyceae*. The demonstration of a reservoir
in which the scales are already arranged, and the suggestion that the reservoir
membrane becomes the cell surface membrane (MOESTRUP and THOMSEN
1974) provide a plausible arrangement mechanism in these organisms. It is
likely that the reservoir membrane is derived from Golgi cisternae, so perhaps
the scales never completely detach from the cisternal membrane and arrive
at the reservoir already arranged. Thecal formation in *Platymonas* is the
most extreme example of the arrangement problem, with fusion of small
particles into a solid theca occurring outside the cell membrane, presumably
beyond any cytoplasmic control. However, this process does take place within
the restricted space of the parent theca, which may even provide a template
for the formation of the new theca.

Another interesting feature of scale formation is the precise movements of
cisternal membranes. The unfolding of the radial network in *Pleurochrysis*,
and the apparent templating seen in more elaborate scale types where the
cisterna exhibits the shape of the scale before synthesis begins (LEADBEATER
1972 a, MANTON 1972 b), are two examples of this. What is the motive force
responsible for these changes in shape, and how is the force applied? In
Pleurochrysis, colchicine does not prevent scale synthesis, ruling out involvement of the prominent microtubular bundles that flank the Golgi apparatus.
There is no information from other studies of scale formation that even applies
to this question.

All of the above considerations are included in the basic question of how
genetic control is exerted over the process of scale formation in the Golgi
apparatus. It is clear that the cisternal membrane plays a crucial role in
determining scale morphology; and, according to the membrane flow hyo-
thesis, cisternal membrane arises from the nuclear envelope and the endoplasmic reticulum. So it may be inferred that information relating to scale
morphology is incorporated into the structure of the cisternal membrane, as in
the arrangement of synthetic enzymes. BROWN and ROMANOVICZ (1976),
observing apparent radial precursor pools that were still attached to the endoplasmic reticulum at the forming face of the Golgi apparatus, suggested that
intracisternal differentiation might begin at this point. An even more
elaborate intracisternal differentiation occurs in *Pyramimonas tetrarhynchus*,
where different scale types are synthesized within a single Golgi cisterna,
with each scale type restricted to a specific cisternal region (MOESTRUP and
WALNE 1979). A careful study of the origin of the Golgi cisternae in this
organism might provide further insight. It has been suggested that the amount

of membrane added to the forming face is insufficient to account for the cisternal dilation observed across the Golgi stack (see WHALEY 1975). If additional membrane is added to cisternae already involved in synthesis, further "recognition" is required to achieve correct cisternal structure. The numerous coated vesicles at the Golgi periphery are often invoked as a source of additional cisternal membrane, but it has never been established that these vesicles are fusing with, rather than blebbing from, the cisternae.

The involvement of the Golgi apparatus in scale formation is a particularly vivid example of the general role of the Golgi in synthesizing the cell coat. The discrete nature of the scales, and their relatively large size with respect to typical cell coat components, offer unique advantages to investigators. Specific alterations of the process may be achieved by changing culture conditions, introducing chemical agents, or inducing a change from one phase of the life history to another. One such alteration occurs in *Platymonas convolutae*, a prasinophyte that normally possesses a theca and flagellar scales, but apparently produced neither while in endosymbiotic association with the worm *Convoluta roscoffensis* (PARKE and MANTON 1967). Another advantage already mentioned is the ability to determine the composition and precise site of synthesis of a particular cell product. While much remains to be learned about scale formation itself, it is anticipated that future research into scale formation will contribute as well to the understanding of the Golgi apparatus.

Acknowledgements

The author wishes to acknowledge the numerous contributions of R. MALCOLM BROWN, jr., to the work on *Pleurochrysis*, and his role in stimulating a general interest in scale-forming organisms. Acknowledgment is also due LINDA K. McKEE for typing the manuscript and BETTINA A. BUSSARD for assistance in its preparation.

References

ALBERSHEIM, P., BAUER, W. D., KEEGSTRA, K., TALMADGE, K. W., 1973: The structure of the wall of suspension-cultured sycamore cells. In: Biogenesis of plant cell wall polysaccharides (LOEWUS, F., ed.), pp. 114—147. New York: Academic Press.

ALLEN, D. M., NORTHCOTE, D. H., 1975: The scales of *Chrysochromulina chiton*. Protoplasma 83, 389—412.

BALASUBRAMANIAN, A. S., BACHHAWAT, B. K., 1970: Sulfate metabolism in brain. Brain Res. 20, 341—360.

BELCHER, J. H., 1969: Some remarks on *Mallomonas papillosa* Harris and Bradley and *M. calceolus* Bradley. Nova Hedwigia 18, 257—270.

— SWALE, E. M. F., 1967: *Chromulina placentula* sp. nov. (Chrysophyceae), a freshwater nanoplankton flagellate. Br. Phycol. Bull. 3, 257—267.

— — 1971: The microanatomy of *Phaester pascheri* Scherffel (Chrysophyceae). Br. Phycol. J. 6, 157—169.

BILLARD, C., GAYRAL, P., 1972: Two new species of *Isochrysis* with remarks on the genus *Ruttnera*. Br. Phycol. J. 7, 289—297.

BLANKELY, W. F., 1971: Auxotrophic and heterotrophic growth and calcification in coccolithophorids. Ph. D. dissertation, University of California at San Diego.

BOALCH, G. T., MOMMAERTS, J. P., 1969: A new punctate species of *Halosphaera*. J. mar. biol. Ass. U.K. 49, 129—139.

BONEY, A. D., 1967: Experimental studies on the benthic phases of *Haptophyceae*. II. Studies on "aged" benthic phases of *Pleurochrysis scherffelii* E. G. Pringsh. J. Exp. Mar. Biol. and Ecol. 1, 7—33.

— BURROWS, A., 1966: Experimental studies on the benthic phases of *Haptophyceae*. I. Effects of some experimental conditions on the release of coccolithophorids. J. mar. biol. Ass. U.K. 46, 295—319.

BOUCK, G. B., 1971: The structure, origin, isolation, and composition of the tubular mastigonemes of the *Ochromonas* flagellum. J. Cell Biol. 50, 362—384.

BRADLEY, E. E., 1964: A study of *Mallomonas, Synura,* and *Chrysosphaerella* of Northern Iceland. J. Gen. Microbiol. 37, 321—333.

BROWN, R. M., jr., 1969: Observations on the relationship of the Golgi apparatus to wall formation in the marine Chrysophycean alga *Pleurochrysis scherffelii* Pringsheim. J. Cell Biol. 41, 109—123.

— FRANKE, W. W., 1971: A microtubular crystal associated with the Golgi field of *Pleurochrysis scherffelii*. Planta 96, 354—363.

— — KLEINIG, H., SITTE, P., 1969: Cellulosic wall component produced by the Golgi apparatus of *Pleurochrysis scherffelii*. Science 166, 894—896.

— — — FALK, H., SITTE, P., 1970: Scale formation in Chrysophycean algae. I. Cellulosic and noncellulosic wall components made in the Golgi apparatus. J. Cell Biol. 45, 246—269.

— HERTH, W., FRANKE, W. W., ROMANOVICZ, D., 1973: The role of the Golgi apparatus in the biosynthesis and secretion of a cellulosic glycoprotein in *Pleurochrysis:* a model system for the synthesis of structural polysaccharides. In: Biogenesis of plant cell wall polysaccharides (LOEWUS, F., ed.), pp. 207—257. New York: Academic Press.

— MONTIZENOS, D., 1976: Cellulose microfibrils: Visualization of biosynthetic and orienting complexes in association with the plasma membrane. Proc. Nat. Acad. Sci. U.S.A. 73, 143—147.

— ROMANOVICZ, D. K., 1976: Biogenesis and structure of Golgi-derived cellulosic scales in *Pleurochrysis*. I. Role of the endomembrane system in scale assembly and exocytosis. Applied Polymer Symposium 28, 537—585.

— WILLISON, H. M., RICHARDSON, C. L., 1976: Cellulose biosynthesis in *Acetobacter xylinum:* Visualization of the site of synthesis and direct measurement of the *in vivo* process. Proc. Nat. Acad. Sci. U.S.A. 73, 4565—4569.

CAUFIELD, D. F., 1971: Crystallite sizes in wet and dry *Valonia ventricosa*. Textile Res. J. 41, 267—269.

CHAMBERLAIN, A. H. L., 1980: Cytochemical and ultrastructural studies on the cell walls of *Thraustochytrium* spp. Bot. Mar. 13, 669—677.

CHRÉTIENNOT, M.-J., 1973: The fine structure and taxonomy of *Platychrysis pigra* Geitler (*Haptophyceae*). J. mar. biol. Ass. U.K. 53, 905—914.

CLARKE, K. J., PENNICK, N. C., 1972: Flagellar scales in *Oxyrrhis marina* Dujardin. Br. Phycol. J. 7, 357—360.

— — 1975: *Syncrypta glomerifera* sp. nov., a marine member of the *Chrysophyceae* bearing a new form of scale. Br. Phycol. J. 10, 363—370.

— — 1976: The occurrence of body scales in *Oxyrrhis marina* Dujardin. Br. Phycol. J. 11, 345—348.

COLVIN, J. R., 1972: The structure and biosynthesis of cellulose. In: CRC critical reviews of macromolecular sciences, Vol. 1, pp. 47—81.

DARLEY, W. M., PORTER, D., FULLER, M. S., 1973: Cell wall composition and synthesis via Golgi-directed scale formation in the marine eukaryote *Schizochytrium aggregatum*, with a note on *Thraustochytrium*. Arch. Mikrobiol. 90, 89—106.

DAUWALDER, M., WHALEY, W. G., KEPHART, J. E., 1969: Phosphatases and differentiation of the Golgi apparatus. J. Cell Sci. 4, 455—497.

DYKSTRA, M. J., 1976: Wall and membrane biogenesis in the unusual labyrinthulid-like organism *Sorodiplophrys stercorea*. Protoplasma 87, 329—346.

FOTT, I. B., 1955: Scales of *Mallomonas* observed in the electron microscope. Preslia 27, 280—287.

FOURNIER, R. O., 1969: Observation of the flagellate *Diacronema vlkianum* Prauser (*Haptophyceae*). Br. Phycol. J. **4**, 185—190.

— 1971: *Chrysocampanula spinifera* gen. et sp. nov., a new marine haptophyte from the Bay of Chaleurs, Quebec. Phycologia **10**, 89—92.

FRANKE, W. W., 1970: Central dilations in maturing Golgi cisternae—a common structural feature among plant cells? Planta **90**, 370—373.

— BROWN, R. M., jr., 1971: Scale formation in chrysophycean algae. III. Negatively stained scales of the coccolithophorid *Hymenomonas*. Arch. Mikrobiol. **77**, 12—19.

FURTADO, J. S., OLIVE, L. S., 1971: Ultrastructure of the protostelid *Ceratiomyxella tahitiensis* including scale formation. Nova Hedwigia **21**, 537—576.

— — 1972: Scale formation in a primitive mycetozoan. Trans. Amer. Micros. Soc. **91**, 594—596.

GIDDINGS, T. H., jr., BROWER, D. L., STAEHELIN, L. A., 1980: Visualization of particle complexes in the plasma membrane of *Micrasterias denticulata* associated with the formation of cellulose fibrils in the primary and secondary cell walls. J. Cell Biol. **84**, 327—339.

GOLDFISCHER, S., ESSNER, E., NOVIKOFF, A. B., 1964: The localization of phosphatase activities at the level of ultrastructure. J. Histochem. Cytochem. **12**, 72—95.

GOODAY, G. W., 1971: A biochemical and autoradiographic study of the role of the Golgi bodies in thecal formation in *Platymonas tetrahele*. J. Exp. Botn. **22**, 959—971.

GREEN, J. C., 1973: Studies in fine-structure and taxonomy of flagellates in the genus *Pavlova*. II. A freshwater representative, *Pavlova grenifera* (Mack) comb. nov. Br. Phycol. J. **8**, 1—12.

— 1975: The fine-structure and taxonomy of the Haptophycean flagellate *Pavlova lutheri* (Droop) comp. nov. (= *Monochrysis lutheri* Droop). J. mar. biol. Ass. U.K. **55**, 785—793.

— 1976: *Corymbellus aureus* gen. et sp. nov., a new colonial member of the *Haptophyceae*. J. mar. biol. Ass. U.K. **56**, 31—38.

— 1980: The fine structure of *Pavlova pinguis* Green and a preliminary survey of the order *Pavlovales* (*Prymnesiophyceae*). Br. Phycol. J. **15**, 151—191.

— HIBBERD, D. J., 1977: The ultrastructure and taxonomy of *Diacronema vlkianum* (*Prymnesiophyceae*) with special reference to the haptonema and flagellar apparatus. J. mar. biol. Ass. U.K. **57**, 1125—1136.

— JENNINGS, D. H., 1967: A physical and chemical investigation of the scales produced by the Golgi apparatus within and found on the surface of the cells of *Chrysochromulina chiton* Parke et Manton. J. Exp. Botn. **18**, 359—370.

— LEADBEATER, B. S. C., 1972: *Chrysochromulina parkeae* sp. nov. (*Haptophyceae*) a new species recorded from S.W. England and Norway. J. mar. biol. Ass. U.K. **52**, 469—474.

— MANTON, I., 1970: Studies in the fine structure and taxonomy of flagellates in the genus *Pavlova*. I. A revision of *Pavlova gyrans*, the type species. J. mar. biol. Ass. U.K. **50**, 1113—1130.

— PARKE, M., 1974: A reinvestigation by light and electron microscopy of *Ruttnera spectabilis* Geitler (*Haptophyceae*), with special reference to the fine structure of the zoids. J. mar. biol. Ass. U.K. **54**, 539—550.

— — 1975: New observations upon members of the genus *Chrysotila* Anand, with remarks upon their relationships within the *Haptophyceae*. J. mar. biol. Ass. U.K. **55**, 109—121.

— PIENAAR, R. N., 1977: The taxonomy of the order *Isochrysidales* (*Prymnesiophyceae*) with special reference to the genera *Isochrysis* Parke, *Dicrateria* Parke, and *Imantonia* Reynolds. J. mar. biol. Ass. U.K. **57**, 7—17.

GRELL, K., BENWITZ, G., 1966: Die Zellhülle von *Paramoeba eilhardi* Schaudinn. Z. Natur. **216**, 600—601.

GROVE, S. N., BRACKER, C. E., MORRÉ, D. J., 1968: Cytomembrane differentiation in the endoplasmic reticulum—Golgi apparatus—vesicle complex. Science **161**, 171—173.

HANKER, J. S., THORNBURG, L. P., YATES, P. E., ROMANOVICZ, D. K., 1975: The demonstration of arylsulfatases with 4-nitro-1,2-benzenediol mono(hydrogen sulfate) by the formation of osmium *blacks* at the site of copper capture. Histochemistry **41**, 207—225.

Harris, K., 1966: The genus *Mallomonopsis*. J. Gen. Micro. **42**, 175—184.

— 1967: Variability in *Mallomonas*. J. Gen. Micro. **46**, 185—191.

— 1970 a: Imperfect forms and taxonomy of *Mallomonas*. J. Gen. Micro. **61**, 73—76.

— 1970 b: Species of the Torquata group of *Mallomonas*. J. Gen. Micro. **61**, 77—80.

— Bradley, E. E., 1956: Electron microscopy of *Synura* scales. Discovery **17**, 329—332.

— — 1958: Some unusual *Chrysophyceae* studied in the electron microscope. J. Gen. Micro. **18**, 71—83.

Herth, W., Franke, W. W., Stadler, J., Bittiger, H., Keilich, G., Brown, R. M., jr., 1972: Further characterization of the alkali-stable material from the scales of *Pleurochrysis scherffelii:* A cellulosic glycoprotein. Planta **105**, 79—92.

Heyn, A. N. J., 1965: Crystalline state of cellulose in fresh and dried mature cotton fiber from unopened bolls as studied by X-ray diffraction. J. Polym. Sci. **A 3**, 1251—1265.

Hibberd, D. J., 1973: Observations on the ultrastructure of flagellar scales in the genus *Synura*. Arch. Mikrobiol. **89**, 291—304.

— 1976: The ultrastructure and taxonomy of the *Chrysophyceae* and *Prymnesiophyceae* (*Haptophyceae*): A survey with new observations on the structure of the *Chrysophyceae*. Botn. J. Linn. Soc. **72**, 55—80.

— 1978: The fine structure of *Synura sphagnicola* (Korch.) Korsh. (*Chrysophyceae*). Br. Phycol. J. **13**, 403—412.

Hori, T., Chihara, M., 1974: Studies on the fine structure of *Prasinocladus ascus* (*Prasinophyceae*). Phycologia **13**, 307—315.

Isenberg, H. D., Douglas, S. D., Lavine, L. S., Spicer, S. S., Weissfellner, H., 1966: A protozoan model of hard tissue formation. Ann. N. Y. Acad. Sci. **136**, 155—190.

Jones, E. B. G., Alderman, D. J., 1970: *Althornia crouchii*, gen. et sp. nov., a marine biflagellate fungus. Nova Hedwigia **21**, 381—400.

DeJong, L. W., Dam, W., Westbroek, P., Crenshaw, M. A., 1976: Aspects of calcification in *Emiliana huxleyi* (unicellular alga). In: The mechanisms of mineralization in the invertebrates and plants (Watabe, N., Wilbur, K. M., eds.), pp. 135—153. Columbia, S.C.: University of South Carolina Press.

Kavookjian, A. M. A., Brown, R. M., jr., 1978: Antigenic properties of the Golgi-derived scales of *Pleurochrysis scherffelii*. Cytobios **23**, 45—70.

Keegstra, K., Talmadge, K., Bauer, W. D., Albersheim, P., 1973: The structure of plant cell walls. III. A model of the walls of suspension-cultured sycamore cells based on the interconnections of the macromolecular components. Plant Physiol. **51**, 188—196.

Kiermayer, O., Dobberstein, B., 1973: Membrankomplexe dictyosomaler Herkunft als Matrizen für die extraplasmatische Synthese und Orientierung von Microfibrillen. Protoplasma **77**, 437—451.

— Sleytr, U. B., 1979: Hexagonally ordered "rosettes" of particles in the plasma membrane of *Micrasterias denticulata* Breb. and their significance for microfibril formation and orientation. Protoplasma **101**, 133—138.

Klaveness, D., 1972 a: *Coccolithus huxleyi* (Lohmann) Kamptner. L. Morphological investigations in the vegetative cell and the process of coccolith formation. Protistologica **8**, 335—346.

— 1972 b: *Coccolithus huxleyi* (Lohmann) Kamptner. II. The flagellate cell, aberrant cell types, vegetative propagation and life cycles. Br. Phycol. J. **7**, 309—318.

— 1973: The microanatomy of *Calyptrosphaera sphaeroidea*, with some supplementary observation on the motile stage of *Coccolithus pelagicus*. Nor. J. Botn. **20**, 151—162.

— Paasche, E., 1971: Two different *Coccolithus huxleyi* cell types incapable of coccolith formation. Arch. Mikrobiol. **75**, 382—385.

Lamport, D. T. A., 1965: The protein component of primary cell walls. In: Advances in botanical research, Vol. 2 (Preston, R. D., ed.), pp. 151—218. New York: Academic Press.

— 1970: Cell wall metabolism. Ann. Rev. Plant Physiol. **21**, 235—270.

— 1973: The glycopeptide linkages of extensin: O-D-galactosyl serine and O-L-arabinosyl hydroxyproline. In: Biogenesis of plant cell wall polysaccharides (Loewus, F., ed.), pp. 149—164. New York: Academic Press.

LEADBEATER, B. S. C., 1970: Preliminary observations on differences of scale morphology at various stages in the life cycle of *"Apistonema-Syracosphaera"* sensu von Stosch. Br. Phycol. J. **5**, 57—69.

— 1971: Observations on the life history of the haptophycean alga *Pleurochrysis scherffelii* with special reference to the microanatomy of the different types of motile cells. Ann. Botn. **35**, 429—439.

— 1972 a: Fine structural observations on six new species of *Chrysochromulina* (*Haptophyceae*) from Norway with preliminary observations of scale production in *C. microcylindra* sp. nov. Sarsia **49**, 65—80.

— 1972 b: Identification, by means of electron microscopy, of flagellate nanoplankton from the coast of Norway. Sarsia **49**, 107—124.

— 1974: Ultrastructural observations of nanoplankton collected from the coast of Jugoslavia and the Bay of Algiers. J. mar. biol. Ass. U.K. **54**, 179—196.

— 1977: Observations on the life history and ultrastructure of the marine choanoflagellate *Choanoeca perplexa* Ellis. J. mar. biol. Ass. U.K. **57**, 285—301.

— 1979 a: Developmental studies on the loricate choanoflagellate *Stephanoeca diplocostata* Ellis. I. Ultrastructure of the non-dividing cell and costal strip production. Protoplasma **98**, 241—262.

— 1979 b: Developmental studies on the loricate choanoflagellate *Stephanoeca diplocostata* Ellis. II. Cell division and lorica assembly. Protoplasma **98**, 311—328.

— MANTON, I., 1969 a: New observations on the fine structure of *Chrysochromulina strobilus* Parke and Manton with special reference to some unusal features of the haptonema and scales. Arch. Mikrobiol. **66**, 105—120.

— — 1969 b: *Chrysochromulina camella* sp. nov. and *C. cymbium* sp. nov., two relatives of *C. strobilus* Parke and Manton. Arch. Mikrobiol. **68**, 116—132.

— — 1971: Fine structure and light microscopy of a new species of *Chrysochromulina* (*C. acantha*). Arch. Mikrobiol. **78**, 58—69.

— MORTON, C., 1973: Ultrastructural observations on the external morphology of some members of the *Haptophyceae* from the coast of Jugoslavia. Nova Hedwigia **24**, 207—233.

LEWIN, R. A., 1958: The cell walls of *Platymonas*. J. Gen. Micro. **19**, 87—90.

LIANG, C. Y., MARCHESSAULT, R. H., 1959: Infrared spectra of crystalline polysaccharides. I. Hydrogen bonds in native celluloses. J. Polym. Sci. **37**, 385—395.

LLOYD, A. G., DODGSON, K. S., PRICE, R. G., ROSE, F. A., 1961: Infrared studies on sulfate esters. I. Polysaccharide sulfates. Biochim. Biophys. Acta **46**, 108—115.

LOEBLICH, A. R., jr., TAPPAN, H., 1971: Annotated index and bibliography of the calcareous nanoplankton. VI. Phycologia **10**, 315—339.

LUCAS, I. A. N., 1967: Two new marine species of *Paraphysomonas*. J. mar. biol. Ass. U.K. **47**, 329—334.

— 1968: A new member of the *Chrysophyceae* bearing polymorphic scales. J. mar. biol. Ass. U.K. **48**, 437—441.

McBRIDE, G. E., 1968: Ultrastructure of the *Choleochaete scutata* zoospore. J. Phycol. **4**, 6.

McCALL, E. R., MORRIS, N. M., TRIPP, V. V., O'CONNOR, R. T., 1971: Low temperature infrared absorption spectra of cellulosics. Appl. Spectrosc. **25**, 196—200.

McLACHLAN, J., PARKE, M., 1967: *Platymonas impellucida* sp. nov. from Puerto Rico. J. mar. biol. Ass. U.K. **47**, 723—733.

MANTON, I., 1964: Observations with the electron microscope on the division cycle in the flagellate *Prymnesium parvum* Carter. J. Roy. Micro. Soc. **83**, 317—325.

— 1966 a: Observations on scale production in *Prymnesium parvum*. J. Cell Sci. **1**, 375—380.

— 1966 b: Observations on scale production in *Pyramimonas amylifera* Conrad. J. Cell Sci. **1**, 429—438.

— 1967 a: Further observations on the fine structure of *Chrysochromulina chiton* with special reference to the haptonema, "peculiar" Golgi structure and scale production. J. Cell Sci. **2**, 265—272.

MANTON, I., 1967 b: Further observations on scale formation in *Chrysochromulina chiton*. J. Cell Sci. **2**, 411—418.

— 1968 a: Further observations on the microanatomy of the haptonema in *Chrysochromulina chiton* and *Prymnesium parvum*. Protoplasma **66**, 35—53.

— 1968 b: Observations on the microanatomy of the type species of *Pyramimonas* (*P. tetrarhynchus*). Proc. Linn. Soc. Lond. **179**, 147—152.

— 1972 a: Preliminary observations on *Chrysochromulina mactra* sp. nov. Br. Phycol. J. **7**, 21—35.

— 1972 b: Observations on the biology and microanatomy of *Chrysochromulina megacylindra* Leadbeater. Br. Phycol. J. **7**, 235—248.

— 1975: Observation on the microanatomy of *Scourfieldia marina* Throndsen and *Scourfieldia caeca* (Korsch.) Belcher et Swale. Arch. Protistenk. **117**, 358—368.

— 1978 a: *Chrysochromulina hirta* sp. nov. a widely distributed species with unusual spines. Br. Phycol. J. **13**, 3—14.

— 1978 b: *Chrysochromulina tenuispina* sp. nov. from arctic Canada. Br. Phycol. J. **13**, 227—234.

— ETTL, H., 1965: Observations on the fine structure of *Mesostigma viride* Lauterborn. J. Linn. Soc. Botn. **59**, 175—184.

— HARRIS, K., 1966: Observations on the microanatomy of the brown flagellate *Sphaleromantis tetragona* Skuja with special reference to the flagellar apparatus and scales. J. Linn. Soc. Botn. **59**, 397—403.

— LEEDALE, G. F., 1961 a: Further observations on the fine structure of *Chrysochromulina ericina* Parke and Manton. J. mar. biol. Ass. U.K. **41**, 145—155.

— — 1961 b: Further observations on the fine structure of *Chrysochromulina minor* and *C. kappa* with special reference to the pyrenoids. J. mar. biol. Ass. U.K. **41**, 519—526.

— — 1961 c: Observations on the fine structure of *Paraphysomonas vestita*, with special reference to the Golgi apparatus and the origin of scales. Phycology **1**, 37—57.

— — 1963 a: Observations on the fine structure of *Prymnesium parvum* Carter. Arch. Mikrobiol. **45**, 285—303.

— — 1963 b: Observations on the microanatomy of *Crystallolithus hyalinus* Gaarder and Morkali. Arch. Mikrobiol. **47**, 115—136.

— — 1969: Observations on the microanatomy of *Coccolithus pelagicus* and *Cricosphaera carterae* with special reference to the origin and nature of coccoliths and scales. J. mar. biol. Ass. U.K. **49**, 1—16.

— OATES, K., 1975: Fine-structural observations on *Papposphaera* Tangen from the Southern Hemisphere and on *Pappomonas* gen. nov. from South Africa and Greenland. Br. Phycol. J. **10**, 93—109.

— — COURSE, P. A., 1981: Cylinder-scales in marine flagellates from the genus *Chrysochromulina* (*Haptophyceae-Prymnesiophyceae*) with a description of *C. pachycylindra* sp. nov. J. mar. biol. Ass. U.K. **61**, 17—26.

— — SUTHERLAND, J., 1981: Cylinder-scales in marine flagellates from the genus *Chrysochromulina* (*Haptophyceae-Prymnesiophyceae*): Further observations from *C. microcylindra* Leadbeater and *C. cyathophora* Thomsen. J. mar. biol. Ass. U.K. **61**, 27—33.

— — PARKE, M., 1963: Observations on the fine structure of the *Pyramimonas* stage of *Halosphaera* and preliminary observations on three species of *Pyramimonas*. J. mar. biol. Ass. U.K. **43**, 225—238.

— PARKE, M., 1960: Further observations on small green flagellates with special reference to possible relatives of *Chromulina pusilla* Butcher. J. mar. biol. Ass. U.K. **39**, 275—298.

— — 1962: Preliminary observations on scales and their mode of origin in *Chrysochromulina polylepis* sp. nov. J. mar. biol. Ass. U.K. **42**, 565—578.

— — 1965: Observations on the fine structure of two species of *Platymonas* with special reference to flagellar scales and the mode of origin of the theca. J. mar. biol. Ass. U.K. **45**, 743—754.

MANTON, I., PETERFI, L. S., 1969: Observations on the fine structure of coccoliths, scales and the protoplast of the freshwater coccolithophorid *Hymenomonas roseola* Stein with supplementary observations on the protoplast of *Cricosphaera carterae*. Proc. Roy. Soc. **172**, 1—15.

— RAYNS, D. G., ETTL, H., PARKE, M., 1965: Further observations on green flagellates with scaly flagella: The genus *Heteromastix* Korshikov. J. mar. biol. Ass. U.K. **45**, 241—255.

— SUTHERLAND, J., 1975: Further observations on the genus *Pappomonas* Manton et Oates with special reference to *P. virgulosa* sp. nov. from West Greenland. Br. Phycol. J. **10**, 337—385.

— — McCULLY, M., 1976 a: Fine structural observations on coccolithophorids from Alaska in the genera *Papposphaera* Tangen and *Pappomonas* Manton et Oates. Br. Phycol. J. **11**, 225—238.

— — OATES, K., 1976 b: Arctic coccolithophorids: Two species of *Turrisphaera* gen. nov. from West Greenland, Alaska, and the Northwest Passage. Proc. Roy. Soc. **B 194**, 179—194.

— — — 1977: Arctic coccolithophorids: *Wigwamma arctica* gen. et sp. nov. from Greenland and arctic Canada, *W. annulifera* sp. nov. from South Africa and S. Alaska and *Calciarcus alaskensis* gen. et sp. nov. from S. Alaska. Proc. Roy. Soc. **B 197**, 145—168.

MARKER, A. F. H., 1965: Extracellular carbohydrate liberation in the flagellates *Isochrysis galbana* and *Prymnesium parvum*. J. mar. biol. Ass. U.K. **45**, 755—772.

MARRINAN, H. J., MANN, J., 1956: Infrared spectra of the crystalline modifications of cellulose. J. Polym. Sci. **21**, 301—311.

MICHELL, A. J., SCURFIELD, G., 1970: An assessment of infrared spectra as indicators of fungal cell wall composition. Aust. J. Biol. Sci. **23**, 345—360.

MILLS, J. T., 1975: *Hymenomonas coronata* sp. nov., a new coccolithophorid from the Texas coast. J. Phycol. **11**, 149—154.

MOESTRUP, Ø., 1970: The fine structure of mature spermatozoids of *Chara corallina* with special reference to microtubules and scales. Planta **93**, 295—308.

— 1974 a: Scale formation: A special function of the Golgi apparatus in algae. In: Proc. Eighth Internat. Congress on Electron Microscopy (SANDERS, J. V., GOODCHILD, D. J., eds.). Canberra: Australian Acad. Sci.

— 1974 b: Ultrastructure of the scale-covered zoospores of the green alga *Chaetosphaeridium*, a possible ancestor of the higher plants and bryophytes. Biol. J. Linn. Soc. **6**, 111—125.

— ETTL, H., 1979: A light and electron microscopic study of *Nephroselmis olivacea* (*Prasinophyceae*). Opera Bot. **49**, 1—39.

— THOMSEN, H. A., 1974: An ultrastructural study of the flagellate *Pyramimonas orientalis* with particular emphasis on Golgi apparatus activity and the flagellar apparatus. Protoplasma **81**, 247—269.

— WALNE, P. L., 1979: Studies on scale morphogenesis in the Golgi apparatus of *Pyramimonas tetrarhynchus* (*Prasinophyceae*). J. Cell Sci. **36**, 437—459.

MORRÉ, D. J., MOLLENHAUER, H. H., BRACKER, C. E., 1971: Origin and continuity of Golgi apparatus. In: Origin and continuity of cell organelles (REINERT, J., URSPRUNG, H., eds.), pp. 82—126. New York: Springer.

MUELLER, S. C., BROWN, R. M., jr., 1980: Evidence for an intramembrane component associated with a cellulose microfibril-synthesizing complex in higher plants. J. Cell Biol. **84**, 315—326.

MÜHLETHALER, K., 1969: Fine structure of natural polysaccharide systems. J. Polym. Sci. **C 28**, 305—316.

NORRIS, R. E., PEARSON, B. R., 1975: Fine structure of *Pyramimonas parkeae*, sp. nov. (*Chlorophyta, Prasinophyceae*). Arch. Protistenk. **117**, 192—213.

NORTHCOTE, D. H., 1971: The Golgi apparatus. Endeavour **30**, 26—30.

NOVIKOFF, A. B., NOVIKOFF, P. M., 1977: Cytochemical contributions to differentiating GERL from the Golgi apparatus. Histochem. J. **9**, 525—551.

Novikoff, P. M., Novikoff, A. B., Quintana, N., Hauw, J. J., 1971: Golgi apparatus, GERL, and lysosomes of neurons in rat dorsal root ganglia, studied by thick section and thin section cytochemistry. J. Cell Biol. 50, 859—886.

Olive, L. S., 1975: The mycetozoans. New York: Academic Press.

Outka, D. E., Williams, D. C., 1971: Sequential coccolith morphogenesis in *Hymenomonas carterae*. J. Protozool. 18, 285—297.

Paasche, E., 1962: Coccolith formation. Nature (Lond.) 193, 1094.

— 1968: Biology and physiology of coccolithophorids. Ann. Rev. Microbiol. 22, 71—86.

Paradies, H. H., Goke, L., Werz, G., 1977: On the spatial structure of a plant cell wall protein. Secondary structure of a cell wall protein from *Acetabularia*. Protoplasma 93, 249—265.

Parke, M., Adams, I., 1960: The motile (*Crystallolithus hyalinus* Gaarder and Markali) and non-motile phases in the life history of *Coccolithus pelagicus* (Wallich) Schiller. J. mar. biol. Ass. U.K. 39, 263—274.

— Boalch, G. T., Jowett, R., Harbour, D. S., 1978: The genus *Pterosperma* (*Prasinophyceae*): Species with a single equatorial ala. J. mar. biol. Ass. U.K. 58, 239—276.

— Dixon, P. S., 1976: Check-list of British marine algae—third revision. J. mar. biol. Ass. U.K. 56, 527—594.

— Green, J. C., Manton, I., 1971: Observations on the fine structure of zoids of the genus *Phaeocystis* (*Haptophyceae*). J. mar. biol. Ass. U.K. 51, 927—941.

— Hartog-Adams, I. den, 1965: Three species of *Halosphaera*. J. mar. biol. Ass. U.K. 45, 537—557.

— Lund, J. W. G., Manton, I., 1962: Observations on the biology and fine structure of the type species of *Chrysochromulina* (*C. parva* Lackey) in the English Lake District. Arch. Mikrobiol. 42, 333—352.

— Manton, I., 1962: Studies on marine flagellates, VI. *Chrysochromulina pringsheimii* sp. nov. J. mar. biol. Ass. U.K. 42, 391—404.

— — 1965: Preliminary observations on the fine structure of *Prasinocladus marinus*. J. mar. biol. Ass. U.K. 45, 525—536.

— — 1967: The specific identity of the algal symbiont in *Convoluta roscoffensis*. J. mar. biol. Ass. U.K. 47, 445—464.

— — Clarke, B., 1955: Studies on marine flagellates. II. Three new species of *Chrysochromulina*. J. mar. biol. Ass. U.K. 34, 579—609.

— — — 1956: Studies on marine flagellates. III. Three further species of *Chrysochromulina*. J. mar. biol. Ass. U.K. 35, 387—414.

— — — 1958: Studies on marine flagellates. IV. Morphology and microanatomy of a new species of *Chrysochromulina*. J. mar. biol. Ass. U.K. 37, 209—228.

— — — 1959: Studies on marine flagellates. V. Morphology and microanatomy of *Chrysochromulina strobilus* sp. nov. J. mar. biol. Ass. U.K. 38, 169—188.

— Rayns, D. G., 1964: Studies on marine flagellates. VII. *Nephroselmis gilva* sp. nov. and some allied forms. J. mar. biol. Ass. U.K. 44, 209—217.

Pennick, N. C., Clarke, K. J., 1972: *Paraphysomonas butcheri* sp. nov., a marine colourless, scale-bearing member of the *Chrysophyceae*. Br. Phycol. J. 7, 45—48.

— — 1973: *Paraphysomonas corbidifera* sp. nov., a marine, colourless, scale-bearing member of the *Chrysophyceae*. Br. Phycol. J. 8, 147—151.

— — 1977: The occurrence of scales in the peridinian dinoflagellate *Heterocapsa triquetra* (Ehrenb.) Stein. Br. Phycol. J. 12, 63—66.

Perkins, F. O., 1973: A new species of marine labyrinthulid *Labyrinthuloides yorkensis* gen. nov. spec. nov.—cytology and fine structure. Arch. Mikrobiol. 90, 1—17.

Pickett-Heaps, J. D., 1968: Ultrastructure and differentiation in *Chara* (*fibrosa*). IV. Spermatogenesis. Aust. J. Biol. Sci. 21, 655—690.

PIENAAR, R. N., 1969 a: The fine structure of *Cricosphaera carterae*. I. External morphology. J. Cell Sci. 4, 561—567.

— 1969 b: The fine structure of *Hymenomonas* (*Cricosphaera*) *carterae*. II. Observations on scale and coccolith production. J. Phycol. 5, 321—331.

— 1971: Coccolith production in *Hymenomonas carterae*. Protoplasma 73, 217—224.

— 1976 a: The microanatomy of *Hymenomonas lacuna* sp. nov. (*Haptophyceae*). J. mar. biol. Ass. U.K. 56, 1—11.

— 1976 b: The rhythmic production of body covering components in the Haptophycean flagellate *Hymenomonas carterae*. In: The mechanisms of mineralization in the invertebrates and plants (WATABE, N., WILBUR, K. M., eds.), pp. 203—229. Columbia, S.C.: University of South Carolina Press.

— 1976 c: The microanatomy of *Sphaleromantis marina* sp. nov. (*Chrysophyceae*). Br. Phycol. J. 11, 83—92.

PRESTON, R. D., 1964: Structural and mechanical aspects of plant cell walls with particular reference to synthesis and growth. In: Formation of wood in forest trees (ZIMMERMAN, M. H., ed.), pp. 169—188. New York: Academic Press.

— 1971: Negative staining and cellulose microfibril size. J. Microsc. 93, 7—13.

— 1974: The physical biology of plant cell walls. London: Chapman and Hall.

RAY, P. M., SHININGER, T., RAY, M., 1969: Isolation of β-glucan synthetase particles from plant cells and identification with Golgi membranes. Proc. Nat. Acad. Sci. 64, 605—612.

RAYNS, D. G., 1962: Alternation of generations in a coccolithophorid *Cricosphaera carterae*. J. mar. biol. Ass. U.K. 42, 481—484.

REES, A. J. J., LEEDALE, G. F., CMIECH, H. A., 1974: *Paraphysomonas faveolata* sp. nov. (*Chrysophyceae*), a fourth marine species with meshwork body scales. Br. Phycol. J. 9, 273—283.

REYNOLDS, N., 1974: *Imantonia rotunda* gen. et sp. nov., a new member of the *Haptophyceae*. Br. Phycol. J. 9, 429—434.

ROBINSON, D. G., PRESTON, R. D., 1972: Plasmalemma structure in relation to microfibril biosynthesis in *Oocystis*. Planta 104, 234—246.

— WHITE, R. K., PRESTON, R. D., 1972: Fine structure of swarmers of *Cladophora* and *Chaetomorpha*. Planta 107, 131—144.

ROMANOVICZ, D. K., 1975: Cell wall biogenesis in *Pleurochrysis scherffelii*. Ph. D. dissertation, University of North Carolina at Chapel Hill.

— BROWN, R. M., jr., 1976: Biogenesis and structure of Golgi-derived cellulosic scales in *Pleurochrysis*. II. Scale composition and supramolecular structure. Applied Polymer Symposium 28, 587—610.

SCHNEPF, E., DEICHGRÄBER, G., 1969: Über die Feinstruktur von *Synura petersenii* unter besonderer Berücksichtigung der Morphogenese ihrer Kieselschuppen. Protoplasma 68, 85—106.

SLABY, F., SLABY, M., 1979: A new theory of the structure and function of the Golgi complex in protein secreting cells: each stack of cisternae is a rotating, spiralled membrane-bounded tube. J. Cell Biol. 83, 431 a.

SWALE, E. M. F., 1973: A third layer of body scales in *Pyramimonas tetrarhynchus* Schmarda. Br. Phycol. J. 8, 95—99.

— BELCHER, J. H., 1968: The external morphology of the type species of *Pyramimonas* (*P. tetrarhynchus* Schmarda) by electron microscopy. Proc. Linn. Soc. Lond. 179, 77—81.

TAKAHASHI, E., 1976: Studies on genera *Mallomonas* and *Synura* and other plankton in freshwater with the electron microscope. X. The genus *Paraphysomonas* (*Chrysophyceae*) in Japan. Br. Phycol. J. 11, 39—48.

TANGEN, K., 1972: *Papposphaera lepida* gen. nov., n. sp., a new marine coccolithophorid from Norwegian coastal waters. Nor. J. Botn. 19, 171—178.

THOMSEN, H. A., 1975: An ultrastructural survey of the chrysophycean genus *Paraphysomonas* under natural conditions. Br. Phycol. J. 10, 113—127.

— 1977: *Chrysochromulina pyramidosa* sp. nov. (*Prymnesiophyceae*) from Danish coastal waters. Botaniska notiser 130, 147—154.

Thomsen, H. A., 1980: *Wigwamma scenozonion* sp. nov. (*Prymnesiophyceae*) from West Greenland. Br. Phycol. J. **15**, 335—342.

— Oates, K., 1978: *Balaniger balticus* gen. et sp. nov. (*Prymnesiophyceae*) from Danish coastal waters. J. mar. biol. Ass. U.K. **58**, 773—779.

Tsuboi, M., 1957: Infrared spectrum and crystal structure of cellulose. J. Polym. Sci. **25**, 159—171.

Turner, F. R., 1968: An ultrastructural study of plant spermatogenesis. Spermatogenesis in *Nitella*. J. Cell Biol. **37**, 370—393.

van der Veer, J., 1976: *Pavlova calceolata* (*Haptophyceae*), a new species from the Tamar Estuary, Cornwall, England. J. mar. biol. Ass. U.K. **56**, 21—30.

Whaley, W. G., 1975: The Golgi apparatus. (Cell Biology Monographs, Vol. 2.) Wien-New York: Springer.

— Dauwalder, M., Kephart, J., 1971: Assembly, continuity, and exchanges in certain cytoplasmic membrane systems. In: Origin and continuity of cell organelles (Reinert, J., Ursprung, H., eds.), pp. 1—45. New York: Springer.

Wright, K., Northcote, D. H., 1976: Identification of β 1-4 glucan chains as part of a fraction of slime synthesized within the dictyosomes of maize root caps. Protoplasma **88**, 225—239.

Wilbur, K. M., Watabe, N., 1963: Experimental studies on calcification in molluscs and the alga *Coccolithus huxleyi*. Ann. N. Y. Acad. Sci. **109**, 82—112.

Zimmermann, B., 1977: Flagellar and body scales in the chrysophyte *Mallomonas multiunca* Asmund. Br. Phycol. J. **12**, 287—290.

Morphogenesis and Biochemistry of Diatom Cell Walls

Anna-Maria M. Schmid, M. A. Borowitzka, and B. E. Volcani

Botanical Institute, University of Salzburg, Salzburg, Austria
Roche Research Institute of Marine Pharmacology, Dee Why, N.S.W., Australia
Scripps Institution of Oceanography, University of California San Diego, La Jolla, California, U.S.A.

With 8 Figures

Contents

I. Introduction

Within the plants diatoms occupy a special position because of their diploidy and their dependence on silicon [1] for growth (Richter 1906, Lewin 1955) which they require not only for formation of their siliceous cell wall, one of the most fascinating creations of single-celled organisms, but for cellular metabolism in general (for reviews see Werner 1977, 1978, Volcani 1978), including DNA replication (Darley and Volcani 1969, Sullivan and Volcani 1973, 1976, Okita and Volcani 1973, 1980).

Because of the recent upsurge of interest in silicon as an essential trace element and pathogenic agent in mammals, diatoms have become the model system for studying silicon metabolism.

The cell wall, or *frustule*, surrounds the protoplast as a small, bipartite box consisting of an *epitheca* (the "top") that overlaps the *hypotheca* (the "bottom"). Each theca is composed of a *valve* ("Schale" in German) and an accompanying *girdle* which constitutes the overlapping zone of the two thecae (Anonymous 1975) (Fig. 1 *a*). The relative proportion of valve to girdle varies widely and ranges from species in which the valve predominates (*e.g., Campylodiscus, Surirella*) to those in which the girdle predominates (*e.g., Cylindrotheca, Rhizosolenia, Tabellaria*). The degree of silicification also varies widely, and although most species are heavily silicified, there are a number in which the frustule consists chiefly of organic material.

The girdle itself is composed of open bands or short segments which are generally uniform and more simple in structure than the valve. In some species one or more so-called *intercalary bands (copulae)* are found between girdle and valve. They differ structurally from the girdle bands by their more valve-like micro-architecture, or by the possession of septae that protrude into the cell lumen [*e.g., Thalassiosira, Mastogloia, Tabellaria, Rhabdonema* (Hustedt 1930, Round 1972 a, von Stosch 1975)].

The architecture of the valves is considered to be species-specific and differences among diatoms in their symmetry have given rise to their separation into two groups: the *centric* diatoms, with tri- to omniradiate symmetry and the *pennate* diatoms with bipolar, very occasionally tripolar, symmetry. The valve pattern is always related to a "center" from which it fans out: in centric diatoms this is generally identical with the midpoint of the valve, while in the pennates it lies along the median of the bilateral valve. The highly complicated yet delicately structured architecture of the diatom cell wall is expressed in an enormous variety of geometrically ordered shapes that have thrilled scientists since the early days of microscopy.

In most species the rigidity of the cell wall causes a progressive reduction in size during successive cell divisions. This is generally accompanied by a change in the allometric proportions (von Stosch 1967) and the loss of struc-

---

[1] Silicon (Si) is used to denote the element, and as a generic term when the nature of the specific compound is not known. Silicic acid in nonionized form $Si(OH)_4$, occurs in aqueous solution at pH 8. Silica $(SiO_2)_n$ also annotated $SiO_2 \cdot nH_2O$, refers to amorphous polymerized silicic acid with unknown molecular weight: it is also referred to as silica gel. In diatoms it is sometimes referred to as "opaline" and as "opal" in other organisms.

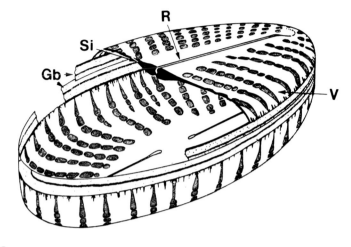

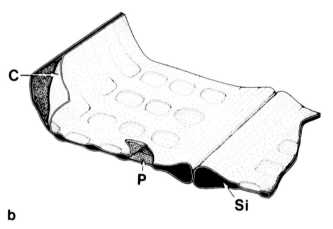

Fig. 1. *a* Drawing of a cell wall of *Navicula pelliculosa*. Part of the hypotheca is removed to show the inside of the epitheca. *V* valve, *Gb* girdle bands, *R* raphe slit, *Si* silica shell. *b* Drawing of magnified portion of the valve showing the silica shell (*Si*) interlocked with the casing (*C*); *P* pore. From Reimann et al. 1966.

tural elements (Geitler 1932, Holmes and Reimann 1966, Holmes 1967, Drebes 1967, Hostetter and Hoshaw 1972, Lowe and Crang 1972, Hostetter and Rutherford 1976). Both types of change can also be triggered by a change in environmental conditions (Drum 1964, Stoermer 1967, Geissler 1970 a, b, Hasle et al., 1971, Schultz 1971, Paasche et al. 1975, Schmid 1976 b, 1979 b, Li and Chiang 1979). This tendency to polymorphism is even greater in weakly silicified species, e.g., *Phaeodactylum tricornutum* (for references see Borowitzka and Volcani 1978) and *Bellerochea malleus* (von Stosch 1977).

Many species have been able to avoid the gradual reduction in size during successive cell divisions by developing a special morphology or arrangement of the girdle bands (Round 1972 b) that compensates for the rigidity of the valves and, without a change in proportions, maintains the valve size and architecture over a period of several months or even years in culture.

II. Morphology of the Frustule

A. Heavily Silicified Walls

Each individual part of a diatom frustule is composed of an inorganic silica component—an amorphous silica gel, *i.e.*, opaline ($SiO_2 \cdot nH_2O$), that is enveloped by an organic layer (Figs. 1 *b* and 2 *a–c*). The basic investigations of the frustule's gross structure were started in the last century (van Heurck 1899, Mangin 1908, Liebisch 1928, 1929) but it was not until the development of the transmission electron microscope (TEM), that the ultrastructural organization of the frustule could be demonstrated in detail (Drum and Pankratz 1964, Reimann 1964).

The great variety in architectural patterns of the valve is expressed in the *pennates* merely as a series of modifications of one basic geometrical form—a branching pattern (Fig. 1 *a*), similar to that of bird-feathers. A very heavily silicified rib extends along the median of the valve (apical axis); in the motile pennates this rib is pervalvarly penetrated, and therefore split into two ribs, the *raphe ribs,* by a longitudinal slit, the *raphe fissure* that is interrupted only in the center by the *central nodule.* From the raphe ribs the more lightly silicified, regularly spaced *transapical ribs (interstriae)* extend either perpendicularly or at a more or less acute angle. Neighbouring transapical ribs are interconnected by various kinds of thin apical extensions, leaving small pores between them.

In the simplest case this rib system constitutes a more or less monolayered valve, as is seen in *Gomphonema parvulum* (Drum and Pankratz 1964, Dawson 1973), in *Amphipleura pellucida* (Stoermer et al. 1965), or in *Navicula pelliculosa* (Fig. 1 *a*) (Reimann et al. 1966). The valve structure is more complex in a diatom such as *Pinnularia* sp. in which the transapical ribs protrude septa-like into the cell lumen (Pickett-Heaps et al. 1979 a).

In *centric* diatoms the valves are generally exceedingly complex, bearing a great variety of areolations and processes. In *Thalassiosira eccentrica,* for example (Figs. 5 *a* and 6 *d*), the valve is constructed of "*loculate areolae*", hexagonal chambers in the wall occluded by a "porous" plate at the proximal part of the valve facing the protoplast; a foramen is open to the exterior. The areolae are arranged in tangential striae. The frustule has two sorts of tubular processes that project through the valve: a single *labiate process* per valve, the "rimoportula"—a structure whose function is quite unknown, although Hasle (1974) suggested that it is the predecessor of the pennate raphe—and the *strutted tubuli,* the "fultoportulae". In the living *T. eccentrica* cells chitinous threads extend through the strutted tubuli, connecting the sister cells after cell division, to form four-celled colonies (Fryxell and Hasle 1972, Schmid and Schulz 1979) (Fig. 5 *a*).

Whereas the areolae are morphological units in *T. eccentrica,* this is not the case in *Melosira nummuloides* in which the rectangular network of the vertical areolae walls shows no correlation with the porous pattern in the proximal layer of the valve, indicating that in *M. nummuloides,* two quite independent deposition systems are involved (CRAWFORD 1974 a).

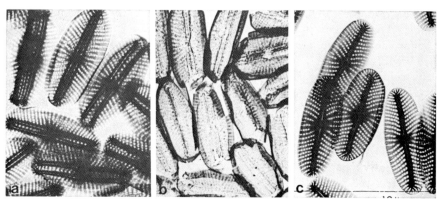

Fig. 2. Electron micrographs of *N. pelliculosa* valves and their components. *a* Intact valves, *b* organic casing, *c* silica shell. ×4,200. Micrographs by B. E. F. REIMANN.

B. Weakly Silicified Walls

Light microscopy can be used for morphological evaluation of highly silicified cell walls, but this is impossible in weakly silicified cells where the wall material is predominantly organic. Therefore the investigation of weakly silicified diatoms and also the discovery that they are in fact diatoms, has become closely associated with the development of electron microscopy and the improvement of preparation techniques.

Cylindrotheca fusiformis is typical of diatoms with lightly silicified valves. Only the raphe system and two adjacent strips are silicified. The unsilicified organic valve extends laterally from the strips to the girdle region, which consists of numerous silicified girdle bands (REIMANN and LEWIN 1964, REIMANN et al. 1965). The only transapically oriented silica structures are the *fibulae* connecting the raphe ribs; the whole cell is twisted along its apical axis, a torsion that would not be possible in the presence of rigid transapical ribs. However, such torsion does not necessarily require a silicon-free area of the valve, for *C. gracilis* is twisted to a much greater degree though the valve is completely surrounded by silicified longitudinal bands (SCHMID 1976 a).

In *Subsilicea fragilarioides* (VON STOSCH and REIMANN 1970) the predominantly organic component envelops a very thin silica layer, which resembles the developmental stage of a rapheless pennate that has highly silicified porous plates at the apices. The organic envelope, however, is structurally and chemically highly differentiated.

In the polymorphic pennate *Phaeodactylum tricornutum,* the only siliceous structures are found in the girdle bands (BOROWITZKA and VOLCANI 1978)

(Fig. 8 c) in the triradiate and fusiform morphotypes; in the oval morphotype, they appear only on one valve which has the structure of a *Cymbella* (LEWIN 1958). The other valve in the oval form, and both valves of the fusiform and triradiate morphotypes, are wholly organic, showing a three-layered structure (REIMANN and VOLCANI 1968).

The changes from one morphotype to another in *P. tricornutum* presumably occur in response to certain as yet little-investigated environmental circumstances. For example, COOKSEY and COOKSEY (1974) report that a lack of calcium can result in the transformation of the oval to the fusiform type, however BOROWITZKA and VOLCANI (1978) were unable to reproduce this. Furthermore, as yet unknown environmental changes can interrupt the normal development of the cleavage furrow in *Phaeodactylum;* this causes the newly forming organic walls to fuse on both sides of the furrow and results in the formation of chains (BOROWITZKA et al. 1977).

C. Auxospore Walls and Naked Cells

The relative percentage of organic material and inorganic silica in the diatom frustule seems to be species-specific, but two factors are known to affect the silica content of the frustule, one environmental, the other endogenous. The latter one is correlated with certain stages in the life cycle. During the sexual phase the haploid gametes are either completely naked (*pennates*) or partially so (*centrics*) (for references see DREBES 1977). In both groups, after nuclear fusion, the zygote develops into an *auxospore*, usually several times larger than the parent cell, and during this process the cell wall has to be very flexible. In centric diatoms, e.g., *Melosira varians* (REIMANN 1960), *M. nummuloides* (CRAWFORD 1974 b), and *Cyclotella meneghiniana* (HOOPS and FLOYD 1979), the auxospore "membrane" is mainly a thick, layered organic wall, beneath which are overlapping perforated silica scales. In pennates, e.g., *Anomoeoneis sphaerophora* (SCHMID, unpublished), observations in the light microscope as well as in the scanning electron microscope (SEM), show that during the enlargement from zygote to auxospore the new cell wall material is added to the equatorial region of the bipolar, expanding cell. The two halves of the former zygote "membrane" become the polar caps of the auxospore's organic cover; beneath this relatively thin organic sheath, the *perizonium*, a "corset" of rather weakly silicified open girdle bands is deposited over the whole cell. Within the perizonium of the mature auxospore the initial cell for a new generation is formed. In the special case of *A. sphaerophora*, at least, the perizonium does not substitute for the auxospore membrane, as summarized by DREBES (1977), but forms an additional layer inside it, similar to the layer of silica scales in the centrics.

These naked or weakly silicified stages occur only during a very short period of the normal life cycle. There is, however, quite another reason for the occurrence of naked or weakly silicified diatom cells—environmental conditions that cause a reduction or complete loss of the silica shell. As has been seen in *P. tricornutum*, this can be accompanied by a change from one morphotype to another, where each of the morphotypes is found to occupy a different ecological niche (for references see VON STOSCH 1967). The silica content of the frustule can be also affected by the silicic acid supply of the medium or by inter- or intraspecific competition for silicon (LEWIN 1962, FUHRMAN et al. 1978, BOOTH and HARRISON 1979). One of the most intriguing situations of environmentally influenced silica content is presented when the cell has become an endosymbiont. Such endosymbiosis has been observed e.g. between diatoms and foraminifera in which the host completely represses the formation of the frustule. The previously endosymbiontic diatoms proved capable of producing a normal frustule in a vitamin-enriched medium (for references see LEE et al. 1979).

Under certain circumstances naked protoplasts can appear during culturing. This is especially likely to occur in marine centric diatoms and has often been falsely interpreted as auxospore formation (e.g., GROSS 1940). BACHRACH and LEFÉVRE (1929) report this phenomenon in marine pennates grown on agar. GEITLER (1932) was able to show that as soon as these pennate protoplasts were transferred from the agar plate into a liquid medium of identical composition, the cells were able to recover and build frustules. The protoplasts can leave the frustule and swell to giant spheres, but how and why is completely unknown and may be a response to any one or several of a variety of conditions, e.g., osmotic changes, as it seems to be the case in T. eccentrica. Exposure to the microtubule inhibitors, amiprophos-methyl (APM; see p. 178 f.) or colchicine produces a similar response. As it is accompanied by an inhibition of mitosis and cytokinesis the possibility was raised that it is a manifestation of a microtubule disturbance (SCHMID 1979 c, d). In some species the phenomenon of forming protoplasts occurs as a result of an ionic imbalance of the medium leading to a relative or absolute deficiency in certain minerals. In Nitzschia putrida, for example, protoplasts were produced following sodium-starvation (RICHTER 1911), and in N. alba after calcium-deprivation (HEMMINGSEN 1971), while VON STOSCH (1942) found that Achnantes longipes cells in culture lose their ability to form valves when manganese is lacking.

III. Morphogenesis

Cell division and valve formation were first studied in the light microscope, and although very fine detail is lacking, it allows the entire sequence of morphogenesis to be followed in the living cell. In the electron microscope, despite its many great advantages, the developmental processes cannot be followed, and this often results in misinterpretation of phenomena, or even failure to perceive them at all, as pointed out by GEITLER (1979). Some of the most striking alterations during morphogenesis, that are best followed in vivo in the LM, are "contractions" of the whole protoplast or local defined morphological changes of the cytoplasmic surface. Already LAUTERBORN (1896) described very precisely how in Surirella calcarata the cleavage furrow becomes extended, especially in the region of the broad pole in connection with the formation of the valvar wings. In 1907 PERAGALLO reported that in the centric diatom, Biddulphia mobiliensis, the two daughter cells retract after cytokinesis, leaving an "empty" gap between them that is later closed when the cell again expands. This so-called "spontaneous plasmolysis" has been reported in many species in connection with valve formation (for references see von Stosch 1967, Drebes 1977) but more recent investigations suggest that this retraction might, at least in the case of *Attheya decora*, be the result of a local secretion of "swelling mucilage" rather than a real plasmolysis (Schnepf et al. 1980).

It was discovered in the course of early light microscopic investigations that deposition of the new valve does not occur simultaneously over the whole surface, nor is it random; rather it is a "directed" growth that starts at a "primary silicification site" in the center of the new cell surface and gradually proceeds to the margins until the whole cell is covered with the new wall (Peragallo 1907, Geitler 1932). The entire silicification process is detailed below.

A. Relation Between Nucleus and Primary Silicification Site

Diatom valves are generally formed only after cytokinesis, but the occurrence of *"internal valves"* in some species, deposited without an apparent cell division, led one to believe in a wall forming mechanism independent from the mitotic cycle. After extensive investigations Geitler (1963) formulated the rule that each valve formation requires a prior, even if only reduced or "rudimentary" mitosis; Geitler's theorem was confirmed by von Stosch and Kowallik (1969) and Hoops and Floyd (1979). However, Oey and Schnepf (1970) found that colchicine, an inhibitor of mitosis, does not prevent valve formation in *Cyclotella cryptica*, although due to the inhibition of cytokinesis, the newly formed valve is deposited onto the protoplasmic surface as a *"lateral valve"* in the girdle region. Microspectrophotometric analysis demonstrated that the building of the lateral valves is dependent on DNA-replication which precedes mitosis (Oey and Schnepf 1970).

Normally the valve is formed after mitosis and cytokinesis, and as the nucleus usually lies very close to the valve formation site, this suggests that the nucleus is intimately involved in formation of the new cell wall. Such a possibility has been substantiated in species in which cell division and the deposition of a new valve is accompanied by nuclear migration. For example, the interphase nucleus of *Coscinodiscus wailesii* (Schmid and Volcani, in preparation) lies in a cone-shaped cytoplasmic "pocket" attached to the center of the epivalve. Prior to mitosis the nucleus migrates into the girdle region where it divides. The daughter nuclei move concurrently with the invading cleavage furrow in its progress from the margin to the center, beneath the newly created cytoplasmic surfaces. The nucleus remains in this area until the new hypovalve is formed, and then returns to its interphase position in the center of the epivalve. A similar process has been described for *Melosira varians* by Tippit et al. (1975).

In pennate diatoms the nucleus generally lies within a central pervalvar plasma column (CPPC), beneath or close to the central nodule of the valve, and remains in this area during mitosis. In the *Surirellaceae* however, the central nodule is translocated onto one pole, the headpole in heteropole species such as *S. peisonis* and *S. ovalis* (Schmid 1979 a, b). In the interphase cell the nucleus is situated halfway between the headpole and the footpole; during mitosis it migrates close to the headpole where silicification starts. The nucleus returns to the interphase position as soon as valve formation has been completed (LAUTERBORN 1896, TIPPIT and PICKETT-HEAPS 1977).

The microtubule inhibitors colchicine and APM, produce dislocation of the CPPC containing the nucleus and associated organelles. When post-telophase stages of *T. eccentrica* (SCHMID 1979 c) were exposed to these compounds the valve pattern center was dislocated showing quite clearly that the position of the primary silicification site is dependent upon the position of the nucleus. Similar experiments with *A. sphaerophora* and *S. peisonis* resulted in dislocation of the central nodule (SCHMID 1980).

B. Primary Silicification Sites

1. Centric Diatoms

In *B. mobiliensis* it was found that silicification starts just above the nucleus in the center of the new hypovalve and expands centrifugally to the margin (PERAGALLO 1907). The same location for the primary silicification site has since been found also in other centric species, e.g., *Cyclotella meneghiniana* (HOOPS and FLOYD 1979), *Thalassiosira eccentrica* (SCHMID and SCHULZ 1979), *Attheya decora* (SCHNEPF et al. 1980), *Melosira roeseana* (ROEMER and ROSOWSKI 1980), and *C. wailesii* (SCHMID and VOLCANI, in preparation). There are some indications (VON SYDOW and CHRISTENHUSS 1972) that growth of the forming valve is centrifugal in *Actinocyclus ehrenbergi*, *Aulacodiscus argus*, and *Coscinodiscus divisus*, as well; CRAWFORD (1974 a) reports that in *M. nummuloides*, also, the valve probably develops centrifugally.

On the other hand there are reports that valve formation is centripetal in *M. varians* (REIMANN 1960) and in germanium-treated *Cyclotella nana* (CHIAPPINO et al. 1977). It is of great taxonomic and evolutionary interest to clarify this situation, especially if there are two quite different modes of valve formation in centric diatoms, even within the same genus as in *Melosira* or in *Cyclotella*.

2. Pennate Diatoms

Investigations carried out in the light microscope (GEITLER 1932) and especially in the electron microscope have revealed that the primary silicification site in the motile pennate diatoms is the raphe area, while in the rapheless pennates it is the "pseudoraphe" area from which the transapical ribs gradually grow to the valve margin (REIMANN 1960, 1964, DRUM and PANKRATZ 1964, REIMANN et al. 1965, 1966, STOERMER et al. 1965, DRUM et al. 1966, COOMBS et al., 1968, DAWSON 1973, PICKETT-HEAPS et al 1975, SCHMID 1976 b, CHIAPPINO et al. 1977, CHIAPPINO and VOLCANI 1977, PICKETT-HEAPS et al. 1979 b, SCHMID 1979 a, b). This development of the silicified structures appears to be general in the pennate diatoms and appears to obey a mechanism that, in its spatial and temporal course, is strictly controlled by a non-reversible sequence of instructions.

Once silicification has started silica is added only to nascent siliceous structures. However, DAWSON (1973) stated that in *Gomphonema parvulum* the outer edges of the valve have already been deposited before deposition occurs in the area between the edge and the central raphe. This deposition sequence requires reexamination and the study of isolated walls, especially since LAURITIS *et al.* (1968) also reported that in *N. alba* the central raphe is formed last, a conclusion, however, that CHIAPIINO *et al.* (1977) showed to be incorrect.

C. Relation Between Silica Deposition, Silica Deposition Vesicle, and Silicalemma

Early studies showed that silica deposition occurs within the cell in a *"silica deposition vesicle"* (SDV) (DRUM and PANKRATZ 1964). In studies on the wall structure of *N. pelliculosa*, REIMANN *et al.* (1966) termed the limiting membrane of this vesicle the *"silicalemma"*. Numerous studies both of pennate diatoms (REIMANN 1964, REIMANN *et al.* 1965, 1966, STOERMER *et al.* 1965, LAURITIS *et al.* 1968, DAWSON 1973, PICKETT-HEAPS *et al.* 1975, CHIAPPINO and VOLCANI 1977, BOROWITZKA and VOLCANI 1978, PICKETT-HEAPS *et al.* 1979 b, SCHMID 1980) and of centric diatoms (CRAWFORD 1974 a, TIPPIT *et al.* 1975, HOOPS and FLOYD 1979, SCHMID and SCHULZ 1979, SCHNEPF *et al.* 1980) demonstrated that silica deposition in diatoms consistently occurs in the SDV. Silica deposition in chrysophytes (*e.g.*, SCHNEPF and DEICHGRÄBER 1969), in radiolarians (ANDERSON 1977, 1981) and sponges (SIMPSON and VACCARO 1974) also occurs in vesicles that are bounded by a silicalemma.

A comprehensive electron microscope study using ultrathin sections as well as isolated walls was done with the pennate *Navicula pelliculosa* (CHIAPPINO and VOLCANI 1977). The initial evidence of a developing valve, after cytokinesis, is a minute SDV, delimited by the silicalemma (Fig. 4 *a*). The SDV extends the length of the cell along the apical axis and beneath the plasmalemma. At this stage polymerized silica already fills the vesicle as a continuous band, the "primary central band", in which the nascent central nodule is already evident (Fig. 3 *A*), within which an array of "knolls" can be seen (Figs. 8 *a* and *b*). Except for the formation of "knolls" in the prospective central nodule the early stages of valve formation in *Pinnularia maior* (Fig. 7 *a*) (PICKETT-HEAPS *et al.* 1979 b) are the same as in *N. pelliculosa*. In *N. pelliculosa* the "primary central band" curves back at the apices towards the central nodule and fuses with "secondary arms" that have developed from this central region, as it has been demonstrated in *Epithemia zebra* var. *saxonica* by GEITLER (1932). The raphe system, *i.e.*, the raphe ribs that enclose the raphe fissure (Figs. 3 *B* and *C*), is thus completed. At the same time the SDV gradually expands laterally as do the transapical ribs forming within it and which have begun to develop along the edges of the raphe ribs in equidistant intervals (first-order branching) (Figs. 3 *C–E*). As the transapical ribs continue to develop cross extensions appear between them at very regular intervals (second-order branching) beginning near the central nodule; they eventually fuse in the apical direction, as was found also in other diatoms (SCHMID 1976 b, 1979 a, PICKETT-HEAPS *et al.* 1979 b). Within the resulting rounded open areas the sieve plates are formed (Fig. 4 *b*) (REIMANN *et al.*

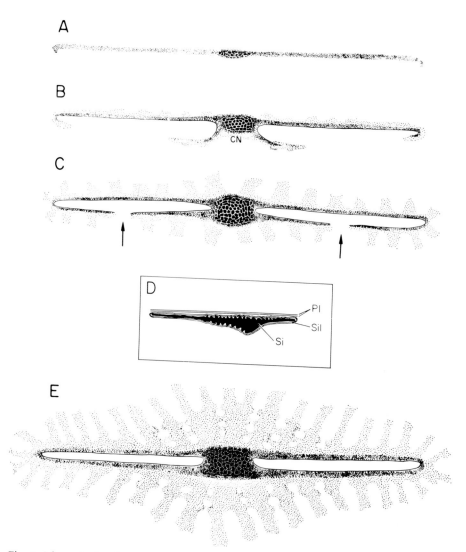

Fig. 3. Schematic drawing summarizing valve development of *N. pelliculosa*. *A* First stage: the primary central band. *B* Second stage: the in-turning of the primary central band and differentiation of the central nodule (*CN*). *C* Third stage: approach of the primary central band, extensions and secondary arms which ultimately fuse forming the raphe ribs; growth of transapical ribs. *D* Cross section of central nodule at the third stage, with mounds or knolls on inner and outer surface. *Si* silica, *Sil* silicalemma, *Pl* plasmalemma. *E* Fourth stage: growth of cross bridges and sieve plates and loss of mounds of central nodule. From CHIAPPINO and VOLCANI 1977.

1966). The spaces between the central nodule "knolls" gradually fill with silica until the "knolls" can no longer be distinguished (Figs. 3 *B–E*); in mature valves no evidence can be found of their existence. Similar steps in valve formation have been observed in the oval morphotype of *P. tricornutum* (BOROWITZKA and VOLCANI 1978) and also in some, though not all of the stages in *Nitzschia alba* (CHIAPPINO *et al.* 1977).

The site, where the "primary central band" fuses with the "secondary arms" can easily be found in the completed valve in some diatoms, as they are manifested as a discontinuity of the transverse pattern. This relationship between the "Voigt-discontinuity"—VOIGT (1943, 1956) described first the very localized pattern aberrations—and the sequence of silica deposition was recently uncovered by MANN (1981).

In *N. pelliculosa*, the transapical ribs seem to develop concurrently on both sides of the raphe ribs, whereas in other species the asymmetric deposition of the two raphe ribs causes a temporary shift in the growth of the transapical ribs as well. A remarkably detailed drawing of this shift, based on light microscopy of *Epithemia zebra* var. *saxonica* (GEITLER 1932), shows that the transapical ribs extending from the primarily deposited raphe rib already reach the valve margin, although on the other side of the slit the raphe rib has not as yet fully developed and fused. A similar asymmetric deposition was observed in SEM investigations of acid-cleaned developing frustules of *Anomoeoneis sphaerophora* (SCHMID 1979 a) and in ultrathin sections of *Pinnularia maior* (PICKETT-HEAPS *et al.* 1979 b) (Figs. 7 *b* and *c*). In both species the asymmetric development of the silica structures occurs in mirror image in the two sister cells.

Longitudinal sections of *Diatoma vulgare* (PICKETT-HEAPS *et al.* 1975) and, in the SEM, *A. sphaerophora* (SCHMID 1980), show that the forming ribs are surrounded by the silicalemma, not as fingers are sheathed in a glove, but as fingers are enclosed in a mitten. In valves of greater structural complexity, e.g., *Amphipleura pellucida* (STOERMER *et al.* 1965), the close attachment of the silicalemma to the developing siliceous structure is more evident. It has therefore been suggested that the silicalemma acts as a "mold" for the forming valves (STOERMER *et al.* 1965, DAWSON 1973, SCHNEPF *et al.* 1980) just as it does for the forming chrysophycean scales, e.g., *Synura* (SCHNEPF and DEICH-GRÄBER 1969).

Despite much investigation the origin of the silicalemma and hence of the SDV is still not known. There have been suggestions that it may derive from Golgi vesicular activity (REIMANN 1964, REIMANN *et al.* 1965, 1966, DAWSON 1973, SCHNEPF *et al.* 1980), or possibly from the endoplasmic reticulum (DAWSON 1973, HOOPS and FLOYD 1979), or even from a combined inter-action between Golgi vesicles, endoplasmic reticulum, and plasmalemma (CHIAPPINO and VOLCANI 1977). In view of the fact that Golgi vesicles have been shown to be capable of manufacturing intricately formed scales and structures in other algae such as *Chrysophyceae* (MANTON and HARRIS 1966, SCHNEPF and DEICHGRÄBER 1969, BROWN and ROMANOVICS 1976, PIENAAR 1976, HIBBERD 1978), there is also the suggestion that the SDV itself may be equivalent to a giant, highly envolved Golgi vesicle, not necessarily func-

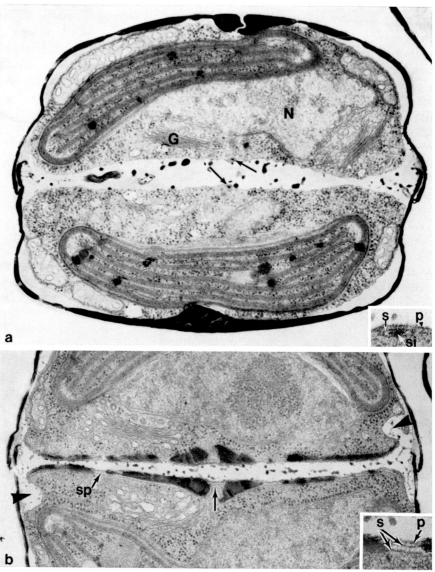

Fig. 4. Electron micrographs of cross sections of *N. pelliculosa* during valve formation. *a* Cell containing two protoplasts in which the initiation of the raphe occures in a minute silica-deposition vesicle (arrows). *N* nucleus, *G* Golgi apparatus. ×30,000. Inset depicts the deposition vesicle bounded by the silicalemma (*s*) containing silica (*si*), lying below the plasmalemma (*p*). ×120,000. *b* Late stage of valve development in which the silica deposition is almost complete. Note the cytoplasmic furrow, developing at opposite sites in the two sister cells (arrowheads). ×34,000. Inset: high magnification of developing sieve plate (*sp*) within the silicalemma. ×120,000. From CHIAPPINO and VOLCANI 1977.

tionally associated with, or derived from, the more typical Golgi bodies (Pickett-Heaps *et al.* 1979 b).

In the centric *T. eccentrica* (Schmid and Schulz 1979) the SDV is formed in the central area beneath the new plasma membrane by the coalescence of small cytoplasmic vesicles, 30–40 nm in diameter (Fig. 5 *b*), which closely resemble in size and content some of the vesicles that have been seen to originate from the Golgi apparatus of this organism (Schulz, in preparation). The SDV extends centrifugally with branching radial tubes; the tubes themselves, by putting out cross extensions, become reticulate and form a two-dimensional system of hexagonal meshes—the outline of the prospective areolae (Figs. 6 *a*, *b*, and *e–1*). The interspace between each mesh—the prospective "porous plate"—is then closed centripetally by fusion of the small vesicles that then constitute the base layer. The silica deposition follows the same sequence within the SDV but with a very small time lag. In the marginal region the elaboration of these structures proceeds somewhat later in each stage of development.

A similar sequence of events has recently been reported for the centric diatom *Attheya decora* (Schnepf *et al.* 1980). Even in the chrysophycean *Ochromonas tuberculata* (Hibberd 1977) cyst formation involves the coalescence of vesicles to form a membranous network which eventually becomes a completely spherical vesicle, between the membranes of which the silica is laid down.

As soon as the base layer has been established in *T. eccentrica* the areolae walls grow vertically outward to the distal side on the previously deposited hexagonal meshes. At the same time the "porous plates" are differentiated by invaginations of the distal and proximal silicalemma (Figs. 6 *c* and *e–2*). Finally, when the walls are built up to their species-specific height, the outer layer of the valve is formed, also by fusion of small vesicles (Schmid and Schulz 1979) (Figs. 5 *c*, *d* and 6 *d*, *e–3*). How the growth of the areolae walls is terminated is completely unknown. As for the mechanism by which the

---

Fig. 5. *Thalassiosira eccentrica: a* SEM micrograph of a whole frustule. Valve structure: tangential striae of loculate areolae; labiate process (white arrows), strutted tubuli (black arrows), tricornate spines (arrowheads). Intercalary band (*IB*), girdle bands (*GB*), valve (*Va*). ×900. *b* Transverse ultrathin section at stage 1 of valve formation: growing edge of the developing valve. Small vesicles (ca. 30–40 nm in diameter), some of them already fused (arrows), form a continuation of the silica deposition vesicle (*SDV*). Chloroplast (*CH*), mitochondrion (*M*), plasmalemma (*PL*), silicalemma (*SL*). Bar = 0.25 μm. ×52,000. From Schmid and Schulz 1979. *c* Ultrathin section of stage 3 of valve formation: outer layer in development; note difference between compacting zone (*CZ*) at the base of the areolae and the growing zone (*GZ*) at the distal part. Small vesicles lie between the plasmalemma (*PL*) and the silicalemma (*SL*) and close to the lateral SL. ×90,000. From Schmid and Schulz 1979. *d* SEM micrograph of a valve at the same stage as *c*, formation of the outer layer. Aggregation of silica spheres in small columns; further aggregations into superstructures (inset). Areola (*A*). ×24,000; inset ×60,000. From Schmid and Schulz 1979. *e* SEM micrograph of a valve stage 3, formation of the outer layer. Reticulate membranous system (*rMS*): tangential tubes become hexagonally shaped prior to close attachment to (or fusion with?) the distal areolae walls. Note numerous vesicles, which are either fusing or budding off (arrows). *PL* removed during preparation. Areola (*A*). ×17,000.

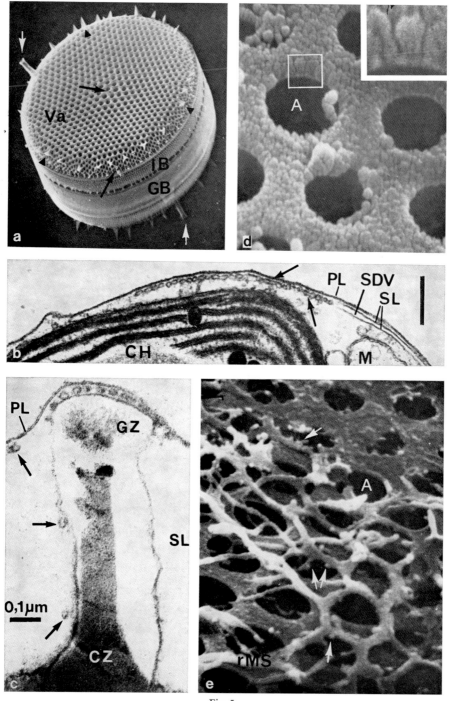

Fig. 5

growth of the entire valve itself is terminated, it has been suggested that a circumferential cytoplasmic furrow in the girdle region of the protoplast acts as a barrier for the expanding SDV (SCHMID 1979 b). As seen from Fig. 4 b, the furrows seem to develop from opposite points (of the apical axis) of the two sister-cells. Whether there is an involvement of filaments or if it is caused by spontaneous plasmolysis is not yet known.

During the formation of the distal areolae walls of *T. eccentrica*, a reticulate membranous system is also involved in valve formation (Fig. 5 e), and may function as a mold for the distal areolae walls. The membranous system develops as tangential tubes between the plasmalemma and the distal silica-lemma; connections form among the tubes and the resulting reticulation becomes hexagonally shaped prior to fusion (if there is indeed fusion) with the SDV. Small vesicles can be seen, but it is not clear whether they are budding off from, or fusing with, the membrane system. The tubular, reticular form of this system suggests a close relationship to the endoplasmic reticulum (ER), (SCHMID, in preparation).

The only indication, so far, for a pattern-related arrangement of the ER during valve formation was observed by PICKETT-HEAPS et al. (1979 b) in *P. maior,* in which the silicalemma, especially in the region of the thickening ribs, is capped by an ER. DAWSON (1973) reported that the ER is in close proximity to the developing valve, though without structural relation. HOOPS and FLOYD (1979) observed a very unusual peripheral ER in the auxospore of *Cyclotella meneghiniana;* it extended completely around the cell and was thought to be associated with the formation of the siliceous scales. However, in auxospores, it has been found that each scale is laid down in its own SDV (CRAWFORD 1974 a). If this completely spherical ER system in the auxospore of *C. meneghiniana* is really the source of the small SDV for each scale, that means it must break up at some point. The role of the ER in diatom valve formation might indeed be similar to its role in chrysophyte scale formation: in all the *Chrysophyceae* so far investigated, the ER always seems to be much involved in the formation of the silica scales, either in molding the SDV, as in *Synura petersenii* (SCHNEPF and DEICHGRÄBER 1969), and *Paraphyso-monas vestita* (MANTON and LEEDALE 1961, LEE 1978) or as the source of the SDV, as has been suggested for *Mallomonas* (WUJEK and KRISTIANSEN 1978) and *Paraphysomonas* (LEE 1978). However, the non-silicified scales of *Sphaleromantis tetragona* and *S. marina* are produced in Golgi vesicles (MANTON and HARRIS 1966, PIENNAR 1976), as are the unmineralized scales of *Synura sphagnicola* (HIBBERD 1978). In *S. petersenii* (SCHNEPF and DEICH-

---

Fig. 6. *a–e* Schematic drawing of valve morphogenesis in *T. eccentrica;* at all stages differen-tiation of the valve margin lags behind the valve center. *a* Stage 1: development of the base layer. *b* Detail of *a*: radial structures become reticulate constituting the outline of the prospective areolae. The finished base layer is a thin silicified plate, enveloped by the silicalemma (*SL*). *c* Stage 2: development of the areolae walls and the "porous" plates. *d* Stage 3: development of the outer layer. *e* Cross-section of the valve in developmental stages 1–3 and relative positions of silica (*Si*), *SL* and *PL*. Based on SCHMID and SCHULZ 1979.

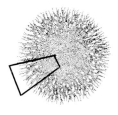

a

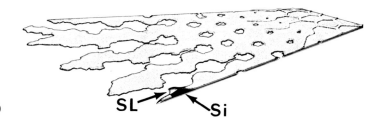

b

SL ← → Si

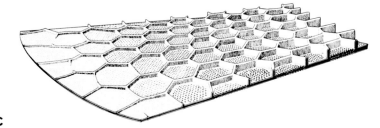

c

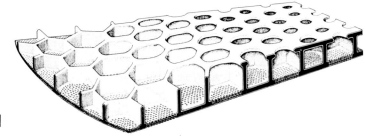

d

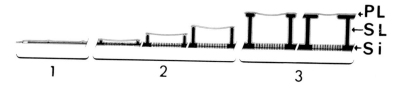

→ PL
← SL
← Si

1   2   3

e

Fig. 6

GRÄBER 1969), and *Mallomonas* (WUJEK and KRISTIANSEN 1978), "coated" vesicles, apparently derived only from the Golgi apparatus have also been found to be incorporated in the SDV.

D. Role of Microtubules, Microtubule Organizing Center, and Microfilaments in Valve Formation

During the earliest stages of valve formation in the pennate diatoms, a row of microtubules (MT) can usually be seen in the cytoplasm beneath, and parallel to, the raphe ribs (STOERMER et al. 1965, DAWSON 1973, CHIAPPINO and VOLCANI 1977, PICKETT-HEAPS et al. 1979 b) (Figs. 7 a and d). This finding agrees with the observation of GREEN (1962), LEDBETTER and PORTER (1963), and HEPLER and NEWCOMB (1964), that the peripheral MT participate in the orientation and organization of wall elements in algae as well as in higher plants; it is highly probable that in diatoms the peripheral MT also govern the orientation of the SDV. It was the extensive work of PICKETT-HEAPS and his associates (1975, 1979 b, TIPPIT et al. 1975, TIPPIT and PICKETT-HEAPS 1977) that elucidated the connection between these MT and the *microtubule organizing center* (MTOC = MC), an organelle similar to a centriole, whose functional implications in diatom mitosis had previously been uncovered by LAUTERBORN (1896).

The MT play an important role in nuclear migration, e.g., *Micrasterias* (KIERMAYER 1968). In diatoms they are probably involved in moving the nucleus closer to the primary silicification site, and hence are probably involved also in centering the valve pattern, as is apparent in *Surirella ovalis* (TIPPIT and PICKETT-HEAPS 1977) and in cells where the nucleus is dislocated by the MT inhibitors, colchicine and APM (SCHMID 1979 c, 1980). In addition the MC itself seems to be highly involved in valve formation as suggested by observations in *P. maior* and *P. viridis* (PICKETT-HEAPS et al. 1979 b); during late cytokinesis the MC moves to the prospective primary silicification site and becomes the focus of increasing numbers of MT which fan out over the

---

Fig. 7. *a Pinnularia viridis*; transverse ultrathin section, half way along the valve between the end and the center. The central sac of the silicalemma is flanked on one side by a group of microtubules (*MT*) and on the other side by a flattened bundle of microfilaments (*MF*). Silica deposition vesicle (*SDV*). ×68,700. From PICKETT-HEAPS et al. 1979 b. *b P. maior*; transverse section half way along the valve between the end and the center. The longitudinal rib of the wall is still single, with few microtubules nearby. The silica deposition vesicle (*SDV*) has extended laterally from the rib asymmetrically (brackets). A bundle of micro-filaments is seen at both the outgrowing edge of the silicalemma and next to the central rib (arrows). ×43,200. (From PICKETT-HEAPS et al. 1979 b). *c P. maior*; transverse section through the central part. Microtubule center (*MC*) on top of the forming central nodule (*CN*). ×13,600. From PICKETT-HEAPS et al. 1979 b. *d P. viridis*; longitudinal section, girdle view; the MC surmounts the thickening central nodule (*CN*) and microtubules extend over the forming wall; note the characteristic structure of the silica in the growing zone (arrows). *M* mitochondria, *MT* microtubules. ×30,600. From PICKETT-HEAPS et al. 1979 b. *e T. eccentrica*; SEM micrograph of an APM-treated cell. Breakdown of the valve (*LaVa*) into scales without areolation (arrows), fragmentation of the intercalary-band and girdle bands into segments of different width. *GB* girdle bands, *IB* intercalary bands. ×3,000. *f A. sphaerophora*; SEM micrograph of an APM-treated cell; inner raphe fissure (*RF*) is fragmented; transapically oriented striae-interstriae pattern (*TS*) undisturbed. ×6,000.

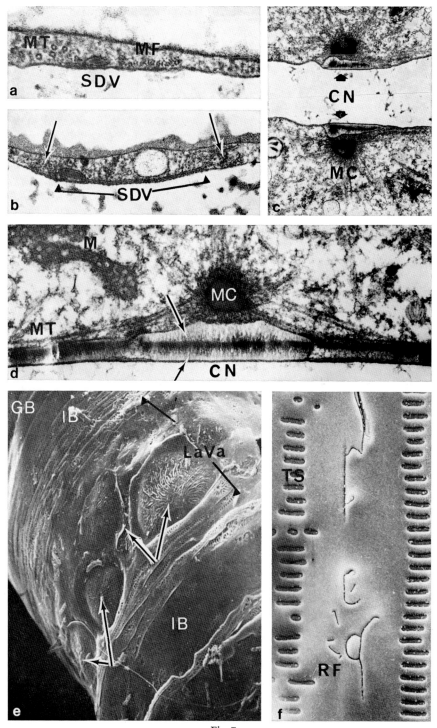

Fig. 7

cleavage furrow. The SDV, which at this stage is oval or triangular in profile and already contains silica, is sandwiched between the plasmalemma and the MT, and throughout the whole process of valve formation the MC remains precisely on the vertex of the forming central nodule (Figs. 7 c and d). A bundle of microfilaments appears, first on one side, then on the other side of this longitudinal wall rib; these microfilaments eventually line the edge of the silicalemma accompanying, and probably causing its extension outwards across the plasmalemma (Pickett-Heaps et al. 1979 b) (Figs. 7 a and b). As soon as the second raphe rib develops, a new structure appears, the "raphe fiber", a bundle of filaments that lies in the cytoplasmic fissure of the raphe running its length and following the curves around the valve poles. The raphe fiber probably functions to prevent silica deposition in the raphe fissure. The MT of the MC have been observed to form a layer over only a part of the wall and only on one side of the raphe (Pickett-Heaps et al. 1979 b), but they seem to be intimately associated with the raphe fiber. In dissected cells of critical-point dried A. sphaerophora and S. peisonis examined with the SEM, bundles of filaments were found (Schmid 1980). After exposure of these two species to colchicine and APM (Schmid 1979 c, 1980) the normal, stretched raphe-fissure was not formed. Instead, all sorts of small, baroquely fragmented fissures appeared along the length of what would have been the raphe fissure (Fig. 7 f). These small fragments indicate that the filaments of the raphe fiber had lost their coordination and/or connection, with consequent disruption of the raphe-fissure. The remaining part of the valve striae-interstriae pattern, however, was disturbed *only* as a consequence of more serious raphe defects. Even if the two raphe ribs are completely dislocated, the transapical ribs retain their normal position vis-à-vis the raphe ribs; hence it is apparent that the raphe system and transapical ribs develop as an entity. Preliminary investigations give the impression that the entire SDV system is suspended within a framework consisting of three sets of filamentous structures oriented along the apical axis—one set in the valve median, and one set on each of the margins; in addition, a number of filamentous structures can be seen extending perpendicular to the apical structures, on top of the interspaces between the transapical ribs (Schmid 1980). Extensive studies are required to elucidate the nature of this framework.

The possibility that the MT might be involved in valve-pattern formation in diatoms is contested by some workers (Schnepf, personal communication, Oey and Schnepf 1970, Schnepf et al. 1980). In the pennates the involvement of MT seems to be confined to raphe formation, as indicated by their reactions to MT-inhibitors. Only at very high concentrations of colchicine and APM, which caused a complete breakdown of the raphe system in A. sphaerophora and S. peisonis, did this destruction cause a disturbance of the striae-interstriae arrangement. However, among the centric diatoms there seem to be species-specific differences in the involvement of MT in valve formation as expressed in the reactions of the cells to MT-inhibitors. When incubated in APM or colchicine, T. eccentrica reacts with the formation of "lateral valves" due to inhibition of cytokinesis, as was found when Cyclotella cryptica was treated with colchicine (Oey and Schnepf 1970). Only when post-telophase cells of *T. eccentrica* were exposed to the MT-inhibitors, were the new walls deposited in the normal valvar plane, but with the same concentration-dependent aberrations as the lateral valves, which, in contrast to the findings in *C. cryptica* climaxed in a complete fragmentation of the valve into scales made up of the base layer, the porous plates of which were irregularly shaped, without even the rudiments of an areolation (SCHMID 1979 c) (Fig. 7 e). On the other hand, *Coscinodiscus wailesii*, similarly treated (SCHMID and VOLCANI, in preparation), forms like pennates "internal valves" instead of "lateral valves", and exhibits only slight disarrangements of the rows of areolae.

Plasmatic MT are generally found in a variety of other silica-depositing organisms where they seem to be associated with the formation of siliceous structures; *e.g.*, the formation of silica scales in the chrysophytes *Synura petersenii* (SCHNEPF and DEICHGRÄBER 1969), and *Mallomonas caudata* (WUJEK and KRISTIANSEN 1978), in choanoflagellates, where the "strip vesicles" are always associated with two parallel MT (LEADBEATER 1979), and in sponge spicules (SIMPSON and VACCARO 1974). In all organisms in which siliceous structures occur, MT seem to be involved; in the heavily silicified radiolarians not only MT, but even an organelle similar to the MC in diatoms, the "podoconus", seems to act as a "morphogenetic organizing center" (ANDERSON 1977).

E. Silica Deposition and Silica Spheres

Transmission electron (TEM) micrographs show that the early stage of silica deposition appears as an electron-dense feathery structure within the SDV, sometimes described as "mottled" (DRUM and PANKRATZ 1964, STOERMER *et al.* 1965, DAWSON 1973). However, such electron-dense "fibrillar material" has also been observed (SCHNEPF and DREBES 1977) in weakly silicified parts of mature valves, *e.g.*, in the tips and the basal layer of the horns of *Attheya decora* which are encrusted with siliceous fibrils, soluble in hydrofluoric acid (HF). During early stages of valve morphogenesis in this species another kind of organized and structured material occurs within the SDV that is not soluble in HF and hence is presumably organic. SCHNEPF *et al.* (1980) have suggested that this material is the matrix for silicification onto which silicon, which passes through the membrane of the SDV, polymerizes. This hypothesis has also been offered to account for the silicification of *Pinnularia* (PICKETT-HEAPS *et al.* 1979 b); in this organism silica appears to be deposited as gradually thickening "microfibrils" that extend progressively from both the distal and the proximal surfaces of the silicalemma (Figs. 7 *d* and 8 *d*). According to the authors the "microfibrils" are the result of silica precipitation on a possibly polysaccharide framework, thought to be secreted by the silicalemma. A number of studies of the areas where these fibrillar siliceous structures develop—scanning electron microscopy (SCHMID 1976 b, 1979 a, b, 1980, SCHMID and SCHULZ 1979), carbon replicas (CHIAPPINO and VOLCANI 1977), and freeze fracture replicas (SCHMID and SCHULZ 1979)—reveal the presence of a loose assemblage of silica spheres of varying sizes (12–50 nm), capable of aggregating into superstructures of 30–40 nm in *N. pelliculosa* and up to

200 nm in *T. eccentrica* (Fig. 7 *d*). Schmid (1976 b) hypothesized that in *A. sphaerophora* silica deposition might occur in small spherical units that are transported by small spherical vesicles whose membranes fuse with the silicalemma and release their silica contents into the SDV. Later observations suggest that indeed this may be the case (Schmid and Schulz 1979). In TEM micrographs of the early development stage in *T. eccentrica*, many electron dense, 30–40 nm diameter vesicles, can be seen lying over the developing

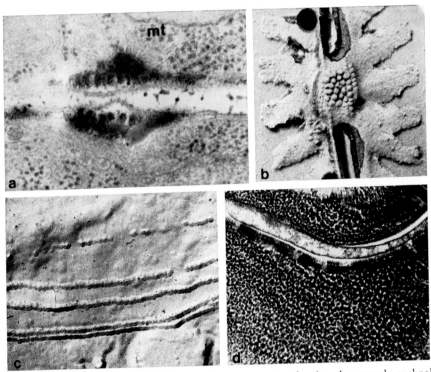

Fig. 8. *a* Central nodule of *N. pelliculosa* in cross section showing the mounds or knolls of silica. Note array of microtubules (*mt*). ×60,000. From Chiappino and Volcani 1977. *b* Carbon-Pt replica of the inner face of a developing valve showing the knolls of silica. ×40,000. From Chiappino and Volcani 1977. *c* Carbon-Pt replica of girdle region of fusiform cell wall of *P. tricornutum*: silica bands appear made up of series of granules. ×100,000. From Borowitzka and Volcani 1978. *d P. maior*, valvar plane section through the growing zone of the forming valve. The siliceous "microfibrils" are sectioned transversely. ×25,000. From Pickett-Heaps *et al.* 1979 b.

"porous" plates and fusing with the SDV; however, when the valve at this stage is cleaned with sulfuric acid the vesicles turn out to be silica spheres. Therefore these small electron dense vesicles were termed "silicon transport vesicles" (Schmid and Schulz 1979). In high power TEM micrographs of the growing-zone similar spherical silica units were observed in ultra-thin sections of *T. eccentrica*. They were also detected in *Ditylum brightwellii* (Kotler, Hewes, and Volcani, unpublished), *Melosira nummuloides*, and

M. varians (CRAWFORD, personal communication); in the latter some of the spheres appear to have a dark "nucleus". Some spherical subunit structures, 30—40 nm in diameter, can also be recognized in the delicate girdle bands of *P. tricornutum* (BOROWITZKA and VOLCANI 1978) (Fig. 8 c). The silica spheres so far reported in diatoms correspond in size to the primary silica particles described in precious opal (DARRAGH et al. 1966), another form of amorphous silica.

At the present time it is very difficult to decide whether the siliceous structures (spheres and microfibrils) are genuine, since artifacts can be introduced in EM procedures during any stage of specimen preparation and as a result of electron bombardment. Nothing is known of how these spheres are assembled and compacted to become the mature wall; the difficulty of investigating the problem is increased by the fact that in some diatoms such as *A. decora* the silicification process does not appear to involve preformed spheres (SCHNEPF et al. 1980). On the other hand, there are convincing indications that formation of spheres does occur during silicification in sponges (GARONNE et al. 1981), in radiolarians (ANDERSON 1981), in chrysophycean cysts (HIBBERD 1977), and in higher plants (see review articles of SANGSTER and PARRY 1981, KAUFMAN et al. 1981); granular in the developing structures, and dense in the older parts.

F. Silicalemma-Plasmalemma Relationship

When the valve is completed it is released exteriorly by fusion of the silicalemma with the plasmalemma (STOERMER et al. 1965, LAURITIS et al. 1968, DAWSON 1973, DUKE and REIMANN 1977, CHIAPPINO and VOLCANI 1977). Although there is no final evidence to date, it is assumed that the proximal membrane of the SDV becomes the functional plasmalemma of the new cell (STOERMER et al. 1965, DAWSON 1973, DUKE and REIMANN 1977, SCHNEPF et al. 1980), a process similar to that occurring during cyst formation in the chrysophyte *Ochromonas tuberculata* (HIBBERT 1977). But it might also be assumed that the lost plasmalemma is replaced by a newly formed one. In *N. pelliculosa* it was demonstrated (REIMANN et al. 1966), that a new plasmalemma is already visible when the valves are still under construction. However in this special case it was not possible to detect a plasmalemma outside the SDV, so caution should be taken in interpreting such pictures. In a reexamination of *N. pelliculosa,* CHIAPPINO and VOLCANI (1977) were not able to establish whether the new plasmalemma was derived from the silicalemma or from an extension of the old plasmalemma surrounding the cell.

In *N. pelliculosa* (CHIAPPINO and VOLCANI 1977) and in *N. alba* (LAURITIS et al. 1968) the distal silicalemma and the old plasmalemma—each of which is well defined up to this stage—are no longer delineated after the release of the valve and are indistinguishable. They now constitute the distal part of the "organic casing" of the mature valve; this casing is visible in HF-treated sections. In the case of *Amphipleura pellucida* the old plasmalemma and the distal silicalemma "slough off" as vesicular remnants (STOERMER et al. 1965), leaving the valves naked of organic material. However, in HF-treated sections of *A. pellucida* the organic casing can be seen of both the new and the parental valves (see Fig. 12, STOERMER *et al.* 1965). Since the SDV breaks down *before* the porous plates of the punctae have been deposited and *before* the developing valve has attained its final thickness (STOERMER *et al.* 1965, p. 1073) the disintegration of the two membranes external to the siliceous valve might be artifactual.

It might be hypothesized that the organic casing is formed *de novo* on the maturing walls after the disintegration of the distal silicalemma-plasmalemma as was suggested for the inner part of the valve envelope by STOERMER *et al.* (1965), DAWSON (1973), DUKE and REIMANN (1977) and SCHNEPF *et al.* (1980). During the complex process of membrane turnover in connection with valve formation, the inner silicalemma is thought to become part of the functional plasmalemma, in this way radically changing its function from that of involvement in silica deposition to that of carrying out the numerous activities of the plasmalemma, *e.g.*, the deposition of organic material.

DUKE and REIMANN (1977) report that in *C. fusiformis* the material constituting the organic wall between the silicified raphe and girdle band is secreted through the plasmalemma, which in this case is the plasma membrane formed during cytokinesis. In *P. tricornutum*, BOROWITZKA and VOLCANI (1978) have described the formation of various types of oval cells, which usually produce a single, lightly silicified valve per cell. As in *C. fusiformis*, the silica component of the valve is formed within the SDV but the organic component is deposited onto the outside of the plasmalemma in those areas not underlaid by the SDV. There must, therefore, be some variation in the secretory activity of the plasmalemma, controlled in conjunction with the development of the SDV. Once formed, the silica portion of the valve is released and its edges attach to the edges of the organic wall, thereby forming the completed wall. In this organism, as in *C. fusiformis*, silicalemma and plasmalemma are highly specialized in their secretory activity. However, there is increasing evidence produced by TEM—and/or LM—studies that in some diatoms organic material is added beneath an existing siliceous valve, constituting a "secondary organic wall" that is secreted through that part of the plasmalemma which is thought to have been the former silicalemma [*e.g.*, *Subsilicea fragilarioides* (VON STOSCH and REIMANN 1970); *Melosira nummuloides* (CRAWFORD 1973); *Attheya decora* (SCHNEPF and DREBES 1977); *Bellerochea malleus*, *B. horologicalis*, *B. yucatanensis*, *Streptotheca tamesis*, *Neostreptotheca subindica* (VON STOSCH 1977); *P. tricornutum* (BOROWITZKA and VOLCANI 1978)]. There is no indication whether there are differences or homologies between the inner part of the "organic casing" of the diatom frustule and the "secondary organic wall". If indeed the proximal silicalemma becomes the functional plasmalemma, in both cases the organic material is secreted through the modified silicalemma, whereas in *C. fusiformis* the silicalemma seems to be restricted in its function to silica deposition.

In all the investigated species the organic layer stains intensively with ruthenium red. These relatively recent studies are in complete agreement with the early investigations of LIEBISCH (1928, 1929), who demonstrated that this organic layer is secreted after deposition of the silica valve in *Triceratium antediluvianum* and *Biddulphia biddulphiana*. As this organic part stained with basic dyes it was thought to consist of pectin, and MANGIN (1908) introduced the term "pectin layer". However, the extensive investigations carried out by VOLCANI (1978) and his associates to characterize the chemical composition of the "pectin" layer revealed a great complexity but no pectin (see section below).

G. Formation of Girdle Bands

Following valve formation the intercalary and girdle bands are formed, each individual band in its own SDV. Whether their formation is associated with mitosis (or with DNA replication) is not as clear as it is with regard to valve formation since girdle bands may be produced throughout the life cycle of the cell (ROUND 1972 a); e.g., in all the morphotypes of P. tricornutum girdle band formation is continuous (BOROWITZKA and VOLCANI 1978), as it is in Striatella unipunctata (ROTH and DE FRANCISCO 1977). In species such as N. pelliculosa, in which the epicingulum consists of 3, but the hypocingulum of only 2 girdle bands during interphase, the "missing" girdle band is deposited prior to mitosis (CHIAPPINO and VOLCANI 1977) (but probably following DNA-replication!), becoming part of the epitheca of the new cell. In A. decora, also, the connecting band of the hypotheca is formed just before cell division (SCHNEPF and DREBES 1977).

Girdle or intercalary band formation has been observed in a number of diatoms. Silica deposition and release of the completed girdle bands seem to be very similar in all the investigated species. For the formation of the girdle band SDV, however, there is only one report (DAWSON 1973) for Gomphonema parvulum. In this organism the flat SDV for each individual band is formed by the coalescence and fusion of small vesicles. Once formed within the vesicle, in all diatoms studied, the completed band is then released as described for the diatom valve (CHIAPPINO and VOLCANI 1977) and becomes located next to the edge of the valve, an adjacent girdle- or an intercalary band to which it becomes attached (DAWSON 1973, ROTH and DE FRANCISCO 1977). In N. pelliculosa only two girdle bands are deposited after each valve formation, thus restoring the 3/2 girdle band configuration of the interphase cell.

The girdle bands may be wholly siliceous, or partly organic and partly siliceous, as in C. fusiformis (REIMANN et al. 1965) and P. tricornutum (BOROWITZKA and VOLCANI 1978). Little is known about the formation of these bands and the mechanism by which they intercalate with, and position themselves in relation to the valve and other girdle and/or intercalary bands. Some ideas as to these complex relationships and their possible development are offered by ROUND (1972 a, b). The recent investigations of A. decora (SCHNEPF and DREBES 1977, SCHNEPF et al. 1980), a very unusual species in many respects, might lead to an understanding of the construction of "normal" cells, and of why the individual siliceous components of a diatom frustule do not fall apart, although the endogenous pressure must be in general higher than that of the surrounding medium. Part of the answer may lie in the fact that the girdle bands seem to be connected by an organic material. In A. decora the protoplast is covered by an organic layer (that stains with ruthenium red) except in the circumferential area of the thecal junction, the regions of the labiate processes, and the area within the horns. Where this layer occurs, the protoplast is retracted, but at the horn bases the organic layer is fused with the silicified basal layer. The interior part of the intercalary bands shows an organic continuation that is connected with the organic cover of the protoplast (SCHNEPF and DREBES 1977). The localized retractions of the protoplast from the valve commence with the formation of the first intercalary band. As in other species, each of these bands is deposited in a separate SDV. The continuation for each band is released simultaneously, but at the distal surface of the plasmalemma in the region when the last retraction has occurred. When the intercalary band is released by fusion of the SDV with the plasmalemma the connection between the interior part of the siliceous band and the organic continuation is established. The proximal membrane of the SDV, now part of the plasmalemma, retracts, and extrudes the organic continuation for the next band. This sequential process is repeated at regular intervals until the new cingulum is established (SCHNEPF et al. 1980).

The appearance of an organic "matrix" between the intercalary bands of *Striatella unipunctata,* where a "dense material" is located even at the distal side of the junctions between two bands (ROTH and DE FRANCISCO 1977), suggests a deposition mechanism that is basically similar to that in *A. decora.* Girdle band formation in *C. fusiformis* (REIMANN et al. 1966) might also be similar.

IV. Biochemistry of the Diatom Cell Wall

Wall formation appears to proceed according to the following sequence of biochemical events: 1. silicon is taken up from the medium; 2. the constituents and macromolecules of the unit membrane bounding the SDV, the silicalemma, are synthesized; 3. silicon (the form is unknown) is translocated from the cytoplasmic pool into the SDV; 4. silicon is polymerized in the SDV; and 5. "secondary" organic material is synthesized for addition to the developing walls. This section of the review will deal almost entirely with the chemical composition of the cell wall. (The assembly of the organic casing, the role of the casing, and more detailed discussion of the silicification mechanism can be found in VOLCANI 1981).

In all diatoms so far investigated the siliceous structures are always enveloped by an organic casing. In most species the mature casing consists of a single, undifferentiated layer 8–10 nm thick. The casing adheres so tightly to the silica valve that the two components cannot be separated by mechanical means; however, HF vapor can be used to dissolve the siliceous portion of the frustule, leaving the casing intact. The latter retains the patterning of the silica valve but appears to be structurally undifferentiated and the macromolecular components (proteinaceous and carbohydrates) are very tightly meshed (VOLCANI 1978). In the fusiform type of *P. tricornutum,* which has an entirely unsilicified valve, the organic material appears under the electron microscope to consist of three layers although they cannot be dissociated (DYCK and VOLCANI, unpublished).

The biochemistry of the diatom frustule can be studied only in isolated cell walls. HECKY et al. (1973) used sonication to free the frustules from the cell contents, but this method cannot completely prevent contamination with cellular material. An alternative method involves mechanically cracking the cells by shaking them with glass beads (COOMBS and VOLCANI 1968); both light and electron microscopy showed the walls to be free of contamination.

A. Siliceous Component

The silica structures of diatoms are always composed of highly pure amorphous silica in varying degrees of hydration from $SiO_2 \cdot 2 H_2O$ to $SiO_2 \cdot nH_2O$; in some species its character is intermediate between gel and opaline silica (KAMATANI 1971). TORRI and VOLCANI (unpublished) investigated the chemical composition of the cell wall in six diatom species (the freshwater *Navicula pelliculosa*, and the marine *Cylindrotheca fusiformis, Nitzschia angularis, N. thermalis, Cyclotella cryptica, Phaeodactylum tricornutum*); in this group silicon content of the frustule ranges from 10% (*P. tricornutum* "fusiform") to 72% (*N. pelliculosa*).

B. Organic Component

In the above diatoms the organic material accounts for 36–90% of the total wall material plus water of hydration. The protein/carbohydrates ratio varies strikingly, ranging from only 0.33 in *P. tricornutum* to 6.5 in *C. cryptica*.

1. Amino Acids Composition

All the 18 common amino acids occur in the proteinaceous material of the casing. In addition a number of unknown ninhydrin-positive compounds are found, but 10 of them appear in only one or more of the species examined. Since these compounds are not present in TCA extracts of the cytoplasm, they are apparently constituents of the organic casing only. Three of them were later identified as 2,3-cis-4-trans-3,4-dihydroxy-L-proline, an analogue of proline (NAKAJIMA and VOLCANI 1969), ε-N-trimethyl-L-δ-hydroxylysine, and its phosphorylated compound (NAKAJIMA and VOLCANI 1970). These compounds are particularly interesting as they are analogs of the 3- and 4-hydroxyproline and hydroxylysine present in collagen.

The proportions of the various amino acids differ markedly in the casing and cytoplasm, as found by HECKY et al. (1973) in a quantitative comparison of 17 amino acids of three freshwater diatoms (*Cyclotella stelligera, Melosira granulata* var. *angustissima, N. pelliculosa*) and 2 estuarine species (*Cyclotella cryptica, Nitzschia brevirostris*). In terms of amino acid residues/1000, serine (especially), threonine, and glycine are significantly higher in the casing. In the estuarine *C. cryptica* and the freshwater *C. stelligera*, the serine content is, respectively, 147 and 157% greater in the casing than in the cytoplasm and accounts for 23 and 25% of total wall amino acids; the latter amount, according to the authors, is the highest yet reported for any biological material. On the other hand, glutamic (especially) and aspartic acids are much lower in the casing, as are tyrosine, phenylalanine, methionine, and cystine.

2. Carbohydrates

Carbohydrate material of the casing in the diatoms studied by Torri and Volcani (unpublished) consists of 5 sugars: rhamnose, fucose, xylose, mannose, and glucose, as well as a number of unidentified sugars including deoxy-sugars, glucuronic- and galacturonic acid (the latter only in *N. angularis*), and unidentified uronic acids. Ten sulfated polysaccharide fractions were found in the wall of *N. pelliculosa* (Banerji and Volcani, unpublished), two of which were isolated and found to be homogeneous. One is a fucosan sulfate containing fucose-mannose-galactose (a trace of the latter) and a sugar—unidentified but presumably a deoxy-sugar; this polysaccharide is similar to a previously isolated polysilicate-binding polysaccharide detected in the cells of *N. pelliculosa* (Katsumata and Volcani, unpublished). The other, a glucuronomannan (glucuronic acid-mannose) sulfate (GMS), comprised more than 70% of the total polysaccharides content of the casing. It is similar to the GMS containing 27% glucuronic acid and 7% sulfate (bound to mannose) reported by Ford and Percival (1965) from whole cells of *P. tricornutum*. Dyck and Volcani (unpublished) found that in *P. tricornutum* and *N. pelliculosa* the wall material remaining after a series of extractions consists of a mat of very fine fibrils composed entirely of GMS; indeed Percival and McDowell (1967) have suggested that GMS might be a characteristic cell wall component in all diatoms.

In comparing the sugar composition of casing and cytoplasm, Hecky *et al.* (1973) found ribose and arabinose in some of the diatoms studied, in addition to the sugars found by Torri and Volcani; however, the Hecky group did not detect any uronic acids or deoxy sugars. Compared to the cell contents, the casing is enriched in xylose and, to a somewhat greater degree, in mannose, but the most salient difference is the markedly lower glucose content of the casing, ranging from 14.5% to as much as 64.2% less than in the cytoplasm. Of interest, also, is the finding that fucose content is lower in the casing of the freshwater species than in that of the estuarine, whereas the glucose content is higher; it may be that in the estuarine environment fucose substitutes for glucose in the organic wall material.

3. Lipids

Kates and Volcani (1968) studied the casing lipids in the 6 species of the Torri and Volcani study; in 5 of the diatoms, the lipids constitute only a minor fraction (0.8–5.0%) of the total casing constituents, but in *C. cryptica* the lipid content is 13%. The lipid composition is unusual consisting mostly of non-phosphatides (neutral lipids, free fatty acids, sulfolipid, and mono- and digalactosyl diglycerides) with only a small proportion of phosphatides, mostly phosphatidyl glycerol. The major fatty acids are oleic, palmitoleic, and palmitic, with small-to-trace amounts of C_{16}-, C_{18}-, and C_{20}-polyenoic acids.

C. The Silicification Mechanism

Since the polymerization of silicic acid takes place within a vesicle bounded by the silicalemma, this membrane is presumably permeable to silicon, but whether as free silicate, organo-silicon, or a silicate complexed with organic molecules is not known, nor is the mechanism of transport known. However, there is evidence (COOMBS and VOLCANI 1968) that the state or the form of the silicon species changes during wall formation in *N. pelliculosa;* in early stages it is bound or polymerized, whereas in later stages a portion of it is extractable with all solvents and hence free. Changes also occur in the silicon pool during wall formation (SULLIVAN 1979).

Polymerization of the silicic acid could be produced by various factors such as changes in pH and/or concentration (see ILER 1979) or binding to specific sites in the silicalemma with subsequent polymerization. The binding, in turn, could be the result of hydrogen bonding or ionic interaction (HOLT and YATES 1953), or could be produced by condensation on hydroxyl groups (COTTON and WILKINSON 1966). HECKY *et al.* (1973) propose the latter—*i.e.,* that silicification is protein-mediated—and they present a chemical model in which the hydroxyl-containing amino acids, serine and threonine, provide a template on which condensed silicic acid molecules are fixed in a geometric arrangement favoring further polymerization.

This model is based on the assumption that the composition of the casing and of the silicalemma is the same. But the embryonic walls, in which silica deposition is actually taking place, differ in important respects from the casing, notably in having much less of the hydroxyl-containing amino acids, serine and threonine (AZAM and VOLCANI, unpublished). Nevertheless, this does not exclude the possibility of a higher content of these amino acids in the silicalemma *per se,* and hence does not rule out the validity of the HECKY *et al.* model. PICKETT-HEAPS *et al.* (1979 b) on the other hand, suggest that the silicalemma creates a frame-work of fine strands of template of polysaccharide upon which the siliceous matrix is precipitated.

Almost all the questions as to the mechanism of silicification in diatoms remain to be answered and research in this area should be richly rewarding.

Acknowledgement

A. M. SCHMID wishes to thank the Max Kade Foundation for the grant of a fellowship.

References

ANDERSON, O. R., 1977: Cytoplasmic fine structure of nassellarian Radiolaria. Mar. Micropaleontol. 2, 251—264.
— 1981: Radiolarian fine structure and silica deposition. In: Silicon and siliceous structures in biological systems (SIMPSON, T. L., VOLCANI, B. E., eds.). New York: Springer. In press.
ANONYMOUS, 1975: Proposals for a standardization in diatom terminology and diagnoses. Nova Hedwigia Beih. 53, 323—354.
BACHRACH, E., LEFÉVRE, M., 1929: Contribution à l'étude du rôle de la silice chez les êtres vivants. Observations sur la biologie des Diatomées. J. Phys. Path. gén. 27, 241—249.
BOOTH, B., HARRISON, P. J., 1979: Effect of silicate limitation on valve morphology in *Thalassiosira* and *Coscinodiscus.* J. Phycol. 15, 326—329.
BOROWITZKA, M. A., CHIAPPINO, M. L., VOLCANI, B. E., 1977: Ultrastructure of a chain-forming diatom *Phaeodactylum tricornutum.* J. Phycol. 13, 162—170.
— VOLCANI, B. E., 1978: The polymorphic diatom *Phaeodactylum tricornutum:* Ultrastructure of its morphotypes. J. Phycol. 14, 10—21.

Brown, R. M., Romanovicz, D. K., 1976: Biogenesis and structure of Golgi-derived cellulosic scales in *Pleurochrysis*. I. Role of the endomembrane system in scale assembly and exocytosis. Appl. Polym. Symp. **28**, 537—585.

Chiappino, M. L., Volcani, B. E., 1977: Studies on the biochemistry and fine structure of silica shell formation in diatoms. VII. Sequential cell wall development in the pennate *Navicula pelliculosa*. Protoplasma **93**, 205—221.

— Azam, F., Volcani, B. E., 1977: Effect of germanic acid on developing cell walls of diatoms. Protoplasma **93**, 191—204.

Cooksey, K. E., Cooksey, B., 1974: Calcium deficiency can induce the transformation from oval to fusiform cell in cultures of *Phaeodactylum tricornutum*, Bohlin. J. Phycol. **10**, 89—90.

Coombs, J., Lauritis, J. A., Darley, W. M., Volcani, B. E., 1968: Studies on the biochemistry and fine structure of silica shell formation in diatoms. V. Effects of colchicine on wall formation in *Navicula pelliculosa* (Bréb.) Hilse. Z. Pflanzenphysiol. **59**, 124—152.

— Volcani, B. E., 1968: Studies on the biochemistry and fine structure of silica-shell formation in diatoms. Chemical changes in the wall of *Navicula pelliculosa* during its formation. Planta **82**, 280—292.

Cotton, F., Wilkinson, G., 1966: Advanced inorganic chemistry, p. 468. New York: Interscience.

Crawford, R. M., 1973: The organic component of the cell wall of the marine diatom *Melosira nummuloides* (Dillw.) C. Ag. Br. Phycol. J. **8**, 257—266.

— 1974 a: The structure and formation of the siliceous wall of the diatom *Melosira nummuloides* (Dillw.) C. Ag. Nova Hedwigia, Beih. **45**, 131—141.

— 1974 b: The auxospore wall of the marine diatom *Melosira nummuloides* (Dillw.) C. Ag. and related species. Br. Phycol. J. **9**, 9—20.

Darley, W. M., Volcani, B. E., 1969: Role of silicon in diatom metabolism. A silicon requirement for desoxyribonucleic acid synthesis in the diatom *Cylindrotheca fusiformis* Reimann and Lewin. Exp. Cell Res. **58**, 334—343.

Darragh, P. J., Gaskin, A. J., Terrell, B. C., Sanders, J. V., 1966: Origin of precious opal. Nature **209**, 12—16.

Dawson, P., 1973: Observations on the structure of some forms of *Gomphonema parvulum* Kütz. III. Frustule formation. J. Phycol. **9**, 353—365.

Drebes, G., 1967: *Bacteriastrum solitarum* Mangin, a stage in the life history of the centric diatom *Bacteriastrum hyalinum*. Mar. Biol. **1**, 40—42.

— 1977: Sexuality. In: The biology of diatoms (Werner, D., ed.), pp. 250—283. (Botanical Monographs, Vol. 13.) Oxford: Blackwell.

Drum, R. W., 1964: Notes on Iowa diatoms. VI. Frustular aberrations in *Surirella ovalis*. J. Iowa Acad. Sci. **71**, 51—55.

— Pankratz, H. S., 1964: Post mitotic fine structure of *Gomphonema parvulum*. J. Ultrastruct. Res. **10**, 217—223.

— — Stoermer, E. F., 1966: Electron microscopy of diatom cells. In: Diatomeenschalen im elektronenmikroskopischen Bild, VI (Helmcke, J. G., Krieger, W., eds.). Lehre: Cramer Verlag.

Duke, E. L., Reimann, B. E. F., 1977: The ultrastructure of the diatom cell. In: The biology of diatoms (Werner, D., ed.), pp. 65—109. (Botanical Monographs, Vol. 13.) Oxford: Blackwell.

Ford, C. W., Percival, E., 1965: Carbohydrates of *Phaeodactylum tricornutum*. Part II. A sulphated glucuronomannan. J. Chem. Soc. 7042—7046.

Fryxell, G. A., Hasle, G. R., 1972: *Thalassiosira eccentrica* (Ehrb.) Cl., *T. symmetrica* sp. nov., and some related centric diatoms. J. Phycol. **8**, 297—317.

Fuhrmann, J. A., Chisholm, S. W., Guillard, R. R. L., 1978: Marine alga *Platymonas* sp. accumulates silicon without apparent requirement. Nature **272**, 244—245.

GARRONE, R., SIMPSON, T. S., POTTU-BOUMENDIL, J., 1981: Ultrastructure and deposition of silica in sponges. In: Silicon and siliceous structures in biological systems (SIMPSON, T. L., VOLCANI, B. E., eds.). New York: Springer. In press.

GEISSLER, U., 1970 a: Die Schalenmerkmale der Diatomeen. Ursachen ihrer Variabilität und Bedeutung für die Taxonomie. Nova Hedwigia, Beih. 31, 511—535.

— 1970 b: Die Variabilität der Schalenmerkmale bei den Diatomeen. Nova Hedwigia Beih. 19, 623—773.

GEITLER, L., 1932: Der Formwechsel der pennaten Diatomeen (Kieselalgen). Arch. Protistenk. 78, 1—226.

— 1963: Alle Schalenbildungen der Diatomeen treten als Folge der Zell- oder Kernteilung auf. Ber. dtsch. bot. Ges. 75, 393—396.

— 1979: On some peculiarities in the life history of pennate diatoms hitherto overlooked. Amer. J. Bot. 66, 91—97.

GREEN, P. B., 1962: Mechanism for plant cellular morphogenesis. Science 138, 1404—1405.

GROSS, F., 1940: The development of isolated resting spores into auxospores in *Ditylum brightwelli* (West.). J. Mar. Biol. Ass. U.K. 24, 375—380.

HASLE, G. R., 1974: The "mucilage pore" of pennate diatoms. Nova Hedwigia Beih. 45, 167—186.

— HEIMDAL, B. R., FRYXELL, G. A., 1971: Morphologic variability in fasciculated diatoms as exemplified by *Thalassiosira tumida* (Janisch) Hasle, comb. nov. Antarct. Res. Ser. 17, 313—333.

HECKY, R. E., MOPPER, K., KILHAM, P., DEGENS, E. T., 1973: The amino acid and sugar composition of diatom cell-walls. Mar. Biol. 19, 323—331.

HEMMINGSEN, B. B., 1971: A mono-silicic acid stimulated adenosintriphosphatase from proto-plasts of the apochlorotic diatom *Nitzschia alba*. Ph.D. thesis, University of California, San Diego.

HEPLER, P. K., NEWCOMB, E. H., 1964: Microtubules and fibrils in the cytoplasm of *Coleus* cells undergoing secondary wall deposition. J. Cell Biol. 20, 529—533.

HEURCK, H. VAN, 1899: Traité des Diatomées. Anvers.

HIBBERD, D. J., 1977: Ultrastructure of cyst formation in *Ochromonas tuberculata* (*Chrysophyceae*). J. Phycol. 13, 309—320.

— 1978: The fine structure of *Synura sphagnicola* (Korsh.) Korsh. (*Chrysophyceae*). Br. Phycol. J. 13, 403—412.

HOLMES, R. W., 1967: Auxospore formation in two marine clones of the diatom genus *Coscinodiscus*. Amer. J. Bot. 54, 163—168.

— REIMANN, B. E. F., 1966: Variation in valve morphology during the life cycle of the marine diatom *Coscinodiscus concinnus*. Phycologia 5, 233—244.

HOLT, P. F., YATES, D. M., 1953: Tissue silicon: a study of the ethanolsoluble fraction, using ^{31}Si. Biochem. J. 54, 300—305.

HOOPS, H. J., FLOYD, G. L., 1979: Ultrastructure of the centric diatom *Cyclotella meneghiniana*: vegetative cell and auxospore development. Phycologia 18, 424—435.

HOSTETTER, H. P., HOSHAW, R. W., 1972: Asexual development patterns of the diatom *Stauroneis anceps* in culture. J. Phycol. 8, 289—296.

— RUTHERFORD, K. D., 1976: Polymorphism of the diatom *Pinnularia brebissonii* in culture and a field collection. J. Phycol. 12, 140—146.

HUSTEDT, F., 1930: Die Kieselalgen. In: Rabenhorsts Kryptogamenflora. Leipzig: Akad. Verlagsgesellschaft mbH.

ILER, R. K., 1979: The chemistry of silica: solubility, polymerization, colloid and surface properties, and biochemistry. New York: J. Wiley.

KAMATANI, A., 1971: Physical and chemical characteristics of biogenous silica. Mar. Biol. 8, 89—95.

Kates, M., Volcani, B. E., 1968: Studies on the biochemistry and fine structure of silicon shell formation in diatoms. Lipid components of the cell walls. Z. Pflanzenphysiol. **60**, 19—29.

Kaufmann, P. B., Dayanandan, P., Takeoka, Y., Srinivasan, J., Lau, E., Bigelow, W. C., Jones, J., Iler, R. K., la Croix, J. D., Ghoshah, N. S., 1981: Silica in shoots of higher plants. In: Silicon and siliceous structures in biological systems (Simpson, T. L., Volcani, B. E., eds.). New York: Springer. In press.

Kiermayer, O., 1968: Hemmung der Kern- und Chloroplastenmigration von *Micrasterias* durch Colchicin. Naturw. **55**, 299—300.

Lauritis, J. A., Coombs, J., Volcani, B. E., 1968: Studies on the biochemistry and fine structure of silica shell formation in diatoms. IV. Fine structure of the apochlorotic diatom *Nitzschia alba* Lewin and Lewin. Arch. Microbiol. **62**, 1—16.

Lauterborn, R., 1896: Untersuchungen über Bau, Kernteilung und Bewegung der Diatomeen. Leipzig: W. Engelmann.

Leadbeater, B. S. C., 1979: Developmental studies on the loricate choanoflagellate *Stephanoeca diplocostata* Ellis. I. Ultrastructure of the non-dividing cell and costal strip production. Protoplasma **98**, 241—262.

Ledbetter, M. C., Porter, K. R., 1963: A "microtubule" in plant fine structure. J. Cell Biol. **19**, 239—250.

Lee, J. J., McEnery, M. E., Shilo, M., Reiss, Z., 1979: Isolation and cultivation of diatom symbionts from larger foraminifera (protozoa). Nature **280**, 57—58.

Lee, R. E., 1978: Formation of scales in *Paraphysomonas vestita* and the inhibition of growth by germanium dioxide. J. Protozool. **25**, 163—166.

Lewin, J. C., 1955: Silicon metabolism in diatoms. II. Sources of silicon for growth of *Navicula pelliculosa*. Plant Physiol. **30**, 129—134.

— 1958: The taxonomic position of *Phaeodactylum tricornutum*. J. Gen. Microbiol. **18**, 427—432.

— 1962: Silicification. In: Physiology and biochemistry of algae (Lewin, R. A., ed.), pp. 445—455. New York: Academic Press.

Li, C. W., Chiang, Y. M., 1979: A euryhaline and polymorphic new diatom, *Proteucylindrus taiwanensis* gen. et sp. nov. Br. Phycol. J. **14**, 377—384.

Liebisch, W., 1928: *Amphitetras antediluviana* Ehrb., sowie einige Beiträge zum Bau und zur Entwicklung der Diatomeenzelle. Z. Bot. **20**, 225—271.

— 1929: Experimentelle und kritische Untersuchungen über die Pektinmembran der Diatomeen unter besonderer Berücksichtigung der Auxosporenbildung und der Kratikularzustände. Z. Bot. **22**, 1—65.

Lowe, R. L., Crang, R. E., 1972: The ultrastructure and morphological variability of the frustule of *Stephanodiscus invisitatus* Hohn and Hellermann. J. Phycol. **8**, 256—259.

Mangin, M. L., 1908: Observations sur les Diatomées. Ann. Sci. Nat. Bot. **9**, 117—219.

Mann, D. G., 1981: A note on valve formation and homology in the diatom genus *Cymbella*. Ann. Bot. **47**, 267—269.

Manton, I., Harris, K., 1966: Observations on the microanatomy of the brown flagellate *Sphaleromantis tetragona* Skuja with special reference to the flagellar apparatus and scales. J. Linn. Soc. (Bot.) **59**, 397—403.

— Leedale, G. F., 1961: Observations on the fine structure of *Paraphysomonas vestita*, with special reference to the Golgi apparatus and the origin of scales. Phycologia **1**, 37—57.

Nakajima, T., Volcani, B. E., 1969: 3,4-Dihydroxyproline: a new amino acid in diatom cell walls. Science **164**, 1400—1406.

— — 1970: ε-N-Trimethyl-L-δ-hydroxylysine phosphate and its nonphosphorylated compound in diatom cell walls. Biochem. Biophys. Res. Comm. **39**, 28—33.

Oey, J. L., Schnepf, E., 1970: Über die Auslösung der Valvenbildung bei der Diatomee *Cyclotella cryptica*. Versuche mit Colchicin, Actinomycin-D und Fluordesoxyuridin (FUDR). Arch. Mikrobiol. **71**, 199—213.

OKITA, T. W., VOLCANI, B. E., 1978: Role of silicon in diatoms. IX. Differential synthesis of DNA polymerases and DNA-binding proteins during silicate starvation and recovery in *Cylindrotheca fusiformis*. Biochim. Biophys. Acta 519, 76—86.

— — 1980: Role of silicon in diatom metabolism. X. Polypeptide labelling patterns during the cell cycle, silicate starvation and recovery in *Cylindrotheca fusiformis*. Exp. Cell Res. 125, 471—481.

PAASCHE, E., JOHANSSON, S., EVENSEN, D. L., 1975: An effect of osmotic pressure on the valve morphology of the diatom *Skeletonema subsalsum* (A. Cleve) Bethge. Phycologia 14, 205—211.

PERAGALLO, H., 1907: Sur la division cellulaire du *Biddulphia mobiliensis*. Soc. sci. d'Arcachon Stat. Biol. Trav. des lab. 10, 1—26.

PERCIVAL, E., McDOWELL, R. H., 1967: Chemistry and enzymology of marine algal polysaccharides, p. 188. London-New York: Academic Press.

PICKETT-HEAPS, J. D., McDONALD, K. L., TIPPIT, D. H., 1975: Cell division in the pennate diatom *Diatoma vulgare*. Protoplasma 86, 205—242.

— TIPPIT, D. H., ANDREOZZI, J. A., 1979 a: Cell division in the pennate diatom *Pinnularia*. III. The valve and associated cytoplasmic organelles. Biol. Cellulaire 35, 195—198.

— — — 1979 b: Cell division in the pennate diatom *Pinnularia*. IV. Valve morphogenesis. Biol. Cellulaire 35, 199—203.

PIENAAR, R. N., 1976: The microanatomy of *Sphaleromantis marina* sp. nov. (*Chrysophyceae*). Br. Phycol. J. 9, 273—283.

REIMANN, B., 1960: Bildung, Bau und Zusammenhang der Bacillariophyceenschalen (elektronenmikroskopische Untersuchungen). Nova Hedwigia 2, 349—373.

REIMANN, B. E. F., 1964: Deposition of silica inside a diatom cell. Exp. Cell Res. 34, 605—608.

— LEWIN, J. C., 1964: The diatom genus *Cylindrotheca* Rabenhorst. J. Roy. Microsc. Soc. 83, 283—296.

— — VOLCANI, B. E., 1965: Studies on the biochemistry and fine structure of silica shell formation in diatoms. I. The structure of the cell wall of *Cylindrotheca fusiformis* Reiman and Lewin. J. Cell Biol. 24, 39—55.

— — — 1966: Studies on the biochemistry and fine structure of silica shell formation in diatoms. II. The structure of the cell wall of *Navicula pelliculosa* (Bréb.) Hilse. J. Phycol. 2, 74—84.

— VOLCANI, B. E., 1968: Studies on the biochemistry and fine structure of silica shell formation in diatoms. III. The structure of the cell wall of *Phaeodactylum tricornutum*. J. Ultrastruct. Res. 21, 182—193.

RICHTER, A., 1906: Zur Physiologie der Diatomeen (I. Mitteilung). Sitzungsber. Österr. Akad. Wiss., Math.-naturw. Kl. 115, 27—119.

— 1911: Die Ernährung der Algen. Intern. Revue Ges. Hydrobiol. Hydrograph., Monographien und Abhandlungen 2, 1—193.

ROEMER, S. C., ROSOWSKI, J. R., 1980: Valve and band morphology of some freshwater diatoms. III. Pre- and postauxospore frustules and the initial cell of *Melosira roeseana*. J. Phycol. 16, 399—411.

ROTH, L. E., DE FRANCISCO, A., 1977: The marine diatom, *Striatella unipunctata*. II. Siliceous structures and the formation of intercalary bands. Cytobiologie 14, 207—221.

ROUND, F. E., 1972 a: The formation of girdle, intercalary bands and septa in diatoms. Nova Hedwigia 23, 449—463.

— 1972 b: The problem of reduction of cell size during diatom cell division. Nova Hedwigia 23, 291—303.

SANGSTER, A. G., PARRY, D. W., 1981: Ultrastructure of silica deposition on higher plants. In: Silicon and siliceous structures in biological systems (SIMPSON, T. L., VOLCANI, B. E., eds.). New York: Springer. In press.

SCHMID, A. M., 1976 a: Morphologische und physiologische Untersuchungen an Diatomeen des Neusiedler Sees: I. Methodik der Analyse der Schalenmorphologie von Cylindrotheca gracilis (Bréb.) Grun. Mikroskopie 32, 81—89.

— 1976 b: Morphologische und physiologische Untersuchungen an Diatomeen des Neusiedler Sees: II. Licht- und rasterelektronenmikroskopische Schalenanalyse der umweltabhängigen Zyklomorphose von Anomoneoneis sphaerophora (Kg.) Pfitzer. Nova Hedwigia 28, 309—351.

— 1979 a: The development of structure in the shells of diatoms. Nova Hedwigia Beih. 64, 219—236.

— 1979 b: Influence of environmental factors on the development of the valve of diatoms. Protoplasma 99, 99—115.

— 1979 c: Wall morphogenesis in diatoms: the role of microtubules during pattern formation. Europ. J. Cell Biol. 20, 125.

— 1979 d: Wall morphogenesis in diatoms: the antimicrotubule action of osmotic pressure. Europ. J. Cell Biol. 20, 134.

— 1980: Valve morphogenesis in diatoms: a pattern-related filamentous system in pennates and the effect of APM, colchicine and osmotic pressure. Nova Hedwigia 33 (in press).

— SCHULZ, D., 1979: Wall morphogenesis in diatoms: deposition of silica by cytoplasmic vesicles. Protoplasma 100, 267—288.

SCHNEPF, E., DEICHGRÄBER, G., 1969: Über die Feinstruktur von Synura petersenii unter besonderer Berücksichtigung der Morphogenese ihrer Kieselschuppen. Protoplasma 68, 85—106.

— DREBES, G., 1977: The structure of the frustule of Attheya decora West (Bacillariophyceae, Biddulphiineae) with special reference to the organic compounds. Br. Phycol. J. 12, 145—154.

— DEICHGRÄBER, G., DREBES, G., 1980: Morphogenetic process in Attheya decora (Bacillariophyceae, Biddulphiineae). Plant Syst. Evol. 135, 265—277.

SCHULTZ, M., 1971: Salinity-related polymorphism in the brackish-water diatom Cyclotella cryptica. Canad. J. Bot. 49, 1285—1289.

SIMPSON, T. L., VACCARO, C. A., 1974: An ultrastructural study of silica deposition in the freshwater sponge Spongilla lacustris. J. Ultrastruct. Res. 47, 296—309.

STOERMER, E. F., PANKRATZ, H. S., BOWEN, C. C., 1965: Fine structure of the diatom Amphipleura pellucida. II. Cytoplasmic fine structure and frustule formation. Amer. J. Bot. 52, 1067—1078.

— 1967: Polymorphism in Mastogloia. J. Phycol. 3, 73—77.

STOSCH, H. A. VON, 1942: Form und Formwechsel der Diatomee Achnanthes longipes in Abhängigkeit von der Ernährung. Mit besonderer Berücksichtigung der Spurenstoffe. Ber. dtsch. bot. Ges. 60, 2—15.

— 1967: Diatomeen. In: Vegetative Fortpflanzung, Parthenogenese und Apogamie bei Algen (RUHLAND, W., Hrsg.). (Handbuch der Pflanzenphysiologie, Bd. 18.) Berlin-Heidelberg-New York: Springer.

— 1975: An amended terminology of the diatom girdle. Nova Hedwigia Beih. 53, 1—28.

— 1977: Observations on Bellerochea and Strephtotheca, including descriptions of three new planktonic diatom species. Nova Hedwigia Beih. 54, 113—166.

— KOWALLIK, D., 1969: Der von L. Geitler aufgestellte Satz über die Notwendigkeit einer Mitose für jede Schalenbildung von Diatomeen. Beobachtungen über die Reichweite und Überlegungen zu einer zellmechanischen Bedeutung. Österr. Bot. Z. 116, 454—474.

— REIMANN, B. E. F., 1970: Subsilicea fragilarioides gen. et spec. nov., eine Diatomee (Fragilariaceae) mit vorwiegend organischer Membran. Nova Hedwigia Beih. 31, 1—36.

SULLIVAN, C. W., 1979: Diatom mineralization of silicic acid. IV. Kinetics of soluble Si pool formation in exponentially growing and synchronized Navicula pelliculosa. J. Phycol. 15, 210—216.

SULLIVAN, C. W., VOLCANI, B. E., 1973: Role of silicon in diatom metabolism. III. The effects of silicic acid on DNA polymerase, TMP kinase and DNA synthesis in *Cylindrotheca fusiformis*. Biochim. Biophys. Acta **308**, 212—219.

— — 1976: Role of silicon in diatom metabolism. VII. Silicic acid-stimulated DNA synthesis in toluene-permeabilized cells of *Cylindrotheca fusiformis*. Exp. Cell Res. **98**, 23—30.

SYDOW, B. v., CHRISTENHUSS, R., 1972: Rasterelektronenmikroskopische Untersuchungen der Hohlräume in der Schalenwand einiger zentrischer Kieselalgen. Arch. Protistenk. **114**, 256—271.

TIPPIT, D. H., McDONALD, K. L., PICKETT-HEAPS, J. D., 1975: Cell division in the centric diatom *Melosira varians*. Cytobiologie **12**, 52—73.

— PICKETT-HEAPS, J. D., 1977: Mitosis in the pennate diatom *Surirella ovalis*. J. Cell Biol. **73**, 705—727.

VOIGT, M., 1943: Sur certaines irrégularités dans la structure des Diatomées. Notes Bot. chin. **4**, 1—50.

— 1956: Sur certaines irrégularités dans la structure des Diatomées. Revue algol., N. S. **2**, 85—97.

VOLCANI, B. E., 1978: Role of silicon in diatom metabolism and silicification. In: Biochemistry of silicon and related problems (BENDZ, G., LINDQVIST, I., eds.), pp. 177—204. New York: Plenum Press.

— 1981: Cell wall formation in diatoms: morphogenesis and biochemistry. In: Silicon and siliceous structures in biological systems (SIMPSON, T. L., VOLCANI, B. E., eds.). New York: Springer. In press.

WERNER, D., 1977: The biology of diatoms. (Botanical Monographs, Vol. 13.) Oxford: Blackwell.

— 1978: Regulation of metabolism by silicate in diatoms. In: Biochemistry of silicon and related problems (BENDZ, G., LINDQVIST, I., eds.), pp. 149—176. New York: Plenum Press.

WUJEK, D. E., KRISTIANSEN, J., 1978: Observations on bristle- and scale-production in *Mallomonas caudata* (*Chrysophyceae*). Arch. Protistenk. **120**, 213—221.

Form and Pattern in *Pediastrum*

W. F. MILLINGTON

Department of Biology, Marquette University, Milwaukee, Wisconsin, U.S.A.

With 24 Figures

Contents

I. Introduction

The members of the *Hydrodictyaceae*, which includes such genera as *Hydrodictyon* (the water net), *Sorastrum*, and *Pediastrum*, and also the *Scenedesmaceae* are of morphogenetic interest because the colony is assembled in a specific form from individual cells. The colony is a coenobium; the number of cells of the colony is determined at the time of mitosis and cytokinesis in the parental cell, and is fixed at 2^n (4, 8, 16, 32, 64, etc.). No further cell division occurs until a new daughter colony is generated; growth of the colony is entirely by cell enlargement. Colony formation is asexual; the colony is assembled from spores. Sexual reproduction leads to the production of spores, which form the new colony. While genera of the *Scenedesmaceae* form colonies from non-motile autospores (*Scenedesmus, Coelastrum*), in the *Hydrodictyaceae*, swarming zoospores aggregate to form the colony (*Hydrodictyon, Pediastrum, Sorastrum*). The form of the colony is thus directed by the manner in which the spores assemble and the shapes the cells assume.

Pediastrum has the distinctive form of a cog or gear wheel in which the cells are arranged in a flat disc one cell thick (Fig. 1). The cells occur in concentric rings in a quite regular, least surface, configuration (SULEK 1969, INGOLD 1973). The peripheral cells of the *Pediastrum* colony have characteristic protuberances (Fig. 1), referred to as prongs or horns in the literature, while interior cells are usually less strongly lobed or lacking in protuberances. The species of *Pediastrum* differ in the number of prongs of the peripheral cells (Fig. 2), whether the prongs branch, and in the sculpturing (Figs. 1 and 3) of the cell wall (SULEK 1969, PARRA BARRIENTOS 1979). Spaces ("performations" or "fenestrations") may occur between the interior cells of the colony in some species, and is a constant enough feature in some species to be taxonomically significant. Eight or nine species are recognized (SULEK 1969, PARRA BARRIENTOS 1979).

Fig. 1. Scanning electron micrograph (SEM) of portion of *Pediastrum boryanum* colony. Prongs are suppressed on inner cells (arrows). ×1,300.

Fig. 2. SEM of portion of colony of *P. simplex*, species with single prong per cell and spaces between cells. Note suppressed prongs (arrow). ×1,650.

Fig. 3. Transmission electron micrograph (TEM) of outer wall layer of cells of young and older colonies of *P. boryanum*. Vesicle with zoospores was released through split (arrow), where pattern is evident in lower wall. ×3,800.

Fig. 4. Feulgen stained *P. boryanum* colony shows cells in various stages of nuclear multiplication. ×900.

Figs. 5–8. TEM of cell wall of *P. boryanum*.

Fig. 5. The walls of two adjacent cells fused by electron dense material (arrow). ×47,000.

Fig. 6. The inner fibrillar layer (*i*) varies in thickness: the outer layer (*o*) is generally uniform. ×80,000.

Fig. 7. In older cell outer wall lawer is folded outward at ridges in pattern. Electron dense columns penetrate the outer layer and appear as dots in surface view. ×84,000.

Fig. 8. The outer wall layer is undulated in older cell. ×9,000.

Figs. 1 and 8 from MILLINGTON and GAWLIK 1970; Figs. 2, 3, 5, and 6 from GAWLIK and MILLINGTON 1969; Fig. 4 from MILLINGTON and RASCH 1980.

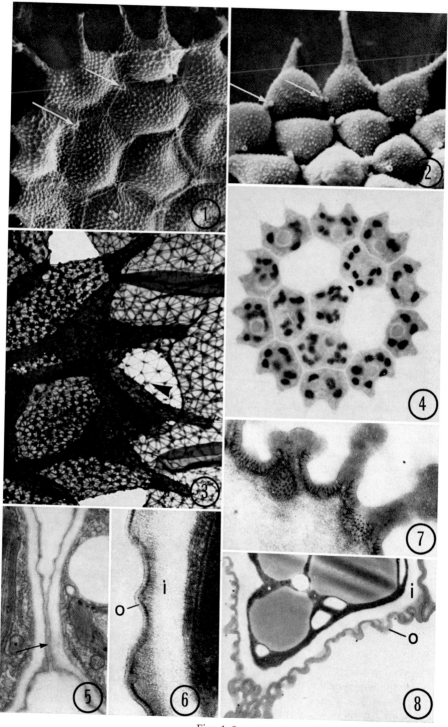

Figs. 1–8

Some aspects of cell shape (*e.g.*, number of prongs) appear to be genetically determined in *Pediastrum* and the genus provides a favorable system for studying the mechanism of cell shape determination. Furthermore, the characteristic sculpturing in the rigid cell wall is of interest from the standpoint of pattern regulation in cells. The electron microscope reveals a variety of patterns (PARRA BARRIENTOS 1979). In *P. boryanum* a network of ridges covers the cell surface in the form of triangles clustered into groups of five and six over the cell surface (Figs. 1 and 3). The cell wall is bilayered, and empty colonies, consisting of just the outer wall layer after release of the zoospores, reveal the pattern in the material of the outer layer (Fig. 3). A thickened disc is present at the points of each triangle, with the ribs of the pattern interconnecting the discs. Growth of the cells is accompanied by elongation or stretching of the ribs (Fig. 3). How the wall pattern is templated has been another of the concerns in the study of *Pediastrum*.

II. Development in *Pediastrum*

A. Colony Formation

Each cell of the colony produces a new colony following mitosis and cleavage and differentiation of the protoplast into uninucleate biflagellate zoospores. The zoospores are released within a vesicle (Fig. 17) which is exuded from the parental cell through a crescent-shaped split in the outer wall layer. The vesicle was reported by BRAUN in 1851 to be the inner layer of the cell wall. Electron microscopy has substantiated that the origin is in the inner wall layer (GAWLIK and MILLINGTON 1969), although additional material may be secreted by the zoospores (MARCHANT 1979). The zoospores swim for about 10 minutes within the vesicle and arrange themselves in a disc. The new colony grows within the vesicle, which is hydrolized away in a few hours, liberating the new colony.

Seldom do all cells of the colony release zoospores at the same time (HARPER 1918); complete release may take a few hours, leaving the parental colony as an empty skeleton. Nor are all the daughter colonies of the same size; an extra mitosis before zoospore formation may double the number of zoospores in some of the cells of the parent colony (Fig. 4).

Before the zoospores are extruded from the parental cell slight twitchings are evident and they begin to show writhing movements (HARPER 1918). With extrusion of the vesicle it expands and flagellar activity increases the movement of the zoospores. Individual zoospores appear to dart in all directions although remaining in a cluster. The swelling vesicle enlarges ahead of the swarming mass, and the zoospores are not confined by the vesicle membrane. Active swarming continues for 3–4 minutes and then slows down and the outline of the colony becomes evident. Writhing of the zoospores continues as the colony flattens into a disc. The complete process takes 10–12 minutes. While the zoospores are initially spherical and then ovoid they suddenly become angular, almost rectangular in outline, as they assume their positions in the colony (HARPER 1918, HAWKINS and LEEDALE 1971, MARCHANT 1979). As movement of the zoospores slows to a vibratory motion the zoospores tend to remain in position. Change in position of one cell may lead to changes in all its neighbors (SMITH 1916). The peripheral cells are the first to cease motion.

If motility of the zoospores is suppressed, as when the colonies are sealed under a cover slip, the zoospores fail to position properly, and the colony has irregular form (HARPER 1918). Under certain conditions the vesicle may not be extruded, the zoospores swarm only feebly and the new colonies are atypical.

Following aggregation the zoospores adhere where the cells are in contact and retract the flagella. Soon the prongs begin to extend from bulges on the peripheral cells, and are fully extended within 5 minutes (HAWKINS and LEEDALE 1971). Rapid growth of the colony continues for an hour of so. The colony of *P. boryanum* within its vesicle is less than 35 μm in diameter and can be filtered from a suspension of the older larger (60–100 μm) colonies with nylon mesh of 37 μm porosity. The colony may double in diameter in less than 90 hours, and is ready to reproduce.

B. Triggering of Colony Formation

Despite reports (SMITH 1916) that colony formation in *Pediastrum* is induced by light, laboratory experiments with a culture of *P. boryanum* (isolated in Iowa by J. S. Davis) show little effect of light conditions on formation of zoospores (MILLINGTON, unpublished, SCHRAUDOLF and FRAUENKRON 1979). Zoospores and colony formation can be induced, however, by transferring mature colonies to fresh culture medium.

The limiting factor in the culture medium in laboratory studies is nitrogen (MILLINGTON and GAWLIK, unpublished). As cultures age, the cells change from green to orange with the accumulation of carotenoids. Orange colonies are dormant and can be stored for long periods. In this resting state *Pediastrum* cells have high survival rates. Colonies on dry herbarium sheets for 12 years were still viable, and experiments in vacuum and at high temperatures revealed as much as 6% survival at 100 °C (DAVIS 1979). Both mature green and resting orange colonies can be stimulated to form new colonies by transfer to fresh medium or by adding nitrogen. If components of the culture medium (Medium C, DAVIS 1962) are added singly to stale medium, only nitrate permits greening and release of new colonies. Similarly, substituting ammonium or the amino acids alanine or glycine in the medium also initiates greening and reproduction. However, leucine does not permit re-greening or sporogenesis.

Although transfer to fresh medium or addition of nitrogen triggers reproduction, the response is not completely synchronous in a culture or among the cells of a colony. While a high percentage of colonies give rise to daughter colonies from every cell, the process may require 4–5 hours, and in 24 hours not all cells in a culture have reproduced. When mature colonies are placed in Medium C (DAVIS 1962) in a controlled environment of 16 hours warm white fluorescent light of 6,500 lux at 21 °C and 8 hours dark at 16 °C, a burst of release of new colonies occurs in about 58 hours (MILLINGTON, unpublished).

It is thus possible to obtain swarming zoospores and early stages in colony formation in large concentrations, but a means of obtaining pure preparations of the swarmer stage has not yet been achieved.

C. The Polyeder in the Life Cycle

In some species of *Pediastrum* (DAVIS 1967, TRYON and DAVIS 1976), and in *Hydrodictyon* (MARCHANT and PICKETT-HEAPS 1972) unusual 4- or 5-pronged cells called polyeders or polyhedra (Fig. 21) occur in the life cycle. Polyeders were observed to develop from zygotes in *P. boryanum* (PALIK 1933) and they gave rise to zoospores which aggregated within a vesicle to form a colony. In *P. simplex* various polyhedra occur. A series of 4-angled, flattened polyeders, 4-spined spherical cells and 4-celled, 4-angled colonies are involved in sexual and asexual cycles that may be conditioned by environmental factors (DAVIS 1967, TRYON and DAVIS 1976). These cells and cycles are of interest from the standpoint or regulation of cell shape. In addition some polyeders lack the cell wall pattern expressed in the cells within a colony that have gone through different phases of development and gene expression, and they are of interest in this regard as well (MILLINGTON *et al.* 1981).

III. Organization of the Cell

A. The Cell Wall

Electron micrographs show the bilayered nature of the cell wall (Figs. 5 and 6). The outer layer, of uniform thickness except at the ridges of the pattern, follows the contours of the thicker inner wall layer with its ridges and valleys (Fig. 6).

The outer wall layer is initially trilaminar, and is homogeneous except for columns that stain with phosphotungstic acid and extend from the inner layer to the exterior where in surface view they appear as minute dots (Fig. 7). Adjacent cells of the colony are cemented together by material which shows in electron micrographs as a dark line between the clear outer wall layers.

The outer wall layer resists degradation and is found in fossil deposits (PIA 1927, WILSON and HOFFMEISTER 1953). It survives acetolysis and thus is assumed to contain sporopollenin (BROOKS and SHAW 1971, GUNNISON and ALEXANDER 1975), a carotenoid derivative. In this regard the wall resembles that of *Chlorella* (ATKINSON, GUNNING, and JOHN 1972), and *Coelastrum* (MARCHANT 1977 b). In addition this layer contains silica (MILLINGTON and GAWLIK 1967). In older cells the outer layer is folded up into the pronounced ridges of the wall pattern (Fig. 8), and becomes undulated within the inner wall layers of adjacent cells.

The inner wall layer is fibrillar or granular in appearance in electron micrographs, and of greater electron density at the ridges where it is thicker (Fig. 6). Separation from the outer layer during colony formation occurs in the region where the inner wall tears away as the zoospore-containing vesicle. The fibrillar nature of the wall is especially evident when it has been discharged as the vesicle. The inner layer has been reported to be a glucan-mannan polymer (PARKER 1964), although a more recent analysis (GUNNISON and ALEXANDER 1975) revealed glucose, xylose, fucose, and galacturonic acid but no mannose in the wall of *P. duplex.*

The properties of the vesicle membrane (inner wall layer) are unique in that it functions as an osmotic membrane, swelling following discharge, but it also retains a degree of form as it swells. Vesicles derived from peripheral cells retain slight bumps at the position where the wall extended into the prongs (Fig. 17), revealing some structural rigidity. Ultimately the vesicle is hydrolized, freeing the new colony. That its disintegration may be enzymatic is suggested by the fact that refrigeration delays hydrolysis, and boiling prevents it (GAWLIK and MILLINGTON, unpublished). Lysis of the vesicle is prevented by exposure of new colonies to actinomycin D and to colchicine (MILLINGTON, unpublished) and to concanavalin A (Con A) (SCHRAUDOLF and FRAUENKRON 1979). Strangely, colchicine in the medium results in the release of the protoplast within the vesicle although it inhibits cleavage and zoospore formation (Fig. 17). Like actinomycin D, colchicine may block successive events in development, possibly preventing the subsequent formation of hydrolytic enzymes. Con A apparently blocks lysis through its effect on the glucan-mannan polymer of the vesicle (SCHRAUDOLF and FRAUENKRON 1979).

Although the prongs of the peripheral cells are initially filled with cytoplasm, the cytoplasm is ultimately displaced by cell wall. Bristles of unknown composition extending from the tips of the prongs have been reported in some species (HARPER 1918, MARCHANT and PICKETT-HEAPS 1974, WILCOX and GRAHAM 1979).

B. The Nucleus and Cytoplasm

A feature of the *Pediastrum* cells is the polarized distribution of organelles, giving the cell a bilateral symmetry in the plane of the colony (Figs. 9 and 14). Despite a gross change in cell shape from the zoospore to the vegetative cell, there is little rearrangement in relative position of major organelles during development (HAWKINS and LEEDALE 1971, MARCHANT 1974, 1979).

1. Nucleus

The nucleus is adjacent to the outward-directed flagella in the zoospores and at the inner side of each cell in the completed colony. At the same time the nucleus changes shape from an angular form in the zoospore to become elongate and quite sausage-shaped in the mature cell. A lobe of the zoospore nucleus extends toward the flagella (HAWKINS and LEEDALE 1971). A perinuclear envelope of endoplasmic reticulum encloses the mitotic nucleus, and within this envelope the nuclear membrane remains intact during mitosis, except for polar fenestrations (MARCHANT 1974). In the growing cell, extensive blebbing of the nuclear envelope along the inner nuclear face appears to contribute vesicles to the forming face of the adjacent Golgi dictyosome (Fig. 9). The nuclear envelope is free of ribosomes in this region though rough elsewhere.

Persistent centrioles are associated with the nucleus (MARCHANT 1974) and flagella of the zoospores. The flagellar bases (basal bodies) have a complex substructure and are interconnected by a bridge (HAWKINS and LEEDALE 1971, MARCHANT 1979). Following flagellar retraction the axonemal microtubules, basal bodies, and rootlet microtubules disintegrate, and new centrioles appear on the nuclear envelope (MARCHANT 1974).

a) DNA Replication and Mitosis

Soon after new colonies are initiated in laboratory cultures and begin to grow, mitosis commences. Binucleate cells in P. boryanum can be detected within 12 hours after colony formation, and by 48 hours 50% of the colonies have one or more binucleate cells (MILLINGTON and RASCH 1980). By 96 hours, when some new colonies are being released, 40% of the colonies have 8-nucleate or 16-nucleate cells (Fig. 4). Microspectrophotometric analysis of colonies during development showed that mitosis follows each DNA replication, since no nuclei show DNA levels above the 20 level. The DNA level in nuclei of P. boryanum before replication is about 0.20 picograms, close to the genome size of other algae such as Pandorina and Eudorina reported in the literature (MILLINGTON and RASCH 1980).

2. Chloroplast

Occupying a large volume of the cell and displaced toward the outer prong side is a single chloroplast containing starch grains and osmiophilic droplets. The chloroplast is bilobed in the zoospore, the lobes interconnected by a narrow bridge in which the pyrenoid lies (HAWKINS and LEEDALE 1971, MARCHANT 1979), and in the cells of the colony of bi-pronged species. Thyla-

---

Figs. 9–11. The Golgi apparatus in P. boryanum.

Fig. 9. The dictyosome (d) lies adjacent to the nucleus (n). Blebs of the nuclear envelope contribute to the dictyosome cisternae. ×39,000.

Figs. 10 and 11. Distribution of Inosine diphosphatase (IDPase) in the dictyosome at successive stages in development (note wall formation). Reaction product from the Gomorri procedure concentrated in inner cisternal layer early, later in vesicles at periphery. ×20,000.

Figs. 12–16. Microtubules.

Fig. 12. Cortical microtubules (arrows) encircle zoospores at aggregation, are aligned in adjacent cells. ×20,500.

Fig. 13. Cortical microtubules (arrow) lie just beneath the plasma membrane in sectioned cells at aggregation. Note material between cells. ×98,000.

Figs. 14–16. Microtubules in P. boryanum are associated with prong extension. Plaques (arrows) at the site of prong initiation and in the tip as prong extends are associated with microtubules

Fig. 14. Early stage in prong formation. ×14,000.

Figs. 15 and 16. Later stages. Microtubules in prong originate at plaque. ×40,000, ×20,000.

Fig. 17. Colchicine treatment of P. boryanum leads to release of the protoplast within the vesicle instead of normal zoospores which aggregate to form colony (inset). The vesicle membrane from a pronged parent cell (inset) retains slight bulges at prong sites. Phase microscopy. ×600.

Figs. 9 and 12 from MILLINGTON and GAWLIK 1970; Figs. 10 and 11 from J. DI ORIO 1977; Fig. 13 from GAWLIK and MILLINGTON 1969; Figs. 14–16 from MARCHANT 1979; Fig. 17 from DI ORIO and MILLINGTON 1978.

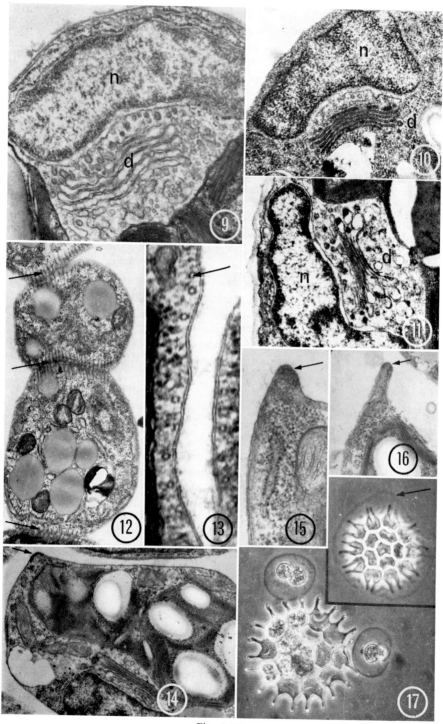

Figs. 9–17

koids occur in interconnected granum-like stacks. The pyrenoid disappears during cleavage, but is complete with new starch plates in young daughter colonies.

3. Golgi Apparatus

The Golgi apparatus is associated with the nucleus (Fig. 9). In *P. boryanum* the single dictyosome lies adjacent to the concave face of the curved nucleus; two to four dictyosomes occupy this position in *P. simplex*. In *P. boryanum* the dictyosome replicates before each mitosis so that at the time of cleavage each nucleus is associated with a dictyosome (DI ORIO and MILLINGTON 1978).

About five to seven flattened cisternae are found in the dictyosome, which appears to be continually supplemented by vesicles blebbing from the nuclear envelope. Vesicles are pinched off at the periphery of the cisternae and from the maturing face (Fig. 9). Small Golgi vesicles of newly released zoospores are replaced by larger and more irregular vesicles during the transition to the vegetative cell (HAWKINS and LEEDALE 1971). At successive stages in development, as indicated by progress of wall formation, three phosphatase enzymes (acid phosphatase, tyrosine pyrophosphatase and inosine diphosphatase), have been shown cytochemically to follow different distribution patterns within the dictyosome (Figs. 10 and 11) (DI ORIO, VIGIL, and MILLINGTON 1977). The role of these Golgi phosphatases in the cell metabolism is not known. However, it has been shown in *P. simplex* that the dictyosome is the source of the inner layer of the cell wall (MILLINGTON and GAWLIK 1975).

4. Endoplasmic Reticulum

The endoplasmic reticulum (ER) also seems to be involved in cell wall formation. Following adhesion and during prong formation when the outer wall layer is being deposited the endoplasmic reticulum is dilated and contains granular material (MARCHANT 1979). Later, when the inner wall layer is degraded leading to separation of the vesicle from the outer wall layer, rough ER is concentrated along the periphery of the cell in *P. tetras* (ROGALSKI et al. 1977), suggesting a possible role in freeing of the vesicle.

5. Microbodies, Lipid Droplets, and Vacuoles

Two contractile vacuoles are present in the zoospore near the flagella and are lost (HAWKINS and LEEDALE 1971). Other clear vacuoles occur in the cytoplasm and disappear at cleavage (ROGALSKI et al. 1977).

Large osmiophilic bodies or lipid droplets (Fig. 12) may correspond to the carotenoid containing droplets seen in older cells in the light microscope. Other large membrane-bound organelles containing a fine granular matrix and associated with ER resemble microbodies, but fail to stain in the DAB reaction, usually diagnostic for microbodies (ROGALSKI et al. 1977).

6. Microtubules

Three systems of microtubules, in addition to the flagellar microtubules, are present in the zoospores or in early colony formation (MARCHANT 1974,

1979). Other microtubules, the phycoplast microtubules, are associated with cleavage of the parental cell (MARCHANT 1974). Within the zoospore, four bands of microtubules (the cruciate system of rootlet microtubules) are associated with the base of the flagella and arise from densely staining material adjacent to the bridge between the basal bodies (MARCHANT 1979). They appear to anchor the flagellar apparatus. Just beneath the plasma membrane of the zoospore is a girdle of cortical microtubules (Fig. 13) apparently involved in regulation of cell shape. They lie in a parallel array circling the cell from side to side, and when the cells are aligned in the colony their microtubules are in alignment (Fig. 12). These cortical microtubules lack discrete organizing centers (MARCHANT 1979).

Another set of microtubules, separate from the cortical girdle, appears to arise from a plaque of dense material at the tip of the developing prong (Figs. 14–16) (MARCHANT 1979).

IV. Regulation of Cell Shape and Colony Form

A. Colony Form

1. Formation of a Disc

It has been suggested that the initially lens-shaped vesicle containing the swarming zoospores spatially limits the swarming mass and establishes the circular outline of the colony (DAVIS 1964). However, other observations (HAWKINS and LEEDALE 1971, MARCHANT 1974, MILLINGTON and GAWLIK 1975) indicate that the vesicle expands in advance of the swarming mass of zoospores and does not limit the periphery of the forming colony. The fact that the cells arrange themselves in a circular layer from a sphere suggests that the circular form must be established by the cells themselves, a conclusion drawn by HARPER in 1918. HARPER attributed form of the colony to cell heredity, which determined basic cell shape and influenced how cells assembled. Contacts between cells then further affected final form. He concluded that cell form determined the character of the colony as a whole, although the form of the colony influenced cell form too, as in the emergence of prongs in peripheral cells. This view has been borne out by more recent studies.

2. Orientation of Cells

The prongs in the peripheral ring of cells of typical disc-shaped colonies always extend outward in the plane of the colony (not perpendicular), and if the colony is fenestrated the major lobes of the inner cells point outward as well. The constancy of prong and lobe orientation and colony symmetry could be accomplished in two ways. One possibility is that prong sites are not predetermined, and prongs emerge from the outward free side of the assembled zoospores. Alternatively, the zoospores orient themselves about an axis so that pre-determined prong sites face outward; *i.e.*, zoospores assume a position in accordance with their cellular organization. The regularity of organelle distribution in the cells of a colony, noted earlier, suggests specific cell orientation and favors the latter idea. Further support is seen in aberrant colonies reported by Harper (1918) where a cell attached in the colony at only one point assumes its specific cell form.

Aggregation of the swarming cells in a disc suggests not only that they are attracted to each other, but adhere at certain lines or points of contact so as to lie in a single layer. The final jostling, squeezing and quivering of the aggregating cells suggests that they may be pulled together tightly. Alignment of the cells indicates that proper cell assembly involves recognition regions and sticky areas at the surface. Marchant and Pickett-Heaps (1974) have suggested that there are discrete adhesive sites, predominantly lateral, about the zoospores. Honda (1973) postulated two presumptive connection sites at the equator of the zoospores of the single-pronged *P. biwae* to account for alignment of cells into a disc. The connection sites on the swarming zoospores have affinity for those on adjacent zoospores and form the cells in a string curled in the plane of the colony. A third site on the zoospore would give rise to the single prong. A computer simulation of the process accounted for arrangement of spheres into a flat plate. Others suggest that the entire surface may be sticky (Hawkins and Leedale 1971), which would seem to favor assembly in a sphere rather than a disc.

That recognition sites are involved is indicated by experiments with lectins. Concanavalin A (Con A), a lectin which binds to the mannoside component of surface polysaccharides and glycoproteins, introduced into the culture medium during colony formation in *P. boryanum* prevents proper aggregation of zoospores and regular assembly of the colony (di Orio 1977, Schraudolf and Frauenkron 1979). Low concentrations of Con A (5×10^{-5} and 5×10^{-6} g/ml) were without effect on colony form, but at higher concentrations (5×10^{-3} and 5×10^{-4} g/ml) zoospore arrangement was highly irregular (Schraudolf and Frauenkron 1979). Con A at the higher level reduced zoospore motility, which could account for the irregular colonies. It also affected cell shape, growth and wall formation, ultimately killing the cells. Use of fluorescein-labelled Con A and a peroxidase method to disclose Con A localization have thus far not revealed discrete adhesive sites on the surface of zoospores, possibly because of limitations in the techniques used and masking of the reaction product by the density of pigmentation of the zoospores (Schraudolf and Frauenkron 1979).

After the cells are assembled in the colony they become tightly adherent. Although they may be dispersed initially by a needle or gentle pressure, in a few hours they become fused and inseparable by mechanical manipulation (Millington, unpublished). Electron micrographs show granular amorphous material between the assembling cells (Gawlik and Millington 1969, Hawkins and Leedale 1971, Marchant 1974). In thin sections of completed colonies a line of electron dense material joining the cells is evident (Fig. 5).

B. Cell Shape

Several workers interested in cell shape in *Pediastrum* have noted the correlation between large organelle distribution and cell shape (Davis 1964, Hawkins and Leedale 1971); flagella are present and the prongs emerge on the outer chloroplast hemisphere of the cells. Although organelle position is consistent, it could be the result of the squeezing of the cells into their wedge shape during assembly of the colony rather than of directed orientation, the large chloroplast being forced to the exterior, the nucleus then toward the center of the colony. Whatever the factors that position the larger organelles, shaping of the cells appears to be regulated by microtubules.

1. Microtubules and Cell Shape

The change in shape of the zoospores during swarming and aggregation is not only correlated with the presence and change in distribution of cytoplasmic microtubules, but treatment with colchicine prevents the normal change in shape of the zoospores and their proper assembly. This effect of colchicine on cell shape has been reported in several algae. Thus *Ochromonas* fails to assume its characteristic form in colchicine (BROWN and BOUCK 1973, BOUCK and BROWN 1973), and *Sorastrum* and *Hydrodictyon* form abnormal colonies because the cells fail to assume the characteristic shape (MARCHANT 1974, MARCHANT and PICKETT-HEAPS 1974). Colchicine also disturbs colony shape in the *Volvocales* where the microtubule effect on cell shape extends to the inversion of the colony in *Eudorina* (MARCHANT 1974 a, GOTTLIEB and GOLDSTEIN 1979) and *Volvox* (VIAMONTES et al. 1979).

Colchicine has an additional effect on *Pediastrum* of interest from the standpoint of successive events in development. Mature colonies in which the cells have undergone several mitoses fail to complete cleavage when placed in colchicine, and they fail to form zoospores (MILLINGTON 1973, MARCHANT and PICKETT-HEAPS 1974, DI ORIO and MILLINGTON 1978). Nevertheless, the undivided protoplast of the parent cell is often extruded within the vesicle (Fig. 17). Thus, even though the events involving microtubules in zoospore formation had been blocked by colchicine, later events, separation of the inner wall layer and vesicle extrusion were already programmed and carried out.

The emergence of prongs also is correlated with the presence of microtubules as is the lobing of cells in fenestrated colonies. Lobing of the interior cells involves formation of a concavity or invagination in the cells near the nucleus, which produces the intercellular space (HAWKINS and LEEDALE 1971). Prongs are suppressed in the presence of colchicine (MARCHANT and PICKETT-HEAPS 1974, MARCHANT 1974). Isopropyl-N-phenyl carbamate, reported to affect microtubule organizing centers, also impaired development of prongs in peripheral cells (see MARCHANT 1979). In some cases it had the effect of inducing 3 prongs on the usually two-pronged *P. boryanum* cells (MARCHANT 1979).

2. Emergence of Prongs

Microtubules thus regulate cell shape, and indirectly colony form. How they direct emergence of prongs is of special interest since the species differ in prong number and prong form. Recently MARCHANT (1979) has demonstrated a pair of densely staining plaques in the cytoplasm of *P. boryanum* associated with the microtubules of the developing prongs. The plaques were first evident as a pair of lamellate bodies near the plasmalemma on the chloroplast side, the prong forming side of the newly released zoospore. The microtubules of the prongs appeared to emanate from the plaques (Figs. 14–16). The number of such microtubule organizing centers may determine the number of prongs in a species.

The number of prongs that actually develop in a cell of *Pediastrum* is affected by contact with other cells. In *P. boryanum* prongs of interior cells are normally suppressed. If a space occurs in a *P. boryanum* colony as it is developing, however, a prong may grow into the space from an adjacent cell (Harper 1918, Davis 1964, Marchant 1974, 1979, Marchant and Pickett-Heaps 1974).

In the same way, a prong of a peripheral cell may be suppressed if the cell is mis-oriented (Fig. 20). All zoospores apparently have the potential to form prongs: when aggregating colonies of *P. boryanum* were mechanically disrupted within the vesicle all the separate cells formed prongs (Marchant and Pickett-Heaps 1974). (It has been reported, on the other hand (Davis 1964), that zoospores of *P. boryanum* released individually from the mother cell swarmed for 45 m and failed to form prongs).

Scanning electron micrographs of *P. boryanum* and *P. simplex* show suppressed prong sites in the wall of interior cells where the cells abut (Fig. 1). That there are several suppressed sites on both surfaces is indicated in cells of a mutant culture of *P. boryanum* (Fig. 18) which spontaneously separates into unicells (Millington et al. 1981). As many as 8 suppressed prong sites are evident in some cells (Fig. 19).

Further indication of prong forming sites is seen in certain unusual situations. Multipronged cells have been induced in *P. boryanum* as a result of mechanically suppressing cleavage in the mother cell (Davis 1964). These multipronged cells arise from multinucleate zoospores having several flagellar pairs. The giant zoospores are released with other typical zoospores within the vesicles. Apparently cleavage was blocked in the manipulation, and the potential for several prongs remained in the protoplast. The multi-pronged polyhedra (Fig. 21), noted earlier, that occur naturally in sexual and asexual reproduction of some species of *Pediastrum* and have been induced in culture (Davis 1967, Tryon and Davis 1976), further suggest the occurrence of several specific prong forming sites.

It will be of interest to determine if these various multipronged cells have multiple plaques of the type described by Marchant (1979) and to see if the interior cells of the colony have dense areas at the putative prong sites which are suppressed from functioning in prong extension by the contact of adjacent cells.

V. Regulation of Pattern in the Wall

A. Emergence of Pattern

The patterned network of the cell wall is templated in the short interval of wall deposition when the zoospores have assembled and retracted their flagella. First evidence of the pattern is the appearance of plaques of material on the plasma membrane in a pattern corresponding to the centers at the

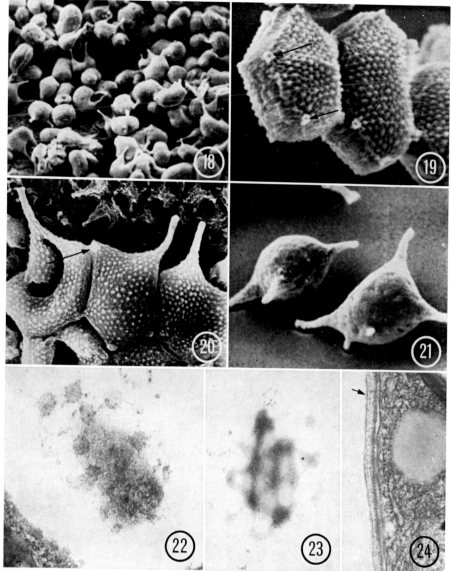

Figs. 18 and 19. Mutant culture of *P. boryanum* in which cells separate after colony formation.

Fig. 19. Isolated internal cells show suppressed prong sites (arrows) along contact faces. ×2,900.

Fig. 20. *P. boryanum* with suppressed prong in peripheral cell (arrow). ×2,100.

Fig. 21. Polyeders of *P. simplex* have multiple prongs. ×1,550.

Figs. 22–24. Initiation of wall pattern in *P. boryanum*.

Fig. 22. Section cut tangent to surface shows plaques at surface in basic pattern of wall. ×47,500.

Fig. 23. Pattern evident in tangential section of inner wall layer. ×37,500.

Fig. 24. Outer wall plaques have extended to complete outer layer (o); inner layer being deposited. ×47,500.

(Figs. 18–21 from MILLINGTON *et al.*, Protoplasma **105**, 1981; Figs. 22–24 from MILLINGTON and GAWLIK 1970.)

triangle junction of the wall pattern (Figs. 9 and 22) (MILLINGTON and GAWLIK 1970, 1975). Faint lines of this material interconnect the plaques. The plaques expand rapidly, forming a continuous wall layer (Fig. 24). The plaques as seen in section are short trilayered pieces (Fig. 9).

No vesicles have been seen in association with plaque deposition; there is no indication of their source in the dictyosome. Rather, the plasma membrane may be the template for wall pattern. How the pattern is directed remains to be discovered. Scattered clear sites in the plasma membrane at the time of deposition, possibly artifact, have been noted (MILLINGTON and GAWLIK 1970, 1975). Furthermore, polysomal configurations occur just beneath the cell surface distributed in a pattern like that of the plaques, but there is no evidence of a direct relationship (MILLINGTON and GAWLIK 1970).

The wall pattern is also evident in the inner wall layer material that is first deposited against the outer layer. A tangential section beneath the outer wall layer (Fig. 23) shows a pattern of more dense material matching the polygonal pattern as if a pattern of deposition was still being imposed during inner wall formation. Still later, the inner wall material deposited is homogeneous. It is the inner layer that seems to be derived from Golgi material (MILLINGTON and GAWLIK 1975). The ridges and valleys of the wall are partly accounted for in the uneven thickness of the inner wall layer, it being thicker at the ridges. However, the outer layer is later folded outward at the ridges, intensifying the pattern.

An earlier explanation for the wall pattern (MONER and CHAPMAN 1963) was based on an interpretation from pioneer electron micrographs that a discontinuous coarse network was distributed through a continuous membrane (MONER 1955). Osmiophilic lipoidal bodies in the cytoplasm appeared to be extruded into the wall. The lipoidal bodies in the young wall were postulated to be drawn out into the connecting strands of the growing wall to explain the expanding pattern (Fig. 3). This attractive idea could relate the carotene-containing lipid droplets with the sporopollenin of the outer wall, reported to be derived from carotenoids (BROOKS and SHAW 1971). However, recent studies have failed to show extrusion of lipids into the wall.

B. Regulation of Pattern in Algae

The occurrence of specific pattern is not uncommon in the walls of algae, and they provide an excellent material for the study of this aspect of gene expression. The origin of pattern in cell walls has been studied extensively in spores and pollen grains (HESLOP-HARRISON 1968; see p. 349 f.) and reveals some of the mechanisms involved. In pollen grains the distribution of endoplasmic reticulum is correlated with the initial basic pattern. Later refinements in pattern seem to be related to other membranes in the cytoplasm near the cell surface.

The diatoms (see p. 63 f.) and dinoflagellates (see p. 27 f.) are perhaps the most complexly, beautifully patterned of the algae. The pattern is established in both groups in the siliceous frustules, thecae, or plates of the cell wall. These are formed within membrane-bound vesicles, apparently Golgi derived, and patterned in the vesicle membrane or silicalemma from which they are deposited at the cell surface. The loricate choanoflagellates (LEADBETER 1979) have a cage (the lorica) of siliceous costae produced in vesicles presumably derived from the Golgi apparatus. The vesicles are distributed in regular pattern about the cell in a manner not yet explained. Patterning in material produced within the Golgi vesicles is well known in the case of the carbohydrate scales that cover the cell body and flagella (see p. 27 f.) of the *Prasinophyceae,* in the silicified scales of *Chrysophyceae* (MANTON and ETTL 1965, NORRIS and PEARSON 1975), and in the cellulosic scales of *Pleurochrysis* (BROWN and ROMANOVICZ 1976).

The initiation of wall patterns at the cell surface, presumably via the plasma membrane, has been observed in a few green algae. Patterns in the cell wall of desmids have been related to patterns in the plasma membrane. Slime pores in the secondary wall of *Cosmarium,* for example, could be related to plasmalemma pattern (PICKETT-HEAPS 1972; see p. 215 f.), and the patterned growth of the primary wall in *Micrasterias* was postulated to be related to specific sites in the plasma membrane where vesicles discharge their contents (KIERMAYER 1970; see p. 181 f.).

Freeze fracture techniques have permitted the correlation of membrane patterns with the deposition of microfibrils of the wall. In *Oocystis* auto-spores, a pattern of granules in the plasmalemma was correlated with the orientation of cellulose microfibrils in the new wall (ROBINSON and PRESTON 1972). It was suggested that the granules might function as synthesis sites of cellulose microfibrils at the surface of the plasmalemma. The lorica of the stalked flagellate *Poteriochromonas* is formed at the plasmalemma in the tip of the growing cell (SCHNEPF *et al.* 1975). The helically arranged fibrils of the lorica coincide with the distribution of microtubules connected to the plasma membrane, apparently directing pattern.

Recently the distribution of "rosettes", hexagonally arranged particles in the plasma membrane of the desmid *Micrasterias denticulata* has been found to be associated with bands of microfibrils in the secondary wall (KIERMAYER and SLEYTR 1979). (See p. 158 f.) In association with the plasma membrane the rosettes were thought to produce and orient the microfibrils (KIERMAYER and DOBBERSTEIN 1973, KIERMAYER 1977). The rosettes are incorporated into the plasma membrane as "flat vesicles" derived from the Golgi apparatus and each bearing 6 globular particles (KIERMAYER and DOBBERSTEIN 1973, KIERMAYER and SLEYTR 1979). The rosettes, each terminating a cellulose microfibril (GIDDINGS *et al.* 1980) function as templates for the production of microfibrils of the cell wall. Similar rosettes of particles in the plasma membrane of maize tissues have now been shown to be associated with microfibril synthesis in higher plants (MUELLER and BROWN 1980).

It will be of interest to determine if the patterned outer sporopollenin wall and the inner fibrillar wall of *Pediastrum* are similarly associated with plasma membrane architecture. Further studies utilizing freeze fracture techniques should contribute to an understanding of the templating of surface patterns.

VI. Conclusion

Pediastrum, and other members of the *Hydrodictyaceae* and *Scenesdesma-ceae* as well, provide excellent material for the study of mechanisms in the regulation of cell differentiation. In most cases they can be cultured in the laboratory in simple media, and development can often be regulated and synchronized. The unique relationship between dictyosome and nucleus in *Pediastrum,* and the presence of a developmental marker in the sequence of wall formation, encourages studies of the role of the Golgi apparatus in the cell. Assembly of the colony from individual cells in these algal groups and the positioning of the aggregating cells is of interest from the standpoint of recognition and communication between cells. The recent studies of micro-tubule organizing centers in relation to emergence of the species-identifying prongs in *Pediastrum* indicate a promising place to look at gene expression in the shaping of cells. Finally, algae with species specific wall patterns, such as *Pediastrum,* provide an opportunity to trace the genetic expression of pattern in cells.

Acknowledgements

I thank JAMES DI ORIO for his considerable help in preparation of this chapter. My sincere appreciation is extended to Dr. J. S. DAVIS and to Dr. H. J. MARCHANT for reading the manuscript and for their valued criticisms. Responsibility for accuracy, of course, rests with the author. In addition, I am grateful to Dr. MARCHANT for providing original camera copies of figures.

Part of this chapter was written while the author was on sabbatical leave from Marquette University in the laboratory of Dr. A. W. ROBARDS, University of York, York, England. My gratitude is expressed to Marquette University for support and to Dr. ROBARDS and the University of York for the hospitality accorded me.

References

ATKINSON, A. W., GUNNING, B. E. S., JOHN, P. C. L., 1972: Sporopollenin in the cell wall of *Chlorella* and other algae: ultrastructure, chemistry, and incorporation of ^{14}C-acetate, studied in synchronous cultures. Planta (Berl.) **107**, 1—32.

BOUCK, G. B., BROWN, D. L., 1973: Microtubule biogenesis and cell shape in *Ochromonas.* I. The distribution of cytoplasmic and mitotic microtubules. J. Cell Biol. **56**, 340—359.

BRAUN, A., 1851: Erscheinungen der Verjüngung in der Natur. Freiburg.

BROOKS, J., SHAW, G., 1971: Recent developments in the chemistry, biochemistry, geochemistry and post-tetrad ontogeny of sporopollemins derived from pollen and exines. In: Pollen: development and physiology (HESLOP-HARRISON, J., ed.). London: Butterworths.

BROWN, D. L., BOUCK, G. B., 1973: Microtubule biogenesis and cell shape in *Ochromonas.* II. The role of nucleating sites in shape development. J. Cell Biol. **56**, 360—378.

BROWN, R. M., ROMANOVICZ, D. K., 1976: Biogenesis of golgi-derived cellulosic scales in *Pleurochrysis.* I. Role of the endomembrane system in scale assembly and exocytosis. Applied Polymer Symposium **28**, 537—585.

DAVIS, J. S., 1962: Resting cells of *Pediastrum.* Amer. J. Bot. **49**, 478—481.

— 1964: Colony form in *Pediastrum.* Bot. Gaz. **125**, 129—131.

— 1967: The life cycle of *Pediastrum simplex.* J. Phycol. **3**, 95—103.

— 1979: Survival of air dried *Pediastrum* in low pressure and high temperature. Pollen et Spores **21**, 499—504.

DI ORIO, J., 1977: Cellular development in *Pediastrum boryanum.* Thesis, Marquette University, Milwaukee, Wisconsin.

Di Orio, J., Millington, W. F., 1978: Dictyosome formation during reproduction in colchicine-treated *Pediastrum boryanum* (*Hydrodictyaceae*). Protoplasma **97**, 329—336.

— Vigil, E., Millington, W. F., 1977: Progressive changes in the Golgi apparatus during development of the alga *Pediastrum boryanum*. Abstract, J. Histochem. Cytochem. **25**, 234.

Gawlik, S. R., Millington, W. F., 1969: Pattern formation and the fine structure of the developing cell wall in colonies of *Pediastrum boryanum*. Amer. J. Bot. **56**, 1084—1093.

Giddings, T. H., Brower, D. L., Staehlin, L. A., 1980: Visualization of particle complexes in the plasma membrane of *Micrasterias denticulata* associated with the formation of cellulose fibrils in primary and secondary cell walls. J. Cell Biol. **84**, 327—339.

Gottlieb, B., Goldstein, M. E., 1977: Colony development in *Eudorina elegans*. J. Phycol. **13**, 358—364.

Gunnison, D., Alexander, M., 1975: Basis for the resistance of several algae to microbial decomposition. Applied Microbiology **29**, 729—738.

Harper, R. A., 1918 a: Organization, reproduction and inheritance in *Pediastrum*. Proc. Amer. Phil. Soc. **57**, 375—438.

— 1918 b: The evolution of cell types and contact and pressure responses in *Pediastrum*. Memoirs of the Torrey Bot. Club **17**, 210—240.

Hawkins, A. F., Leedale, G. F., 1971: Zoospore structure and colony formation in *Pediastrum* spp. and *Hydrodictyon reticulatum*. Ann. Bot. **35**, 201—211.

Heslop-Harrison, J., 1968: The emergence of pattern in the cell walls of higher plants. In: The emergence of order in developing systems (Locke, M., ed.). (27th Sympos. Soc. Devel. Biol.) New York: Academic Press.

Honda, H., 1973: Pattern formation of the coenobial algae *Pediastrum biwae* Negoro. J. Theor. Biol. **42**, 461—481.

Ingold, C., 1973: Cell arrangement in coenobia of *Pediastrum*. Ann. Bot. **37**, 389—394.

Kiermayer, O., 1970: Elektronenmikroskopische Untersuchungen zum Problem der Cytomorphogenese von *Micrasterias denticulata* Breb. Protoplasma **69**, 97—132.

— 1977: Biomembranen als Träger morphogenetischer Information. Naturwiss. Rundschau **30**, 161—165.

— Dobberstein, B., 1973: Membrankomplexe dictyosomaler Herkunft als Matrizen für die extraplasmatische Synthese und Orientierung von Microfibrillen. Protoplasma **77**, 437—451.

— Sleytr, U. B., 1979: Hexagonally ordered "rosettes" of particles in the plasma membrane of *Micrasterias denticulata* Breb. and their significance for microfibril formation and orientation. Protoplasma **101**, 133—138.

Leadbeater, B. S. C., 1979: Developmental studies on the loricate choanoflagellate *Stephanoeca diplocostata* Ellis. Protoplasma **98**, 241—262.

Manton, I., Ettl, H., 1965: Observations on the fine structure of *Mesostigma viride* Lauterborn. J. Linn. Soc. London Bot. **59**, 175—184.

Marchant, H. J., 1974: Cytokinesis and colony formation in *Pediastrum boryanum*. Ann. Bot. **38**, 883—888.

— 1977 a: Colony formation and inversion in the green alga *Eudorina elegans*. Protoplasma **93**, 325—339.

— 1977 b: Cell division and colony formation in the green alga *Coelastrum* (*Chlorococcales*). J. Phycol. **13**, 102—110.

— 1979: Microtubular determination of cell shape during colony formation by the alga *Pediastrum*. Protoplasma **98**, 1—14.

— Pickett-Heaps, J., 1972: Ultrastructure and differentiation in *Hydrodictyon reticulatum*. IV. Conjugation of gametes and the development of zygospores and azygospores. Aust. J. Biol. Sci. **25**, 279—291.

— — 1974: The effect of colchicine on colony formation in the algae *Hydrodictyon*, *Pediastrum*, and *Sorastrum*. Planta **116**, 291—300.

MILLINGTON, W. F., 1973: Regulation of zoospore and colony formation in *Pediastrum* (Abstr.). J. Phycol. **9**, Suppl. 7.

— CHUBB, G., SEED, T., 1981: Cell shape in the alga *Pediastrum* (*Hydrodictyaceae, Chlorophyta*). Protoplasma **105**, 169—176.

— DI ORIO, J., 1978: Dictyosome formation during reproduction in colchicine-treated *Pediastrum boryanum* (*Hydrodictyaceae*). Protoplasma **97**, 329—336.

— GAWLIK, S. R., 1967: Silica in the wall of *Pediastrum*. Nature **216**, 68.

— — 1970: Ultrastructure and initiation of wall pattern in *Pediastrum boryanum*. Amer. J. Bot. **57**, 552—561.

— — 1975: Cell shape and wall pattern in relation to cytoplasmic organization in *Pediastrum simplex*. Amer. J. Bot. **62**, 824—832.

— RASCH, E., 1980: Microspectrophotometric analysis of mitosis and DNA synthesis associated with colony formation in *Pediastrum boryanum* (*Chlorophyceae*). J. Phycol. **16**, 177—182.

MONER, J. G., 1955: Cell wall structure in *Pediastrum* as revealed by electron microscopy. Amer. J. Bot. **42**, 802—806.

— CHAPMAN, G. B., 1963: Cell wall formation in *Pediastrum biradiatum* as revealed by the electron microscope. Amer. J. Bot. **50**, 992—998.

MUELLER, S. C., BROWN, R. M., 1980: Evidence for an intermembrane component associated with a cellulose microfibril synthesizing complex in higher plants. J. Cell Biol. **84**, 315—326.

NORRIS, R. E., PEARSON, B. R., 1975: Fine structure of *Pyramimonas parkae*, sp. nov. (*Chlorophyta, Prasinophyceae*). Arch. Protistenkd. **117**, 192—213.

PALIK, P., 1933: Über die Entstehung der Polyeder bei *Pediastrum boryanum* (*Turpin*) Meneghini. Arch. Protistenk. **79**, 234—238.

PARKER, B. C., 1964: The structure and chemical composition of cell walls of three chlorophycean algae. Phycologia **4**, 63—74.

PARRA BARRIENTOS, O. O., 1979: Revision der Gattung *Pediastrum* Meyen (*Chlorophyta*). Bibliotheca Phycologica **48**, 1—125.

PIA, J., 1927: In: Handbuch der Palaeobotanik, Vol. 1 (HIRMER, M., ed.). München: R. Oldenbourg.

PICKETT-HEAPS, J. D., 1972: Cell division in *Cosmarium botrytis*. J. Phycol. **8**, 343—360.

ROBINSON, D. G., PRESTON, R. D., 1972: Plasmalemma structure in relation to microfibril biosynthesis in *Oocystis*. Planta **104**, 234—246.

ROGALSKI, A. A., OVERTON, J., RUDDAT, M., 1977: An ultrastructural and cytochemical investigation of the colonial green alga *Pediastrum tetras* during zoospore formation. Protoplasma **91**, 93—106.

SCHNEPF, E., RODERER, G., HERTH, W., 1975: The formation of the fibrils in the lorica of *Poteriochromonas stipitata*: Tip growth, kinetics, site, orientation. Planta **125**, 45—52.

SCHRAUDOLF, H., FRAUENKRON, I., 1979: Effects of Concanavalin A on pattern formation in *Hydrodictyaceae*. Protoplasma **98**, 131—138.

SMITH, G. M., 1916: Cytological studies in the Protococcales. II. Cell structure and zoospore formation in *Pediastrum boryanum* (*Turp.*) Menegh. Ann. Bot. **30**, 467—479.

SULEK, J., 1969: Taxonomische Übersicht der Gattung *Pediastrum* Meyen. In: Studies in phycology (FOTT, B., ed.), pp. 197—261. Prag: Academica.

TRYON, E. H., DAVIS, J. S., 1976: A new asexual cycle in a *Pediastrum simplex* strain. Bot. Gaz. **137**, 356—360.

WILCOX, L. W., GRAHAM, L. E., 1979: Fine structure of bristles on *Pediastrum duplex*. J. Phycology **15**, suppl. 27.

WILSON, L. R., HOFFMEISTER, W. S., 1953: Four new species of fossil *Pediastrum*. Amer. J. Sci. **251**, 753—760.

Pattern Formation in *Acetabularia*

H. G. Schweiger and Sigrid Berger

Max Planck Institute for Cell Biology, Ladenburg near Heidelberg, Federal Republic of Germany

With 20 Figures

Contents

I. Introduction

The differing morphologies of the innumerable organisms of the world provide a basis to distinguish and classify them. There are higher multi-cellular or relatively simple unicellular organisms and even subcellular structures. That almost all the different organisms can indeed be distinguished requires that their shape as such, or the individual structural elements into which the organisms can be dissected, deviate to a great extent from basic geometric structures. Such basic structures are, for example, spheres, ellipsoids, or round columns with hemispheres attached to both ends.

The morphological features of the individual organism, within limits and depending on the conditions, result from growth and morphogenesis. The developmental processes that contribute to morphogenesis are nonrandom and regular. The spatial and temporal organization as well as the quantitative regulation of the processes are genetically inherited and follow predetermined lines. The resulting patterns are morphological equivalents of a high level of organization in terms of space. One of the major tasks of modern biology is to reduce these structural phenomena to their molecular and metabolic equivalents. This necessarily leads to the consideration that a basic part of pattern formation is the development of polarity. In this sense, polarity is closely connected to the formation of chemical gradients. Unfortunately, it is not yet clear which substances are primarily involved and how gradients are established.

II. Morphology

For studies on pattern formation, polarity and intracellular gradients, the green marine alga *Acetabularia* is one of the most promising unicellular organisms (for reference see Schweiger and Berger 1979). A total of about 25 species, known under the name *Acetabularia,* belong to the subfamily *Acetabularoideae,* family *Dasycladaceae* (Berger *et al.* 1974). Most of these species are cultured in the Max-Planck Institute for Cell Biology (Ladenburg).

The *Acetabularia* cell exhibits a pronounced species-specific morphology (Solms-Laubach 1895, Valet 1969). The morphological differences between the members of the family *Dasycladaceae* and the subfamily *Acetabularoideae* are significant, and most of the species can be easily distinguished (Fig. 1). All the species of this family share a common basic structure that can be described in a simplified way as being a long cylinder (the stalk) that carries a hemisphere (the rhizoid) at one end and a cap (a disc-like structure) at the other end. The disc-like structure is characteristic for the subfamily *Acetabularoideae* and develops at the end of the vegetative phase. The cap is composed of rays or chambers radiating axially from the apical end to the central stalk (Fig. 2). The stalk carries slim cylinder-shaped processes (hairs) that are circularly arranged and occur at intervals along the long axis of the stalk. Usually, an *Acetabularia* cell has more than one such set of hairs or whorl. The individual hair ramifies into cylinder-shaped processes.

The hemisphere-like basic structure of the rhizoid consists of a number of slim cylinders that look like branches. In the wild, the branches of the rhizoid spread over the substratum and form a holdfast, while in laboratory cultures, they twist together.

III. Life Cycle

The life cycle of *Acetabularia* starts from an almost spherical zygote that has a diameter of about 10 μm (Berger *et al.* 1975). The zygote is formed after the fusion of two isogametes. A few days after the fusion of the gametes, the formation of the rhizoid begins and the stalk grows in an apical direction in a tube-like fashion. Due to this extensive stalk growth, the cell eventually

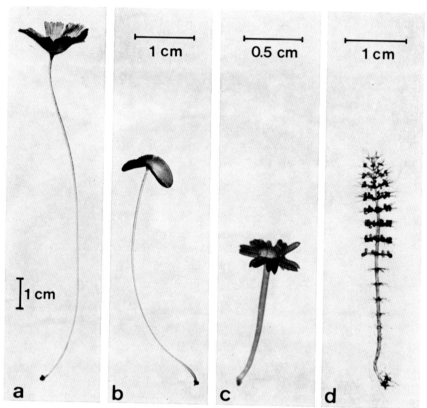

Fig. 1. Four unicellular green algae of the family *Dasycladaceae*. The four algae belong to different species. They are all in an equivalent mature stage. Note the difference in length. Note that *Batophora* does not form an apical cap but rather that the cysts with gametes are contained in small spheres attached to the upper hair whorls. *a Acetabularia major; b A. mediterranea; c A. dentata; d Batophora oerstedii.*

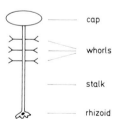

Fig. 2. Schematic drawing of an *Acetabularia* cell. The single primary nucleus normally located in the rhizoid is drawn as a black dot.

reaches a length of 1 to 200 mm or even more, depending on the species. In laboratory cultures in a defined artificial medium (Schweiger *et al.* 1977), the apical growth is completed after three to four months and a cap forms.

When the cap is fully developed, the generative phase of the life cycle begins. During the vegetative phase, the nucleus is located in the rhizoid and is called the primary nucleus (Hämmerling 1931). As soon as the cap reaches its maximum diameter, the primary nucleus starts to divide and numerous secondary nuclei are formed in the rhizoid (Schulze 1939). Together with most of the cytoplasm, the secondary nuclei migrate through the stalk into the cap rays. A thick-walled cyst is formed around each secondary nucleus, together with a certain portion of cytoplasm. In the cyst, the secondary nucleus proceeds to divide, giving rise to 100 to 1,000 secondary nuclei per cyst. Each secondary nucleus and a certain portion

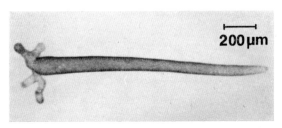

Fig. 3. *A. mediterranea* about 40 days old. The morphogenesis of the stalk and rhizoid formation can be clearly seen.

of cytoplasm within the cyst are surrounded by a membrane and eventually develop into a biflagellated gamete. Meanwhile, the cap structure disintegrates, liberating the cysts. Later, a preformed lid opens in the cyst wall, releasing the gametes, which are positively phototactic and therefore swim toward the light. Each cyst releases only "+" or "—" gametes, but an individual cell produces both types of cysts (Hämmerling 1931, 1934 a, 1944). One "+" and one "—" gamete fuse to give rise to a zygote with four flagellae. In contrast to the gametes, the zygote is negatively phototactic (Hämmerling 1955). It swims away from the light, settles to the substratum, loses its flagellae, and initiates growth of the stalk and rhizoid, thus completing the life cycle.

IV. Morphogenesis

The cylindric stalk grows out of one side and the branched rhizoid out of the other side of the cell, thus manifesting a polarity early in the developmental process (Fig. 3). When the stalk reaches a length of a few millimeters, formation of the first whorl starts. At the initiation of whorl formation, the apical growth of the stalk is stopped. The tip of the stalk flattens and thickens around its outer margin. Out of this margin, a number of processes extrude radially at regular intervals (Fig. 4 a). The processes grow as straight cylinders perpendicular to the stalk and form hairs that later on ramify repeatedly (Fig. 4 b). After a whorl is completed and a bunch of hairs is

Fig. 4. Hair whorls. *a* At the flattened tip of the *A. major* cell, 11 equally spaced processes are formed that later on give rise to a whorl. The older generation of a whorl can partially be seen (BERGER *et al.* 1974). *b* The hairs of a young whorl at the tip of the *A. mediterranea* cell branched once. The hairs are still close together in straight position toward the light. *c* The mature whorl of *Chalmasia antillana* with four-times branched hairs unfolded. This whorl in its basis protects a just forming cap.

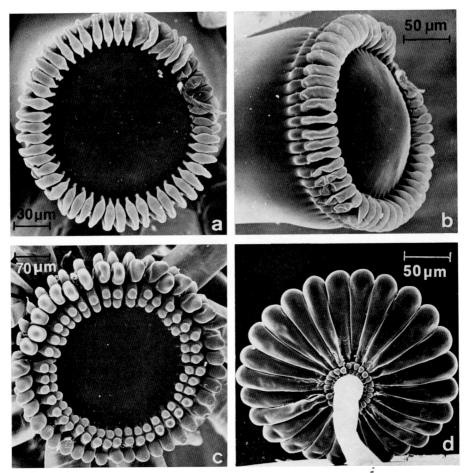

Fig. 5. Cap formation in *A. farlowii*. *a* Equally spaced processes arise from the outer margin of the flattened tip of the cell. Little bulges on their inner sides later form the segments of the *corona superior*. *b* The early cap processes, when viewed from the side, show little bulges beneath them that later give rise to the segments of the *corona inferior*. *c* A young cap already shows distinct cap rays under which, in the left upper side of the micrograph, *corona inferior* segments can be detected. The segments of the *corona superior* carry a species-specific number of hair processes. *d* A fully grown cap, from underneath, shows the segments of the *corona inferior* and the species-specific shape of the cap rays.

formed (Fig. 4 *c*), the flattened end of the stalk is transformed into a tip and resumes growing. These processes are repeated a number of times. Thus, the direction of growth periodically cycles from axial to radial during the vegetative phase.

When the cell has reached a length characteristic of its species, the tip of the stalk flattens again, thickens at its outer margin as was described for whorl formation, and develops a number of radial processes that in top view have an ellipsoid shape (Fig. 5 *a*). This number is not identical with the number of hairs per whorl. In one *Acetabularia* species, *A. mediterranea*, these processes are connected so tightly that instead of several separated processes, one pad of interconnected processes extrudes. The processes eventually will grow out and develop into the cap rays. At the site where each process is attached to the stalk, a bulge is formed (Figs. 5 *a* and *c*) that, depending on the species, has the shape of either a sphere or an ellipsoid. These bulges form the *corona superior*. On each bulge of the *corona superior*, a number of processes develop (Figs. 5 *c*, 6 *a* and *b*). Some of the processes grow into branched hairs, others remain rudimentary. The *corona inferior* (Fig. 5 *d*) is composed of individual bulges that grow out of the outer side of the processes that later form the cap rays (Fig. 5 *b*). The bulges of the

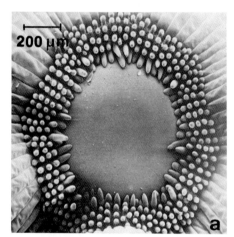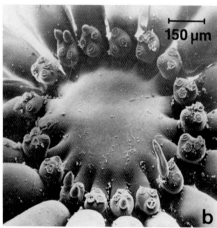

Fig. 6. *Corona superior. a A corona superior* of *A. ryukyuensis* with the species-specific number of five to six hair processes in one row. *b* The *corona superior* of *A. peniculus* having no *corona inferior*, shows a pattern of sphere-like segments with the hair processes arranged in a circle or in two rows. Since this is a mature cap, some of the hair processes have grown out as hairs and fallen off leaving scars at the segments.

corona inferior can be straight or scalloped one to several times and, at their upper end, are smooth or scalloped or toothed (Figs. 7 *a, c*, and *d*). While all species have a *corona superior*, some of them do not form a *corona inferior*. These species, in accordance with an old terminology, were called *Polyphysa* instead of *Acetabularia*.

A vigorous cytoplasmic streaming toward the cap starts when the cap of the cell is fully grown (SCHULZE 1939). Together with the cytoplasmic components, the numerous secondary nuclei are transported into the cap rays. The positions of the secondary nuclei in the cap rays are recognizable as white spots since the area around each secondary nucleus is devoid of chloroplasts (SCHULZE 1939, WOODCOCK and MILLER 1973). The white spots are regularly arranged with equal distances between neighboring spots (Fig. 8 *a*). Around each secondary nucleus, a portion of the plasma separates (Fig. 8 *a*) and is enclosed by the development of the cyst wall. The shape and size of the cysts exhibit species specificity (BERGER *et al.* 1974).

V. Comparative Morphology

A comparison of the morphological features of different species of *Acetabularia* might help elucidate general rules of pattern formation at a cellular level. The shape of the rhizoid exhibits a high degree of individual variability so that it is not generally suited as a marker to distinguish dif-

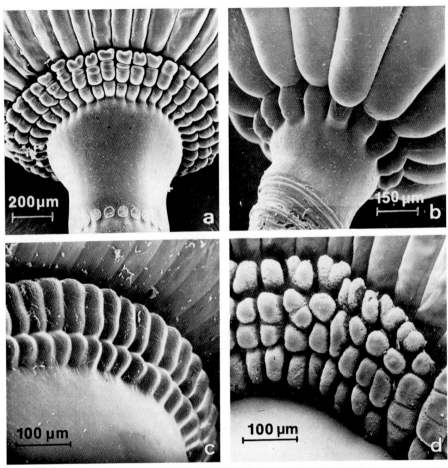

Fig. 7. *Corona inferior. a* The *corona inferior* of *A. crenulata* has bilobed, once to twice invaginated segments. At the bottom of the micrograph, a circle of scars derived from a hair whorl can be seen. *b* The cap rays of *A. parvula*, which has no *corona inferior*, are situated on little stalks (Berger *et al.* 1974). *c* The *corona inferior* of *A. mediterranea* is composed of segments that grow together as do the cap rays. *d* The *corona inferior* segments of *A. ryukyuensis* have many invaginations and are irregularly bilobed at their distal ends.

ferent species. In spite of this fact, a study of the pattern formation of the rhizoid in different species of *Acetabularia* reveals a certain extent of genetic determination. A comparison of the species with a relatively long stalk such as *A. cliftonii* and *A. major* shows that in the first species, the rhizoid branches are short, thick, and not much ramified (Fig. 9 *b*), while in the

Fig. 8. Cap rays and whorls. *a* Part of the cap of a mature *A. mediterranea*. The arrows point to cap rays with "white spots", each one of which represents a secondary nucleus surrounded by a portion of chloroplast-free cytoplasm. Some of the cap rays are in the course of cyst formation, while others already bear cysts. *b* Apical end of a growing *A. mediterranea* cell. The stalk bears three sets of whorls. The older whorls degenerate and fall off. *c* Stalk of *Batophora oerstedii*. The stalk bears numerous branched hairs that do not degenerate.

second species, the rhizoids have long and relatively thin branches with a number of ramifications. In laboratory culture conditions where the branches are not attached to a substrate, the branches form a skein-like structure when they grow around each other (Fig. 9 a). In none of the species a regular pattern is detectable at the site where the basal part of the stalk ramifies into the branches of the rhizoid.

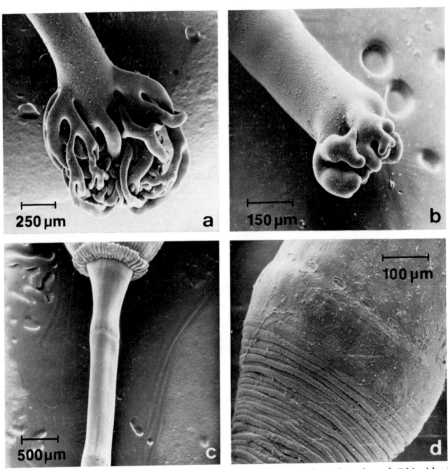

Fig. 9. Rhizoid and stalk. *a* Rhizoid of *A. major* with many long branches. *b* Rhizoid of *A. cliftonii* with short thick branches (Berger *et al* 1974). *c* Stalk of *A. major*. The stalk has a smooth surface. Two circles of scars derive from degenerated whorls. The upper end of the stalk carries the *corona inferior*. *d* Stalk of *A. parvula*. The surface of the stalk has numerous ring formations and one circle of hair scars.

The stalk walls of the large *Acetabularia* species are smooth. The only interruptions are the regularly arranged whorls or corresponding rings of scars where the hairs were lost (Fig. 9 c) the shape of the scars varies and, to a limited extent, is species-specific (Berger *et al.* 1974). In the small *Acetabularia* species such as *A. parvula*, the cell wall of the basal part of the stalk exhibits ring-like structures. The stalk, therefore, looks compressed (Fig. 9 *d*).

Hair whorls are formed by all members of the family *Dasycladaceae*. Usually, the older whorls degenerate and fall off. This means that only two to four of the youngest generations of whorls remain fixed to the stalk

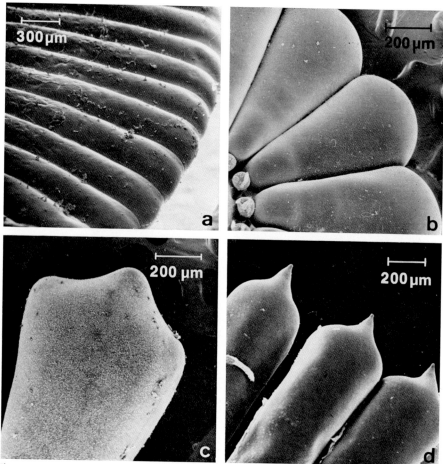

Fig. 10. Cap rays. *a* The cap rays of *A. mediterranea* are fused. They are blunt at their distal ends (BERGER et al. 1974). *b* The cap rays of *A. parvula* are obtusely rounded, close together but not fused. *c* The distal end of a cap ray of *A. polyphysoides* is triangularly stretched. *d* The cap rays of *A. crenulata* bear thorns of varying sizes, although they are usually big enough to be visualized macroscopically.

(Fig. 8 *b*). A few *Acetabularia* species, like *A. kilneri* and *A. calyculus*, retain their whorls for a longer period so that six and more generations of whorls are attached to the stalk at the same time. Members of the family *Dasycladaceae* that do not belong to the subfamily *Acetabularoideae*, e.g., *Batophora* and *Dasycladus*, rarely lose the older whorls. In these species, the whole length of the stalk bears whorls (Fig. 9 *c*). These species also

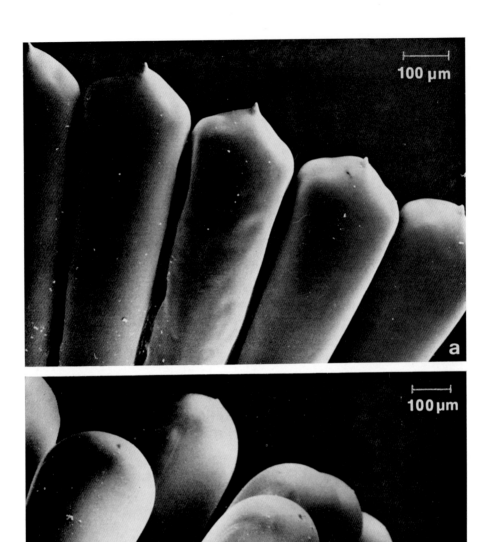

Fig. 11. Cap rays of two cells of *A. kilneri.* *a* The cap rays of this cell bear distinct little thorns at their distal ends. These thorns can be easily visualized by using a dissecting microscope. *b* The cap rays of this cell appear smoothly rounded at their distal ends when observed by using a dissecting microscope. In the scanning electron microscope, it is obvious that these cap rays also bear thorns at their distal ends. The appearance of the thorns is species-specific, whereas the size of the thorns shows individual variation.

Fig. 12. Different species of *Acetabularia*. *a* Cyst-bearing cell of *A. exigua*. This small *Acetabularia* species often forms cysts not only in the cap rays but also in the stalk and even in the rhizoid. *b* Cells of *A. dentata* on a stone as it was found in nature. The cap rays are fused by calcification, which gives them a beaker- or umbrella-shaped appearance. *c* Cells of *A. dentata* grown in the laboratory. In laboratory culture, *Acetabularia* cells usually do not calcify. The cap rays are therefore not fused and may grow in different directions.

show a different pattern in gametangia formation since they do not form apical caps but attach their ball-shaped gametangia to the upper whorls (Fig. 1 *d*) (Liddle et al. 1976).

The cap-shaped gametangia stands of *Acetabularia* have a common basic construction. They are composed of radially arranged rays and a crown at the upper proximal end. While the whorls of the stalk beneath the cap will degenerate and fall off, bunches of hairs grow from this crown, maybe in order to absorb nutrients from the surrounding medium. The cap rays exhibit a high degree of species specificity. In particular, this is true for the distal ends of the cap rays. In one species, *A. mediterranea*, the cap rays are so tightly united that they stick together even without calcification (Fig. 10 *a*).

The distal ends of the cap rays may be blunt or tipped, and they may be of triangular or club-shaped form (Figs. 10 *b*, *c*, and 11 *a*). They may bear thorns that are so tiny they can be detected only with the aid of a microscope (Fig. 11 *b*). The size and shape of the thorns might greatly vary (Figs. 11 *a* and *b*), but their existence is strictly species-specific.

Acetabularia cells are often calcified in nature. The degree of calcification of the cells is influenced by ecological factors. However, to a certain extent it is species-specific, as can be seen when comparing the degree of calcification of cells grown under identical environmental conditions. Calcification influences the apparent morphology of *Acetabularia* cells (Figs. 12 *b* and *c*). Often the caps of cells from the wild appear as discs whose rays stick firmly together. Under laboratory conditions where only little calcification occurs in cells of the same species, the cap rays are separated. Moreover, due to severe calcification, the morphology of the cap rays may change. The calcium carbonate may completely embed the distal ends of the cap rays so that they appear to be blunt or obtusely rounded; nevertheless, some species may indeed still bear little thorns.

The morphological patterns of the *corona superior* segments, as well as the number of hair processes they carry, are strictly species-specific. The *corona superior* segments are either ball shaped or nearly ellipsoid. The hair processes may form one or two rows (Figs. 6 *a* and *b*). In some of the species, only one hair process is found per *corona* segment, like in *Halicoryne* (Solms-Laubach 1895), while in others, there are up to ten processes, like in *A. haemmerlingi* (Schweiger and Berger 1975). As mentioned before, some of the *Acetabularia* species have a *corona inferior*. These species also possess ball-shaped *corona superior* segments. The species specificity of the pattern of the *corona inferior* segments is limited. These segments may be narrow or wide, smooth or one to several times carved. Their distal ends may be blunt or lobed (Figs. 7 *a*, *c*, and *d*). In the species that do not have a *corona inferior*, the cap rays are connected to the stalk by thin segments (Fig. 7 *b*) (Berger et al. 1974).

VI. Polarity

If a unicellular organism produces a species-specific pattern at distinct sites, one may assume that a gradient of chemical substances responsible for the formation of the pattern exists in the cell. However, it would be over-simplifying the problem of pattern formation to believe that it can be explained by a simple chemical gradient inside a closed homogeneous system.

The question of how a gradient originates becomes an obvious problem if one starts with the premise that it originates from a spherical cell with a homogeneous distribution of its content. With such a sphere-symmetrical structure, it is difficult to imagine how, for example, axial growth, which is an extremely simple form of pattern formation, can be initiated.

In order to gain insight into the basic mechanism of pattern formation, it might be useful to resolve the whole process into a sequence of individual steps. Among these steps, the early occurrence of uneven distribution of molecules at the cellular or subcellular level deserves special interest. Although it cannot be ruled out that small molecules might play a part in the formation of polarity, the main search in this context should be focussed on macromolecules because perturbation of an uneven distribution of macromolecules, once it is set up, is much less probable than with small molecules; also, if a gradient of small molecules is to be established, macromolecules may be involved as binding sites, *e.g.*, for inorganic ions and metabolites or as components of compartmentizing structures.

Instead of being a sac filled with a homogeneous soup, the cell contains a high number of compartmentizing structures, like Golgi vesicles, plastids, mitochondria, and certain membrane structures. These will probably be unevenly distributed. Most of the subcellular structures exhibit explicit internal heterogeneity, which again might be the basis for the formation of gradients at the cellular level. Finally, one should keep in mind that recent evidence indicates that transport and receptor sites at the cell surface for a number of substances are unevenly distributed, and the same may be true for other functional molecules.

As has been discussed, the unicellular organism *Acetabularia* exhibits a pronounced pattern formation—stalk at one end and a rhizoid at the other end. These processes are initiated immediately after zygote formation. Stalk and cap formation at opposite ends of the zygote reflect a characteristic polarity and, as underlined by indirect evidence (HÄMMERLING 1934 b, 1936), are probably due to gradients of two corresponding morphogenetic substances.

In the case of *Acetabularia*, it seems obvious how this polarity originates. The gametes, which are highly asymmetric with the flagellae forming one pole and with the nucleus next to it, fuse in a way such that the fused nucleus stays at one end. This end, after losing the flagellae, attaches to the substratum and forms a rhizoid (HÄMMERLING 1955). At the same time, the opposite pole starts to form the stalk and later the whorls and the cap (Fig. 13). The existence of an apical and basal pole of the *Acetabularia* cell, therefore, is derived from a polarity determined by the gamete structure. This primary polarity is necessary but by itself is not sufficient for the performance of all further differentiation events at distinct places of the cell. Under certain experimental conditions, it has been demonstrated that the polarity of the cell may be changed (HÄMMERLING 1934 b, 1936, 1955). The primary polarity is, however, responsible for the development of the gradient of morphogenetic substances and, therefore, for the formation of different structures at different sites of the cell or at different times of the life cycle.

The determining structure for polarity is, in any case, the cell nucleus.

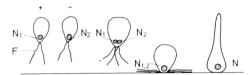

Fig. 13. Zygote formation in *Acetabularia*. The gametes of *Acetabularia* already show a polarity. They have a flagellar pole, and the nucleus is located next to this pole. Two gametes ("+" and "—") fuse along their sides to form a zygote. The zygote settles with its flagellar pole to the substrate. It will lose its flagellae and form a rhizoid at this pole. The tube-like stalk grows out of the opposite pole. *N* nucleus; *F* flagella.

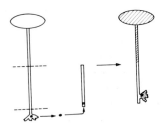

Fig. 14. Role of the nucleus in morphogenesis. If a basal anucleate fragment is cut out of an *Acetabularia* cell and receives an isolated nucleus, it will form a rhizoid in the place where the nucleus is located. Stalk growth and later cap formation occur at the opposite pole. New growth induced by the implanted nucleus is indicated by the striated regions.

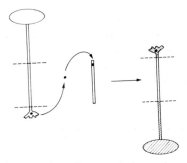

Fig. 15. Change of polarity under the influence of the nucleus. If a basal anucleate fragment is cut out of an *Acetabularia* cell and receives an isolated nucleus at its former apical end, it will reverse its polarity. It will form the rhizoid at the former apical part, and elongation and cap formation occur at the former basal part.

HÄMMERLING was able to show that an anucleate stalk always forms a new rhizoid near the region into which a new nucleus is implanted (Fig. 14). This holds true also when the nucleus is implanted into an apical region; the result is an inversion of the original polarity (Fig. 15). From this observation, HÄMMERLING concluded that the morphogenetic substances responsible for rhizoid formation have a high concentration in the vicinity of the nucleus, while the morphogenetic substances for stalk, whorl, and cap formation, which also must originate in the nucleus, are transported into the apical part of the cell (HÄMMERLING 1955).

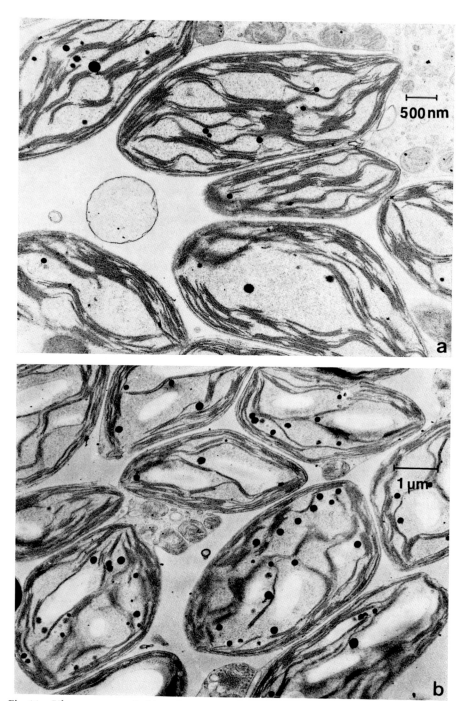

Fig. 16. Ultrastructure of chloroplasts. *a* Section through the apical stalk region. The numerous chloroplasts are elongated and show well-developed grana and no polysaccharide granules. *b* Section through the lower half of the stalk. The chloroplasts are not as elongated as in the apical region. Many of them have several polysaccharide granules and relatively few grana.

VII. Location Dependent Ultrastructure of Chloroplasts

Besides the morphological characteristics of the rhizoid, other ultra-structural differences exist between the apical and basal stalk and cap portions of the cell. The cytoplasm of the *Acetabularia* cell contains numerous chloroplasts that, according to their construction plan, all belong to the same type (Hoursiangou-Neubrun and Puiseux-Dao 1974). The ultra-structure of the chloroplasts is not static, but rather dynamic. During the development of the cell, it is subjected to changes that eventually give rise to a mixed chloroplast population, particularly when the stalk has reached its maximal length. The chloroplasts then show a wide morphological and physiological heterogeneity along an apical-basal gradient (van den Driessche 1974, Puiseux-Dao and Dazy 1970, Boloukhère 1972, Hoursiangou-Neubrun et al. 1977).

The chloroplasts of the apical part of the stalk have nearly no polysac-charide granules (Fig. 16 a), they are primarily small. Chloroplasts of the middle part of the cell usually have more polysaccharide granules (Fig. 16 b). In the rhizoid, two regions can be distinguished. In the vicinity of the cell nucleus, the chloroplasts are similar to those found in the tip of the cell. They have well-organized membranes, form grana, and mostly have no or only small polysaccharide granules (Fig. 17 b). In the rest of the rhizoid, the chloroplasts are usually spherical, have only a few membranes, and are nearly filled with polysaccharides (Fig. 17 a). Moreover, it has been shown by density gradient centrifugation that a hetereogeneity in the chloroplast population exists that reflects the different polysaccharide contents (Lüttke et al. 1976).

The heterogeneity of the chloroplast population has been postulated to be a function of the age of the chloroplasts: the polysaccharide content increases with age, chloroplasts in the tip are presumed to be young and rapidly dividing, chloroplasts in the middle of the cell are mature, and chloroplasts in the basal part are old, except for those in the vicinity of the cell nucleus. These ultrastructural observations are in accordance with results obtained from investigations of the chloroplasts of the whorls (Dujardin et al. 1977). The whorls of the apical part of the cell have a mixed chloroplast population. The older whorls contain a homogeneous population of chloroplasts of the type from the rhizoid cytoplasm that is not in close vicinity to the cell nucleus, *i.e.*, polysaccharide-rich chloro-plasts.

The substantial cytoplasmic streaming that occurs in these cells seems to preclude, at first glance, the existence of stable gradients of the chloro-plast population. Cinematography has demonstrated that the different parts of the cytoplasm stream at different speeds. Within the cytoplasmic strands, some granules move at a high speed while chloroplasts usually move much more slowly (Koop and Kiermayer 1980). Altogether, one gets the impression that the chloroplasts move back and forth over short distances rather than migrating the entire length of the cell.

The fact that chloroplasts in the vicinity of the nucleus look "young"

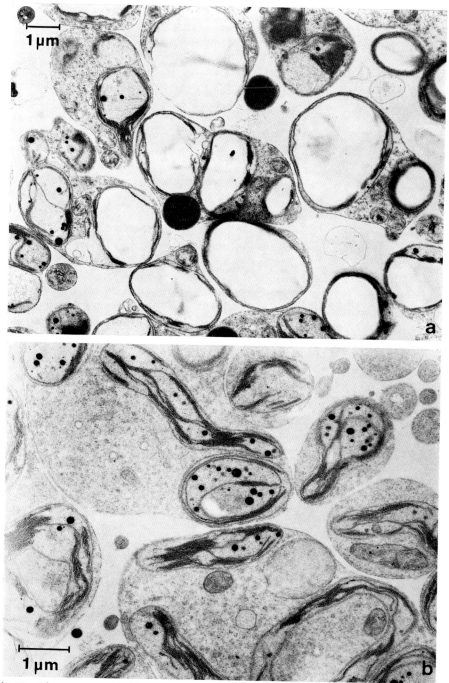

Fig. 17. Ultrastructure of chloroplasts. *a* Section through a rhizoid branch that does not contain the cell nucleus. Many of the chloroplasts are filled with polysaccharides. They therefore look round and do not display grana. *b* Section through the cytoplasm in the vicinity of the nucleus. The chloroplasts are elongated. Only a few of them contain polysaccharide granules. They have well-developed grana.

indicates that a simple age dependency cannot be the reason for the differences in the ultrastructure of the chloroplasts in different parts of the cell. An attractive explanation could be that the heterogeneous distribution of different types of chloroplasts in the cell is a function of the intergenomic cooperation between cell nucleus and chloroplasts. The fact that the chloroplasts in the apex and in the vicinity of the nucleus look similar might indicate that the chloroplasts that take part in this cooperation move directly toward the tip of the cell or that such a cooperation might only be possible in the tip of the cell. Such a hypothesis is supported by the fact that nonregenerating anucleate cell fragments do not contain chloroplasts of the apical type. In both nucleate and anucleate regenerating fragments, the apical-basal gradient of chloroplast types is reestablished.

A relationship between chloroplast heterogeneity and the chloroplast-nucleocytosol cooperation is also indicated by the fact that the chloroplasts in the apex and in the vicinity of the nucleus look similar, presumably because they are in a similar functional state. This situation is paralleled by a high density of 80 S ribosomes in these parts of the cell. One may assume that the chloroplast activity, in particular in terms of capability to divide, is limited by the supply of proteins contributed by the nucleocytosol compartment. However, this requires an explanation of the uneven distribution of ribosomes within the cell. (This problem will be discussed in the section on the intracellular distribution of ribosomes.) The hypothesis that the heterogeneous distribution of different types of chloroplasts within the cell has to do with an uneven distribution of 80 S ribosomes is supported by the observation that in nonregenerating anucleate cell fragments that do not exhibit an uneven distribution of 80 S cytosol ribosomes, the ultrastructure of the chloroplasts is not heterogeneous and does not contain the active type. However, in nucleate and anucleate regenerating fragments, both the uneven distribution of 80 S ribosomes and the heterogeneous distribution of different chloroplast types is observed.

VIII. Chloroplast Migration

Besides the permanently uneven and heterogeneous distribution of different chloroplast types fluctuations in the number of chloroplasts in different parts of the cell are observed. Periods during which a high density of chloroplasts is observed in the apex alternate with periods when the number of chloroplasts in the apex decreases (Koop et al. 1978). The density of the chloroplasts in the rhizoid and in the basal part of the cell fluctuates in a like manner, albeit with an opposite phase. In statistical terms, the chloroplasts are transported by cytoplasmic streaming into the basal part or the apical part, depending on the time of day. During the light phase, the chloroplasts accumulate in the upper part of the cell; during the dark phase, the majority congregate in the basal part. This alternating upward-downward migration is observed even if the cells are kept under constant light. Under constant conditions, the period of this chloroplast migration rhythm is about 24 hours (Fig. 18). It is justified to assume that this endogeneous rhythm is of a circadian nature. The rhythmic pattern is retained even if the nucleus is removed.

Recently, a measuring device was developed that enables the measurement of these chloroplast migration oscillations in an individual cell over several months (BRODA *et al.* 1979). The principle of measurement is based on the variations in light transmission observed when chloroplasts migrate to and from a region of the stalk of a cell held between a pinpoint light source on one side and a light measuring device on the other. The light transmission reflects the density of chloroplasts in the site of measurement.

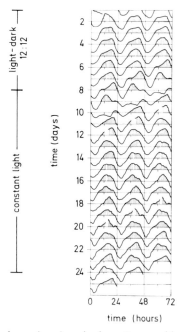

Fig. 18. Endogenous chloroplast migration rhythm. Due to chloroplast migration toward the tip of the cell, the light transmission through the basal part of the cell reaches a maximum in the light phase. During the dark period, chloroplasts migrate toward the basal part of the cell so that transmission reaches a minimum during the dark phase. The pattern repeats periodically and has a phase length of about 24 hours. This rhythmicity is maintained in constant light and is an example of a circadian rhythm.

The rhythmic chloroplast migration is interesting since it is an excellent example of an intracellular gradient that clearly depends on time. Interestingly enough, the chloroplasts not only migrate in an oscillatory way, but also apparently undergo ultrastructural alterations in a circadian manner (VAN DEN DRIESSCHE *et al.* 1976).

IX. Intracellular Distribution of Ribosomes

The heterogeneity of the chloroplast population and the uneven distribution of the different types of chloroplasts within the cell raises the question of whether there are other subcellular structures that exhibit intracellular gradients. As has already been discussed, 80 S ribosomes of the cytosol exhibit a higher density in the tip of the cell and immediately adjacent to the nucleus (Boloukhère 1969). This is also the site where the main part of the protein synthesis takes place. The question of why this is so may be partially answered by a simple type of experiment (Fig. 19). Cells were given a 24-hour pulse of radioactively labelled RNA precursors. At different times after labelling, the distribution of radioactive 80 S ribosomes within the cell was observed. The newly synthesized 80 S ribosomes first appear as unbound monosomes in the basal part of the cell. Within the next three days, more than two-thirds of them are associated with poly- ribosomes in the basal part. Five to seven days after their synthesis, the labelled ribosomes, which now all appear as polysomes, are membrane

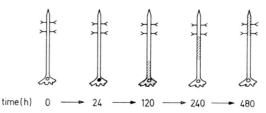

time(h)   0 ⟶ 24 ⟶ 120 ⟶ 240 ⟶ 480

Fig. 19. Intracellular ribosome migration. A growing cell was labelled with ³H-uridine for 24 hours. After this time the RNA produced in the nucleus is highly radioactive and is represented in this schematic drawing by a dark dot. Radioactively labelled rRNA will be released from the nucleus. Its incorporation into ribosomes and their migration through the stalk has been followed over the given time. The distribution of the labelled ribosomes is represented in this drawing as shading.

bound. This binding to membranes occurs simultaneously with the initia- tion of transport of the ribosomes into the apical region (Kloppstech and Schweiger 1975).

The experimental design permits rough estimations of the speed with which the ribosomes are transported from the rhizoid into the apex. This rate is about 2 to 4 mm per day, while the increase in cell length due to stalk growth is only 1 to 2 mm per day. The prerequisite for the uneven distribu- tion of the ribosomes within the cell is the fact that they are bound to membranes. A comparison of the speed of transport of the ribosomes through the cell and the rate of stalk growth indicates that the ribosomes are trans- ported significantly faster than the cell increases its length. Together with the fact that the half-life of ribosomes in *Acetabularia* is on the order of 80 days, *i.e.*, extremely long, the difference in the two rates helps explain why there is an accumulation of 80 S ribosomes in the tip of the cell. On the other hand, one may assume that the high concentration of ribosomes in the vicinity of the nucleus is due to the time lag between the synthesis of the ribosomes within the nucleus and their attachment to membranes. These events probably account for the high density of 80 S ribosomes ob- served in the vicinity of the cell nucleus.

X. The Generative Phase

Comparing the stalks of a young cell and one from a stage with a maximum diameter cap, one realizes that the stalk of the older cell is much paler. This difference is due to the fact that in older cells the chloroplasts, together with other components of the cytoplasm, are transported into the cap. This type of transport does not depend on the time of the day. During the period of transport, the stalk of such cells appears inhomogeneous and the cytoplasm has a striped appearance (SCHULZE 1939). At the same time, numerous secondary nuclei are formed and transported in small groups into the cap rays. At the end of these processes most of the cytoplasm is found in the cap, leaving behind a pale and empty ghost.

The following stages, which include the appearance of white spots and eventually result in cyst formation, include a number of examples for specific pattern formation. They are closely related to an uneven distribution of organelles like chloroplasts, segregation of cytoplasmic droplets, formation of cell walls around the cysts, and introduction of specific structures into the cyst wall that then result in lid formation (WOODCOCK and MILLER 1973, BERGER et al. 1974, BERGER et al. 1975). Although all these processes apparently are highly organized, sometimes irregular events occur, e.g., it has been observed a number of times that cyst formation, which usually is strongly nuclear dependent, occurs even in the absence of a nucleus, i.e., in anucleate cells. Another characteristic though irregular event observed quite frequently in smaller species in particular, is that the transport of secondary nuclei remains incomplete, i.e., the nuclei do not reach the cap but remain in the stalk where cysts are formed (Fig. 12 a). This incomplete migration of secondary nuclei is perhaps due to a weak cytoplasmic streaming during the period of cytosol transport, or the cytoplasm contains components that impair the chloroplast transport.

XI. Bioelectric Pattern Formation

Besides morphological patterns, *Acetabularia* cells are capable of developing functional patterns too. It has been reported that a transcellular electrical potential exists in nucleate cells and also in anucleate apical and basal stalk fragments (NOVAK and BENTRUP 1972; see also p. 390 f.). The electrical potential is oriented parallel to the long axis of the cell where the growing apex is positive relative to the basal region. This transcellular potential is probably due to a transcellular gradient of membrane potentials. By means of a three-part cuvette, it is possible to mechanically and electrically isolate the basal, middle, and apical portions of an *Acetabularia* cell (Fig. 20). If the apical and basal portions are insulted from each other, no membrane potentials are formed. In such a case, there is no regeneration of a cell fragment. Only when the two ends of the cell are in electrical contact with each other will regeneration start (NOVAK and SIRONVAL 1975). In further experiments it was found that about 30 hours after the establishment of an electrical field in a basal anucleate part of an *Acetabularia* cell, electrical polarity can be detected (BENTRUP 1977). This observation indicates that under these conditions, regeneration is initiated since it is known that the tip region of a regenerating cell fragment is continuously hyperpolarized by about 10 mV and depolarized periodically via action potentials. Eight hours after the application of an electrical field, the presumptive tip starts to spontaneously depolarize. At first, action potentials are released at a rate of 4 to 5 per hour and gradually develop a regular pattern that parallels the regeneration of the cell. The action potentials of the *Acetabularia* cells and the maintenance of the electrical potential presumably reflect activity of the ATP-dependent Na/K-pump of the plasmalemma (Gradmann 1976). A coupling of the action potentials with morphogenesis is presumed since they start at the same time regeneration of the cell fragment begins. A similar frequency of depolarizations has been found also in the growth region of pollen tubes of higher

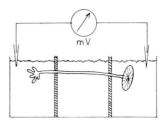

Fig. 20. Electrical potential measuring along an *Acetabularia* cell. The cell is put into a three-part cuvette, which enables the experimentor to electrically isolate the three parts of the cell. The experimental setup is similar to that of Novak *et al.* (1975).

plants. Probably the electrical as well as the morphological pattern formation are the expression of a common functional pattern that develops in the plasmalemma.

XII. The Early Mechanism of Whorl and Cap Formation

When the axial growth of an *Acetabularia* cell stops during formation of a whorl or a species-specific cap, a special regulatory mechanism must function in the stalk tip, which is rather distant from the cell nucleus. By cytochemical and electron microscope investigations, it was possible to gain a little insight into the very early events of differentiation (Werz 1970). One of the first observable changes at the initiation of whorl or cap formation is a characteristic swelling of the stalk tip. Axial growth stops before the swelling is apparent. During the axial growth phase, a characteristic protein is accumulated and evenly distributed in this region. Immediately after the onset of tip swelling, this protein is arranged in segments that all together give rise to a ring-like structure. The segments of this ring divide the cytoplasm of the stalk tip into distinct regions, the number of which corresponds exactly to the number of hairs or cap rays that form later on. In a next step the cell wall is hydrolized from the cytoplasmic side exactly at the sites where this protein had accumulated and where the hairs or cap rays will be formed. This means that the hydrolysis of the cell wall is a characteristic morphogenetic event and is species-specific since the number of cap rays is determined by this event. Electron microscope investigations have revealed that the endoplasmic reticulum is specifically associated with the regions of the cell wall where hydrolysis takes place. When hydrolysis is completed, small vesicles appear and undergo growth and cell wall formation, resulting in the development of whorls or cap rays. At the same time these synthetic processes resume, dictyosomes appear in the stalk tip. The alternating phases of synthesis and hydrolysis demonstrate that a complex system of regulation is involved, even in the earliest events of pattern formation.

XIII. Models of Pattern Formation

In the course of its development, the *Acetabularia* cell develops from a nondifferentiated zygote into a unicellular organism with morphologically and functionally different parts that are characterized by a specific arrangement of the subcellular structures. A number of results suggest that a system of chemical substances that react with each other and are intracellularly distributed by diffusion is responsible for morphogenesis. Simple diffusion or mixing of substances by chance is not sufficient to explain certain morphogenetic events, as for example, the early events in whorl or cap formation mentioned before. To explain this phenomenon of morphogenesis, theories have been developed to link experimental data to predictions based on chemical and mathematical models. Such models discuss biochemically realistic mechanisms for a strictly controlled hydrolysis that produces a pattern in the wall or membrane of the cell. Based on the model of TURING (1952), HARRISON et al. (1980) have developed a theory that tries to explain the formation of hair whorls in *Acetabularia*. Experimental observations have indicated that the spacing between the individual hairs of a whorl within limits is temperature dependent. This temperature dependency together with the apparent activation energy are in accordance with a simple version of a reaction-diffusion theory. Results obtained below 16 °C suggest that the diffusible molecules involved are membrane associated and affected by changes in the membrane fluidity. A combination of two reaction-diffusion models (TYSON and KAUFFMAN 1965, HARRISON and LACALLI 1978) makes it possible to explain the shift of growth activity from the punctuate location in the very tip of the cell to the concentric arrangement at the outer margin of the tip as well as the specific steps of hair morphogenesis (HARRISON et al. 1980).

Acknowledgements

The authors wish to express their gratitude to Dr. ROBERT SHOEMAN for critically reading the manuscript and to Mrs. CHRISTINE FABRICIUS, Mrs. RENATE FISCHER, and Mrs. BRIGITTE NAGEL for technical assistance in preparing the manuscript.

References

BENTRUP, F. W., 1977: Electrical events during apex regeneration in *Acetabularia mediterranea*. In: Progress in *Acetabularia* research (WOODCOCK, C. L. F., ed.), pp. 249—254. New York-London: Academic Press.

Berger, S., Sandakhchiev, L., Schweiger, H. G., 1974: Fine structural and biochemical markers of *Dasycladaceae*. J. Microsc. (Paris) **19**, 89—104.

— Herth, W., Franke, W. W., Falk, H., Spring, H., Schweiger, H. G., 1975: Morphology of the nucleocytoplasmic interactions during the development of *Acetabularia* cells. II. The generative phase. Protoplasma **84**, 223—256.

Boloukhere-Presburg, M., 1969: Ultrastructure de l'algue *Acetabularia mediterranea* au cours du cycle biologique et dans différentes conditions expérimentales. Thesis, Université Libre de Bruxelles.

Boloukhere, M., 1972: Différenciation spatiale et temporelle des chloroplastes d'*Acetabularia mediterranea*. J. Microsc. (Paris) **13**, 401—416.

Broda, H., Schweiger, G., Koop, H.-U., Schmid, R., Schweiger, H. G., 1979: Chloroplast migration: a method for continuously monitoring a circadian rhythm in a single cell of *Acetabularia*. In: Developmental biology of *Acetabularia* (Bonotto, S., Kefeli, V., Puiseux-Dao, S., eds.), pp. 163—167. Amsterdam: Elsevier.

Dujardin, E., Bonotto, S., Sironval, C., 1977: Plastidial heterogeneity during the developmental cycle of *Acetabularia mediterranea*. In: Progress in *Acetabularia* research (Woodcock, C. L. F., ed.), pp. 219—225. New York-London: Academic Press.

Gradmann, D., 1976: "Metabolic" action potentials in *Acetabularia*. J. Membr. Biol. **29**, 23—45.

Harrison, L. G., Lacalli, T. C., 1978: Hyperchirality: a mathematically convenient and biochemically possible model for the kinetics of morphogenesis. Proc. R. Soc. Lond. B **202**, 361—397.

— Snell, J., Verdi, R., Zeiss, G. D., Green, B. R., Lacalli, T. C., 1980: Hair morphogenesis in *Acetabularia*. J. Cell Biol. **22**, 459.

Hämmerling, J., 1931: Entwicklung und Formbildungsvermögen von *Acetabularia mediterranea*. I. Die normale Entwicklung. Biol. Zentralbl. **51**, 633—647.

— 1934 a: Über die Geschlechtsverhältnisse von *Acetabularia mediterranea* und *Acetabularia wettsteinii*. Arch. Protistenkd. **83**, 57—97.

— 1934 b: Über formbildende Substanzen bei *Acetabularia mediterranea*, ihre räumliche und zeitliche Verteilung und ihre Herkunft. Arch. Entwicklungsmech. Org. (Wilhelm Roux) **131**, 1—81.

— 1936: Studien zum Polaritätsproblem I—III. Zool. Jahrb. Abt. Allg. Zool. Physiol. Tiere **56**, 439—486.

— 1944: Zur Lebensweise, Fortpflanzung und Entwicklung verschiedener Dasycladaceen. Arch. Protistenkd. **97**, 7—56.

— 1955: Neuere Versuche über Polarität und Differenzierung bei *Acetabularia*. Biol. Zentralbl. **74**, 545—554.

Hoursiangou-Neubrun, D., Puiseux-Dao, S., 1974: Modifications du gradient apicobasal de la population plastidale chez l'*Acetabularia mediterranea*. Plant Sci. Lett. **2**, 209—219.

— Dubacq, J. P., Puiseux-Dao, S., 1977: Heterogeneity of the plastid population and chloroplast differentiation in *Acetabularia mediterranea*. In: Progress in *Acetabularia* research (Woodcock, C. L. F., ed.), pp. 175—194. New York-London: Academic Press.

Kloppstech, K., Schweiger, H. G., 1975: 80 S ribosomes in *Acetabularia major*. Distribution and transportation within the cell. Protoplasma **83**, 27—40.

Koop, H.-U., Kiermayer, O., 1980: Protoplasmic streaming in the giant unicellular green alga *Acetabularia mediterranea*. I. Formation of intracellular transport systems in the course of cell differentiation. Protoplasma **102**, 147—166.

— Schmid, R., Heunert, H. H., Milthaler, B., 1978: Chloroplast migration: a new circadian rhythm in *Acetabularia*. Protoplasma **97**, 301—310.

Liddle, L., Berger, S., Schweiger, H. G., 1976: Ultrastructure during development of the nucleus of *Batophora oerstedii* (*Chlorophyta; Dasycladaceae*). J. Phycol. **12**, 261—272.

Lüttke, A., Rahmsdorf, U., Schmid, R., 1976: Heterogeneity in chloroplasts of siphonacious algae as compared with higher plant chloroplasts. Z. Naturforsch. **31 c**, 108—110.

Novak, B., Bentrup, F. W., 1972: An electrophysiological study of regeneration in *Acetabularia mediterranea*. Planta (Berl.) **108**, 227—244.

— Sironval, C., 1975: Inhibition of regeneration of *Acetabularia mediterranea* enucleated posterior stalk segments by electrical isolation. Plant Sci. Lett. **5**, 183—188.

Puiseux-Dao, S., Dazy, A. C., 1970: Plastid structure and the evolution of plastids in *Acetabularia*. In: Biology of *Acetabularia* (Brachet, J., Bonotto, S., eds.), pp. 111—122. New York-London: Academic Press.

Schulze, K. L., 1939: Cytologische Untersuchungen an *Acetabularia mediterranea* und *Acetabularia wettsteinii*. Arch. Protistenkd. **92**, 179—223.

Schweiger, H. G., Berger, S., 1975: *Acetabularia haemmerlingi*, a new species. Nova Hedwigia **26**, 33—43.

— — 1979: Nucleocytoplasmic interrelationships in *Acetabularia* and some other *Dasyclada-ceae*. Int. Rev. Cytol. Suppl. **9**, 11—44.

— Dehm, P., Berger, S., 1977: Culture conditions for *Acetabularia*. In: Progress in *Acetabularia* research (Woodcock, C. L. F., ed.), pp. 319—330. New York-London: Academic Press.

Solms-Laubach, H. Graf zu, 1895: I. Monograph of the *Acetabularieae*. Trans. Linn. Soc. London, 2nd Ser. Bot. **5**, 1—39.

Turing, A. M., 1952: The chemical basis of morphogenesis. Philos. Trans. R. Soc. Lond. **B 237**, 37—72.

Tyson, J., Kauffman, S., 1975: Control of mitosis by a continuous biochemical oscillation: synchronization, spatially inhomogeneous oscillations. J. Math. Biol. **1**, 289—310.

Valet, G., 1969: Contribution à l'étude des *Dasycladales* 2 et 3. Nova Hedwigia **17**, 551—644.

Vanden Driessche, Th., 1974: Circadian rhythm in the Hill reaction of *Acetabularia*. In: Proceedings of the third international congress on photosynthesis (Avron, M., ed.), pp. 745—751. Amsterdam: Elsevier.

— Dujardin, E., Magnusson, A., Sironval, C., 1976: *Acetabularia mediterranea*: circadian rhythms of photosynthesis and associated changes in molecular structure of the thylakoid membranes. Int. J. Chronobiol. **4**, 111—124.

Werz, G., 1970: Cytoplasmic control of cell wall formation in *Acetabularia*. Curr. Top. Microbiol. Immunol. **51**, 27—62.

Woodcock, C. L. F., Miller, G. J., 1973: Ultrastructural features of the life cycle of *Acetabularia mediterranea*. II. Events associated with the division of the primary nucleus and the formation of cysts. Protoplasma **77**, 331—341.

Cytoplasmic Basis of Morphogenesis in *Micrasterias*

O. KIERMAYER

Botanical Institute, University of Salzburg, Salzburg, Austria

With 11 Figures

Contents

I. Introduction

The unicellular green alga *Micrasterias* belongs to one of the most favorable organisms in experimental cell biology. This because of its relatively large size, its rapid cell growth and development, its symmetric cell pattern, and because it is easily cultivated under artificial culture conditions. One of the first cell biologists to use this organism for experimental studies, especially for morphogenetic experiments, was the Finnish plant physiologist Hans WARIS. In 1950 he introduced a useful culture medium (Waris-solution) that made experimental studies of growth and development possible for the first time (WARIS 1950). With elegant techniques applied to living cells KALLIO (1951, 1957, 1963) continued and extended these experiments, which were concerned with the influence of the nucleus on cytomorphogenesis in *Micrasterias*. At the same time, the Austrian physiologist K. HÖFLER and his coworkers introduced desmids and especially *Micrasterias* as very useful objects for cell-physiological purposes. In these experiments *Micrasterias* and other desmids had been used for studies of plasmolysis and permeability (HÖFLER 1951, KREBS 1951, 1952 HERRMANN 1966), vital staining (CHOLNOKY and HÖFLER 1950, LOUB 1951, HIRN 1953, KINZEL 1953, KIERMAYER 1954, 1955, KOWALLIK 1965) and centrifugation (KIERMAYER 1954).

In a series of publications, EIBL (1938, 1939, 1941) showed the influence of centrifugation on chloroplast dislocation and restitution and identified plasmatic components (plasmatic strands, "Plasmataue") as key elements in morphogenesis and migration of the chloroplast (see also KOPETZKY-RECHTPERG 1954).

Although the results of WARIS and KALLIO (1964) gave new insight into a possible nuclear control of morphogenesis in *Micrasterias,* very little was known about the cytoplasmic basis of morphogenesis and causal mechanisms of cell wall growth and internal cell architecture. DRAWERT and MIX (1961 a–f, 1962 a–e), in a series of ultrastructural studies, revealed some details on the ultrastructure of *Micrasterias* cells, but fixation procedure did not give satisfactory results at this time. After a series of experiments on osmotic factors on cytomorphogenesis in *Micrasterias* (KIERMAYER 1962, 1964, 1965 a–c, 1967 a, b), the author introduced a special fixation procedure (KIERMAYER 1968 a) with low concentration of glutaraldehyde (0.8% to 1.0%) combined with a short fixation time (maximum 10 minutes). This technique gave satisfactory results on the ultrastructure of protoplasmic components like microtubules and the golgi-system with its various vesicle products (KIERMAYER 1967 a, 1968 a, b, 1970 a, b, 1971, KIERMAYER and DOBBERSTEIN 1973, KIERMAYER and DORDEL 1976, MENGE and KIERMAYER

1977 a, b). In combination with a modified culture method based on the Waris procedure (KIERMAYER 1964, 1970 a, 1980), microcinematographic and physiological studies (KIERMAYER 1965 b, d, 1966 a, b, 1968 c, 1972, 1973 a, 1976 a, b, KIERMAYER and HEPLER 1970, KIERMAYER and FEDTKE 1977, KUNZMANN and KIERMAYER 1978, KIERMAYER and MEINDL 1979, MEINDL and KIERMAYER 1981), and a satisfactory electron microscopical technique specifically worked out for developing *Micrasterias* cells, a number of developmental and morphogenetic events could be elucidated.

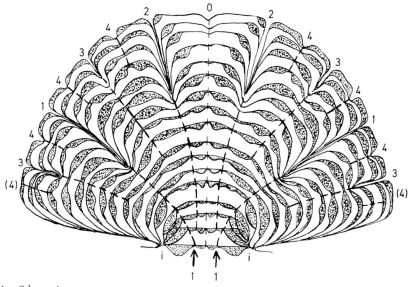

Fig. 1. Schematic representation of the developmental steps of *Micrasterias denticulata*, starting from the septum stage (i–i). The numbers indicate the lobe invaginations corresponding to their formation. The illustration further shows the different pattern of accumulation of cell wall material under turgor reduction. From KIERMAYER 1964.

In addition the works of UEDA (1972), LACALLI (1973, 1974, 1975 a, b, 1976), UEDA and NOGUCHI (1976), UEDA and YOSHIOKA (1976), NOGUCHI (1976, 1978), MAEZAWA and UEDA (1979), NOGUCHI and UEDA (1979), SANO and UEDA (1980), TIPPIT and PICKETT-HEAPS (1974), BROWER and McINTOSH (1980), BROWER and GIDDINGS (1980), and GIDDINGS *et al.* (1980) have given further details about different factors responsible for the highly complex process of cytomorphogenesis in *Micrasterias*. In the following chapter by KALLIO and LETHONEN (see p. 191 f.) special emphasis will be given to the nuclear control of morphogenesis in *Micrasterias*.

II. Cell Development and Morphogenesis

The large ornamental *Micrasterias* cell shows a deep constriction—the isthmus—that divides the cell in two symmetric semicells. Within the isthmus region, a large nucleus (30 µ) is located. Each semicell is composed of a number of lobes (Fig. 1), thus giving the *Micrasterias* cell its typical shape. As can be seen in Fig. 1, a median polar lobe and several lateral lobes (wings) can be distinguished (see also p. 192). Each semicell contains a flat axial chloroplast that extends into the periphery of the lobes. Vegetative growth and morphogenesis of *Micrasterias* is characterized by a number of morphogenetic events that take place in a precise sequence (see p. 193). In Table 1, the time course of developmental processes in *Micrasterias denticulata* is given.

Table 1. *Time course of developmental processes in Micrasterias denticulata.* From KIER-
MAYER 1980

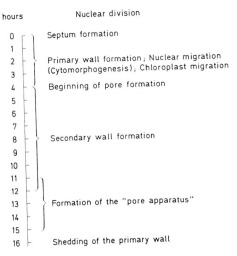

A. Septum Formation

After nuclear division (WARIS 1950), the two semicells are separated by a septum. Septum starts to form at a late anaphase stage and grows inward centripetally like a diaphragm (KIERMAYER 1967 b). Before septum formation, a girdle of wall material in the region of later septum formation is formed (LACALLI 1973). The whole process of septum formation (under 20 °C) takes no more than 15 minutes. In microcinematographic studies (KIERMAYER 1965 d, 1966 a), it was shown that a number of crystals move from lateral positions in the cell into the isthmus region. As long as the isthmus region is not separated by the septum, these crystals move freely within the isthmus. Gradually, their passage is blocked and at last completely prevented by the growing septum. Thus, observation of the moving crystals represents an ideal light microscopic means to distinguish the different stages of septum formation. As will be shown later, this is a very important fact in experimental and electron microscopic studies of the process of septum formation, which seems to be a crucial step for the establishment of a "prepattern" in cell development.

B. Primary Wall Formation

After septum formation, each semicell starts to form a new semicell by a rather rapid growth of the primary wall (WARIS 1950, KALLIO 1954, 1960, KIERMAYER 1964, 1965 c). Time-lapse microcinematography (KIERMAYER 1965 d, 1966 a, MEINDL 1981) shows that each semicell forms a little bulge (bulbus-stage), and approximately 75 minutes after the beginning of primary wall growth, the bulge develops sequentially symmetric invaginations that lead to the typical morphological habitus of the *Micrasterias* cell (Fig. 1). The whole process of primary wall formation, during which shape modelling takes place, lasts approximately 4 hours.

C. Secondary Wall and Pore Formation

After the end of primary wall growth and pattern formation, a thick, rigid secondary wall is formed beneath the primary wall layer. Under the light microscope, the formation of this cell wall layer can be recognized as an increase of contrast of the outer cell contour. Simultaneously with the appearance of the secondary wall, pores are formed (see p. 215 f.). These pores, approximately 200 nm in diameter, penetrate the secondary wall and are distributed in a special hexagonal pattern (NEUHAUS and KIERMAYER 1981). The formation of the secondary wall lasts approximately 8 hours (DOBBERSTEIN 1973). After secondary wall formation, the primary wall is shed ("Häutung", DRAWERT and MIX 1962 e) and disappears (MEINDL 1981).

D. Nuclear Migration and Anchoring

During the growth of the "bulges", the posttelophase nucleus migrates into the growing half-cell and later back into the isthmus region where it is anchored for the whole period of interphase. The postmitotic movements of the nucleus, *e.g.*, the inward movement into the growing half-cell and the backward movement into the isthmus region, had been termed "nuclear migration" (KIERMAYER 1968 c). As will be shown later, nuclear migration represents an interesting microtubule-controlled phenomenon that can be experimentally influenced.

E. Shape Formation of the Chloroplast

During the growth of the bulges, the chloroplast together with the nucleus moves into the growing half-cell (chloroplast migration). At the beginning of this movement (chloroplast immigration), the chloroplast appears as a shapeless mass of material. Soon, however, tips or "tongues" of chloroplast material are formed ending in thin cytoplasmic strands that probably anchor the chloroplast onto special areas of the cortical plasma (chloroplast-anchoring). During shape formation of the chloroplast, cytoplasmic strands apparently pull the chloroplast toward the periphery of the cell (chloroplast spreading). At this stage, the chloroplast of the old and the young semicell is still connected. In a last process of chloroplast development, it is separated in the isthmus region (chloroplast-separation) so that each semicell now contains one chloroplast. The whole postmitotic development of the chloro-

O. KIERMAYER

plast therefore takes place in four different developmental steps, namely (1) chloroplast immigration, (2) chloroplast anchoring, (3) chloroplast spreading, and (4) chloroplast separation (MEINDL 1980, MEINDL and KIERMAYER 1981 c).

F. Osmotic Values

KREBS (1951, 1952) showed that in various species of desmids, the osmotic value of the cell decreases during cell growth. In growing cells of *Micrasterias rotata*, the osmotic value decreases an equivalent of about 0.05 mol glucose under the value of resting cells (KIERMAYER and JAROSCH 1962). During interphase, the cells restore their higher osmotic value. In detailed studies

Table 2. *Osmotic values of developing cells of* Micrasterias thomasiana *and* Micrasterias denticulata

Object	Hours after septum formation	Developmental stages				
		1	2	3	4	5
	Interphase	A B C D	E F G H	I J K L	M N O P	Q R S T
Micrasterias thomasiana *	0.26	0.24 0.22 0.20	0.18	0.16	0.14	
Micrasterias denticulata **	0.22	0.20 0.18	0.16		0.14	0.12

A, B, C . . .: developmental stages in 15 min. sequence (see Fig. 1 and Table 1).

  * Osmoticum: mol concentration of glucose. From HACKSTEIN-ANDERS 1974.

  ** Osmoticum: mol concentration of polyethylenglycol 400 (Lutrol). From KIERMAYER 1964.

on the osmotic values during cell development KIERMAYER (1964) and HACKSTEIN-ANDERS (1974) found that the developmental stages clearly differ in their osmotic values (Table 2). This fact must be seriously taken into consideration in all experimental investigations with growing cells of *Micrasterias*. As will be shown on p. 174, experimental changes of the turgor pressure that depend on the actual osmotic values lead to severe disturbances of morphogenesis (turgomorphosis).

III. Ultrastructural Morphology

A. Septum

As already mentioned, septum growth starts at late anaphase. Already at prophase, a girdle of cell wall material can be observed in the isthmus region (LACALLI 1973). DRAWERT and MIX (1962 b) have shown in osmium tetroxid fixed cells that the septum represents a 3-layered structure. These authors have also demonstrated that an accumulation of vesicles can be found in the vicinity of the growing septum. These vesicles are probably functional in septum formation. In more recent studies on septum formation

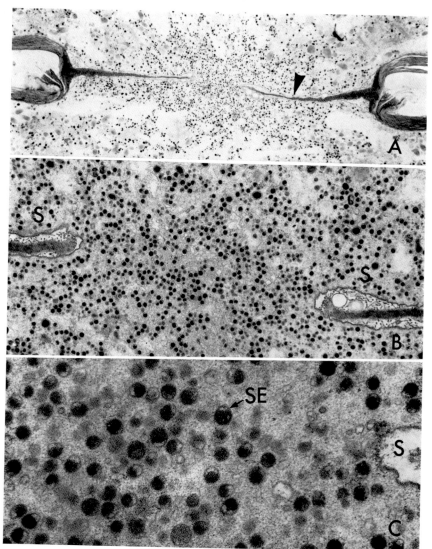

Fig. 2. Ultrastructural aspect of septum formation in *Micrasterias denticulata*. *A* Cross-section of the septum (arrow) that is still not closed. Especially in the plasmatic area between the open septum many septum vesicles (*SE*) are accumulated that are probably involved in septum formation (×1,200). *B* Higher magnification of the area between the open septum (*S*). Septum vesicles with an asymmetric appearance and a diameter of 100 to 160 nm can be recognized (×5,000; *S* ingrowing septum). *C* Higher magnification of septum vesicles (*SE*). *S* ingrowing septum (×20,000).

with glutaraldehyde fixed cells of *Micrasterias denticulata* (KIERMAYER, un-published), these results found by DRAWERT and MIX (1962 b) could be supported. As shown in Figs. 2 *A–C*, a great number of little vesicles are to be found in the vicinity of the growing septum. Especially the open "hole" of the septum is filled with these "septum vesicles" (SEV), which are 100 to 160 nm in diameter. Fig. 2 *C* shows these vesicles at higher magnification; they are characterized by an asymmetric vesicle content. It seems that SE-vesicles are produced by dictyosomes. Although fusion stages of SEV with the plasma membrane surrounding the growing septum could not be found, it seems probable that these vesicles contribute material for both the growing septum wall and the plasma membrane.

B. Primary Wall and the Cytoplasmic Organization
of the Growing Semicell

Primary cell wall growth starts after septum formation, *e.g.*, when each semicells begins to form a bulge. An ultrastructural aspect of a growing bulge of *Micrasterias denticulata* is given in Fig. 3 *A*. The primary wall (*PW*) that surrounds the bulge is composed of a network of thin microfibrils (Fig. 10 *B*, KIERMAYER 1970 a, b). The surface of the primary wall seems to be fringed. As can be seen on Fig. 3 *A,* the primary wall originates in the isthmus region between the (old) secondary wall of the old half-cell (*SW*) and the plasma membrane. In a recent freeze-etch study, GIDDINGS *et al.* (1980) showed that the primary wall consists of randomly oriented microfibrils with a diameter of 6 to 8 nm. These authors could also observe that on the plasma membrane of a growing half-cell, during synthesis of primary wall material, numerous single "rosettes" consisting of six particles can be found. These rosettes of membrane particles might be involved in synthesis of the micro-fibrils (KIERMAYER and SLEYTR 1979, see also p. 160 f.). There are about twice as many membrane particles near the tips of growing lobes as there are in proximal regions of the lobes. Additionally, rosettes consisting of 6 membrane particles are seen predominantly in the distal parts of the lobes (BROWER and GIDDINGS 1980).

During growth of the primary wall, a special kind of vesicle with a dark content (dark vesicles, DV, Fig. 3 *B*) and a diameter of approximately 200 nm appears. These vesicles fuse with the plasma membrane (KIERMAYER 1970 a, b) and seem to deliver wall material (pectins) (UEDA and YOSHIOKA 1976) for the growing primary wall. MENGE (1973, 1976) in her ultracytochemical

Fig. 3. Cross-section of *Micrasterias denticulata* through a young growing half-cell. Within the protoplasm, organelles and different types of vesicles can be seen; however, no cor-relation between the distribution of these components to the later cell pattern can be observed. *SW* secondary wall of the old half-cell; *PW* primary wall; *N* nucleus; *No* nucle-olar material; *D* dictyosomes (×5,400). *B* Dictyosome of the growing half-cell of *Micrasterias denticulata* and different cytoplasmic vesicles—*LV*, large vesicles containing slime material; *DV* dark vesicles containing primary wall material; and *VX* transitorial vesicles carrying spiny (coating) structures (arrows): ER profiles always closely associated with the dictyo-some. The dictyosome shows a simultaneous production of both large vesicles (*LV*) and dark vesicles (*DV*) (×64,000). From KIERMAYER 1970 a.

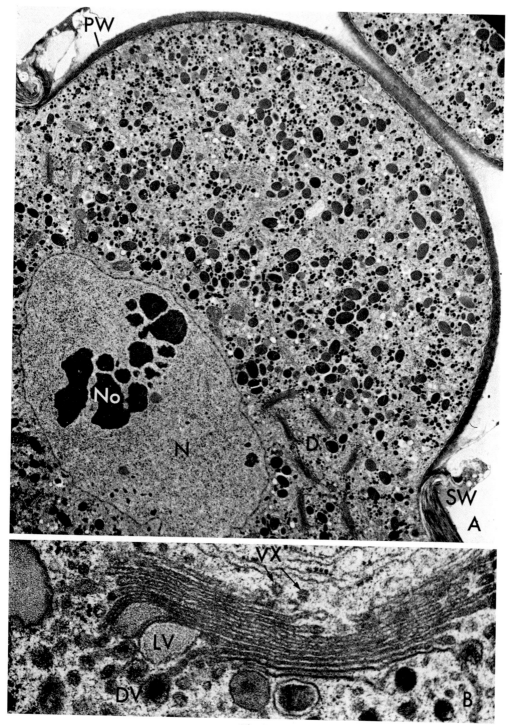

Fig. 3

Table 3. *Survey of different types of cytoplasmic vesicles in* Micrasterias denticulata. *From* MENGE 1973, modified

Vesicle type	Diameter (μm)	Form	Membrane type	Content	Localization	Formation	Function	References
LV	0.3–1.0	round-oval	UM (symmetric)	granulous-fibrillar, low contrast slime	all stages, randomly distributed	dictyosome	slime-transport	KIERMAYER 1968 a, 1970 a, UEDA and NOGUCHI 1976
LV_F	0.3–0.4	round-oval	UM (symmetric)	coil-like fibrils, low contrast slime (?)	all stages, randomly distributed	?	slime-transport (?)	KIERMAYER 1970 a
$PV = LV_h$	0.3	round-oval	thin UM (symmetric)	homogenous, low contrast	cytoplasma during primary wall (randomly) and secondary wall (cortical) formation	?	pore formation	DOBBERSTEIN 1973, MENGE 1976
DV_{as}	0.15–0.25	round-oval	UM (asymmetric)	homogenous, high contrast	all stages, randomly distributed	dictyosome	primary wall growth (?) catalase-transport	KIERMAYER 1970 a, MENGE 1976, UEDA and NOGUCHI 1976
$SEV = DV_K$	0.1–0.16	round	UM (symmetric) with "coat"	high contrast, asymmetric contents ("sphere-cap")	septum formation	dictyosome	transport of material for septum and primary wall growth (?)	DRAWERT and MIX 1962 b, MENGE 1973, LACALLI 1973, KIERMAYER, unpublished
DV_g	0.1–0.3	round-oval	UM (symmetric)	asymmetric, coil-like content, high contrast	growing cells, randomly distributed	fusion products from SEV?	primary wall growth	KIERMAYER 1970 a, MENGE 1973

CV	0.045–0.055	round	UM (asymmetric with "coat")	low contrast	peripheral cytoplasm	from DV_{as} and dictyosome	transport of enzymes (?)	KIERMAYER 1970, BURGSTALLER-GETZINGER and KIERMAYER 1981
SV	0.1–0.2	unregularly	membrane with many thin layers	low contrast, rod-like structure with high contrast	cytoplasm, accumulated in vicinity of the nucleus	nuclear membrane (?)	nucleo-cytoplasmic exchange (?)	KIERMAYER 1970 a, 1971
VX	0.05–0.06	round	UM with spiny (rough) coat	low contrast	transitorial (shuttle) vesicles between ER and dictyosome	ER	material transport for proximal face of the dictyosome cisterna	KIERMAYER 1970 a
FV	0.4 thickness: 0.04–0.06	flat with a sack-like appendage	sack-like appendage with UM; thick membrane (16–20 nm) with particles in rows (15–20 nm)	low contrast	cortical protoplasm, specifically during secondary wall formation	dictyosome	synthesis and orientation of secondary wall microfibrils	DOBBERSTEIN and KIERMAYER 1972, KIERMAYER and DOBBERSTEIN 1973, UEDA and NOGUCHI 1976, PIHAKASKI and KALLIO 1978

LV large vesicles; LV_F large vesicle with coil-like fibrils; PV pore vesicles (= LV_h large vesicle with homogenous content); DV_{as} dark vesicles with asymmetric membrane; SEV septum vesicles (= DV_K dark vesicles with "cap"); DV_g dark vesicles with coil-like contents; CV coated vesicles; SV "rod-containing" vesicles; VX transitorial (shuttle) vesicles; FV flat vesicles; UM unit membrane.

work demonstrated that the content of D-vesicles gives a positive reaction for acid polysaccharides. With freeze-etching Giddings et al. (1980) also observed vesicles, probably D-vesicles, in the cytoplasm of the semicell engaged in primary wall formation. On the membrane of these vesicles, they observed single rosettes that had also been found in the plasma membrane of cells undergoing primary wall synthesis. Therefore it can be assumed that D-vesicles by fusing with the plasma membrane deliver not only wall material (pectins) but also membrane areas carrying the typical membrane rosettes for microfibril production.

The cytoplasmic organization within the growing half-cell reveals a great diversity of ultrastructural elements. Fig. 3 A shows two types of vesicles that are especially prominent: (1) large electron lucent vesicles (LV, large vesicles); and (2) smaller electron dense vesicles (DV, dark vesicles, see also Table 3). In the cytoplasm, these vesicles are randomly distributed, and no relation of their distribution to cytomorphogenesis can be observed. Between the DV and LV, cisternae of the rough endoplasmic reticulum are found. Close to the large nucleus (N), many dictyosomes (D), mitochondria, and microbodies occur. In young growing semicells, therefore, a cytoplasmic zonation exists, e.g., an internal zone around the nucleus with dictyosomes, ER, mitochondria, and microbodies and a peripheral zone consisting of vesicles and ER profiles but without larger organelles (Kiermayer 1970 a). Light-microscopic and microcinematographic studies show that the zone surrounding the nucleus exhibits no or a weak protoplasmic streaming, while the cortical protoplasm with the vesicles is actively streaming (Kiermayer 1970 a).

C. Secondary Wall

Approximately 4 hours after the beginning of primary wall growth and modelling of cell shape (Table 1), an additional cell wall layer is formed—the secondary wall. The structure and formation of this wall differs from that of the flexible primary wall. Mix (1966, 1968, 1969, 1973) in her studies on cell wall structure of desmids showed that in *Micrasterias* and other desmids, the secondary wall is composed of parallel oriented microfibrils that form layers of crossed bands. In freeze-etch studies, Kiermayer and Staehelin (1972) found that the outer surface of the secondary wall is covered with a continuous, thin, fibril-free layer with cleaving properties resembling those of multiple lipid bilayers. Two to 15 microfibrils aggregate laterally to form the characteristic bands of the secondary wall (Fig. 4 A). Each microfibril (20 to 30 nm in diameter) is composed of "elementary fibrils" (4 to 5 nm in diameter; Fig. 4 A).

1. "Flat Vesicles"

Dobberstein and Kiermayer (1972) and Kiermayer and Dobberstein (1973) in their ultrastructural studies on *Micrasterias denticulata* found that during secondary wall formation, a particular type of disc-like vesicles in the cytoplasm of glutaraldehyde fixed cells appears that is produced by the dictyosomes. These vesicles, which had been termed "flat vesicles"

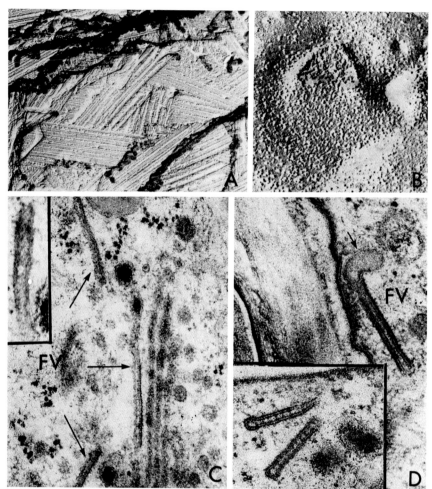

Fig. 4. Secondary wall formation in *Micrasterias denticulata*. *A* The secondary wall is composed of bands of parallel-oriented microfibrils. This picture shows a freeze-etched fracture through the secondary wall with its layers of microfibril bands. Subunits of the microfibrils (elementary fibrils) are clearly visible ($\times$45,000). From KIERMAYER and SLEYTR 1979). *B* Fracture through a freeze-etched cell at the stage of secondary wall formation. A field of small "rosettes" composed of particles and arranged in a hexagonal pattern can be recognized ($\times$97,000). From KIERMAYER and SLEYTR 1979. *C* Dictyosome of a cell at the stage of secondary wall formation. At the distal pole of the dictyosome, flat vesicles (*FV*) are produced (arrows). ($\times$60,000). From DOBBERSTEIN and KIERMAYER 1972). Inled of Fig. 4 C; tangential section through a F-vesicle shows a helical striation ($\times$70,000). From MEINDL, unpublished. *D* A F-vesicle (*FV*) in the vicinity of the plasma membrane and the secondary wall. The F-vesicle is composed of an area with a thick membrane (160 to 200 Å) and a sack-like appendix with a normal unit membrane (arrow) ($\times$100,000). From KIERMAYER and DOBBERSTEIN 1973. Inled of Fig. 4 D: F-vesicles in which globular subunits of the membrane in rows with a spacing of approx. 25 or 50 nm can be observed ($\times$80,000). From KIERMAYER 1977.

(FV; Table 3), carry sack-like structures at their edges (Fig. 4 D) and are incorporated into the plasma membrane during the period of secondary wall formation. The flat areas of the vesicles are characterized by an unusually thick membrane (16 to 20 nm; Fig. 4 D) that contains special globular particles of about 20 nm on the inner membrane surface (Fig. 4 D inled). On tangential sections through F-vesicles, a helical striation, possibly a microfilament, can be observed (Fig. 4 C inled; MEINDL, unpubl.). By fusion of the FV-membrane with the plasma membrane, the globular particles reach the outside of the plasma membrane. KIERMAYER and DOBBERSTEIN (1973) emphasized that the specifically differentiated membranes of these vesicles are functional as templates for the formation of secondary wall microfibrils.

Fig. 4 C reveals that F-vesicles are formed on the distal pole of the dictyosomes. From there, these vesicles, which develop globular subunits oriented in rows on the vesicle membrane after being shed from the dictyosome (membrane differentiation), move to the cell periphery (template transfer) and become incorporated. The new membrane areas, carrying the membrane particles, seem to be involved in the production of the patterned arrays of highly ordered microfibrils (template realization). The globuli on the plasma membrane are thought to be the microfibril synthesizing apparatus (Bildungs-apparat; KIERMAYER and DOBBERSTEIN 1973).

It should be mentioned that formation of a structurally similar secondary wall in desmids is apparently built in another way: MIX and MANSHARD (1977) in their investigation on the fine structure of a strain of *Penium spirostriolatum var. amplificatum* observed cytoplasmic vesicles that contain an amorphous matrix with a bundle of eight, ten, or 12 fibrils. Longitudinal sections through such a vesicle revealed a parallel orientation of the fibrils and their linear expansion throughout the whole vesicle. It seems probable that these fibrils, which are formed within a vesicle, are used for the bands of microfibrils of the secondary wall of this alga.

2. "Rosettes" of Particles Within the Plasma Membrane

When *Micrasterias denticulata* cells are freeze-etched at a stage of secondary wall formation, the plasma membrane shows typical particles. In special areas, a complex of ordered rosettes of small particles can be observed (SLEYTR and KIERMAYER 1979, KIERMAYER and SLEYTR 1979, GIDDINGS *et al.* 1980). These arrays of densely packed particles (rosettes) occur on the plasma membrane in the form of oval fields. Each rosette is apparently composed of six particles. The rosettes are arranged in a hexagonally ordered lattice with a centre-to-centre spacing of approximately 25 nm (Fig. 4 B). The same periodicities can be found in the bands between the parallel oriented microfibrils of the secondary wall (Fig. 4 A). From these observations, it has been concluded that the fields of hexagonally arranged rosettes are areas of incorporated "flat vesicles". The rosettes correspond to the globuli; the pore in the globulus (KIERMAYER and DOBBERSTEIN 1973) seems to be the centre of a rosette. Since the same periodicities can be found in both the lattice of rosettes and the bands of microfibrils, the rosettes seem to be structures responsible for the formation and orientation of microfibrils (KIERMAYER and SLEYTR 1979, GIDDINGS *et al.* 1980).

Similar particles of the plasma membrane have also been found in other plant cells (MOOR and MÜHLETHALER 1963, PRESTON 1964, BARNETT and PRESTON 1970, ROBINSON and PRESTON 1971, ROBENEK and PEVELING 1977, MUELLER and BROWN 1980) and in a sponge (GARRONE *et al.* 1980) so that these structures seem to represent a general component for the formation of microfibrils or the extracellular matrix.

Hypothetically, GIDDINGS *et al.* (1980) assume that each rosette produces one 5 nm elementary fibril. By the action of a row of rosettes, several 5 nm fibrils aggregate laterally to form a 35 nm microfibril. The thickened area of the membrane of F-vesicles, or of incorporated areas of F-vesicles within the plasma membrane (KIERMAYER and DOBBERSTEIN 1973), may serve as a "membrane-associated layer" to hold the rosettes together in a hexagonal array (GIDDINGS *et al.* 1980).

D. Golgi Complex

1. Structure of the Dictyosome

The Golgi complex in *Micrasterias* is a dominant organelle with large and clearly developed dictyosomes. Already PALLA (1894) described structures (Karyoide) in *conjugatae* that seem to be identical with dictyosomes. In a series of light and electron microscopic studies, DRAWERT and MIX (1961 e, 1962 b–d, 1973) and KIERMAYER (1967 a, 1970 a, 1977) gave a detailed description of dictyosomes in *Micrasterias*.

In a growing half-cell, many dictyosomes (D) are to be found in the vicinity of the nucleus (N, Fig. 3 *A*). The single dictyosome in *Micrasterias denticulata* consists of a rather constant number of cisternae, namely 11. The single stack of cisternae exhibits a pronounced polarized configuration showing wide cisternae at the proximal side and narrower cisternae at the distal side (Fig. 3 *B*). The dictyosomes can be found in close spatial relationship to the endoplasmic reticulum (ER). From the ER, small vesicles with a rough spiny surface are pinched off (Fig. 3 *B*, VX), which probably fuse and give rise to the proximal cisterna (KIERMAYER 1970 a). All cisternae actively pinch off vesicles of different kind at their margins. In Fig. 3 *B* it can be seen that one dictyosome simultaneously produces at least two types of vesicles during the stage of primary wall formation: (1) D-vesicles containing primary wall material (Table 3); and (2) L-vesicles containing slime material (Table 3).

Although the number of cisternae in a dictyosome is quite constant (11), the diameter of the single dictyosome varies greatly. Dictyosomes, therefore, can undergo growth in diameter. After a process of diameter increase, the dictyosomes "divide" by a central splitting in each cisterna (KIERMAYER 1967 a). KIERMAYER (1970 b), MENGE (1973), and MENGE and KIERMAYER (1977 a, b) demonstrated that the total diameter of a dictyosome changes in accordance with the stage of development. Growth occurs during interphase followed by division during mitosis (MENGE and KIERMAYER 1977 a, b).

According to their localization in the stack, the cisternae are, to different degrees, organized in central plates, fenestrated areas, and vesicles. The outermost distal cisternae are completely fenestrated and free of vesicles. A three-dimensional model of the dictyosome is given by MENGE and KIERMAYER (1977 b).

Freeze-etch studies by STAEHELIN and KIERMAYER (1970) showed that the dictyosomes of *Micrasterias denticulata* exhibit a gradual increase in the density of particles on the membranes of successive cisternae from the forming to the maturing face. At the maturing face, a drop in particle numbers can be observed in the peripheral regions of the cisternae where fenestration and the packaging of the secretory products occurs.

Investigation on *Micrasterias americana* by UEDA and NOGUCHI (1976) showed that in resting cells, dictyosomes consisted of 11 cisternae. During the development of the new semicell, they produced "dark vesicles" and "large vesicles"; the diameter and number of the cisternae decreased. Dictyosomes with six or seven cisternae of small diameter were found at the end of primary wall growth. At this stage they produced F-vesicles (p. 158), but thereafter they recovered in size and number of cisternae. These authors speculated that lysosomes divide the dictyosomes, a result that has been doubted in a recent work by BURGSTALLER-GETZINGER and KIERMAYER (1981).

2. Transformation

Depending on the special function of the dictyosomes in *Micrasterias,* they are transformed during the different developmental phases. During septum formation, the dictyosomes produce SE-vesicles (see p. 152 f.). When growth of the primary wall begins, the D-vesicles (together with L-vesicles) are prominent. At the end of primary wall growth, the F-vesicles, which are functional in secondary wall formation, are produced. During interphase the dictyosomes increase in diameter and produce hypertrophied L-vesicles (KIERMAYER 1967 a, 1970 a, MENGE 1973, UEDA and NOGUCHI 1976, MENGE and KIERMAYER 1977 a, b, NOGUCHI 1978, Fig. 5).

3. Information Transfer via Dictyosomes

The observations showing that F-vesicle membranes contain elemental particles involved in the formation and orientation of secondary wall microfibrils indicate that biomembranes may carry morphogenetic information (KIERMAYER 1970 b, 1977). In Fig. 6, the possible information-transfer for secondary wall formation via the Golgi system is shown. There are two mobile vesicle phases within this transfer system: (1) the transitorial vesicles (VX) and (2) the flat vesicles (FV). Any change or dislocation of either type of vesicles must lead to a severe disturbance of the information chain (Fig. 6).

E. Cell Wall Pores

It has been known for a long time that *Micrasterias* cells possess pores, penetrating the secondary wall, which are functional in slime excretion and phototactic movements (KLEBS 1885, KOL 1927, URL and KUSEL-

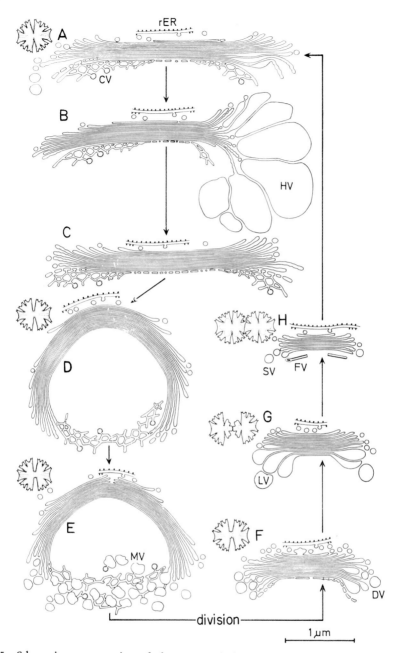

Fig. 5. Schematic representation of the structural changes (transformation) of dictyosomes during the cell cycle of *Micrasterias crux-melitensis*. In resting cells, dictyosomes consist of 11 cisternae (*A–C*). The distal cisterna is a network to which coated vesicles (*CV*) are attached. Just prior to the nuclear division, dictyosomes become elliptical and curve along their long axes (*D*). The dictyosomes then produce middle-sized vesicles (*MV*) from their distal networks and, finally, divide into two from the proximal to the distal face along their short axes (*E*). In metaphase cells of nuclear division, dictyosomes still consist of 11 cisternae and produce dark vesicles (*DV*) from their distal cisternae (*F*). In actively growing cells after cell division, dictyosomes consist of nine or eight cisternae and produce large vesicles (*LV*) from their distal cisternae (*G*). In cells just after full growth, dictyosomes reach their smallest size and produce flat vesicles (*FV*) and small vesicles (*SV*) from their distal cisternae (*H*). In cells cultured for 48 hours after full growth, dictyosomes fully recover their size (*A*) and produce hypertrophied vesicles (*HV*) from their distal cisternae (B). The HV-production lasts for about 10 hours and then ceases (C). From NOGUCHI 1978.

FETZMANN 1973, HADER and WENDEROTH 1977). In her electron microscopic studies, DOBBERSTEIN (1973) demonstrated that these pores are produced by a special kind of vesicle (pore vesicles, PV, Table 3) that form plugs during the formation of the secondary wall layer. The pattered arrangement of pores seems to be controlled by a membrane recognition mechanism operating

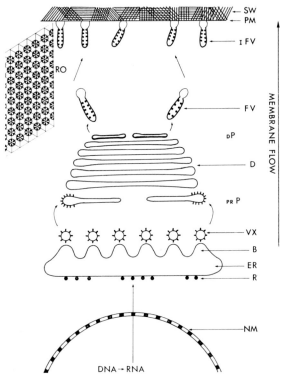

Fig. 6. Schematical representation of the information-transfer for secondary wall formation in *Micrasterias* via the Golgi system. *NM* nuclear membrane; *R* ribosomes; *ER* endoplasmic reticulum; *B* blebs of the ER; *VX* transitorial vesicles with a spiny coat; *prP* proximal pole of the dictyosome with incorporated transitorial vesicles; *D* dictyosome; *dP* distal pole with premature F-vesicles; *FV* typical flat vesicle with a thickened membrane carrying rows of globular subunits and a sack-like appendix; *iFV* flat vesicles incorporated into the plasma membrane (*PM*); *SW* secondary wall with its bands of parallel oriented microfibrils; *RO* schematic drawing of the hexagonally ordered "rosettes" of particles in the plasma membrane, which are formed within the flat vesicles and are transferred and incorporated into the plasma membrane. Modified after KIERMAYER and DOBBERSTEIN 1973, KIERMAYER 1977, KIERMAYER and SLEYTR 1979. (See also Fig. 13, p. 41.)

between the PV-membrane and the plasma membrane during the phase of secondary wall formation (PICKETT-HEAPS 1972, KIERMAYER 1977, NEUHAUS and KIERMAYER 1981). For further detail, see p. 215 f.

Table 4. *Results of cytochemical reactions on polysaccharides in Micrasterias denticulata.* $LV_{diet.}$ = LVs still on the dictyosome; $DV_{diet.}$ = DV still on the dictyosome (other vesicles see Table 3); + weak reaction; ++ medium reaction; +++ strong reaction; n.i. not to identify; * with few exceptions. From MENGE 1976

Reaction	Type of vesicles											
	$LV_{diet.}$	LV_{free}	$LV_h(PV)$	LV_F	$DV_{diet.}$	DV_g	$DV_K(SEV)$	DV_{as}	SV	CV	FV	VX
PAS	n.i.	++	n.i.	n.i.	+	(++)	(++)	(−)?	n.i.	n.i.	−	−
Thororast	+	+++	−	−	−?	++	−	−*	−	−	−	−
Ruthenium brown	+	+	−	+	++	+++	+++	−*	−	−	−	−

F. Various Cytoplasmic Vesicles and Their Ultracytochemical Analysis

In *Micrasterias*, a considerable number of different cytoplasmic vesicles that are functional in diverse morphological and physiological processes could be detected. Table 3 shows a survey of the vesicles so far known in the cytoplasm of *Micrasterias* cells.

Table 4 shows cytochemical results of reactions on polysaccharides in various vesicles of *Micrasterias denticulata* (MENGE 1973, 1976, see also TUTUMI and UEDA 1975). Phosphatase activity and osmium reduction in cell organelles of *Micrasterias americana* had been studied by NOGUCHI (1976).

From the cytochemical and topographical electron microscopic studies, it can be concluded that the vesicle population DVg (Table 4) and DVk (Table 4) are candidates for primary wall formation, while LV (Table 4) seem to be involved in slime excretion.

G. Microtubule Systems

In differentiating cells of *Micrasterias denticulata*, four distinct systems of cytoplasmic microtubules have been found (KIERMAYER 1968 a). In Fig. 7, a schematic drawing of these systems is given.

Isthmus Band of Microtubules

As can be seen in Fig. 8 A, numerous microtubules are clustered in the region of the isthmus. Similar bands of microtubules in the isthmus region were also found in other desmids (*Closterium*, PICKETT-HEAPS and FOWKE 1970, *Cosmarium*, PICKETT-HEAPS 1972, *Micrasterias rotata*, LACALLI 1973; *Euastrum*, NEUHAUS, unpublished). The number of microtubules decreases as the distance from the isthmus increases, and they gradually disappear. A band of microtubules has been found in posttelophase cells where the nuclei have moved out of the isthmus. Thus, it is evident that the band of microtubules remains in its position even though the location of the nucleus changes. In the isthmus region of young half-cells, very few or no microtubules can be found. In resting cells, the region of the isthmus is always occupied by many microtubules. The function of this band is not yet clear. It might be functional in determining the location of septum formation, anchoring of the nucleus and in chloroplast separation (MEINDL 1980).

Microtubules near the Plasmalemma of the Old Half-Cell

This system is characterized by links between the individual microtubules and the plasmalemma (Fig. 8 C). The links may endow this complex with some degree of stability. The function of this system, which is often found in the vicinity of pores of the secondary wall (KIERMAYER 1968 a), is unknown.

Posttelophase Complex of Microtubules and Microfilaments

This transient microtubule system (Figs. 8 B, D, and E), consisting of a massive bundle of microtubules with weaker "stainability" than the other systems, surrounds the posttelophase nucleus. When the cell is sectioned for an edge view, long microtubules can be seen emerging from the isthmus region and running close to the nucleus. Within the array of microtubules, a central complex is prominent where rod-like structures, possibly microfilaments, can be identified. These rod-like structures are often paired with the microtubules and oriented parallel to them (Figs. 8 *B* and *E*, KIERMAYER 1968 a, NEUHAUS-URL 1980, MEINDL, unpublished).

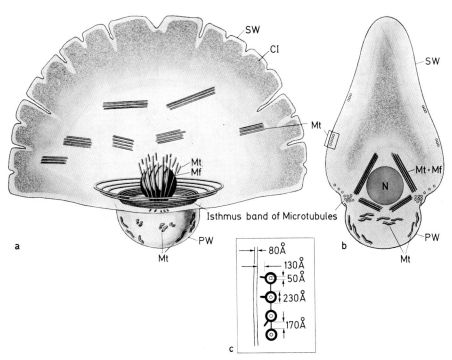

Fig. 7. Drawing of the different microtubule systems in *Micrasterias denticulata*. *a* Median plane view showing the "isthmus band of microtubules", microtubules near the plasmalemma of the old half-cell, the posttelophase microtubule-microfilament complex, and microtubules in the growing half-cell. *b* Edge view. *c* Schematic drawing, at higher magnification, of the microtubule system in the cortical protoplasm of the old half-cell (demonstrated in *a*, *b*) showing both cross-bridges between the microtubules and extensions toward the plasma membrane. *SW* secondary wall; *CL* chloroplast; *Mt* microtubules; *Mf* microfilaments; *PW* primary wall of the new growing half-cell. From KIERMAYER 1968 a, modified.

This system that surrounds the posttelophase nucleus like a "cage" and cannot be found in resting cells seems to participate in postmitotic nuclear migration (KIERMAYER 1968 b, see p. 151).

However, the way in which this highly interesting microtubule-microfilament-system works, and how the two filamentous elements cooperate so that motion of the nucleus and perhaps the chloroplast is evoked, is not clear. (Compare a similar problem in *Acetabularia*) (KOOP and KIERMAYER 1980 a, b).

Microtubules in the Central and Cortical Protoplasm of Growing Half-Cells

In the central protoplasm of young growing half-cells, microtubules are found oriented mostly at right angle to the median plane of the cell. In older stages, however, microtubules are oriented mostly parallel to the median plane. Very few microtubules occur in close proximity to the plasmalemma; furthermore, no special arrangement of them in relation to shape formation has been detected. Since anti-microtubule drugs in general do not influence cytomorphogenesis in *Micrasterias* (KIERMAYER 1968 c, 1970 b, 1972, 1973 a, KIERMAYER and FEDTKE 1977), it seems that microtubules are not functional in shape formation in *Micrasterias* (KIERMAYER 1970 b, TIPPIT and PICKETT-HEAPS 1974).

It should be mentioned that in another desmid, *Closterium littorale*, PICKETT-HEAPS and FOWKE (1970) described a microtubule organizing center (MTOC) that aggregates at telophase and later migrates along the old semicell with microtubules extending from it back toward the nucleus. The nucleus follows the MTOC into a cleavage in the chloroplast. This migration may be mediated by the microtubules. In *Micrasterias* (KIERMAYER 1968 a) and *Cosmarium* (PICKETT-HEAPS 1972), the existence of a similar MTOC is not obvious. As the latter author demonstrated, the posttelophase nucleus of *Cosmarium botrytis* moves deep into the new semicell and then migrates back into the isthmus. No microtubules could be found near the nucleus during this return movement in *Cosmarium*.

IV. Pattern Formation in Turgor-Reduced Cells

From studies on various plant cells, it is known that elongation growth depends on a critical turgor pressure (see p. 234). Decreasing the turgor pressure below a critical value causes cessation of elongation growth (REINHARDT 1899, CLELAND 1958, FREY-WYSSLING 1959, RAY 1961, 1962, SCHRÖTER and SIEVERS 1971).

A. The "Septum Initial Pattern"

When *Micrasterias* cells are placed into an osmotic solution immediately after septum formation, elongation growth is blocked. Instead of elongation growth, wall thickenings are formed showing a distinct pattern. As can be seen on Figs. 9 *A* and *B*, this pattern of deposited primary wall material,

---

Fig. 8. Microtubule systems in *Micrasterias denticulata*, partly associated with microfilaments and cross-bridges. *A* Longitudinal section through the isthmus region of a non growing cell showing the "isthmus band of microtubules"; *SW* secondary wall ($\times$66,000). *B* Longitudinal section through a young growing half-cell showing microtubules near the posttelophase nucleus. Between the microtubules intertubular structures, probably microfilaments can be seen ($\times$50,000). *C* Cortical microtubules in a non growing half-cell; rod-like cross-bridges between the microtubules and between the microtubules and the plasma membrane can be observed ($\times$152,000). From KIERMAYER 1968 a, 1970 a. *D, E* Microtubules near the posttelophase nucleus (*N*). Arrow points toward a possible intermicrotubular microfilament ($\times$40,000). From MEINDL, unpublished.

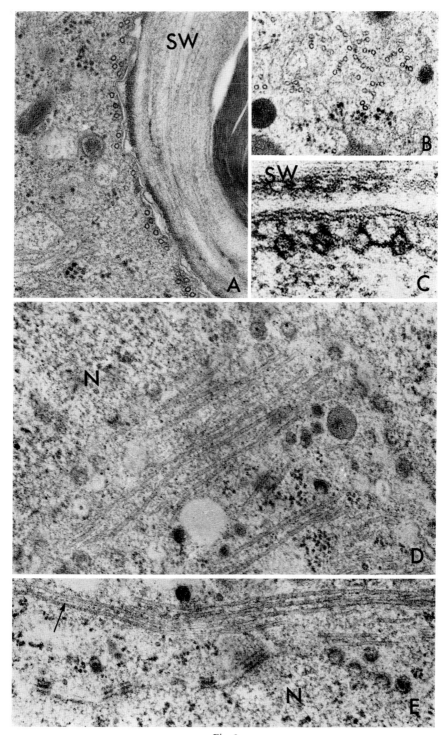

Fig. 8

which has been called "septum initial pattern" (KIERMAYER 1964, 1967 b, 1970 b), is characterized by two lateral minimum zones (1) and one central minimum zone (0). The schematic drawing in Fig. 9 A shows that the lateral zones (1) correspond to the first lobe invaginations, whereas the central zone (0) corresponds to the invagination of the polar lobe formed late in the developmental sequence. This experiment clearly shows that the basic symmetry of the *Micrasterias* cell (a three-lobed structure) is already determined at this very early developmental stage.

If *Micrasterias* cells in mitosis (prophase, grape-stage, WARIS 1950) are placed into 0.20 to 0.22 mol glucose, nuclear division shows no changes from turgor reduction, but septum growth exhibits (1) a decrease in growth velocity (from normally 15 to 20 minutes to 210 minutes in 0.22 mol glucose) and (2) an increased thickening with an accumulation of wall material in a pattern as described in the previous paragraph. This demonstrates that already during septum formation, a patterned deposition of primary wall material on the partly developed septum takes place. From these findings it was concluded that the basic developmental pattern for the *Micrasterias* cell is formed sequentially during the growth of the septum (KIERMAYER 1967 b). The shape of the isthmus may exert some morphogenetic influences on septum growth (LACALLI 1976).

As Figs. 9 C and D demonstrates, wall thickening formed in an osmotic solution suddenly changes its pattern within seconds. This phenomenon has been explained by a sudden rupture of protoplasmic strands within the accumulated wall material (KIERMAYER 1965 a). It further shows that the accumulated wall thickening consists of a rather liquid fusible material (see p. 174, "turgomorphosis").

B. The "Accumulation Pattern" of the Primary Wall

If growing *Micrasterias* cells of different developmental stages are placed into osmotic solution, each stage produces a specific pattern of wall thickening (KIERMAYER and JAROSCH 1962, JAROSCH and KIERMAYER 1962, KIERMAYER 1964, 1965 a, c, TIPPIT and PICKETT-HEAPS 1974, UEDA and YOSHIOKA 1976). In Fig. 1, a graphic representation of accumulation patterns at different developmental stages is shown. As can be seen, certain zones of the cell periphery are distinguished by the potential to accumulate great amounts of primary wall material (maximum zones), whereas other zones accumulate little or no material (minimum zones). From the drawing it can be recognized that the minimum zones represent the prospective zones of cell invaginations, whereas the maximum zones represent the actual growing areas. Thus, these experiments reveal "that local secretion of wall material at the surface of the protoplast takes place at different intensities, according to a predetermined pattern" (KIERMAYER 1970 b). In Fig. 10 A, the accumulation pattern of an older developmental stage is shown; Fig. 10 B shows the electron microscopic aspect of a wall thickening exhibiting a loose, spongy material. Within the thickening cavities (arrow) with membraneous elements, probably remnants of cytoplasmic strands can be recognized (KIERMAYER and DORDEL 1976, UEDA and YOSHIOKA 1976). The latter authors, working with

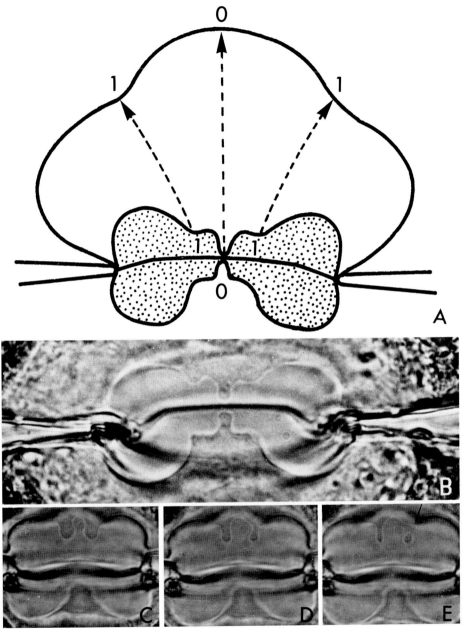

Fig. 9. Septum formation under the influence of turgor reduction in *Micrasterias denticulata*. *A, B* "Septum initial pattern" that developed in a 0.22 m solution of glucose after 2 hours and 30 minutes. The schematic drawing shows the connection of the "septum initial pattern" with shape formation. Thus, the minimum zones (1-1) of the septum pattern correspond to the first invagination zones (1-1) of the growing half-cell. *C–E* In a series of pictures (sequence 1 minute), a rapid change of the septum initial pattern can be seen demonstrating the flexible and semiliquide nature of the accumulated cell wall material. (*D* Arrow); cells 32 minutes in 0.22 mol glucose. From KIERMAYER 1967 b, 1970 b.

Micrasterias americana, studied cell wall development, especially in isotonic and hypertonic solutions. From digestion experiments with pectinase and cellulase and from fluorescence spectra in Calcofluor and Coriphosphin solutions, the authors concluded that pectic substances can be found as a main constituent of primary cell walls, whereas cellulose is formed after the semicell has fully developed (secondary wall formation).

Accumulated wall material, which has been termed "pectic lenses", where also observed in enucleate *Micrasterias* cells and in cells treated with glucose (4%) or different amino acids (WARIS 1951). Cell wall thickenings produced by turgor-reduction has been observed also in developing zygotes of *Micrasterias papillifera* (KIES 1968).

C. Plasmolysis Pattern

When growing half-cells of *Micrasterias denticulata* are placed into hypertonic solutions, plasmolysis occurs and growth is immediately stopped. Plasmolysis is not even since the protoplasm adheres more tightly on special areas. These zones of "negative plasmolysis" (WEBER 1929) have been found to be the areas of the prospective or already visible lobe invaginations. On the areas of lobe invagination the protoplasm adheres much stronger than on areas where active growth occurs, *e.g.,* zones of minimal growth of the primary wall are characterized by strong protoplasmic adhesion to the cell wall. Since the zones of minimal growth (lobe invaginations) are distributed according to a species and stage specific pattern, also a typical "plasmolysis pattern" occurs (KIERMAYER 1964). It seems that in zones of "negative plasmolysis", not only is plasma adhesion higher than in other zones but also protoplasmic strands ("Hecht'sche Fäden") formed during plasmolysis between the cell wall and the plasmalemma are less elastic.

It could be further shown that in plasmolyzed cells, secretion of primary wall material continues so that the space between cell wall and protoplast is gradually filled with wall material. Hanging onto plasmatic strands of different extensibility, the protoplast exhibits various shapes of plasmolysis. During the process of wall material secretion, protoplasmic strands are gradually embedded into primary wall material. The strands can be visualized

---

Fig. 10. Effects of physical and chemical treatments on cells of *Micrasterias denticulata.* *A* Accumulation pattern of the primary wall as a result of turgor reduction during the growth of the primary wall; cells 90 minutes in 0.12 mol Lutrol then plasmolyzed in 0.25 mol Lutrol (polyethylenglycol 400). From KIERMAYER 1964. *B* Electron microscopic aspect of a thickening of the primary wall induced by turgor reduction (isotonic solution of glucose). The wall thickening appears as a spongy material of thin fibrils in which cavities with membraneous elements (arrow) can be recognized ($\times$18,000. From KIERMAYER and DORDEL 1976. *C* Inhibition of nuclear and chloroplast migration by an anti-microtubule substance (10^{-6}% Amiprophos-methyl, *APM*). The dislocated nucleus (*N*) is clearly visible. From KIERMAYER and FEDTKE 1977. *D* Morphogenetic effect by an inhibition of protein synthesis during growth of the primary wall (Cycloheximide 10^{-3}%). From KIERMAYER and MEINDL 1980 a. *E* Electron microscopic aspect of a cell area after centrifugation (7,938 $\times$ g) during growth of the primary wall. The accumulated material contains large vesicles (*LV*) deposited on the centrifugal side of the cell wall ($\times$18,000). From KIERMAYER and DORDEL 1976.

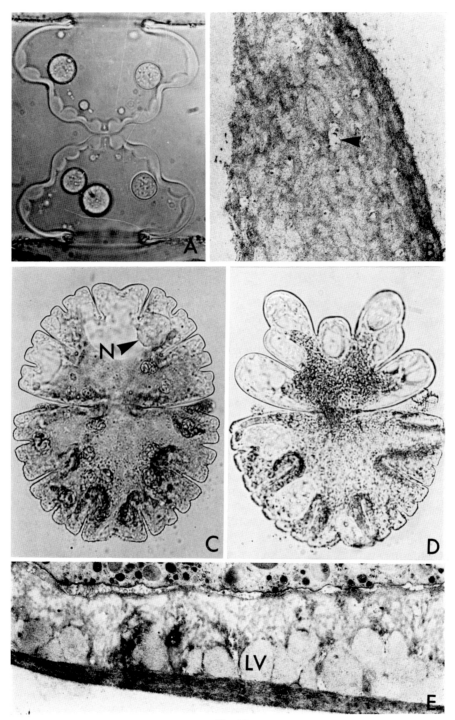

Fig. 10

with a special light microscopic staining technique after CRAFTS (1931), with JJK followed by gentianaviolet. Using this staining method, it has been shown that wall thickenings that form in isotonic and hypertonic solutions are traversed by fibrilar structures, probably protoplasmic strands (KIERMAYER 1965 a). On electronmicrographs, remnants of these strands can be seen in cavities of wall thickenings (Fig. 10 B).

Plasmolyzed cells that had ceased primary wall growth started growing again when placed into a hypotonic solution (deplasmolysis). KIERMAYER (1964) showed that such deplasmolyzed cells accomplish a rather normal morphogenesis. This astonishing fact has been explained by an intact system of protoplasmic strands that is not destroyed during plasmolytical contraction.

UEDA and YOSHIOKA (1976) in their electron microscopic studies of *Micrasterias americana* cells treated with isotonic solutions showed zones of "negative plasmolysis" but absolutely no ultrastructural indications for the stronger plasma adhesion. The wall material of plasmolyzed half-cells appeared loose and spongy.

D. Turgomorphosis

When developing cells of *Micrasterias* are placed into a hypotonic solution, growth of the half-cell is temporarily inhibited and wall thickenings are formed. After a certain time, however, growth starts again, which leads to malformations called "turgomorphosis" (KIERMAYER and JAROSCH 1962, KIERMAYER 1964, NEUHAUS-URL 1980). The reason for the aberrant cell growth, which takes place even at a low concentration range of 0.07 to 0.12 mol (Table 2), seems to be a temporary blockade of elongation growth and an accumulation of cell wall material. After a certain time, during which turgor pressure increases, the outer layer of the primary wall partially ruptures in certain areas, and there growth starts again. During this growth, the accumulated wall material is used up. Growth velocity is significantly increased at the areas of aberrant cell wall growth ("stored growth", RAY 1961). Turgomorphosis is characterized by a strong reduction of cell pattern and the formation of 3-dimensional bubble-like lobes (trifoliate-forms, KIERMAYER 1964). Malformation of that kind had already been described by BARG (1942) in *Micrasterias truncata* and *M. papillifera*. As KIERMAYER (1964) points out, turgomorphosis of cells should be seriously taken into consideration in all experiments in which osmotically active concentrations of various substances are used.

E. Malformations of the Secondary Wall

When *Micrasterias denticulata* cells are treated with low concentrated glucose solutions (0.14 to 0.16 mol), besides the abnormal accumulation of primary wall material, the secondary wall shows malformations. Electronmicrographs of treated young half-cells show an abnormally thickened secondary cell wall with centripetal protuberances, which are formed especially in the region of the isthmus (NEUHAUS-URL 1980, Fig. 11). The formation of the protuberances is probably induced by an irregularity on the surface area of the thickened secondary cell wall. More F-vesicles seem to be associated with the regions of increased secondary wall deposition. In turgor reduced cells, the secondary wall remains surrounded by the primary wall, which is normally shed. DOBBERSTEIN (1973) showed the abnormal formation of a secondary wall around plasmolytically contracted protoplasts.

Fig. 11. Malformation of the secondary wall of *Micrasterias denticulata* under the influence of turgor reduction by 0.14 mol glucose (after 2 days, ×11,000). From NEUHAUS-URL 1980.

V. Other Physical and Chemical Influences on Cytomorphogenesis

A. Effect of Temperature on Shape Formation

Not many facts are known about the influence of temperature on shape formation in *Micrasterias*. In experiments on optimal culture conditions of *Micrasterias denticulata*, KIERMAYER (1964) noted that changes in temperature of about a few degrees over 20 to 21 °C lead to malformations that had been termed "thermomorphosis". LACALLI (1975 b), in his studies of morphogenesis in *Micrasterias rotata*, found striking alterations in lobe form when cells developed at increased temperature. Up to 35 °C, shape formation has been found to be unaffected. At 38 °C, growth of the half-cell was arrested, while at a temperature range between 36 to 37.5 °C, the lobes elongated substantially which lead to abnormal cell shape. Also, low temperature (2 °C) caused malformations (MARČENKO 1966). Further experiments are necessary to elucidate the morphogenetic effect of temperature changes.

B. Effect of Vital Centrifugation on Developing Cells

Centrifugation of developing cells of *Micrasterias* had been used as an experimental tool for studies on cytomorphogenesis (Kallio 1949, 1951, see p. 196 f.). Tippit and Pickett-Heaps (1974) showed that centrifugation makes it possible to dislocate organelles in developing half-cells leading to severe morphological changes of the cell. Dordel (1973) and Kiermayer and Dordel (1976) in their light and electron microscopical studies of centrifuged young stages of *Micrasterias denticulata* demonstrated a local inhibition of cell wall growth, on both the centrifugal and centripetal side of the cell leading to highly asymmetric cell forms. Inhibition of growth of the primary wall on the centrifugal side of the cell had been explained by a prevention of the incorporation of D-vesicles (see p. 154 f.) caused by an abnormal layer of L-vesicles. Fig. 10 E shows an electronmicrograph of the cortical centrifugal area of a developing half-cell in which a compact layer of L-vesicles (slime vesicles) is visible; this apparently prevents D-vesicle incorporation and, thus, growth of this side of the cell. Inhibition of growth also occurs on the centripetal side of the cell; this can be explained as a direct influence of the centrifugal force on vesicle transport and incorporation. Also, secondary wall formation is influenced by centrifugation probably caused by a dislocation of "flat-vesicles" (see p. 158 f.; Kiermayer and Dordel 1976). In addition to cell wall growth, the Golgi system is strongly influenced by centrifugation (Dordel 1973, Burgstaller-Getzinger and Kiermayer 1981). Centrifugation experiments thus demonstrate that dislocation of vesicles leads to severe disorders of cytomorphogenesis and organelle functions.

C. Effect of Electric Fields

Brower and McIntosh (1980) showed that applied electric fields of approximately 14 V/cm^{-1} have profound effects on shape formation in *Micrasterias denticulata*. The lobes show a galvanotropism toward the cathode; growth of lobes parallel to the field is inhibited to varying degrees, depending on their orientation. In osmotically treated cells, wall material accumulates along the cathode-facing sides of lobes oriented perpendicular to the electric field. Under the influence of applied electric fields, radioactively labelled glucose and the methyl groups from methionine (Lacalli 1975 a) are incorporated along the cathode-facing sides of lobes as well as the lobe tips. The experiments indicate that the wall-depositing machinery can be altered by the field. However, no gross cytoplasmic abnormities nor an asymmetric distribution of cytoplasmic organelles or vesicles could be found in growing cells oriented perpendicular to the electric field (Brower and Giddings 1980).

Freeze-fracture studies of Brower and Giddings (1980) showed that applied electric fields cause a large asymmetry in the distribution of membrane particles (Table 5). Larger numbers of rosettes could be found on the cathode-facing sides of lobes oriented perpendicular to the fields. The applied electric fields cause a clear redistribution of membrane particles with the greater concentration on the cathode-facing side where extra wall material is deposited.

As BROWER and GIDDINGS emphasize, the experiment with applied electric fields support the hypothesis of KIERMAYER that "the plasma membrane itself is an important structure in the control of the formation and ultimate shape of the wall" (KIERMAYER 1970 b).

Table 5. *Distribution of cytoplasmic vesicles (DV, LV) and plasma membrane particles of electric field-treated growing lobes of Micrasterias denticulata.* After BROWER and GIDDINGS 1980

Distribution of cytoplasmic vesicles

	Dark vesicles (DV)	Large vesicles (LV)
Percent in cathode-facing half of the lobe (mean, n = 13)	47.9	50.8
Standard deviation	6.0	5.7
Significance	$P > 0.05$	$P > 0.05$

Distribution of plasma membrane particles

	E-face (n = 8)	P-face (n = 6)	Combined (n = 14)
Percent in cathode-facing half of the lobe (mean)	66.8	62.8	65.1
Standard deviation	8.7	10.6	9.4
Significance	$P < 0.001$	$P < 0.05$	$P < 0.001$

D. Effect of Ethanol on Septum Growth and Subsequent Differentiation

When cells of *Micrasterias denticulata* are treated with 10% ethanol during septum growth and afterwards placed into nutrient solution, changes of septum growth and subsequent cell development can be observed (KIERMAYER 1966 b). An 1 hour treatment of the cell leads to an irreversible inhibition of septum growth. Since subsequent cell elongation growth (bulge formation) remained uneffected, binuclear double-cells with a teratogenic middle section were formed (see also p. 196 f.). The degree of differentiation of this middle section was found to depend on the extent of septum formation. Thus, a direct relationship between the degree of differentiation of the middle section and septum formation exists. These experiments support the view of a template function of the septum (see p. 181).

Septum formation seems to be a sensitive process that can be influenced rather unspecifically by various effects such as osmotic factors (KIERMAYER 1967 b) and chemical and physical factors (KALLIO 1951, 1963, MARČENKO 1966, MEINDL 1980) leading to double-cells. Even in nature, double-cells can be frequently found induced by environmental conditions, *e.g.,* changes of temperature, light, and nutritional factors.

E. Effect of Drugs Inhibiting RNA or Protein Synthesis

SELMAN (1966) has been the first to show that the RNA blocking drug, Actinomycin D, has a pronounced influence on morphogenesis when administered in the young developmental stages of *Micrasterias* cells (see also KALLIO and HEIKKILÄ 1972, TIPPIT and PICKETT-HEAPS 1974). Cells that develop in the presence of this drug form new semicells with a simplified cell pattern including a number of undifferentiated, bubble-like lobes. This aberrant shape formation has been termed "anuclear type of development" (SELMAN 1966) because similar malformation had been found in anucleate cells produced after centrifugation (WARIS 1951, SELMAN 1966) and after treatment with ultraviolet light (KALLIO 1959, 1963, KALLIO and HEIKKILÄ 1969, SELMAN 1966; see p. 191 f.). The same malformation is formed under the influence of ethidium bromide (HACKSTEIN-ANDERS 1975). In treated cells, the degree of malformation depends on the age of the semicell at the beginning of the treatment. The treated cells exhibit at least a minimum pattern that corresponds with the "septum initial pattern". HACKSTEIN-ANDERS (1975) further emphasizes that malformation of semicells often leads to cell death by a rupture of the cell (plasmoptysis) caused by an abnormally increased turgor pressure; also, secondary wall formation is prevented in these cells. Following the removal of actinomycin D, treated cells during subsequent divisions developed normally. Puromycin had no effect on cytomorphogenesis (TIPPIT and PICKETT-HEAPS 1974).

Similar morphogenetic aberrations could be obtained with the protein synthesis inhibitors cycloheximide (TIPPIT and PICKETT-HEAPS 1974, KIERMAYER and MEINDL 1980 a, MEINDL 1981; Fig. 10 D) and gougerotin (KIERMAYER and MEINDL 1980 b). TIPPIT and PICKETT-HEAPS (1974) found that developing cells treated with a combination of cycloheximide and D-mannitol showed cell wall accumulation that lacked the characteristic pattern of accumulated wall material. Cycloheximide also exerts strong influence on the ultrastructure of the cytoplasm, especially the dictyosomes (NOGUCHI and UEDA 1979). KIERMAYER and MEINDL (1980 a) showed effects of this drug on the ultrastructure of the growing primary wall; it appears that normal incorporation of D-vesicles into the growing wall is prevented by cycloheximide; thus, an uneven distribution of wall material within the primary wall results. Similar disorders in the fine structure of primary walls had been observed in UV-irradiated malformed cells (PIHAKASKI and KALLIO 1978). In time-lapse studies of cycloheximide and gougerotin treated cells, a prolonged elongation growth could be observed leading to undifferentiated bubble-like lobes that finally burst by plasmoptysis (MEINDL 1981).

F. Effect of Anti-Microtubule Substances

Since different microtubule systems are found in developing *Micrasterias* cells, anti-microtubule drugs might reveal the underlying function of these cellular structures. When developing *Micrasterias* cells are treated with colchicine, little or no morphogenetic influence could be found (KIERMAYER 1968 c). Many other anti-microtubule drugs listed in Table 6 also had no special effect on cytomorphogenesis (KIERMAYER 1968 c, KIERMAYER and

HEPLER 1970, TIPPIT and PICKETT-HEAPS 1974, MEINDL and KIERMAYER 1981 a, b). From these findings, it had been assumed that microtubules might not play an important role in shape formation of *Micrasterias* cells (KIERMAYER 1970 b). Although shaping is not affected, anti-microtubule drugs like colchicine and vinblastine (KIERMAYER 1968 c, 1970 b), IPC and CIPC (KIERMAYER and HEPLER 1970, KIERMAYER 1973 a), trifluralin (KIERMAYER 1972), amiprophos-methyl (APM) (KIERMAYER and FEDTKE 1977), lindane,

Table 6. *Anti-microtubule activity of various substances in Micrasterias denticulata in relation to the concentration of the drugs. As a criterion for anti-microtubule activity, the specific inhibition of nuclear and chloroplast migration has been taken* ($\oplus$ lowest active concentration). *From* MEINDL and KIERMAYER 1981 a

Substance	10^{-1}	10^{-2}	10^{-3}	10^{-4}	10^{-5}	10^{-6}	10^{-7}	%C
Colchicine	$\oplus$							
Colcemid		$\oplus$						
Vinblastine				$\oplus$				
Podophyllotoxin			$\oplus$*					
Griseofulvin			$\oplus$					
Bacitracin **		$\oplus$						
Gougerotin **	$\oplus$							
Nocodacole				$\oplus$				
IPC					$\oplus$			
CIPC					$\oplus$			
Trifluralin					$\oplus$			
APM						$\oplus$		
Lindane				$\oplus$				
Dacthale				$\oplus$				
MBC	−							
EPTC	−							
EGTA	−							
DMSO	−							

* Strong inhibition of chloroplast migration.
** Variable results.

dacthale, nocodacole (KIERMAYER and MEINDL 1979), and gougerotin (KIERMAYER and MEINDL 1980 b) inhibit postmitotic migration of the nucleus and the chloroplast. In Table 6, all active anti-microtubule agents so far tested in *Micrasterias* are given (MEINDL and KIERMAYER 1981 a). From this table, it becomes evident that the most active substances on nuclear and chloroplast migration are various herbicides. The most active compound found so far is the organophosphorous herbicide amiprophos-methyl (APM) which is active at concentrations of 10^{-3} to 10^{-6} %. As far as active concentration ranges are concerned, anti-microtubule substances like colchicine, vinblastine, or griseofulvin are less active (Table 6). It has been suggested that the microtubules surrounding the posttelophase nucleus (Figs. 8 *B, D,* and *E*) are disorganized by these drugs with the consequential loss of normal migratory ability (KIERMAYER 1968 a–c, TIPPIT and PICKETT-HEAPS 1974).

12*

Chloroplast migration, which may be controlled by microtubules, too, is also inhibited by anti-microtubule substances (KIERMAYER 1968 c, MEINDL 1980). The dislocation of the nucleus and the chloroplast after treatment with anti-microtubule drugs can easily be watched under the light microscope (Fig. 10 C) and serves as a specific criterion for anti-microtubule activity. Thus, developing *Micrasterias* cells represent an ideal object for physiological testing of anti-microtubule activity of various compounds (MEINDL and KIERMAYER 1981 a). MEINDL (1980) in her recent electron microscopic investigations demonstrated that the postmitotic bundles of microtubules surrounding the posttelophase nucleus are no longer present in APM-treated cells. This observation supports the hypothesis of an involvement of the microtubule system surrounding the posttelophase nucleus in nuclear migration. When cells treated with anti-microtubule drugs are placed in nutrient solution, in many cases a complete recovery from the inhibitory effect results (MEINDL 1980).

G. Effect of Antibiotics and Other Drugs on Shape Formation of the Chloroplast

KUNZMANN and KIERMAYER (1978) in their investigations on the action of antibiotics in developing *Micrasterias* cells have shown a disturbance of chloroplast migration by penicillin G and cephalosporines. In many cases, a strong "thinning" associated with the formation of a "hole" in the chloroplast of the old half-cell as well as a strong chloroplast contraction within the young half-cell have been observed. Since chloroplast migration might be controlled by microtubules, MEINDL and KIERMAYER (1979, 1981 c) tested the influence of the anti-microtubule agents APM and CIPC on penicillin G induced disturbance. An antagonistic effect of APM and CIPC on penicillin G induced disturbances have been observed; therefore, it has been suggested that the action of penicillin G is in some unknown way related to the microtubules system controlling chloroplast migration. Malformations had also been observed in cells treated with streptomycin and nystatin (KUNZMANN and KIERMAYER 1978, BURGSTALLER-GETZINGER 1981).

An interesting inhibiting action of phenyl-α-naphtylamine (a component of the psychopharmacon Vinydan) and valinomycin has been observed by MEINDL (1980) on chloroplast separation (see p. 151), which demonstrates the autonomous character of this process.

H. Effect of Various Substances on Cytomorphogenesis

A number of different substances not mentioned so far have been found to be specifically active in changing cytomorphogenesis of *Micrasterias*: WARIS and KALLIO (1964) report about a morphogenetic action of ribonuclease (see also TIPPIT and PICKETT-HEAPS 1974), sodium 2.4-dinitrophenol, and mercaptoethanol that cause cell forms similar to those of enucleate or hypohaploid cells (KALLIO 1963). TIPPIT and PICKETT-HEAPS (1974) studied the effect of a number of different substances: caffeine caused the expanding primary wall to burst and stopped deposition of new wall material; cytochalasin B stopped cytoplasmic streaming and expansion of the primary wall and severely suppressed deposition of new wall material; glusulase caused expanding cell wall to burst. When a mixture of glusulase and D-mannitol was used, wall material remained on the future site of lobe invaginations, *i.e.*, these sites appeared more resistant to this enzyme.

VI. Hypothetical Aspects on Cytomorphogenesis: Membranes and Pattern Formation

From the data given in this article on light and electron microscopic studies of developing *Micrasterias* cells, some general aspects about the cytoplasmic basis of cytomorphogenesis deserve attention. One of the most important structures for the control of morphogenesis seems to be the plasma membrane overlying the septum. As had been shown pattern formation under turgor reduction occurs already during septum formation in late anaphase. A special population of Golgi-derived vesicles (SEV) seem to fuse with the plasma membrane to form a membrane layer that apparently contains the "septum initial pattern". It can be speculated that the plasma membrane of the septum contains a kind of "membrane recognition pattern" (specific receptors) responsible for an ordered incorporation of another kinds of vesicles—the D-vesicles. Corresponding to a species-specific template for membrane recognition, incorporation of D-vesicles containing primary wall material takes place only at specific zones of the plasma membrane where local affinities between the plasma membrane and D-vesicle membranes may exist. By such a membrane controlled mechanism, differential growth of the primary wall, *e.g.*, pattern formation (lobes, spines), could be explained. In this hypothesis, membrane recognition is considered to be the major tool for cytomorphogenesis (KIERMAYER 1970 b, 1973 b). If such a process is considered to be the means for ordering vesicle incorporation, it is not necessary to assume the participation of structures such as microtubules. The experiments with anti-microtubule drugs have indeed shown that microtubules do not play an important role in the patterned growth of the primary wall. However, protoplasmic streaming as a means for vesicle transportation is essential for wall growth, as had been demonstrated in experiments with cytochalasin B.

A major role for cytomorphogenesis is postulated for the Golgi system. The dictyosomes are producing different kinds of vesicles, and the production of these vesicles is strictly correlated with developmental events. Thus, at the beginning of cell growth, septum vesicles (SEV) are formed that seem to build the first template for later membrane recognition. After the formation of SE-vesicles, the dictyosomes "switch" to produce another kind, the D-vesicles, which contain wall material (pectins) and deliver membrane areas with particle rosettes responsible for microfibril formation. These D-vesicles are incorporated into the plasma membrane, probably corresponding to the membrane template formed by SE-vesicles. After primary wall growth, dictyosomes "switch" again. Now "flat vesicles" (FV) responsible for secondary wall formation are pinched off by the dictyosomes. The membrane of F-vesicles is highly differentiated and contains arrays of hexagonally arranged rosettes of particles that are active in microfibril formation and orientation when incorporated into the plasma membrane. Membrane particles formed on dictyosome-derived vesicle-membranes thus give rise to the highly ordered bands of microfibrils of the secondary wall.

In addition to F-vesicles, "pore vesicles" (PV) are produced, the origin of which is still obscure but might be also the dictyosome. During secondary wall formation, these P-vesicles form "plugs" on special areas of the plasma membrane where local deposition of wall material is thus prevented. It

Table 7. *Hypothetical representation of a possible "switching" of the dictyosomes to produce different vesicles during the course of cytodifferentiation*

Septum growth	Primary wall growth	Secondary wall growth	Pore formation
Template formation, "initial pattern" of the septum membrane	Membrane recognition, template realization	Incorporation of membrane "rosettes"	Membrane recognition
↑	↑	↑	↑
SE-vesicles	D-vesicles	F-vesicles	P-vesicles
↑	↑	↑	↑
dictyosome	dictyosome	dictyosome	dictyosome
↑	↑	↑	↑
VX (transitorial vesicles)	VX (transitorial vesicles)	VX (transitorial vesicles)	VX (transitorial vesicles)
↑	↑	↑	↑
ER	ER	ER	ER
↑	↑	↑	↑
R	R	R	R
↑	↑	↑	↑
RNA	RNA	RNA	RNA
		Nucleus (DNA)	

might be assumed that the distribution of pore vesicles is again controlled by a membrane recognition phenomenon, *e.g.*, the membrane of pore-vesicles exhibits a specific affinity to local areas of the plasma membrane.

There is no doubt that dictyosomes sequentially change their vesicle products. The "switching" of dictyosomes from one product to the other might be controlled by nuclear factors. A chain of information transfer can be assumed in which membranes of the ER receive information by a sequence DNA-RNA-ribosomes (R)-protein. From the ER, "transitorial vesicles" carry information to the membranes of the Golgi system. From the dictyosomes, special vesicles are formed that finally transfer morphogenetic information to the plasma membrane where it is realized (Table 7). It might be speculated that various structures associated with the vesicle membranes such as "coats" and "spines" (see Table 3) may function as special markers for membrane recognition; membrane associated microfilaments may be involved in the process of vesicle formation (blebbing etc.).

Sequential "switching" of the dictyosomes strictly correlated to specific developmental events might be controlled by nuclear factors (RNA) that are formed corresponding to a precise time-scheduled program (inner clock) of the nucleus (Table 7). Any change of nuclear function might thus be transferred via the Golgi system to the plasma membrane and might induce morphological aberrations (see also p. 210).

In addition to nuclear and membrane factors establishing a cortical template that causes uneven patterned distribution of wall material, some other physiological and physical factors must be operating to maintain normal cytomorphogenesis:

(1) osmotic pressure, representing a hydrostatic "abutment" for primary wall growth (KIERMAYER 1964)

(2) continuous production of wall material, its transport toward the periphery of the cell by protoplasmic streaming and deposition into the primary wall

(3) maintenance of protein-synthesis during cell growth (TIPPIT and PICKETT-HEAPS 1974), since it is possible that certain proteins are involved in a proper vesicle incorporation during primary wall growth (KIERMAYER and MEINDL 1980 a).

It should be emphasized that normal cell growth and morphogenesis greatly depends on a continuous flow of mobile vesicles, *i.e.*, of transitorial vesicles (shuttle vesicles, VX, Fig. 6) from the ER to the dictyosomes and the wall forming vesicles (SEV, DV, and FV), from the dictyosomes to the plasma membrane. If the movement of these vesicles is inhibited or blocked either chemically or physically (centrifugation), severe disturbances of the Golgi function and cell wall growth can be observed. These experimental data strongly support the hypothesis on the function of different vesicles during the process of cell wall growth and morphogenesis.

As far as formation and maintenance of intracellular architecture of the *Micrasterias* cell is concerned, cytoskeletal structures such as microtubules and microfilaments may be functional. Thus, fixation of the nucleus at a central cell position is achieved by bundles of microtubules. A special bundle of microtubules and microfilaments seems to control postmitotic nuclear migration. Time-lapse studies showed that anchoring of the chloroplast is achieved in the presence of anti-microtubule drugs, also (MEINDL 1980). Therefore, the resulting ornamental chloroplast shape does not depend on microtubules but seems to be controlled by filaments that anchor the chloroplast on special local areas of the cortical protoplasma or the plasma membrane. It can be speculated that in the case of shape modelling of the chloroplast, the plasma membrane contains a pattern of species-specific adhesion zones for microfilaments that pull and anchor the chloroplast in the cortical protoplasma at the proper location.

In my hypothesis on morphogenesis of *Micrasterias*, biomembranes play the major role (KIERMAYER 1964, 1970 b, 1977). Thus, the rapidly increasing knowledge of membrane structure, biochemistry and physiology in both plant and animal cells will also greatly enhance understanding of the basic principles for cytomorphogenesis.

Acknowledgement

This work has been supported by the "Fonds zur Förderung der wissenschaftlichen Forschung", Austria, Projekt Nr. 2783 and 3660. I am most grateful to Professor Dr. Peter K. Hepler for his help in translating this manuscript. I am also very thankful to Dr. Ursula Meindl and my secretary Susanne Hampl for assistence in the preparation of the manuscript.

References

Barg, T., 1942: Beiträge zur Cytomorphologie der Desmidiaceen. Arch. Protistenkunde **95**, 391—432.

Barnett, J. R., Preston, R. D., 1970: Arrays of granules associated with the plasmalemma in swarmers of *Cladophora*. Ann. Bot. N.S. **38**, 1011—1017.

Brower, D. L., McIntosh, J. R., 1980: The effects of applied electric fields on *Micrasterias*. I. Morphogenesis and the pattern of cell wall deposition. J. Cell Sci. **42**, 261—277.

— Giddings, T. H., 1980: The effect of applied electric fields on *Micrasterias*. II. The distribution of cytoplasmic and plasma membrane components. J. Cell Sci. **42**, 279—290.

Burgstaller-Getzinger, Ch., Kiermayer, O., 1981: Einfluß der Zentrifugierung und chemischer Stoffe auf sich differenzierende Zellen von *Micrasterias denticulata* Bréb. mit besonderer Berücksichtigung des Golgi-Systems. Sitzungsber. Österr. Akad. Wiss. (in press).

Cholnoky, B. von, Höfler, K., 1950: Vergleichende Vitalfärbeversuche an Hochmooralgen. Sitzungsber. Österr. Akad. Wiss., math.-nat. Kl., Abt. I, **159**, 143—182.

Cleland, R., 1958: Effects of osmotic concentration on auxin-action and on irreversible and reversible expansion of *Avena* coleoptile. Physiol. Plant. **11**, 599—603.

Crafts, A. S., 1931: A technic for demonstrating plasmodesma. Stain Technol. **6**, 127—129.

Dobberstein, B., 1973: Einige Untersuchungen zur Sekundärwandbildung von *Micrasterias denticulata* de Brébisson (*Desmidiaceae*). Nova Hedwigia **42**, 83—90.

— Kiermayer, O., 1972: Das Auftreten eines besonderen Typs von Golgivesikeln während der Sekundärwandbildung von *Micrasterias denticulata* Bréb. Protoplasma **75**, 185—194.

Dordel, S., 1973: Licht- und elektronenmikroskopische Untersuchungen über den Einfluß der Zentrifugalkraft auf *Micrasterias denticulata* Bréb. Thesis (Köln).

Drawert, H., Mix, M., 1961 a: Licht- und elektronenmikroskopische Untersuchungen an Desmidiaceen. III. Mitt. Der Nucleolus im Interphasekern von *Micrasterias rotata*. Flora **150**, 185—190.

— — 1961 b: Licht- und elektronenmikroskopische Untersuchungen an Desmidiaceen. IV. Mitt. Beiträge zur elektronenmikroskopischen Struktur des Interphasekernes von *Micrasterias rotata*. Z. Naturf. **16 b**, 546—551.

— — 1961 c: Licht- und elektronenmikroskopische Untersuchungen an Desmidiaceen. V. Mitt. Über die Variabilität der Chloroplastenstruktur bei *Micrasterias rotata*. Planta **56**, 648—665.

— — 1961 d: Licht- und elektronenmikroskopische Untersuchungen an Desmidiaceen, VI. Mitt. Der Einfluß von Antibiotika auf die Chloroplastenstruktur von *Micrasterias rotata*. Planta **57**, 51—70.

— — 1961 e: Licht- und elektronenmikroskopische Untersuchungen an Desmidiaceen. VII. Mitt. Der Golgi-Apparat von *Micrasterias rotata* nach Fixierung mit Kaliumpermanganat und Osmiumtetroxyd. Mikroskopie **16**, 207—212.

— — 1961 f: Licht- und elektronenmikroskopische Untersuchungen an Desmidiaceen. VIII. Mitt. Die Chondriosomen von *Micrasterias rotata*. Flora **151**, 487—508.

— — 1962 a: Licht- und elektronenmikroskopische Untersuchungen an Desmidiaceen. IX. Mitt. Die Struktur der Pyrenoide von *Micrasterias rotata*. Planta **58**, 50—74.

— — 1962 b: Zur Funktion des Golgi-Apparates in der Pflanzenzelle. Planta **58**, 448—452.

— — 1962 c: Zur Frage von Struktur und Funktion des "Golgi-Apparates" in Pflanzenzellen. Sitzungsber. d. Ges. zur Förderung der gesamten Naturwissenschaften zu Marburg, 361—382.

— — 1962 d: Zur Frage der Identität von Karyoiden und Golgi-Apparat bei den Conjugaten. Naturwiss. **15**, 353—354.

DRAWERT, H., MIX, M., 1962 e: Licht- und elektronenmikrospopische Untersuchungen an Desmidiaceen. X. Mitt. Beiträge zur Kenntnis der „Häutung" von Desmidiaceen. Arch. Mikrobiol. **42** (1), 96—109.

— — 1973: Licht- und elektronenmikroskopische Untersuchungen an Desmidiaceen. XI. Mitt. Die Struktur von Nucleolus und Golgi-Apparat bei *Micrasterias denticulata* Bréb. Portugaliae Acta Biologica **VII**, 17—28.

EIBL, K., 1938: Kontraktion der Chromatophoren bei *Micrasterias rotata*. Protoplasma **32**, 251—264.

— 1939: Das Verhalten der *Spirogyra*-Chloroplasten bei Zentrifugierung. Protoplasma **33**, 73—102.

— 1941: Die Restitution der Chromatophorenform bei *Micrasterias rotata* nach Schleuderung. Protoplasma **35**, 595—617.

FREY-WYSSLING, A., 1959: Die pflanzliche Zellwand. Berlin-Göttingen-Heidelberg: Springer.

GARRONE, R., LETHIAS, C., ESCAIG, J., 1980: Freeze-fracture study of sponge cell membranes and extracellular matrix. Preliminary results. Biol. Cell. **38**, 71—74.

GIDDINGS, T. H., BROWER, D. L., STAEHELIN, L. A., 1980: Visualization of particle complexes in the plasma membrane of *Micrasterias denticulata* associated with the formation of cellulose fibrils in primary and secondary cell walls. J. Cell Biology **84**, 327—339.

HACKSTEIN-ANDERS, CH., 1974: Untersuchungen zur Wirkung von Actinomycin D und Ethidiumbromid auf die Cytomorphogenese und Ultrastruktur von *Micrasterias thomasiana* und *Micrasterias denticulata* Bréb. unter besonderer Berücksichtigung des Golgiapparates. Thesis (Köln).

— 1975: Untersuchungen zur Cytomorphogenese von *Micrasterias thomasiana* und *Micrasterias denticulata* Bréb. unter Einfluß von Actinomycin D und Ethidiumbromid. I. Lichtmikroskopische Untersuchungen. Protoplasma **86**, 83—105.

HÄDER, D. P., WENDEROTH, K., 1977: Role of three basic light reactions in photomovement of Desmids. Planta **137**, 207—214.

HERRMANN, R., 1966: Zur Frage des Plasmalemmas bei Desmidialen. Protoplasma **61**, 12—59.

HIRN, I., Vitalfärbestudien an Desmidiaceen. Flora **140**, 453—473.

HÖFLER, K., 1951: Plasmolyse mit Natriumkarbonat. Zur Frage des Plasmalemmas bei Süßwasseralgen und bei Gewebszellen von Landblütenpflanzen. Protoplasma **40**, 426—460.

JAROSCH, R., KIERMAYER, O., 1962: Die Formdifferenzierung von *Micrasterias*-Zellen nach lokaler Lichteinwirkung. Planta **58**, 95—112.

KALLIO, P., 1949: Artificially produced binuclear, diploid and anuclear Desmids. Arch. Soc. Zoo. Bot. Fenn. Vanamo **2**, 42—44.

— 1951: The significance of nuclear quantity in the genus *Micrasterias*. Ann. Bot. Soc. Zool. Bot. Fenn. Vanamo **24**, 1—122.

— 1954: Morphogenetic studies in *Micrasterias rotata* (Grev.) Ralfs var. *evoluta* Turner (*Desmidiaceae*). Arch. Soc. Zool. Bot. Fenn. Vanamo **8**, 118—122.

— 1957: Studies on artificially produced diploid forms of some *Micrasterias* species (*Desmidiaceae*). Arch. Soc. Zool. Bot. Fenn. Vanamo **11**, 193—204.

— 1959: The relationship between nuclear quantity and cytoplasmic units in *Micrasterias*. Ann. Acad, Sci. Fenn. IV, **44**, 1—44.

— 1960: Morphogenetics of *Micrasterias americana* in clone culture. Nature **187**, 164—166.

— 1963: The effect of ultraviolet radiation and some chemicals on morphogenesis in *Micrasterias*. Ann. Acad. Sci. Fenn. **70**, 5—39.

— HEIKKILÄ, H., 1969: UV-induced facies change in *Micrasterias torreyi*. Österr. bot. Z. **116**, 226—243.

— — 1972: On the effect of elimination of nuclear control in *Micrasterias*. In: Biology and radiobiology of anuclate systems. II. Plant cells (BONOTTO, S., GOUTIER, R., KIRCHMANN, R., MAISIN, J. R., eds.), pp. 145—164. New York-London: Academic Press.

KIERMAYER, O., 1954: Die Vakuolen der Desmidiaceen, ihr Verhalten bei Vitalfärbe- und Zentrifugierungsversuchen. Sitzungsber. Österr. Akad. Wiss., math.-nat. Kl., Abt. I, **163**, 175—222.

— 1955: Über die Reduktion basischer Vitalfarbstoffe in pflanzlichen Vakuolen. Sitzungsber. Österr. Akad. Wiss. math.-nat. Kl., Abt. I, **164**, 275—302.

Kiermayer, O., 1962: Die Rolle des Turgordruckes bei der Formbildung von *Micrasterias*. Ber. Dtsch. Bot. Ges. **75**, 78—81.

— 1964: Untersuchungen über die Morphogenese und Zellwandbildung bei *Micrasterias denticulata* Bréb. Protoplasma **59**, 382—420.

— 1965 a: Plasmatische Strukturen in experimentell hervorgerufenen Zellwandverdickungen von *Micrasterias denticulata* Bréb. Planta **66**, 216—220.

— 1965 b: *Micrasterias denticulata* (*Desmidiaceae*) — Morphogenese bei reduziertem Turgor. Film E 869, Inst. Wiss. Film, Göttingen.

— 1965 c: Zur Mechanik der Formbildung von *Micrasterias*. Ber. Dtsch. Bot. Ges. **78**, 1.

— 1965 d: *Micrasterias denticulata* (*Desmidiaceae*) — Morphogenese. Film E 868, Inst. Wiss. Film, Göttingen.

— 1966 a: Differenzierung und Wachstum von *Micrasterias denticulata* (*Conjugatae*). Film C 924, Inst. Wiss. Film, Göttingen.

— 1966 b: Septumbildung und Cytomorphogenese von *Micrasterias denticulata* nach der Einwirkung von Äthanol. Planta **71**, 305—313.

— 1967 a: Dictyosomes in *Micrasterias* and their "division". J. Cell. Biol. **35**, 68 A.

— 1967 b: Das Septum-Initialmuster von *Micrasterias denticulata* und seine Bildung. Protoplasma **64**, 481—484.

— 1968 a: The distribution of microtubules in differentiating cells of *Micrasterias denticulata* Bréb. Planta **83**, 223—236.

— 1968 b: Mikrotubuli um den Posttelophasekern von *Micrasterias* und ihre mögliche Funktion. Ber. Dtsch. Bot. Ges. **81** (7), 319.

— 1968 c: Hemmung der Kern- und Chloroplastenmigration von *Micrasterias* durch Colchicin. Naturwissenschaften **55** (6), 299—300.

— 1970 a: Elektronenmikroskopische Untersuchungen zum Problem der Cytomorphogenese von *Micrasterias denticulata* Bréb. I. Allgemeiner Überblick. Protoplasma **69**, 97—132.

— 1970 b: Causal aspects of cytomorphogenesis in *Micrasterias*. Ann. N.Y. Acad. Sci. **175**, 686—701.

— 1971: Elektronenmikroskopischer Nachweis spezieller cytoplasmatischer Vesikel bei *Micrasterias denticulata* Bréb. Planta **96**, 74—80.

— 1972: Beeinflussung der postmitotischen Kernmigration von *Micrasterias denticulata* Bréb. durch das Herbizid Trifluralin. Protoplasma **75**, 421—426.

— 1973 a: Störung der Kernmigration von *Micrasterias denticulata* (*Desmidiaceae*) durch eine die Mikrotubuli beeinflussende Substanz Chlor-isopropyl-N-phenylcarbamat (CIPC). Film B 1070, Inst. Wiss. Film, Göttingen.

— 1973 b: Feinstrukturelle Grundlagen der Cytomorphogenese. Ber. Dtsch. Bot. Ges. **86**, 287—291.

— 1976 a: Formbildung des Chloroplasten von *Micrasterias denticulata* (*Desmidiaceae*). Film B 1106, Inst. Wiss. Film, Göttingen.

— 1976 b: Rhythmische Kontraktionen des Chloroplasten von *Micrasterias denticulata* Bréb. dargestellt durch kinematographische Zeitraffung. Mikroskopie **32**, 301—304.

— 1977: Biomembranen als Träger morphogenetischer Information. Naturwiss. Rundschau **30** (5), 161—165.

— 1980: Control of morphogenesis in *Micrasterias*. In: Handbook of phycological methods. Developmental and cytological methods (Gantt, E., ed.), pp. 6—12. Cambridge: University Press.

— Dobberstein, B., 1973: Membrankomplexe dictyosomaler Herkunft als „Matrizen" für die extraplasmatische Synthese und Orientierung von Mikrofibrillen. Protoplasma **77**, 437—451.

— Dordel, S., 1976: Elektronenmikroskopische Untersuchungen zum Problem der Cytomorphogenese von *Micrasterias denticulata* Bréb. II. Einfluß von Vitalzentrifugierung auf Formbildung und Feinstruktur. Protoplasma **87**, 179—190.

— Fedtke, C., 1977: Strong anti-microtubule action of amiprophos-methyl (APM) in *Micrasterias*. Protoplasma **92**, 163—166.

— Hepler, P., 1970: Hemmung der Kernmigration von Jochalgen (*Micrasterias*) durch Isopropyl-N-phenylcarbamat. Naturwissenschaften **5**, 252.

KIERMAYER, O., JAROSCH, R., 1962: Die Formbildung von *Micrasterias rotata* Ralfs. und ihre experimentelle Beeinflussung. Protoplasma 54, 382—420.

— MEINDL, U., 1979: Anti-microtubule action of various substances in *Micrasterias denticulata*. Europ. J. Cell Biol. 20, 130.

— — 1980 a: Elektronenmikroskopische Untersuchungen zum Problem der Cytomorphogenese von *Micrasterias denticulata* Bréb. III. Einfluß von Cycloheximid auf die Bildung und Ultrastruktur der Primärwand. Protoplasma 103, 169—177.

— — 1980 b: Cytomorphogenetic and anti-microtubule action of the antibiotic gougerotin in *Micrasterias denticulata* Bréb. Protoplasma 104, 175—179.

— SLEYTR, U., 1979: Hexagonally ordered "rosettes" of particles in the plasma membrane of *Micrasterias denticulata* Bréb. and their significance for microfibril formation and orientation. Protoplasma 101, 133—138.

— STAEHELIN, L. A., 1972: Feinstruktur von Zellwand und Plasmamembran bei *Micrasterias denticulata* Bréb. nach Gefrierätzung. Protoplasma 74, 227—237.

KIES, L., 1968: Über die Zygotenbildung bei *Micrasterias papillifera* Bréb. Flora, Abt. B, 157, 301—313.

KINZEL, H., 1953: Untersuchungen über die Chemie und Physikochemie der Gallertbildungen von Süßwasseralgen. Österr. bot. Z. 100, 25—79.

KLEBS, G., 1885: Über die Bewegung und Schleimausbildung der Desmidiaceen. Biolog. Zentralblatt V, 12.

KOL, E., 1927: Über die Bewegung mit Schleimbildung von einigen Desmidiaceen aus der Hohen Tatra. Fol Krypt. 1, 5.

KOOP, H. U., KIERMAYER, O., 1980 a: Protoplasmic streaming in the giant unicellular green alga *Acetabularia mediterranea*. I. Formation of intracellular transport systems in the course of cell differentiation. Protoplasma 102, 147—166.

— — 1980 b: Protoplasmic streaming in the giant unicellular green alga *Acetabularia mediterranea*. II. Differential sensitivity of movement systems to substances acting on microfilaments and microtubules. Protoplasma 102, 295—306.

KOPETZKY-RECHTPERG, O., 1954: Beobachtungen an Protoplasma und Chloroplasten der Alge *Netrium digitus* (Ehrenberg) bei Kultur unter Lichtabschluß. Protoplasma 44, 322—331.

KOWALLIK, K., 1965: Vergleichende cytomorphologische und cytochemische Vitalfärbeversuche an Hochmooralgen. Protoplasma 60, 243—301.

KREBS, I., 1951: Beiträge zur Kenntnis des Desmidiaceen-Protoplasten. I. Osmotische Werte. II. Plastidenkonsistenz. Sitzungsber. Österr. Akad. Wiss., math.-nat. Kl., Abt. I, 160, 579—613.

— 1952: Beiträge zur Kenntnis des Desmidiaceen-Protoplasten. III. Permeabilität für Nichtleiter. Sitzungsber. Österr. Akad. Wiss., math.-nat. Kl., Abt. I, 161, 291—328.

KUNZMANN, R., KIERMAYER, O., 1978: Über die Wirkung verschiedener Antibiotika auf sich differenzierende Zellen von *Micrasterias denticulata*. Sitzungsber. Österr. Akad. Wiss., math.-nat. Kl., Abt. I, 187, 233—255.

LACALLI, T. C., 1973: Cytokinesis in *Micrasterias rotata*. Problems of directed primary wall deposition. Protoplasma 78, 433—442.

— 1974: Composition of the primary wall in *Micrasterias rotata*. Protoplasma 80, 269—272.

— 1975 a: Morphogenesis in *Micrasterias*. I. Tip growth J. Embryol. exp. Morph. 33, 95—115.

— 1975 b: Morphogenesis in *Micrasterias*. II. Patterns of morphogenesis. J. Embryol. exp. Morph. 33 (1), 117—126.

— 1976: Morphogenesis in *Micrasterias*. III. The morphogenetic template. Protoplasma 88, 133—146.

LOUB, W., 1951: Über die Resistenz verschiedener Algen gegen Vitalfarbstoffe. Sitzungsber. Österr. Akad. Wiss., math.-nat. Kl., Abt. I, 160, 829—866.

MAEZAWA, R., UEDA, K., 1979: Formation and multiplication of double cells in a green alga, *Micrasterias crux melitensis* (Ehrbg.) Hass. Cytologia 44, 849—859.

MARČENKO, E., 1966: Über die Wirkung der Gammastrahlen auf Algen (Desmidiaceen). Protoplasma 62, 157—183.

Meindl, U., 1980: Licht- und elektronenmikroskopische Untersuchungen zur Kern- und Chloroplastenmigration von *Micrasterias denticulata* Bréb. Thesis, Salzburg.

— 1981: Störung der Cytomorphogenese bei *Micrasterias denticulata* durch Hemmung der Proteinsynthese. Film D 1425. Inst. Wiss. Film, Göttingen.

— Kiermayer, O., 1979: Antagonistic effect of anti-microtubule substances on penicillin G-induced changes of chloroplast migration in *Micrasterias*. Europ. J. Cell Biol. **20** (1), 131.

— — 1981 a: Biologischer Test zur Bestimmung der Antimikrotubuli-Wirkung verschiedener Stoffe mit Hilfe der Grünalge *Micrasterias denticulata*. Mikroskopie (in press).

— — 1981 b: Über die Kern- und Chloroplastenmigration von *Micrasterias denticulata* Bréb. I. Licht- und elektronenmikroskopische Untersuchungen der Kernmigration nach Behandlung mit Antimicrotubuli-Substanzen. Phyton (Austria) (in press).

— — 1981 c: Über die Kern- und Chloroplastenmigration von *Micrasterias denticulata* Bréb. II. Die Chloroplastenmigration und ihre Veränderung durch verschiedene Stoffe. Phyton (Austria) (in press).

Menge, U., 1973: Elektronenmikroskopische und cytochemische Untersuchungen der Dictyosomen und Vesikel von *Micrasterias denticulata* Bréb. Thesis, Köln.

— 1976: Ultracytochemische Untersuchungen an *Micrasterias denticulata* Bréb. Protoplasma **88**, 287—303.

— Kiermayer, O., 1977 a: Dictyosomen von *Micrasterias denticulata* Bréb. — ihre Größenveränderung während des Zellzyklus. Protoplasma **91**, 115—123.

— — 1977 b: Beobachtung zur Struktur der Dictyosomen von *Micrasterias denticulata* Bréb. Mikroskopie **33**, 168—176.

Mix, M., 1966: Licht- und elektronenmikroskopische Untersuchungen an Desmidiaceen. XII. Zur Feinstruktur der Zellwände und Mikrofibrillen einiger Desmidiaceen vom *Cosmarium*-Type. Arch. Mikrobiol. **55**, 116—133.

— 1968: Zur Feinstruktur der Zellwände in der Gattung *Penium* (*Desmidiaceae*). Ber. Dtsch. Bot. Ges. **80**, 715—721.

— 1969: Zur Feinstruktur der Zellwände in der Gattung *Closterium* (*Desmidiaceae*) unter besonderer Berücksichtigung des Porensystems. Arch. Mikrobiol. **68**, 306—325.

— 1973: Die Feinstruktur der Zellwände der Conjugaten und ihre systematische Bedeutung. Nova Hedwigia **42**, 179—194.

— Manshard, E., 1977: Über Mikrofibrillen-Aggregate in langgestreckten Vesikeln und ihre Bedeutung für die Zellwandbildung bei einem Stamm von *Penium* (*Desmidiales*). Ber. Dtsch. Bot. Ges. **90**, 517—526.

Moor, H., Mühlethaler, K., 1963: Fine structure of frozen-etched yeast cells. J. Cell Biol. **17**, 609—628.

Mueller, S. C., Brown, R. M., 1980: Evidence for an intramembrane component associated with a cellulose microfibril-synthesizing complex in higher plants. J. Cell Biol. **84**, 315—326.

Neuhaus, G., Kiermayer, O., 1981: Rasterelektronenmikroskopische Untersuchungen an Desmidiaceen: Die Poren und ihre Verteilungsmuster. Nova Hedwigia (in press).

Neuhaus-Url, G., 1980: Über die Wirkung von Turgoränderungen auf die Cytomorphogenese und Ultrastruktur von *Micrasterias denticulata* Bréb. Thesis, Salzburg.

Noguchi, T., 1976: Phosphatase activities and osmium reduction in cell organelles of *Micrasterias americana*. Protoplasma **87**, 163—178.

— 1978: Transformation of the Golgi-apparatus in the cell cycle, especially at the resting and earliest developmental stages of a green alga, *Micrasterias americana*. Protoplasma **95**, 73—88.

— Ueda, K., 1979: Effect of cycloheximide on the ultrastructure of cytoplasm in cells of a green alga, *Micrasterias crux melitensis*. Biol. Cell **35**, 103—110.

Palla, E., 1894: Über ein neues Organ der Conjugatenzelle. Ber. Dtsch. Bot. Ges. **12**, 153—162.

Pickett-Heaps, J. D., 1972: Cell division in *Cosmarium botrytis*. J. Phycol. **8**, 343—360.

— Fowke, L. C., 1970: Mitosis, cytokinesis, and cell elongation in the desmid *Closterium littorale*. J. Phycol. **6**, 189—215.

PIHAKASKI, K., KALLIO, P., 1978: Effect of denucleation and UV-irradiation on the subcellular morphology in *Micrasterias*. Protoplasma 95, 37—55.

PRESTON, R. D., 1964: Structural and mechanical aspects of plant cell walls with particular reference to synthesis and growth. In: Formation of wood and forest trees (ZIMMERMANN, M. H., ed.), pp. 169—188. New York: Academic Press.

RAY, P. M., 1961: Hormonal regulation of plant cell growth. In: Control mechanism in cellular processes (BONNER, D. M., ed.), pp. 928—939. New York: Ronald Press Comp.

— 1962: Cell wall synthesis and cell elongation in oat coleoptile tissue. Amer. J. Bot. 49, 928—939.

ROBENEK, H., PEVELING, E., 1977: Ultrastructure of the cell wall regeneration of isolated protoplasts of *Skimmia japonica* Thunb. Planta 136, 135—145.

ROBINSON, D. G., PRESTON, R. D., 1971: Fine structure of swarmers of *Cladophora* and *Chaetomorpha*. I. plasmalemma and Golgi apparatus in naked swarmers. J. Cell Sci. 9 581—601.

SANO, M., UEDA, K., 1980: Relationship between nucleo-cytoplasmic ratio and final cell volume in *Micrasterias crux-melitensis* (*Desmidiaceae, Chlorophyta*). J. Phycol. 16, 52—56.

SCHRÖTER, K., SIEVERS, A., 1971: Wirkung der Turgorreduktion auf den Golgi-Apparat und die Bildung der Zellwand bei Wurzelhaaren. Protoplasma 72, 203—211.

SELMAN, G. G., 1966: Experimental evidence for the nuclear control of differentiation in *Micrasterias*. J. embryol. exp. Morph. 16, 469—485.

SLEYTR, U., KIERMAYER, O., 1979: Hexagonal lattice of "rosettes" in the plasma membrane of *Micrasterias* and its relation to microfibril formation. Europ. J. Cell Biol. 20, 126.

STAEHELIN, L. A., KIERMAYER, O., 1970: Membrane differentiation in the Golgi-complex of *Micrasterias denticulata* Bréb. visualized by freeze-etching. J. Cell Sci. 7, 787—792.

TIPPIT, D. H., PICKETT-HEAPS, J. D., Experimental investigations into morphogenesis in *Micrasterias*. Protoplasma 81, 271—296.

TUTUMI, T., UEDA, K., 1975: A cytochemical study on polysaccharides in cells of *Micrasterias americana*. Cytologia 40, 113—118.

UEDA, K., 1972: Electron microscopical observation on nuclear division in *Micrasterias americana*. Bot. Mag. (Tokyo) 85, 263—271.

— NOGUCHI, T., 1976: Transformation of the Golgi-apparatus in the cell cycle of a green alga *Micrasterias americana*, Protoplasma 87, 145—162.

— YOSHIOKA, S., 1976: Cell wall development of *Micrasterias americana*, especially in isotonic and hypertonic solutions. J. Cell Sci. 21, 617—631.

URL, W., KUSEL-FETZMANN, E., 1973: *Desmidiaceae:* Fortbewegung durch Schleimausscheidung. Film E 1913, Inst. Wiss. Film, Göttingen.

WARIS, H., 1950: Cytophysiological studies on *Micrasterias*. I. Nuclear and cell division. Physiol. Plant. 3, 1—16.

— 1951: Cytophysiological studies on *Micrasterias*. III. Factors influencing the development of enucleated cells. Physiol. Plant. 4, 387—409.

— KALLIO, P., 1964: Morphogenesis in *Micrasterias*. Advan. Morphogen. 4, 45—80.

WEBER, F., 1929: Plasmolyse-Ort. Protoplasma 7, 583—601.

Nuclear Control of Morphogenesis in *Micrasterias*

P. KALLIO and J. LEHTONEN

Department of Biology, Botanical Laboratory, University of Turku, Turku, Finland

With 7 Figures

Contents

I. Introduction

Morphogenesis lies under the control of the nucleus. But there must also be something to be controlled—the cytoplasm with its organelles. The roles played in morphogenesis by these two main cell components are to some degree independent of each other.

The nuclear control of the development of form has its own mechanisms, the discovery of which has been the real beginning of potential understanding of morphogenesis. The mechanism of information—consisting mainly of the genetic code, transcription, translation, and the formation of specific enzymes—controls events in the cytoplasm. Consequently, knowledge of the structure and function of the cytoplasmic units is most important for an understanding of the mutual relationships.

In multicellular organisms, differentiation is complicated: in addition to the genetic control in the individual cells, control also occurs at other levels. The environment affects the cells in different ways, and the neigh-

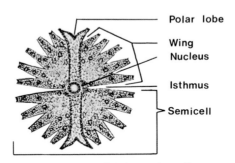

Fig. 1. Normal *Micrasterias* cell.

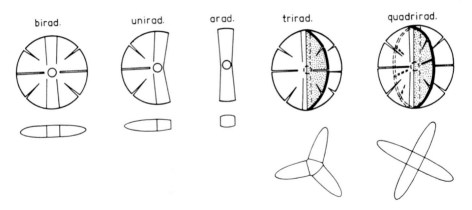

Fig. 2. Different facies in *Micrasterias*.

bouring cells with their connecting plasmodesmata are part of the environment. For these reasons a unicellular, eukaryotic organism, which has its own photosynthetic apparatus and is thus independent of other cells or organisms, is an ideal object for morphogenetic studies at cellular level. This kind of organism can also be cultivated in a synthetic, inorganic medium in controlled conditions.

Ever since the first studies of Hämmerling (*e.g.,* 1943 a, b, 1963), *Acetabularia* has been a favorite object of morphogenetic research, especially in studies on nuclear control using denucleated cells. The splendid results presented at the Mol-symposium in 1972 (Bonotto *et al.* 1972) and in studies and reviews (*e.g.,* those by Werz 1965, Zetsche 1966, and Brachet 1970) form a firm basis for the study of cellular morphogenesis. *Acetabularia*

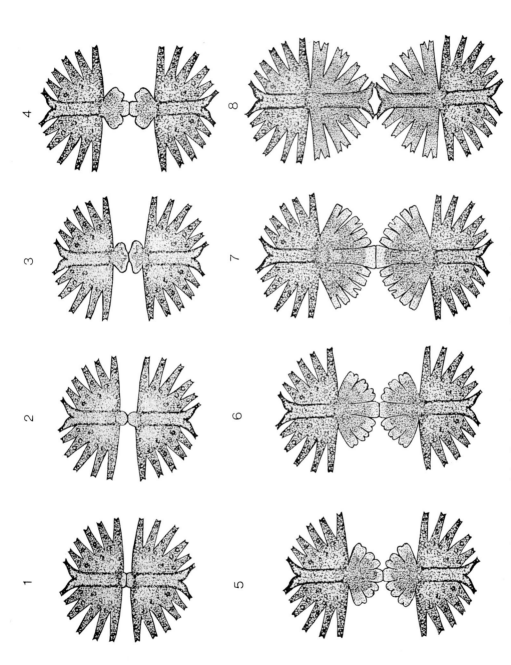

Fig. 3. Normal development of *Micrasterias* cell. About one hour between each stage.

has, however, some very exceptional features, *e.g.*, the unique behavior of its nucleus. Therefore, other unicellular organisms are also needed for the elucidation of morphogenetic problems. *Stentor* and other ciliates have proved to serve this purpose, as the numerous studies of Tartar, Sonneborn, and others have shown.

Another very suitable object is the alga *Micrasterias* (*Desmidiaceae*). The *Micrasterias* cell is complicated enough morphologically to provide many different parameters for morphogenetic analysis, and yet, it is a fairly typical plant cell as regards its functions. The cell is big enough to be manipulated, mitosis can be followed *in vivo* without staining, and development and differentiation are rapid, so the results of experiments or the effects of treatments can be seen within a few hours. The main taxonomical characters of the genus *Micrasterias* were described by Krieger (1937) and morphological analyses have been made by, for instance, Teiling (1950) (*cf.*, also Waris and Kallio 1964). Fig. 1 shows the main features and the terminology used in this paper; Fig. 2 illustrates the different facies in *Micrasterias;* and Fig. 3 shows the normal development of the cell.

Cytomorphogenetic laboratory studies of *Micrasterias* were started in the 1940s in the university of Turku, Finland, by Waris (earlier Warén), after his contacts with Pringsheim and his laboratory (Waris 1950 a, b, 1951), and these studies have been continued in the same laboratory for 40 years. The starting point was long-term studies on culture techniques (Warén 1926, Pringsheim 1930). The next stage was detailed analysis of the division of the cell *in vivo*, and these studies indicated that the beginning of mitosis can be predicted, which is very important for experimental morphogenetic research (Waris 1950). Waris also discussed the extraordinary plasmatic continuity of the structure of the *Micrasterias* cell, paying particular attention to the spontaneously formed uniradiate cells. The progress made by him allowed induction of phenotypic and genotypic variation, such as polyploidy, complex cells and facies changes (Kallio 1949, 1951, 1953 a, b, 1954, 1957, 1959, 1960, 1961, 1963, Waris and Kallio 1964, Kallio and Heikkilä 1969, 1972, Kallio and Lehtonen 1973).

The light microscope, used in all the studies mentioned above, did not allow a proper understanding of certain mechanisms. For example, a plasmatic factor causing the structural continuity of the cell was demonstrated in different ways in those studies, but studies made with the transmission electron microscope by Waddington (1962), Kiermayer with his school (Kiermayer 1970), and also Pihakaski and Kallio (1978) failed to show any structural basis for this; no "plasmatic axes" were to be found. The abundant data provided by TEM studies of the ultrastructure of *Micrasterias* have done much to promote understanding of various morphogenetic details and functions. Drawert and his school studied the ultrastructure without particular reference to the morphogenesis (Drawert and Mix 1960, 1961 a–e, 1962 a–d, 1963). Mix also extended her studies to many other families of *Conjugatae* (Mix 1967, 1973 and the literature cited), as did also Pickett-Heaps (1975). Kiermayer approached morphogenesis through a study of turgor reduced cells. He and his school also paid attention to the dictyosomes and the vesicles formed by them, the formation of the cell wall and the role of the plasmalemma in it, and the microtubules and the role of these organelles in morphogenesis (JAROSCH and KIERMAYER 1962, KIERMAYER and JAROSCH 1962, KIERMAYER 1962, 1964, 1965 a, b, 1966, 1967, 1968, 1970 a, 1970 b, 1972, 1973, KIERMAYER and DOBBERSTEIN 1973, KIERMAYER and FEDTKE 1977, MENGE and KIERMAYER 1977; see p. 147 f.). LACALLI has also contributed greatly to the understanding of the ultrastructure and the theory of morphogenesis, for example, in laser and autoradiographical studies (LACALLI and ACTON 1974, LACALLI 1975 a, b, 1976). UEDA and NOGUCHI (1976) have concentrated on the dictyosome cycle, TOURTE (1972) and PIHAKASKI and KALLIO (1978) have studied the influence of the nucleus on the ultrastructure of the cell, and LEHTONEN (1977) has analysed the morphogenesis by such means as UV-microbeam irradiation, cytochalasin B, which acts on the microfilaments, and also chemical manipulation of microtubules. The present study, survey of cytomorphogenesis in *Micrasterias* with special reference to the influence of nuclear quantity, is based mainly on the studies of the authors at the university of Turku.

II. Increase of Nuclear Control

The increase in cell size attending an increase in nuclear mass has been so well known since the time of HERTWIG (*cf., e.g.,* DARLINGTON 1937), that there is very little to add concerning the constancy of the mass relationship between the nucleus and the cytoplasm. This usually means that few or no changes occur in ploidization. A more fruitful subject of study is changes in the nuclear mass per morphological unit—in *Micrasterias* the wings, lobes, and their subunits. These changes result in new qualities and reveal new relationships.

In the laboratory the usual ways to increase nuclear influence are the production of multinucleate cells and ploidization. The techniques of changing the nuclear mass, *e.g.,* by splitting the nucleus, are difficult to control (WARIS 1958). On the other hand, the effect of the quantity of the nucleus on the development of the morphology can easily be examined by applying chemical or physical inhibitors at some point in the information chain from the nucleus to the site of development. In *Micrasterias,* a further specific method of changing the quantitative relation between the nucleus and the cytoplasm is to induce the formation of complex cells of different types and thus obtain different quantities of nuclear mass per morphological unit.

A. Binucleate Cells (see also p. 177)

In principle there are three different types of binucleate cells in *Micrasterias* (Fig. 4): (1) an almost normally shaped cell with two nuclei in the isthmus; (2) a double cell that has formed after unsuccessful cytokinesis (when the septum has not formed normally) with two sets of morphological units and a nucleus in each of its two isthmuses; (3) a double cell, as in case 2, with the two nuclei situated in the same isthmus, the other isthmus being anucleate.

Although binucleate cells may be produced in *Micrasterias* by such procedures as exposure to continuous light (KALLIO 1953 b), more accurate methods have been developed. Cold shock (KALLIO 1951, 1959, 1963), DNP, and a number of similar chemicals (KALLIO, unpublished), colchicine, and trifluralin (LEHTONEN 1977), and griseofulvin (LEHTONEN, in preparation) have been used to disturb septum formation and induce formation of symmetric binucleate cells with a middle section and one nucleus in each isthmus. The division of these cells leads to development of two normal haploid cells, the binucleate cell being preserved. Centrifugation of the cell is a further very useful and accurate method of producing binucleate cells.

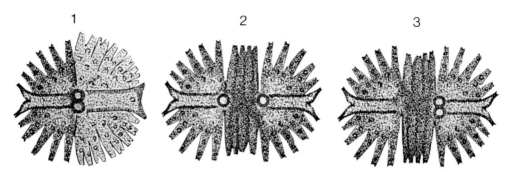

Fig. 4. Different types of binucleate cells: *1* almost normally shaped cell with two nuclei in the isthmus; *2* double cell with a middle section, one nucleus in each isthmus; *3* asymmetric double cell; the one isthmus binucleate and the other anucleate.

Centrifugation must take place at metaphase. A chloroplast bulge then protrudes into the isthmus and pushes the nucleus away from its normal position. After centrifugation, the chloroplast withdraws. The nucleus is normally fastened by microtubules to the isthmus (KIERMAYER 1968; see p. 166), and after centrifugation it will move back to its original position. If centrifugation is correctly timed, however, the septum, which in *Micrasterias* can form only in the centre of the isthmus, develops before the nucleus is in the equatorial plane, thus dividing the cell into two different daughter cells. All the nuclear mass is then in the one cell, and the other is anucleate. On the other hand, if the septum begins to develop when the chloroplast bulge is still in the isthmus, the formation of the septum is disturbed and a middle section develops between the original semicells. The result is a double cell with two isthmuses, one of which contains both nuclei.

When a binucleate and an anucleate cell have formed, the two new semicells develop in quite different ways (Fig. 5)—one under the control of two nuclei and the other without any controlling nucleus. The new semicell of the binucleate cell has some hypertrophic features, compared with a normal, haploid semicell, but they are much less pronounced than in a typical diploid cell, which is clearly larger and more differentiated than a normal haploid cell. Actually, the only effect of the binucleate condition in this case is the slight widening of the wing angle (angle formed by the longitudinal axes

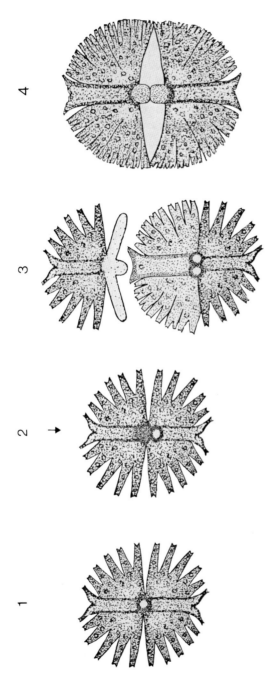

Fig. 5. Development of a cell after centrifugation: *1* normal, haploid *Micrasterias* cell; *2* the cell centrifuged at metaphase; *3* an anucleate and a binucleate cell formed by centrifugation; *4* typical diploid cell formed later from the binucleate cell at early dividing stage.

of the cell and the sinus edge). Why the two nuclei have so little effect on the morphogenesis in this case is not exactly known. But many facts point to the significance of the limiting effect of the structure and quantity of the cytoplasm in the original, haploid mother semicell. When the binucleate cell then divides, the two nuclei unite and two diploid cells are formed. The new semicells are larger than the normal cells but do not attain the size or degree of differentiation typical of a diploid cell.

The different complex cells demonstrate clearly the effects of the changed ratio between the quantities of the nucleus and the cytoplasm. In the asymmetric double cell the nuclear control is not the same in all parts of the cell due to the differences in distances. When this kind of *M. torreyi* cell divides, septum formation normally takes place in the anucleate isthmus as well. But when the new semicells develop, the significance of the distance from the nucleus is seen: the new semicell formed in the anucleate end is always smaller and less differentiated. At that end of the double cell an anucleate, separate cell is formed, and at the other end there develops a diploid cell (Fig. 6). Even this diploid cell has not reached the size and characteristics of a typical diploid type. And when the double cell divides a second time, a facies change usually happens at the anucleate end: the new daughter semicell formed by the middle section is uniradiate with the polar lobe and only one wing (Fig. 7). The number of morphological units is thus diminished.

Further treatment of these types of binucleate cells may result in even more complicated cells. For example, centrifugation of the double cells may result in cells with more than one middle section or a double cell with one nucleus in the one isthmus and more in the other (Kallio 1961).

B. Diploid and Polyploid Cells

The first semicells formed by binucleate cells were found to be smaller than an average diploid cell. The development needs some "mother biomass", which is correlated with the size of the mother semicell. This is also readily observed in cell cultures where, for many reasons, the cells may differ in size. It is seen most clearly in an asymmetric complex cell in which septum formation in the anucleate isthmus has failed. The new semicell formed in the binucleate end is larger than normal. This indicates that a developing semicell can utilize the cytoplasmic resources of the whole structural combination, not only those in the nearest semicell. This kind of situation is typical in *M. thomasiana* var. *notata*.

The characteristic features of the diploid cell are seen in Fig. 5. If the length of the haploid cell is taken as 1.0, the relative length of the diploid cell in, e.g., *M. thomasiana* var. *notata* is 1.24. The width has increased more than the length, so that the ratio of the width to the length is 0.96 in the diploid as against 0.92 in the haploid. Further changes have occurred in the morphology: the wing angle in the haploid cell is approx. 90° or less, while in the diploid cell it exceeds 90°, sometimes even reaching 120°. This means that the widths of the wings react more strongly to increased nuclear influence and also that the edges of the semicells in a

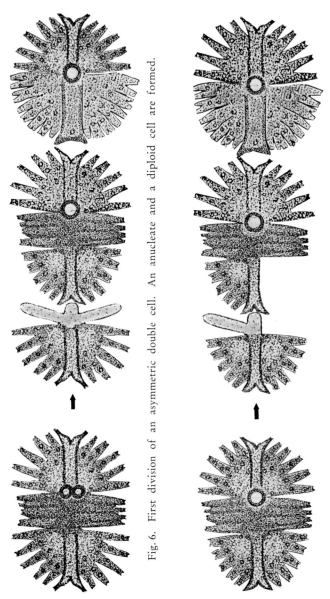

Fig. 6. First division of an asymmetric double cell. An anucleate and a diploid cell are formed.

Fig. 7. Next division of the double cell in Fig. 6. Facies change has occurred at the anucleate end.

diploid cell overlap. The ratio between the total cell volumes of the haploid and diploid cells is 1 : 2.3 in *M. thomasiana* var. *notata* (KALLIO 1951), and the number of the teeth in the lobes of a typical diploid cell is 32, as compared with 16 in a haploid cell. The number of pyrenoids in the diploid cell is also doubled.

The diploid cell has one further typical morphological feature that varies greatly with the species but occurs in all the species studied. The morphology of some tertiary lobes in the wings is often simpler than normal, but at the same time, they show some hypertrophy. This is called "lobe doubling" because the lobes are on two levels when seen from the front. When the physiology of haploid and diploid (*M. thomasiana* var. *notata*) cells is studied and compared, differences are found in, for example, the temperature range in which the cells are able to divide. A temperature of $+ 25\ °C$ is often too high for diploids but not for haploid cells, whereas the lower limit is higher for diploids than haploids. Thus, the temperature range is narrower for diploid cells. Another difference is that division is clearly more rapid in haploid clones.

The increase in nuclear influence in diploid clone cultures may lead to facies changes, the number of wings increasing. A diploid biradiate form may produce a triradiate form. The latter facies arises spontaneously in *M. thomasiana*, is the most constant facies in diploid *M. torreyi*, and has also been found in *M. rotata* and *M. sol*. These changes have also been observed in nature (*e.g.*, HEIMANS 1942). Diploid quadriradiate *M. thomasiana* cells are known as well (KALLIO 1951).

In the quadriradiate form, with a diploid nucleus and four wings, the relation between the nuclear and cytoplasmic quantities has not changed greatly, compared with that in the haploid biradiate cell. As a result, the wings are very similar in the two forms in both size and shape. In triradiate diploid cells, the size and shape of the wings are often intermediate between those of the biradiate diploid and biradiate haploid cells. These forms indicate that in *Micrasterias* the nuclear control of morphogenesis can largely be understood on the basis of the quantity of nucleus per cytoplasmic unit. Diploidy has not any specific qualitative effect at this level of morphogenesis. However, comparison of the rate of division in biradiate and triradiate diploid cells shows that it is more rapid in the latter. This may mean that the facies change has some evolutional value (KALLIO 1961).

Further manipulation of diploid *Micrasterias* cells can produce triploid and tetraploid forms. This can be achieved by centrifuging (KALLIO 1951) or trifluralin treatment (LEHTONEN, unpublished). These forms show increased hypertrophic features when compared with diploid cells, but they have never been found to be capable of division.

III. Decrease of Nuclear Control

A. Anucleate Cells

In the anucleate cells produced by centrifuging as described above, the connection of the cell with the nucleus is broken before telophase, and the only possible nuclear control is that released to the cytoplasm before septum formation. The nucleus has already given some morphogenetic information (*e.g.,* information needed for the first developmental events, for the formation of the "growth centres") (LEHTONEN 1977), but after septum formation, the semicell grows without a controlling nucleus. In this situation almost all the *Micrasterias* species develop in a similar way, forming only three cylindrical lobes corresponding to the main morphological units—the polar lobe and the wings (Fig. 5). The polar lobe tends to be shorter than the wing lobes, which are symmetric. The chloroplast is not usually able to move into a new semicell, and no secondary wall is formed. Anucleate cells typically survive for a very short period, living approx. one day and mostly bursting after this.

In *M. thomasiana* var. *notata,* the new semicell developing on the anucleate cell has three or five lobes, three being more common. The occurrence of five lobes has been held to prove that some nuclear "after-effect" is involved (KALLIO 1951). If the anucleate cell is exposed to UV-radiation after denucleation, only three lobed semicells are formed, and even normal cell divisions produce three-lobed semicells after UV-irradiation. Some "aftereffect" has also been demonstrated in other experiments (KALLIO and HEIKKILÄ 1972). Five-lobed anucleate semicells are known in *M. denticulata* also (KALLIO 1954).

The morphology and formation of anucleate cells have been described in *M. thomasiana* (KALLIO 1949, WARIS 1951), *M. rotata, M. denticulata* (KALLIO 1953), *M. torreyi* (KALLIO 1959), and *M. sol* (KALLIO 1968) (*cf.,* also WARIS and KALLIO 1972).

B. Effect of Some Chemicals

A direct way to diminish nuclear control is to split the nucleus (WARIS 1958), but it is very difficult to control the changes in the chromosome set or the deletions of the chromosomes (see, *e.g.,* KASPRIK 1973). Chemical methods affecting part of the chain of nuclear control, transcription, translation, or protein synthesis are easier to use. Typical inhibitors are RNase, which degrades RNAs, AMD, which prevents the synthesis of RNAs, and puromycin, which prevents translation, causing loosening of the incomplete polypeptide chains from the surface of the ribosomes. The use of these chemicals has resulted in forms with the shape of anucleate cells, and it has proved possible to diminish nuclear control to the desired degree.

KALLIO (1963) showed that RNase is a useful tool in studying morphogenesis in *Micrasterias* (as shown already in *Acetabularia* by STICH and PLAUT 1958), but, unfortunately, there was a misprint in his paper: the concentrations of RNase were printed as mg/litre when they should have been mg/ml. This caused much confusion later. AMD treatments were performed first by SELMAN (1966), and later by KALLIO and HEIKKILÄ (1972). In these experiments, anucleate-shaped forms were not found because the treatments were started after septum formation at the beginning of the development of the new semicells. Since AMD affects transcription, the treatment must be started earlier to eliminate nuclear control completely. Later it was shown that the denucleation caused by centrifugation or UV-irradiation of the nucleus can be imitated very closely with AMD (Lehtonen, in preparation). The course of transcription can also be followed with AMD. The morphogenetic "timetable" of transcription revealed by AMD is similar to that shown by UV-irradiation of the nucleus (Lehtonen 1977).

The results of studies with puromycin are analogous to those obtained with AMD, and because of the reversible effect of the drug, different sequences of the control can be "cut away" during the development while the result is seen in the final form. These results confirm the observation that the nuclear control of morphogenesis proceeds without interruption from the beginning to the end, without any feed-back from the cytoplasm to the nucleus, and independently of the developing form (Lehtonen 1977).

Mercaptoethanol is another chemical that inhibits protein synthesis and can be used in morphogenetic studies. Its effects are similar to those described above (Kallio 1963).

C. Decrease of Nuclear Influence in Complex Cells

As already mentioned (Section II. A), complex cells may also produce anucleate cells, especially in *M. torreyi*. This is seen in the double cells obtained by centrifugation that have two nuclei in one isthmus and none in the other. When such cells divide, a septum normally forms in the anucleate isthmus as well, and this is followed by the development of an anucleate cell (Fig. 6). The nuclei are thus able to induce the same reaction in both isthmuses in spite of the distance between them and the anucleate isthmus. In that isthmus, however, the septum forms a little later, and the semicell developing on the anucleate side of the middle section is smaller than normal. This is due to the weaker influence of the nucleus at the anucleate end of the double cell; the strength of the influence is dependent on the distance of the nucleus from the developing semicell.

The weakened influence of the nucleus may also be evident as a change of facies at the anucleate end of the cell. This never occurs in the first division but usually in the next one (Fig. 7). Thus, this is the opposite situation to the increase in wings caused by increased nuclear influence, *e.g.*, in diploid cells (see Section II. B). The complex cell with a middle section and a uniradiate semicell at the anucleate isthmus does not always continue to divide, but if this occurs, the cell formed at the anucleate and produces a uniradiate, anucleate-shaped semicell.

D. Effect of Irradiation

Since the nucleic acids specifically absorb radiation at the UV-wavelength 253.7 nm, and this deactivates the nuclear information, this method is one of the physical ways of eliminating nuclear control in the cell. Gamma irradiation also has an effect similar to the denucleation of the cell, if the nucleus is irradiated in the division stage (Marčenko 1968, Waris and Rouhiainen 1970).

The effects of irradiation on the nucleus and the nuclear information may be of two kinds. First, the irradiation may eliminate the information, and if the cell is irradiated during development, the treatment diminishes the degree of differentiation. Second, the irradiation may have mutagenic effects.

By adjusting the dose, it is possible to obtain new semicells that are shaped completely like anucleate cells and not even viable. This is achieved by exposing the cells to the standard lethal dose (SLD) described by KALLIO (1951). The anucleate-shaped semicells are, in this case, a little smaller than those in denucleated cells. This has been attributed to disturbance of the RNAs in the cytoplasm as well. In most cases these cytoplasmic RNAs have no effect on the degree of differentiation, as has been demonstrated by irradiating denucleated cells. Moreover, if centrifuged cells are irradiated, the binucleate daughter semicell is larger than the anucleate one (KALLIO 1959). This shows that some nuclear influence is left after irradiation, but this affects only the size and not the differentiation.

Irradiation can be limited to the nucleus by using an UV-microbeam irradiation microscope. In this case the daughter semicells formed by these cells are larger than those formed by totally irradiated cells. They can also be more differentiated, which indicates recovery of the nucleus after UV irradiation due to the nonirradiated cytoplasm (KALLIO and HEIKKILÄ 1972). With the UV-microbeam, it is also possible to study the course of the transcription in detail by irradiating the nucleus in different phases, as mentioned in Section III. B.

Sublethal UV-irradiation applied to interphase cells may also decrease the wing number, *i.e.*, cause facies change. This is analogous to the situation in complex cells in which facies changes may occur in the anucleate end due to decreased nuclear influence. Sublethal irradiation applied to *M. sol* (KALLIO 1968) and *M. torreyi* (KALLIO and HEIKKILÄ 1969) produced uniradiate cells and even clones. On the other hand, irradiation during division does not lead to this result, the facies of the new semicell is always determined before cell division.

This may be illustrated by some experiments (KALLIO and LEHTONEN 1973). If the cytoplasm of both semicells in a dividing cell receives a strong UV dose but the nuclei are not damaged, the daughter semicells sometimes fail to reach normal size or full differentiation, although they may survive. These daughter semicells are symmetric, biradiate; but in the next division, a facies change may take place. Further, using the UV-microbeam, it is possible to irradiate only one wing in the mother semicell during early cell division. Then, too, the new semicell is always symmetric. But in the following cell division, the daughter semicell often produces a uniradiate semicell in which the wing corresponding to the irradiated wing is lacking. This phenomenon agrees well with early observations and theories on the plasmatic continuity of the main cell structure and the theory of plasmatic macro-templates developed to explain this (KALLIO 1963, KALLIO and HEIKKILÄ 1969, KALLIO and LEHTONEN 1973). The fact that the symmetry of the new semicell is determined before differentiation of the main morphological units is connected with the plasmatic structural continuity and may be explained by the theory of "growth centres". The growth of the new semicell is obviously tip growth (LACALLI 1975 a, LEHTONEN 1977), and the initially only hypothetical growth centres seem to be real structures, the first growing points formed for tip growth. These may also be damaged by UV, and their development depends on various factors such as microtubules (KALLIO and LEHTONEN 1973, LEHTONEN 1977).

IV. Aneuploid Chromosome Numbers

Few detailed analyses have been made of aneuploid chromosome numbers and their effect on morphogenesis. In KASPRIK's (1973) cytological study of some morphologically deviating cells, a typical "cross form" appeared to have extra chromosomes. This type of a cell is frequent among the hyperhaploid cells produced when nuclei are split by centrifuging (WARIS 1958). The form of these cells is always somewhat unstable. Another form with abnormal branching of the side lobes and hypertrophic features in *M. thomasiana* had 46 chromosomes instead of the normal 37 to 39, but this form was also apparently unstable (KASPRIK 1973). This type of cell has likewise been found as a result of "unsuccessful" centrifugation by KALLIO. Similar side lobe branching is seen in Fig. 7 B, in KALLIO (1959), and in KASPRIK (1973). WARIS (1958) stresses the quantitative correlation between nuclear control and structural units in the hyperhaploid cells.

V. Facies Change and the Shape of Morphological Units in the Cell

After facies change, the cells are well able to live and divide. In algae collections of the Department of Biology, University of Turku, there are some uniradiate clones dating back to the year 1939 (*M. thomasiana* var. *notata*). An aradiate strain of *M. americana* arose spontaneously in treatment with continuous light (KALLIO 1960) and was cultured for a considerable time, and an aradiate strain that arose in *M. torreyi* in 1967 is still in the collections. A morphological comparison of an aradiate *M. torreyi* and a normal haploid cell must necessarily be restricted to the polar lobes. In the aradiate cell, the size of the polar lobe has increased so much that the volume of the cell may reach the volume of a normal haploid uniradiate cell. In addition, an aradiate cell has some extra spines, "protuberances", at the ends of the lobes. The chloroplast shows some abnormalities, not being distributed so evenly as in biradiate cells. Uniradiate facies are known in *M. thomasiana* var. *notata*, *M. fimbriata*, *M. sol*, *M. americana*, and *M. torreyi*. These all have certain features in common and also have structural features similar to those in biradiate diploid cells. The wing angle is greater than in biradiate cells. This is easily seen, as the polar lobes are not in the same line but form an angle. The increase of nuclear influence is sometimes also evident in lobe doubling in the wings of the uniradiate cells, occurring in the lobes nearest to the polar lobe, as in diploid cells.

A different situation is found in the H-strain, which is some kind of genetic mutation (KALLIO and HEIKKILÄ 1969). The cells vary between uniradiate and aradiate facies and there are some hypohaploid features: the wings are less differentiated than in normal uniradiates, and the wing angle has not increased. The H-strain also has a diploid uniradiate facies. This is not stable and changes to the biradiate form, but this form shows no hypertrophy in the structure and no lobe doubling like that found in normal diploid biradiates. However, in the H-strain, too, the size of the polar lobe is much greater in the aradiate facies.

VI. Differences Between the Polar Lobe and the Wings

There is hardly any experimental evidence concerning morphogenetic differences between the two main lobe types, *i.e.*, the polar lobe and the wings. KALLIO (1951) and SELMAN (1966) have presented some evidence that they differ in nature. On the other hand, in some particular cases, the polar lobe may closely resemble a sublobe of a wing, and certain chemicals can cause the wings to develop some of the main features of the polar lobe (WARIS and KALLIO 1957). The effect of some substances actually differs with the morphological units; with certain treatments, only the polar lobes show a tendency to double (GRÖNBLAD and KALLIO 1954, WARIS and KALLIO 1957, LACALLI 1975 b), although it is usually the wings that react more sensitively.

In the experiments of LACALLI (1975 b) the "veining pattern" of the lobe development was different in the two lobe types. This difference is further reflected in the development after leasing of the two types. It has not been possible to remove the growth centre of the polar lobe and obtain a cell permanently consisting solely of wings. Although such semicells have sometimes been formed (*e.g.*, in complex cells), this feature has not been transmitted to the daughter semicells. And also the change of facies is an indication of some difference in the two lobe types.

VII. Interaction of the Nucleus and the Cytoplasm

When two nuclei in a complex cell divide, mitosis commences simultaneously. This well-known fact, observed in many experiments and reported in the literature, shows that the cytoplasm is the mitogenic factor and determines the time of mitosis, doing so with very great accuracy. New information on this point was obtained from the double cell with one haploid and one diploid nucleus (KALLIO 1961). As already mentioned, the diploid and haploid cells normally have their own rates of division, the haploids being faster. In a cell with one haploid nucleus in one isthmus and a diploid in the other, it is interesting to note that in this case, too, the two nuclei divide simultaneously.

The quantitative relationship between the cytoplasm and the nucleus is clearly reflected in the division rates in different types of complex cells. Especially interesting are the different rates in diploid biradiate and triradiate cells (apparently with idential genomes), the biradiates being much slower (KALLIO 1961) (Section II. B). This may prove that the facies change has some ecological and evolutional significance. The explanation of this fact could lie in differences in the surface—volume ratios.

As is shown by the development of an asymmetric double cell with two nuclei in one isthmus and none in the other, the nucleus is the factor that induces septum formation at division. This is well illustrated in *M. torreyi* cells in which septum formation takes place at nearly the same time in both isthmuses, and they develop completely. On the other hand, other species develop in different ways. In *M. thomasiana,* for instance, the nuclear influence is obviously weaker, and a septum only rarely forms in the anucleate isthmus. Even in cases when septum formation is induced and started, formation is not complete and a middle section develops. If, however, the nuclear influence is very strong, as in triploid cells, complete septum formation is possible (Kallio 1951).

Interaction of the cytoplasm and nucleus has also been discussed in the case of reactivation of the cells after SLD of UV-irradiation (Kallio 1968, Kallio and Heikkilä 1972). Marčenko (1968) has suggested that the nucleus is the site of photoreactivation, which is an enzymatic system in which the nucleic acids are repaired after UV-irradiation (Saito and Werbin 1969 a, b). This has been shown in *Micrasterias,* also, the most effective way to cause photoreactivation is to illuminate the damaged cells with blue light soon after irradiation. In this way it is possible to reactivate cells so that the DRF value recorded is about 0.5. The effect of the photoreactivation is also seen in the form of the developing irradiated cells (Kallio and Heikkilä 1972). But experiments have shown that undamaged cytoplasm may also be able to reactivate the nucleus; cells in which only the nucleus has been irradiated may survive, while cells that have been completely irradiated will die (Kallio and Heikkilä 1972). These experiments also show that semicells with a typical anucleate morphology may still be able to function as mother semicells in normal cell divisions. The shape of a new semicell thus indicates only the nuclear physiological condition during cell division, and the anucleate-shaped form shows that the semicell has developed without nuclear control and that in normal cases the cell will not survive. Unfortunately transplantation of the nucleus has not succeeded in *Micrasterias,* and it is not known whether a denucleated cell can recover after obtaining a new nucleus.

VIII. Nuclear Influence and the Ultrastructure of the Cell

The above studies are concerned with the influence of the nucleus on morphogenesis in *Micrasterias,* but between the studied nuclear information chain and the final form, there is a wide gap, the details of which are not exactly known and not studied in this Chapter. The nucleus has specific effects on the cytoplasmic components and structures, and these and their possible significance can be studied in this connexion.

The morphogenesis necessitates a specific ultrastructure. The most important components in this respect are the dictyosomes, the microtubules, possibly the microfilaments, and the cell membranes functioning in cell wall formation. A key role is played by the dictyosomes, which produce vesicles containing cell wall material. Kiermayer (1970 a, b) has described several different types of vesicles, the most important in morphogenesis being the dark vesicles and the flat vesicles. The former contain primary wall material and transport it to the growing points, and the latter function in formation of the secondary wall (see p. 154 f.).

The real relations between the nucleus and the dictyosomes and the degree of independence of the dictyosomes are not well known, but denucleation has effects on the dictyosomes. Their normal cycle, described by UEDA and NOGUCHI (1976), and their ultrastructure is disturbed (PIHAKASKI and KALLIO 1978). At the beginning of cell division, there are no marked differences in the amounts of vesicles between normal and anucleate cells. It actually seems that UV-irradiation may even activate the production of dark vesicles (PIHAKASKI and KALLIO 1978). But the vesicle types normally appearing later in the development are lacking in denucleated cells, and the wall material production and transport slow down and eventually stop, so that growth and differentiation are interrupted and wall material does not collect much inside the cell, as, for example, in turgor reduced cells (KIERMAYER 1962, KIERMAYER and JAROSCH 1962). The expansion of the cell is thus prevented, though the information is not affected. The dictyosomes are, thus, a very important link between the specific enzymes formed by the nucleus and the final form. Their number and capacity to produce cell wall material seem to depend on the volume of the cytoplasm in the cell. A certain volume may be a feature of the "mature state" of the cytoplasm (described by KALLIO 1959), which is needed for cell division, and this may also partly explain why the volume of the first daughter semicells formed by binucleate cells is not much greater than in haploid cells, though older diploid cells are much greater and more differentiated (Section II. *A*).

The microtubules do not play a key role in the actual differentiation, but they affect the transport of the cytoplasm from the mother semicell to the daughter semicell and can move and anchor the nucleus (KIERMAYER 1970). Evidently they also influence septum formation and growth centre formation (LEHTONEN 1977). Denucleation affects the microtubules, and in anucleate cells, their number soon diminishes (PIHAKASKI and KALLIO 1978), but this does not play a role in the determination of the structure of an anucleate cell. The cell membranes may also be affected by denucleation; for instance, it may change their properties and their function in cell wall formation and morphogenesis described by KIERMAYER (1965 a, 1970 b, 1973, KIERMAYER and DOBBERSTEIN 1973), but these changes are apparently slow and may not be significant during morphogenesis. On the other hand, local destruction of the membranes, *e.g.*, by laser beam (LACALLI 1975 a) may have drastic effects on the morphogenesis, showing the importance of the membranes. The microfilaments obviously act as some kind of mediators between the nuclear information and the final form, as shown, for instance, by LEHTONEN (1977, in preparation). The microfilaments cause cellular streaming and can evidently guide the transport of the vesicles in the cell, under the control of the nucleus. Cytochalasin B has strong effects on the morphogenesis, and denucleation obviously also affects the functioning of the microfilaments, though this has not been studied.

As regards the other organelles, the ribosomes are affected by denucleation. Their number is not readily changed, but their grouping shows alterations. No typical spiral organizations are found in denucleated or in UV-irradiated cells (Pihakaski and Kallio 1978). Some other organelles, such as the chloroplast and mitochondria, are affected only slightly and change slowly due to their possessing their own DNAs and being largely independent (Tourte 1972, Pihakaski and Kallio 1978).

The events between the nuclear information and the final form are complicated and not easily studied, but some information can be obtained by studying the influence of the nucleus on the organelles and the ultrastructure of the cell.

IX. Comparison of the Nucleo-Cytoplasmic Interaction in *Micrasterias* and *Acetabularia*

As *Acetabularia* is another very thoroughly studied cell (see p. 119 f.), comparison between these two cells might reveal some general tendencies. The *Acetabularia* cell deviates considerably from a typical plant cell in its nuclear relations and functions as well as in its size and life cycle. But in some respects, *e.g.*, in studies of denucleated cells, *Acetabularia* gives information that can be obtained only from a cell of this kind.

One of the most important and marked differences between *Micrasterias* and *Acetabularia* is the very long life span of the genetic information in *Acetabularia*, compared with *Micrasterias*. *Micrasterias* has a cell cycle of 3 to 5 days, *Acetabularia* some months. But in *Micrasterias*, no species-specific control is seen approx. one hour after denucleation, while in *Acetabularia* the normal morphogenetic control functions for many weeks. In *Acetabularia* the species-specific form is reached even when no nucleus is present, while in *Micrasterias* differentiation needs the continuous presence of a controlling nucleus. This explains why such treatments as exposure to RNase and UV irradiation have different effects on the morphogenesis of the two species (Kallio 1963).

In *Acetabularia* the morphological development (Hutbildung) affects the nucleus and causes the division of the secondary nucleus; some feed-back occurs. In *Micrasterias* no specific morphological features are correlated with the onset of the mitotic processes; the maturity of the cell is not clearly evident from the morphology, although many features seem to indicate it (Waris 1950 a). There are, however, strong arguments for the theory that some abrupt change in the cytoplasm determines the beginning of mitosis in *Micrasterias*, *e.g.*, the evidence provided by the double cells.

X. Main Ideas on the Nucleo-Cytoplasmic Relationships in *Micrasterias* Morphogenesis

The presented relationships between the nucleus and the cytoplasm in *Micrasterias* morphogenesis can be roughly and schematically presented as follows:

— The basis for the morphogenesis of a semicell is created during the preceding interphase of the mother semicell. Some kind of maturity in the mother semicell is needed for mitosis and for the structural continuity of the main morphological units of the cell, *i.e.*, the functioning of the macro-templates. This continuity can be disturbed by such treatments as UV-irradiation, with resultant elimination of morphological units.

— Mediating between the plasmatic macrotemplates and the units of the new semicell are the growth centres. These and their formation are also sensitive to UV. These are the primary growing points of the lobes, the growth of which is tip growth, resembling, for instance the growth of fungal hyphae.

— When the growth centres have formed, the daughter semicell can develop three lobes without the control of the nucleus. The lobes represent the wings and the polar lobe, and in anucleate cells they remain quite un-differentiated.

— The normal, species-specific growth and differentiation—controlled tip growth with branching after determined periods—requires continuous nuclear control. This control proceeds without interruption from the beginning to the end without any feed-back between the cytoplasm and the nucleus. Cells may therefore occur in which, for instance, differentiation has stopped at an early phase but the nucleus is not damaged, or is reactivated, and which have developed a secondary wall and are viable.

— The wings and the polar lobe develop in different ways under the control of the nucleus, *i.e.*, they react differently to the control. This is possibly due to differences in the structure of the growth centres or growing points.

— Increased nuclear influence in a cell (*e.g.*, after ploidization) increases the complexity and size of the new semicells developed by it. The number of the wings may also be increased. A reduction of nuclear influence has the opposite effect on the morphology.

— When the morphology of the wings in cells with different wing numbers (different facies) are compared, it appears that in *Micrasterias* it is possible and useful to study the nuclear control expressed as the amount per morpho-logical unit. The importance of this quantitative relationship is revealed for instance, by comparison between biradiate haploid and quadriradiate diploid cells. In this case the nuclear influence per wing is the same, and the units are similar. In a biradiate diploid cell, the form and size of the wings show clear hypertrophic features, and triradiate diploids have some features intermediate between biradiate diploids and biradiate haploids. Uniradiate haploid cells also show hypertrophy.

— Polyploid cells also have some typical differences from haploid ones in their physiology; differences in the amount of influence per cytoplasmic unit thus have some physiological consequences. The triradiate and biradiate diploid cells differ clearly in the rate of division and clone growth, the triradiate being faster. The quantitative nucleo-cytoplasmic relationship may thus have some ecological and evolutionary significance.

— The cytoplasm controls the commencement of mitosis in the nucleus. The normal prerequisites being maturity of the cytoplasm and the plasmatic templates.

— The nucleus entering mitosis controls the initiation of the septum, the site of which is determined by the plasmatic structure.

— The nucleus controls (to a greater or lesser extent) the structure and function of various cytoplasmic organelles, *e.g.*, the cycle and division of the dictyosomes and their vesicle production, the occurrence of microtubules, the functioning of the microfilaments, ribosomal behavior, the support of the membranes, and the structure of the mitochondria. All these are affected by denucleation of the cell and are thus components of the mechanism controlled by the nucleus and needed for normal function and development (see also p. 181 f.). Disturbance of this mechanism causes abnormal development. But it is also evident from the development of the *Micrasterias* cell that the cytoplasm and its organelles have some degree of independence.

References

Bonotto, S., Goutier, R., Kirchmann, R., Maisin, J.-R. (eds.), 1972: Biology and radio-biology of anucleate systems II. Plant cells. New York-London: Academic Press.

Brachet, J., 1970: Concluding remarks. In: Biology of *Acetabularia* (Brachet, J., Bonotto, S., eds.), pp. 273—291. New York-London: Academic Press.

Darlington, C. D., 1937: Recent advances in cytology. London-Philadelphia.

Drawert, H., Mix, M., 1960: Licht- und elektronenmikroskopische Untersuchungen an Desmidiaceen. III. Mitteilung: Der Nucleolus im Interphasekern von *Micrasterias rotata*. Flora 150, 185—190.

— — 1961 a: Licht- und elektronenmikroskopische Untersuchungen an Desmidiaceen. II. Mitteilung: Hüllgallerte und Schleimbildung bei *Micrasterias, Pleurotaenium* und *Hyalotheca*. Planta 56, 237—261.

— — 1961 b: Licht- und elektronenmikroskopische Untersuchungen an Desmidiaceen. IV. Mitteilung: Beiträge zur elektronenmikroskopischen Struktur des Interphasekerns von *Micrasterias rotata*. Z. Naturforsch. 16 b, 546—551.

— — 1961 c: Licht- und elektronenmikroskopische Untersuchungen an Desmidiaceen. V. Mitteilung: Über die Variabilität der Chloroplastenstruktur bei *Micrasterias rotata*. Planta 56, 648—665.

— — 1961 d: Licht- und elektronenmikroskopische Untersuchungen an Desmidiaceen VII. Mitteilung: Der Golgi-Apparat von *Micrasterias rotata* nach Fixierung mit Kaliumpermanganat und Osmiumtetroxid. Mikroskopie 16, 207—212.

— — 1961 e: Licht- und elektronenmikroskopische Untersuchungen an Desmidiaceen. VIII. Mitteilung: Die Chondriosomen von *Micrasterias rotata*. Flora 151, 487—508.

— — 1962 a: Licht- und elektronenmikroskopische Untersuchungen an Desmidiaceen. IX. Mitteilung: Die Struktur der Pyrenoide von *Micrasterias rotata*. Planta 58, 50—74.

— — 1962 b: Licht- und elektronenmikroskopische Untersuchungen an Desmidiaceen. X. Mitteilung: Beiträge zur Kenntnis der „Häutung" von Desmidiaceen. Arch. Mikrobiol. 42, 96—109.

— — 1962 c: Zur Frage von Struktur und Funktion des Golgi-Apparates in Pflanzenzellen. Sitzber. Ges. Beförd. ges. Naturwiss. Marburg 83/84, 361—382.

— — 1962 d: Zur Funktion des Golgi-Apparates in der Pflanzenzelle. Planta 58, 448—452.

— — 1963: Licht- und elektronenmikroskopische Untersuchungen an Desmidiaceen. XI. Mitteilung: Die Struktur von Nucleolus und Golgi-Apparat bei *Micrasterias denticulata* Bréb. Portugaliae Acta Biologica Sér. A 8, 17—28.

Grönblad, R., Kallio, P., 1954: A new genus and a new species among the desmids. Botaniska Notiser 1954, 167—178.

Hämmerling, J., 1943 a: Entwicklung und Regeneration von *Acetabularia crenulata*. Zschr. Ind. Abst. Vererb. 81, 85—113.

HÄMMERLING, J., 1943 b: Ein- und zweikernige Transplantate zwischen *Acetabularia mediter-ranea* und *A. crenulata*. Zschr. Ind. Abst. Vererb. **81**, 114—180.
— 1963: The role of the nucleus in differentiation especially in *Acetabularia*. Symp. Soc. Exper. Biol. **17**, 127—137.

HEIMANS, J., 1942: Triquetroes forms of *Micrasterias*. Blumea, Suppl. **2**, 52—63.

JAROSCH, R., KIERMAYER, O., 1962: Die Formbildung von *Micrasterias*-Zellen nach lokaler Lichteinwirkung. Planta **58**, 112.

KALLIO, P., 1949: Artificially produced binuclear, diploid and anuclear desmids. Arch. Soc. "Vanamo" **2**, 42—44.
— 1951: The significance of nuclear quantity in the genus *Micrasterias*. Ann. Bot. Soc. "Vanamo" **24**, 1—120.
— 1953 a: On the morphogenetics of the desmids. Bull. Torrey Bot. Club **80**, 247—263.
— 1953 b: The effect of continued illumination on the desmids. Arch. Soc. "Vanamo" **8**, 58—74.
— 1954: Morphogenetic studies in *Micrasterias rotata* var. *evoluta* Turner (*Desmidiaceae*). Arch. Soc. "Vanamo" **8**, 118—122.
— 1957: Studies on artificially produced diploid forms of some *Micrasterias* species (*Desmidiaceae*). Arch. Soc. "Vanamo" **11**, 193—204.
— 1959: The relationship between nuclear quantity and cytoplasmic units in *Micrasterias*. Ann. Acad. Sci. fenn. A, IV Biologica **44**, 1—44.
— 1960: Morphogenetics of *Micrasterias americana* in clone culture. Nature **187**, 164—166.
— 1961: Mitotic cycles and mitogenesis in *Micrasterias*. Cytologia **26**, 155—169.
— 1963: The effects of ultraviolet radiation and some chemicals on morphogenesis in *Micrasterias*. Ann. Acad. Sci. fenn. A, IV Biologica **70**, 1—39.
— 1968: On the morphogenetic system of *Micrasterias sol*. Ann. Acad. Sci. fenn. A, IV Biologica **124**, 1—23.
— HEIKKILÄ, H., 1969: UV-induced facies change in *Micrasterias torreyi*. Österr. Bot. Z. **116**, 226—243.
— — 1972: On the effect of elimination of nuclear control in *Micrasterias*. In: Biology and radiobiology of anucleate systems II. Plant cells (BONOTTO, S., GOUTIER, R., KIRCH-MANN, R., MAISIN, J.-R., eds.), pp. 145—164. New York-London: Academic Press.
— LEHTONEN, J., 1973: On the plasmatic template system in *Micrasterias* morphogenesis. Ann. Acad. Sci. fenn. A, IV Biologica **199**, 1—6.

KASPRIK, W., 1973: Beiträge zur Karyologie der Desmidiaceen-Gattung *Micrasterias Ag*. Beitr. Nova Hedwigia **42**, 115—137.

KIERMAYER, O., 1962: Die Rolle des Turgordrucks bei der Formbildung von *Micrasterias*. Ber. Deutsch. Bot. Gesellschaft **75**, 77—81.
— 1964: Untersuchungen über die Morphogenese und Zellwandbildung bei *Micrasterias denticulata* Bréb. Protoplasma **59**, 76—132.
— 1965 a: Zur Mechanik der Formbildung von *Micrasterias*. Ber. Dtsch. Bot. Gesellschaft **78**.
— 1965 b: Plasmatische Strukturen in experimentell hervorgerufenen Zellwandverdickungen von *Micrasterias denticulata* Bréb. Planta (Berl.) **66**, 216—220.
— 1966: Differenzierung und Wachstum von *Micrasterias denticulata* (*Conjugatae*). Institut für den wissenschaftlichen Film, Wissenschaftlicher Film C 924/1966 (Göttingen).
— 1967: Dictyosomes in *Micrasterias* and their division. J. Cell Biol. **35**, 68 A.
— 1968: The distribution of microtubules in differentiating cells of *Micrasterias denticulata* Bréb. Planta (Berl.) **83**, 223—236.
— 1970 a: Elektronenmikroskopische Untersuchungen zum Problem der Cytomorphogenese von *Micrasterias denticulata* Bréb. Protoplasma **69**, 97—132.
— 1970 b: Causal aspects of cytomorphogenesis in *Micrasterias*. Ann. N.Y. Acad. Sci. **175**, 686—701.
— 1971: Elektronenmikroskopischer Nachweis spezieller cytoplasmatischer Vesikel bei *Micrasterias denticulata* Bréb. Planta (Berl.) **96**, 74—80.

14*

Kiermayer, O., 1972: Beeinflussung der postmitotischen Kernmigration von *Micrasterias denticulata* Bréb. durch das Herbizid Trifluralin. Protoplasma 75, 421—426.

— 1973: Feinstrukturelle Grundlagen der Cytomorphogenese. Ber. Dtsch. Bot. Ges. 86, 287—291.

— Dobberstein, B., 1973: Membrankomplexe dictyosomaler Herkunft als „Matrizen" für die extraplasmatische Synthese und Orientierung von Mikrofibrillen. Protoplasma 77, 437—451.

— Fedtke, C., 1977: Strong anti-microtubule action of amiprophos-methyl (APM) in *Micrasterias*. Protoplasma 92, 163—166.

— Jarosch, R., 1962: Die Formbildung von *Micrasterias rotata* Ralfs. und ihre experimentelle Beeinflussung. Protoplasma 54, 382—420.

Krieger, W., 1937: Die Desmidiaceen. Rabenhorts Kryptogamenflora XIII/2.

Lacalli, T. C., 1975 a: Morphogenesis in *Micrasterias* I. Tip growth. J. Embryol. exp. Morph. 33, 95—115.

— 1975 b: Morphogenesis in *Micrasterias* II. Patterns of morphogenesis. J. Embryol. exp. Morph. 33, 117—126.

— 1976: Morphogenesis in *Micrasterias* III. The morphogenetic template. Protoplasma 88, 133—146.

— Acton, A. B., 1974: Tip growth in *Micrasterias*. Science 183, 665—666.

Lehtonen, J., 1977: Morphogenesis in *Micrasterias torreyi* Bail. and *M. thomasiana* Arch. studied with UV microbeam irradiation and chemicals. Ann. Bot. Fennici 14, 165—190.

Marčenko, E., 1968: The site of photoreactivation of reproductive ability in the alga *Netrium* as determined by exposure to a microbeam of visible light. Rad. Bot. 8, 325—338.

Menge, U., Kiermayer, O., 1977: Dictyosomen von *Micrasterias denticulata* Bréb. — ihre Größenveränderung während des Zellzyklus. Protoplasma 91, 115—123.

Mix, M., 1965: Zur Variationsbreite von *Micrasterias swainei* Hastings und *Staurastrum leptocladum* Nordst. sowie über die Bedeutung von Kulturversuchen für die Taxonomie der Desmidiaceen. Archiv für Mikrobiologie 51, 168—178.

— 1967: Zur Feinstruktur der Zellwände in der Gattung *Penium* (*Desmidiaceae*). Ber. dtsch. Bot. Ges. Bd. 80, 715—721.

— 1973: Die Feinstruktur der Zellwände der Conjugaten und ihre systematische Bedeutung. Beitr. Nova Hedwigia 42, 179—194.

Pickett-Heaps, J. D., 1975: The green algae. Sunderland, Ma.: Sinauer Assoc.

Pihakaski, K., Kallio, P., 1978: Effect of denucleation and UV-irradiation on the subcellular morphology in *Micrasterias*. Protoplasma 95, 37—55.

Pringsheim, E. G., 1930: Die Kultur von *Micrasterias* and *Volvox*. Arch. Protistenkunde 72, 1—47.

Saito, N., Werbin, H., 1969 a: Evidence for a photoreactivating enzyme in higher plants. Photochem. Photobiol. 9, 389—393.

— — 1969 b: Action spectrum for a DNA-photoreactivating enzyme isolated from higher plants. Rad. Bot. 9, 421—425.

Selman, G. G., 1966: Experimental evidence for the nuclear control of differentiation in *Micrasterias*. J. Embryol. exp. Morph. 16, 469—485.

Stich, H., Plaut, W., 1958: The effect of ribosomes on protein synthesis in nucleated and enucleated fragments of *Acetabularia*. J. Biophys. Biochem. Cytol. 4, 119—121.

Teiling, E., 1950: Radiation of Desmids, its origin and its consequences as regards taxonomy and nomenclature. Bot. Not. 150, 299—327.

Tourte, M., 1972: Ultrastructural investigations on anucleate cells in *Micrasterias fimbriata*. Comparison to three-lobed actinomycin D treated cells. In: Biology and radiobiology of anucleate systems II. Plant cells (Bonotto, S., Goutier, R., Kirchmann, R., Maisin, J.-R., eds.), pp. 193—223. New York-London: Academic Press.

UEDA, K., NOGUCHI, T., 1976: Transformation of the Golgi-apparatus in the cell of a green alga, *Micrasterias americana*. Protoplasma **87**, 1–3, 145—162.

WADDINGTON, C. H., 1962: New patterns in genetics and development. New York: Columbia University Press.

WARÉN, H., 1926: Nahrungsphysiologische Versuche an *Micrasterias rotata*. Comm. Biol. Soc. Sci. Fenn. II, **8**, 1—42.

WARIS, H., 1950 a: Cytophysiological studies on *Micrasterias* I. Nuclear and cell division. Phys. Plant. **3**, 1—16.

— 1950 b: Cytophysiological studies on *Micrasterias* II. The cytoplasmic framework and its mutation. Phys. Plant. **3**, 236—246.

— 1951: Cytophysiological studies on *Micrasterias* III. Factors influencing the development of enucleate cells. Phys. Plant. **4**, 387—409.

— 1958: Splitting of the nucleus by centrifuging in *Micrasterias*. Ann. Acad. Sci. fenn. A, IV, Biologica **40**, 1—20.

— KALLIO, P., 1957: Morphogenetic effects of chemical agents and nucleo-cytoplasmic relations in *Micrasterias*. Ann. Acad. Sci. fenn. A, IV, Biologica **37**, 3—16.

— — 1964: Morphogenesis in *Micrasterias*. Adv. in Morphogenesis **4**, 45—80.

— — 1972: Effects of enucleation on *Micrasterias*. In: Biology and radiobiology of anucleate systems II. Plant cells (BONOTTO, S., GOUTIER, R., KIRCHMANN, R., MAISIN, J.-R., eds.), pp. 137—144. New York-London: Academic Press.

— ROUHIAINEN, I., 1970: Permanent and temporary changes in *Micrasterias* induced by gamma rays. Ann. Acad. Sci. fenn. A, IV, Biologica **167**, 1—13.

WERZ, G., 1965: Determination and realization of morphogenesis in *Acetabularia*. In: Genetic control of differentiation. Brookhaven Symposia in Biology **18**, 185—203.

ZETSCHE, K., 1966: Regulation der zeitlichen Aufeinanderfolge von Differenzierungs-vorgängen bei *Acetabularia*. Zeitschrift für Naturforschung **21 b**, 375—379.

Formation and Distribution of Cell Wall Pores in Desmids

G. Neuhaus and O. Kiermayer

Max-Planck Institute for Cell Biology, Ladenburg near Heidelberg, Federal Republic of Germany, and Botanical Institute, University of Salzburg, Salzburg, Austria

With 19 Figures

Contents

I. Introduction

Desmids offer a fascinating system for cytological research. This chapter deals with a certain aspect of desmidiology—the formation and distribution of cell wall pores.

The placoderm desmids (Desmidiales) can be subdivided into two distinct groups: the *Closteriinae* and the *Desmidiinae*. The *Closteriinae* possess three kinds of cell wall layers in the interphase stage: (1) the compact outer layer; (2) the primary wall, and (3) the secondary wall (Mix 1966, 1969, 1972, Pickett-Heaps and Fowke 1970). In this group the cell wall pores travers only the outer layer. Under these pores the inner cell wall layers (primary wall and secondary wall) only show a less dense wall material in contrast to the pore-surrounding area (Mix 1969). The *Desmidiinae* have pores in their secondary wall (Drawert and Mix 1961, Mix 1966, 1973, Kiermayer 1970 a, b, Pickett-Heaps 1972, Kiermayer and Staehelin 1972, Chardard 1977). The primary wall, which is layed down during morphogenesis, is shed after the secondary wall material is fully deposited (see p. 151). The outer surface of the secondary wall is covered with a thin fibril-free layer, that is also penetrated by pores (Kiermayer and Staehelin 1972). The cell wall pores and their function of slime excretion was studied earlier by several investigators (Hauptfleisch 1888, Lütkemüller 1902, Schröder 1902, Krieger 1937). However, their formation and distribution over the cell wall first became clear through investigations carried out employing transmission electron microscopy (Lhotský 1949, Drawert and Mix 1961, Mix 1969, Kiermayer 1970 a, 1977, Kiermayer and Staehelin 1972, Pickett-Heaps 1972, Chardard 1977) and scanning electron microscopy (Pickett-Heaps 1973, 1974, Lacalli and Harrison 1978, Neuhaus and Kiermayer 1979, Neuhaus 1980, Neuhaus and Kiermayer 1981).

This article attempts a comprehensive description of (1) the formation of cell wall pores during morphogenesis of desmids, (2) the shaping of pores on the cell surface, and (3) the distribution and arrangement of these pores.

II. Formation of Cell Wall Pores

The process of formation of pores in *Micrasterias denticulata* starts at the beginning of secondary wall material deposition (Kiermayer and Staehelin 1972, Dobberstein 1973), about 4 hours after septum formation (see pp. 150, 158). In this stage of morphogenesis a special type of vesicle, which is described as the "LVh" (= large vesicle with homogeneous contents; see p. 156 f.), can be observed near the plasma membrane (Fig. 1/stage *A*; Fig. 5). This vesicle, also known as the "pore vesicle" (PV; Dobberstein 1973, Kiermayer 1977), cytochemically differs from other intracellular LVs. Menge (1976) in her cytochemical investigations showed that the PV contain neutral polysaccharids. In contrast to the other large vesicles— LVf, LVfrei, and LVdict (see p. 156 f.)—this type of vesicle shows lower contrast in transmission electron micrographs and does not respond with an expansion after longer glutaraldehyde fixation. The round or elliptical-shaped PV possess a diameter of approximately 0.30 µm (Menge 1976; Fig. 5).

The approach of the pore vesicle toward and its incorporation into the plasma membrane is probably determined by a membrane recognition mechanism (Pickett-Heaps 1972, Kiermayer 1977). The latter author was able to demonstrate a special "coat" of such vesicles that may be functional for membrane recognition.

After fusion of the pore vesicle membrane with the plasma membrane, the contents of the P-vesicle builds up a plug that is situated between the primary wall and the plasma membrane (Fig. 1/stage *B;* Fig. 6). This initial pore stage shows the same cytochemical reactions as the PV (Menge 1976). Bands of microfibrils of the growing secondary wall are forced to surround the pore plugs (Fig. 8; Kiermayer and Staehelin 1972). Such pore initials were never present during primary wall deposition. Pickett-Heaps (1972) also observed similar plugs during morphogenesis of *Cosmarium botrytis*.

In the following stage the plug is being formed, as can be seen in Fig. 1/ stage C. This stage represents the phase following the completion of secondary wall deposition. The thick (0.38 to 0.45 µm) secondary wall is disrupted in the area of the pore plug (diameter about 0.20 µm). The transition from pore plug to the differentiated cell wall pore is quite unclear. This process

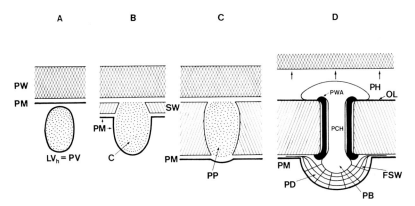

Fig. 1. Developmental stages of formation of cell wall pores in desmids; stage A: LV_h = PV (large vesicle with homogeneous contents = pore vesicle), PM plasma membrane; PW primary wall; stage B: contents (C) of the incorporated pore vesicle, PM plasma membrane, SW secondary wall; stage C: the secondary wall is interrupted by the pore plug (PP), PM plasma membrane; stage D: the differentiated cell wall pore ("pore apparatus") after shedding of the primary wall (arrows), FSW fibrilous "spider web", OL outer layer, PB pore bulb, PCH pore channel, PD pore depression, PH pore head, PM plasma membrane, PWA pore wall.

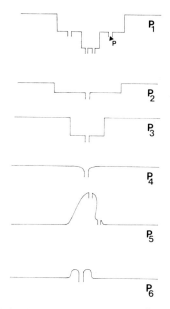

Fig. 2. Schematic drawing of the six pore types (P_{1-6}); cell wall pore (P). From NEUHAUS and KIERMAYER 1981.

Table 1. *Values of Pores in Different Desmids*. From Neuhaus and Kiermayer 1981

Algae	l. × w.	Pore-φ	Pore type	Pore systems	Pore patterns
Enastrum oblongum	135 × 75	0.14–0.16	1, 2, 3	CPS, R	hex., tetr.
Enastrum insigne	100 × 48	0.12–0.15	2, 3, 4	CPG, R	hex., tetr.
Enastrum denticulatum	25.1 × 19.5	0.9–0.14	(2), 4, 5	CPP	SDP
Enastrum ansatum	100 × 48	0.11–0.15	1, 2, 3, 4	CPS, R	hex., tetr.
Enastrum didelta	148 × 70	0.14–0.16	1, ?	CPS, R	—
Enastrum sinnosum	49 × 28	0.12–0.14	1, 2, 3	CPS	hex.
Enastrum intermedium	64 × 36	0.13–0.15	2, 3, 4	CPG, R	SDP
Enastrum bidentatum	54 × 29	0.14–0.16	4, 5	CPP	SDP
Enastrum pectinatum		0.14–0.16		AP	—
Cosmarium subcucumis	70.5 × 41	0.16–0.19	4	AP, IP	hex.
Cosmarium pseudopyramidatum	56.4 × 31.5	0.14–0.17	2, (3)	AP, IP	hex.
Cosmarium cucurbita	37.8 × 18.3	0.10–0.13	3	AP, IP	hex.
Cosmarium ornatum	27.3 × 31.2	0.10–0.12	4	AP, IP	SDP
Cosmarium spec.	27.1 × 21.3	0.07 (0.13, AP)	2, 4	AP	hex.
Staurastrum arachne	20.2 × 34.4	0.11–0.13	4	AP, IP	SDP
Staurastrum teliferum	55.5 × 35	0.11–0.13	4	AP, IP	SDP
Staurastrum scabrum	29.3 × 26	0.11–0.13	4	AP, IP	SDP
Tetmemorus brébissonii	68.1 × 15	0.12	3, 4 (IP)	IP	hex.
Tetmemorus granulatus	157.2 × 30.1	0.12–0.14	4	IP, R	hex.
Pleurotaenium minutum	80.3 × 11.5	0.13–0.18	4	AP, IP	tetr.
Xanthidium armatum	100 × 75	0.19–0.21	4	AP, IP	hex., tetr.
Micrasterias thomasiana	180 × 194	0.19–0.23	4	—	hex. (only on PL)
Desmidium cylindricum	21 × 33	0.14	6	—	—

AP = apical pore system; CPG = central pore group; CPP = central pore protuberance system; CPS = central pore system—pore type 1; IP = isthmical pore system; PL = polar lobe; R = pore row; SDP = species determined pore distribution; l. × w. = length × width of the alga, in μm; pore-φ = pore diameter, in μm; hex. = hexagonal; tetr. = tetragonal.

takes place before shedding of the primary wall (Drawert and Mix 1962). The cytochemical investigations of Menge (1976) demonstrate that the differentiated pore apparatus reacts variably to the PV; the pore apparatus is built up by acid polysaccharids. A similar acid reaction can be seen in the LVfrei, which are responsible for intracellular slime transport (Kiermayer 1968, 1970 a, Menge 1976).

Stage *D in* Fig. 1 demonstrates the differentiated pore in the secondary wall after shedding of the primary wall. In contrast to the observations of Mix (1966), the results of Kiermayer and Staehelin (1972) clearly show

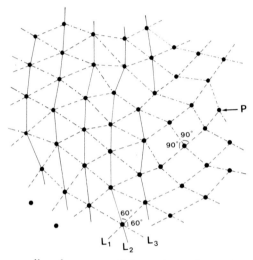

Fig. 3. Pore pattern on a cell surface sector of *Cosmarium subcucumis;* every point represents a cell wall pore (P). The three lines (L_1, L_2, L_3) connect the pores of the hexagonal pattern; on the right side one line is interrupted and a tetragonal pore pattern is established. From Neuhaus and Kiermayer 1981.

that only the topmost fibril layer of the secondary wall is disrupted by the pore apparatus; the fibrils underneath surround the pores (Fig. 8). The structure of a cell wall pore was postulated by Lütkemüller (1902) and Schröder (1902) and established by the transmission electron microscopic observations of Chardard and Rouiller (1957), Drawert and Metzner-Küster (1961), Drawert and Mix (1961), Mix (1966), Kiermayer (1970 a, b), Kiermayer and Staehelin (1972), and Chardard (1977). As shown in Fig. 1/stage D, the pore consists of a pore apparatus that can be subdivided into the pore bulb on the inner face of the secondary wall, the pore head on the outer face, and the connecting pore channel. Under the cell wall pore, the plasma membrane shows a depression (Porenhof) in which the contents of the LVf can be detected (Figs. 7 and 9; Kiermayer 1970 a, Kiermayer and Staehelin 1972). In these pore depressions, two different types of fibrilous material can be observed (Fig. 7). One type represents the contents of the incorporated LVf (Kiermayer 1970 a). This material always forms balls of fibrils (Fig. 7). The other type builds up a regular "spider web" (Chardard

1977; Fig. 1/stage *D*; Fig. 7). The fibrils of this web, which are parallel to the plasma membrane, reach into the area beside the pore depression (Fig. 1/ stage *D*). The vertical fibril connects the pore bulb with the plasma membrane and crosses the horizontal fibrils (Fig. 1/stage *D*).

III. Morphology of Pore Openings

Scanning electron microscopic studies by Neuhaus (1980) and Neuhaus and Kiermayer (1981) clearly showed the close morphological relationship between the pore opening and the surrounding cell surface. Therefore,

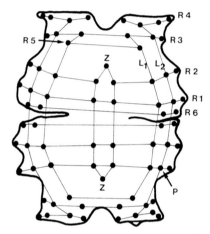

Fig. 4. Pore distribution on the cell surface of *Euastrum denticulatum*. The pores are connected by vertical lines (L_1, L_2) and horizontal rows (R_{1-6}); the central pore (Z) cannot be arranged in these lines. *P* pore. From Neuhaus and Kiermayer 1981.

the technical term "pore type" (Mix 1973) had to be redefined (Neuhaus 1980). The pore type characterizes the functional unity of the pore opening and the differentiated surrounding area of the cell surface.

Fig. 5. *Micrasterias denticulata*; the approach of a pore vesicle (*PV*) to the plasma membrane (*PM*); the structures that surround the PV represent the "expanded" contents of other LV. *PW* primary wall ($\times 105,000$). From Dobberstein 1972.

Fig. 6. *Micrasterias denticulata*; the secondary wall (*SW*) is interrupted by a pore plug (*PP*), *PM* plasma membrane, *SW* secondary wall ($\times 100,000$). From Dobberstein 1972).

Fig. 7. *Micrasterias denticulata*; the differentiated cell wall pore (*P*) in cross-section. Underneath this pore, a pore depression (*PD*) can be observed with balls of fibrils (*BF*) and a fibrilous "spider web" (arrow). *SW* secondary wall, *PM* plasma membrane ($\times 36,000$). From Kiermayer 1970 a.

Fig. 8. *Micrasterias denticulata*; freeze-etch view of microfibrils curved around the cell wall pore (arrow) and a cell wall pore and its pore wall (*PWA*). *OL* outer layer ($\times 45,000$). From Kiermayer and Staehelin 1972.

Fig. 9. *Micrasterias denticulata*; two pore depressions (*PD*) of a freeze-etched cell. The arrows mark the holes that probably represent the incorporated large vesicle. *PM* plasma membrane ($\times 34,000$). From Kiermayer and Staehelin 1972.

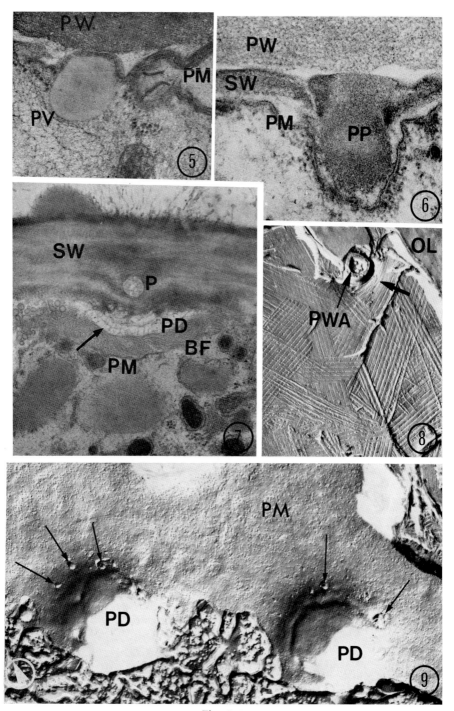

Figs. 5–9

Within this new definition it was possible to identify six different pore types in the investigated desmids (Neuhaus and Kiermayer 1981):

Pore type 1, which is only differentiated in some species of the genus *Euastrum,* is a sunken pore system build up at two different levels (Fig. 18). This pore type was called "central mucus porus"; if many of these types are present they are called *"Scrobiculae"* (Krieger 1937). The nearly round cell wall excavation of pore type 1 (diameter 6.5 to 8.5 µm) usually possesses five pores at the first level. At the centre of this level is another depression (diameter 2.4 to 3.2 µm), the bottom of which is also penetrated by some pores. The pores show a constant diameter of about 0.14 to 0.16 µm. A profile of this pore type 1 can be seen in Fig. 2.

Pore type 2 is characterized by pore openings that are situated in the center of a smooth cell wall depression (Fig. 2) called a pore field (Neuhaus and Kiermayer 1981). Sometimes the central pore opening is on an elevation that looks like a truncated cone. This pore type occurs in various desmids (Figs. 2, 10, 18; Table 1).

Pore type 3 is differentiated similarly to pore type 2, but the pore field is deeper and smaller in diameter (Fig. 2; Neuhaus 1980). This pore type was found on the lobes and swellings, especially in the genus *Euastrum* (Fig. 18; Table 1).

Pore type 4 is characterized by pore openings located at the same level with the cell surface, without any special differentiated cell area surrounding it (Figs. 2, 11, 12, 15, and 16). Only sometimes the pore opening is found in a very smooth cell wall depression (Fig. 14). This pore type was observed on various desmids (Table 1).

Pore type 5 is characterized by pores that are always combined with cell protuberances. The pore opening of this pore type is situated on the bottom

Fig. 10. *Euastrum insigne;* pores of the central pore group that show pore type 2 (P_2); in the area of the isthmus, some pores of pore type 4 (P_4) can be seen. *PF* pore field (×2,500). From Neuhaus and Kiermayer 1981.

Fig. 11. *Euastrum denticulatum;* morphology of pore type 5 (P_5); the arrow marks the channel of the protuberance (×12,800). From Neuhaus and Kiermayer 1981.

Fig. 12. *Pleurotaenium minutum;* apical area with an apical pore system that consists of a larger central pore (*CP*) and a circle of pores (arrows); the pores show pore type 4 (P_4) (×4,400). From Neuhaus and Kiermayer 1981.

Fig. 13. *Cosmarium subcucumis;* triangularly formed apical pore system (thick marked lines); all pores of the apex are arranged in a hexagonal pore pattern (×10,000). From Neuhaus and Kiermayer 1981.

Fig. 14. *Staurastrum teliferum;* apical view of one semicell with the apical pore system consisting of one central pore (*CP*) and two circles of pores (arrows) surrounding the central pore; the pores demonstrate pore type 4 (×4,000).

Fig. 15. *Cosmarium subcucumis;* pores of pore type 4 (P_4) are connected by lines that intersect each other at an angle of about 60° (arrows) and build up a hexagonal pore pattern (×2,900).

Fig. 16. *Cosmarium cucurbita;* near the isthmus, the pores (P_4) are arranged in an isthmical pore system; the circles of pores on each semicell are marked by arrows (×3,100).

Fig. 17. *Micrasterias thomasiana;* hexagonal pore pattern on the end of a polar lobe; the pores show pore type 4 (P_4) (×1,800). From Neuhaus and Kiermayer 1981.

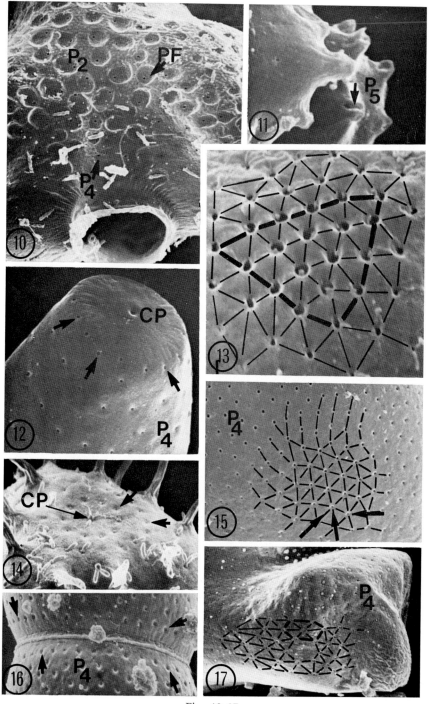

Figs. 10–17

or on the top of the protuberance (Figs. 2 and 11). Sometimes the cell protuberances show a differentiated channel that appears to be a continuation of the pore opening as can be seen in Fig. 11.

Pore type 6 is characterized by pore openings that are surrounded by a cell wall rampart (Fig. 2). This pore type could only be observed in the genus *Desmidium* (Table 1).

In Table 1, the different pore types of several desmids are listed. In addition, it can be seen that the pore diameter is related to the size of the cell (NEUHAUS 1980), as was already supposed by MIX (1966).

IV. Pore Systems and Distribution of Pores

On the basis of variable or constant distances between pores, some specific groups of cell wall pores can be recognized; they are called pore systems (NEUHAUS 1980, NEUHAUS and KIERMAYER 1981). A pore system is defined as a specific small group of pores on the cell surface that can be sharply outlined against the other pores of the cell by their position and by their closeness to each other.

A preferred location for differentiating pore systems is the apex of some desmids. Usually, the apical pore system consists of a central pore, surrounded by a certain number of pores (Figs. 12, 14, and 19). *Cosmarium subcucumis* (Fig. 13) possesses an apical pore system; its pores are arranged in the shape of a triangle (Fig. 9).

In the genus *Euastrum*, some species exhibit a central situated pore system that also can be called pore type 1 on the basis of its morphology (Fig. 18). This pore system (= pore type 1) was described in Section III of this Chapter. Cells of the genus *Euastrum* also show a special group of pores that are arranged in the form of a line. This line of pores starts at the apical incision and ends in the area of the central pore system (Fig. 18). In this genus, some cells that never possess a central pore system as described above only show a group of a certain number of pores that can be distinguished from the other pores of the cell by their pore type or by their special arrangement (Table 1).

Especially in the genera *Cosmarium, Staurastrum, Tetmemorus, Pleurotaenium,* and *Xanthidium,* an isthmical pore system is differentiated. As can be seen in Fig. 16, a circle of cell wall pores is situated around the isthmus at a constant distance on each semicell (NEUHAUS and KIERMAYER 1979).

The different pore systems in various desmids are summarized in Table 1.

Fig. 18. *Euastrum oblongum;* semicell with different pore types. In the centre pore type 1 (P_1); sunken pore system in two levels, pore type 3 (P_3) is differentiated on the lobes of the cell, the rest of the pores show pore type 2 (P_2); the arrows mark the row of pores that starts at the apical incision to the central pore system ($\times 4,700$). From NEUHAUS and KIERMAYER 1981.

Fig. 19. *Staurastrum arachne;* the apex of this alga possesses an apical pore system with one central pore (*CP*) and a circle of pores (arrows); the pore openings are still covered with pore heads ($\times 3,700$).

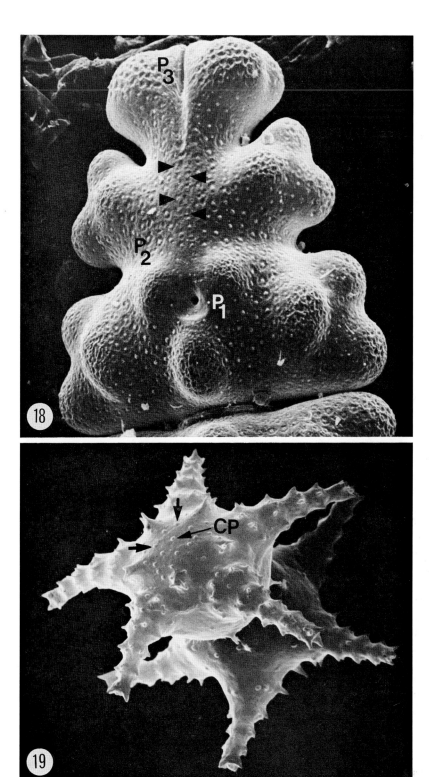

Figs. 18 and 19

Conversely to the results of Lacalli and Harrison (1978), in various desmids a regular distribution of pores could be found (Neuhaus 1980, Neuhaus and Kiermayer 1981). Especially in desmids with pores all over the cell surface, a hexagonal pattern of cell wall pores could be found (Figs. 13, 15, and 17). Sometimes this regular pattern is interrupted by disturbances in pore distribution. In these regions, a tetragonal pore distribution can be recognized. As demonstrated in Fig. 3, the three pore-connecting lines intersect each other at an angle of 60°. In cell regions with a tetragonal pattern, only two pore-connecting lines, which form an angle of about 90°, can be drawn (Fig. 3). Fig. 13 demonstrates the triangularly formed apical pore system of *Cosmarium subcucumis*, whose pores are included in the hexagonal pore pattern of the surrounding cell wall pores. The same alga also shows, in a central cell area, a hexagonal pore distribution (Fig. 15). *Micrasterias thomasiana* var. *notata* only possesses a clear hexagonal pore-distribution at the ends of the lobes, especially at the polar lobes (Fig. 17), while in central areas of the cell, no regular patterns can be observed.

Desmids of various genera (*Euastrum denticulatum*, *Euastrum bidentatum*, *Cosmarium ornatum*, *Staurastrum arachne*, *Staurastrum teliferum*, *Staurastrum scabrum*) show a clear reduction in the number of differentiated cell wall pores (Fig. 19). Some of these pores are arranged in pore systems, but the pores never form patterns as in other demids (Neuhaus and Kiermayer 1981). The distribution of cell wall pores on those desmids is specific for each species; *e.g.*, on every cell of *Euastrum denticulatum* the cell wall pores are arranged in the same pattern as can be seen in Fig. 4.

The formation of pore-distribution patterns, such as hexagonal or tetragonal arrangement and pore systems on specific areas of the cell wall (apical pore systems, central pore systems, isthmical pore systems), can probably be explained as a result of a membrane recognition mechanism operating between the plasma membrane and the P-vesicle membrane (Kiermayer 1970 b, 1977). However, further morphogenetic information must be necessary for modelling the shape of the pores (pore-type) and determining their dimensions. At present no facts are known about these questions.

Defects in pore distribution and formation were observed in *Micrasterias denticulata* under the influence of a specific group of antibiotics (cephalosporines) by Kunzmann and Kiermayer (1979). Although a secondary wall was formed, these drugs specifically inhibited pore formation. Also, a turgor reduction by glucose during the development of the secondary wall prevents the formation of differentiated cell wall pores (Neuhaus-Url 1980).

A strong influence of ethanol and diethylether on pore formation of developing *Micrasterias denticulata* cells was shown by Exner (1978) who demonstrated a considerable increase of the pore diameter of cells grown at higher concentrations of these substances.

Acknowledgements

The work has been supported by the Fonds zur Förderung der wissenschaftlichen Forschung, Vienna (project 3660).

We are very grateful to Dr. M. TEMPLE for reading the manuscript. We also thank Miss Ing. D. PINEGGER for making the drawings and Miss C. SKRINJAR for the preparation of the manuscript.

References

CHARDARD, R., 1977: La sécrétion de mucilage chez quelques desmidiales. I. Les pores. Protistologica 13, 241—251.
— ROUILLER, C., 1957: L'ultrastructure de trois algues Desmidiée. Etude au microscope électronique. Rev. Cytol. et Biol. végét. 18, 153.
DOBBERSTEIN, B., 1972: Zur Sekundärwandbildung von Micrasterias denticulata Bréb. Thesis, Köln.
— 1973: Einige Untersuchungen zur Sekundärwandbildung von Micrasterias denticulata de Brébisson. Nova Hedwigia 42, 83—90.
— KIERMAYER, O., 1972: Das Auftreten eines besonderen Typs von Golgivesikeln während der Sekundärwandbildung von Micrasterias denticulata Bréb. Protoplasma 75, 185—194.
DRAWERT, H., METZNER-KÜSTER, I., 1961: Licht- und elektronenmikroskopische Untersuchungen an Desmidiaceen. I. Mitt.: Zellwand- und Gallertstrukturen bei einigen Arten. Planta 56, 213—228.
— MIX, M., 1961: Licht- und elektronenmikroskopische Untersuchungen an Desmidiaceen. II. Mitt.: Hüllgallerte und Schleimbildung bei Micrasterias, Pleurotaenium und Hyalotheka. Planta 56, 237—261.
— MIX, M., 1962: Licht- und elektronenmikroskopische Untersuchungen an Desmidiaceen. X. Beiträge zur Kenntnis der „Häutung" von Desmidiaceen. Archiv für Mikrobiologie 42, 96—109.
EXNER, CH., 1978: Die Wirkung von Äthylalkohol und Diäthyläther auf Wachstumsprozesse und Ultrastruktur von Micrasterias denticulata Bréb. Thesis, Salzburg.
HAUPTFLEISCH, P., 1888: Zellmembran und Hüllgallerte der Desmidiaceen. Mitt. nat. Verein f. Neuvorpomm. u. Rügen 20, 59—139.
KIERMAYER, O., 1968: The distribution of microtubules in differentiating cells of Micrasterias denticulata Bréb., Planta 83, 223—236.
— 1970 a: Elektronenmikroskopische Untersuchungen zum Problem der Cytomorphogenese bei Micrasterias denticulata Bréb. 1. Allgemeiner Überblick. Protoplasma 69, 97—132.
— 1970 b: Causal aspects of cytomorphogenesis in Micrasterias. Ann. N.Y. Acad. Sci. 175, 686—701.
— 1977: Biomembranen als Träger morphogenetischer Information. Naturwiss. Rundschau 30, 161—165.
— STAEHELIN, A., 1972: Feinstruktur von Zellwand und Plasmamembran bei Micrasterias denticulata Bréb. nach Gefrierätzung. Protoplasma 74, 227—237.
KRIEGER, W., 1937: Die Desmidiaceen Europas mit Berücksichtigung der außereuropäischen Arten. Rabenh. Kryptofl. 13, Abt. 1, 1. und 2. Teil. Leipzig: Akad. VerlagsgesmbH.
KUNZMANN, R., KIERMAYER, O., 1978: Über die Wirkung verschiedener Antibiotica auf sich differenzierende Zellen von Micrasterias denticulata Bréb. Sitzungsber. Österr. Akad. Wiss. math.-nat. Kl. Abt. I, 178, 233—255.
LACALLI, C. T., HARRISON, L. G., 1978: Development of ordered arrays of cell wall pores in desmids: a nucleation model. J. theor. Biol. 74, 109—138.
LHOTSKÝ, O., 1949: The pore-system of the Desmidiaceae. Exper. V, 158.
LÜTKEMÜLLER, J., 1902: Die Zellmembran der Desmidiaceen. Beitr. Biol. d. Pflanz. 8, 347.
MENGE, U., 1976: Ultracytochemische Untersuchungen an Micrasterias denticulata Bréb. Protoplasma 88, 287—303.
MIX, M., 1966: Licht- und elektronenmikroskopische Untersuchungen an Desmidiaceen. XII. Zur Feinstruktur der Zellwände und Mikrofibrillen einiger Desmidiaceen vom Cosmarium-Typ. Arch. Mikrobiol. 55, 116—133.
— 1969: Zur Feinstruktur der Zellwände in der Gattung Closterium (Desmidiaceae), unter besonderer Berücksichtigung des Porensystems. Arch. Mikrobiol. 68, 306—325.

Mix, M., 1972: Die Feinstruktur der Zellwände bei *Mesotaeniaceae* und *Gonatozygaceae* mit einer vergleichenden Betrachtung der verschiedenen Wandtypen der Conjugaten und über deren systematischen Wert. Arch. Mikrobiol. **81**, 197—220.

— 1973: Die Feinstruktur der Zellwände der Conjugaten und ihre systematische Bedeutung. Beih. Nova Hedwigia **42**, 179—194.

— Manshard, E., 1977: Über Mikrofibrillen-Aggregate in langgestreckten Vesikeln und ihre Bedeutung für die Zellwandbildung bei einem Stamm von *Penium* (*Desmidiaceae*). Ber. Dtsch. Bot. Ges., Band **90**, 517—526.

Neuhaus, G., 1980: Raster-elektronenmikroskopische Untersuchungen an Desmidiaceen mit besonderer Berücksichtigung der Poren und ihrer Verteilung. Thesis, Salzburg.

— Kiermayer, O., 1979: Scanning electron microscope-studies in pore-distribution on some desmids. Europ. J. Cell Biol. **20**, 132.

— Kiermayer, O., 1981: Raster-elektronenmikroskopische Untersuchungen an Desmidiaceen: die Poren und ihre Verteilungsmuster. Nova Hedwigia (in press).

Neuhaus-Url, G., 1980: Über die Wirkung von Turgoränderungen auf die Cytomorphogenese und Ultrastruktur von *Micrasterias denticulata* Bréb. Thesis, Salzburg.

Pickett-Heaps, J. D., 1972: Cell division in *Cosmarium botrytis*. J. Phycol. **8**, 343—360.

— 1973: Stereo-scanning electron microscopy of desmids. J. Micros. **99**, 109—116.

— 1974: Scanning electron microscopy of some cultured desmids. Trans. Amer. Micros. Soc. **93**, 1—23.

— Fowke, L. C., 1970: Mitosis, cytokinesis, and cell elongation in the desmid, *Closterium littorale*. J. Phycol. **6**, 189—215.

Schröder, B., 1902: Untersuchungen über die Gallertbildung der Algen. Verh. naturh.-med. Ver. Heidelberg **7**, 139—196.

II. Cytomorphogenesis in Multicellular Plants

Control of Cell Expansion in the *Nitella* Internode

L. Taiz [1], J.-P. Métraux [2], and P. A. Richmond [3]

[1] Division of Natural Sciences, Thimann Laboratories, University of California, Santa Cruz, Calif., U.S.A.
[2] Department of Botany, University of California, Berkeley, Calif., U.S.A.
[3] Department of Biological Sciences, University of the Pacific, Stockton, Calif., U.S.A.

With 9 Figures

Contents

I. Introduction

Members of the giant-celled, freshwater *Characeae,* particularly the ecorticate species of *Chara* and *Nitella,* have figured prominently in classical studies on ion transport, water permeability and cytoplasmic streaming (see Hope and Walker 1975). These algae are characterized by erect, branched coenocytic filaments, differentiated into nodes and internodes, each node bearing a whorl of specialized lateral internodes (Fig. 1). Since the early work of Green on *Nitella (e.g.,* Green 1954, 1958 a, 1960), the usefulness of this alga as a model system for studying cell expansion has been generally recognized. The plant is derived from divisions of a single apical cell and its immediate progeny. The internode cell, initially a shortened disk 20 microns in length, eventually undergoes dramatic elongation and a small amount of lateral expansion to produce a slender cylinder about 6 cm long and 0.5 mm wide at maturity (Green 1958 a, 1963 a). The two fundamental cytomorphogenetic questions which have been studied most extensively are the determination of cell shape and the regulation of cell elongation.

The structure which limits the plant cell's tendency to expand under typical hypoosmotic conditions is the rigid cell wall. It is axiomatic that for cell enlargement to occur the wall must yield to the force of turgor pressure (Cleland 1971, Ray et al. 1972). Wall yielding may be localized, as in the case of the tip growth of root hairs, pollen tubes, fungal hyphae and Nitella rhizoids, or it may be distributed more or less uniformly along the entire cell surface (diffuse growth). Diffuse growth is characteristic of higher plant epidermal and parenchyma cells (Castle 1955, Wardrop 1955, Wilson 1967) as well as the Nitella internode (Green 1954). The physical factors which influence shape in tip-growing and diffuse-growing cells have been described in a review by Green (1969). This chapter will focus on the principles governing the latter type as exemplified by the Nitella internode.

Since an isotropic wall yielding to the non-directional force of turgor pressure would be expected to assume a spherical shape (Green 1963 a), it follows that the walls of diffuse-growing cylindrical cells must be reinforced circumferentially. The walls of Nitella internodes are structurally, physically and optically anisotropic, conforming to these expectations (Green 1960, Gertel and Green 1977). It has also been established that the arrangement of cellulose microfibrils in Nitella walls is consistent with the "multinet growth" model suggested by Houwink and Roelofson (Green 1960). The predominantly transverse orientation of the microfibrils correlates with the anisotropic growth. Microtubule toxins, such as colchicine, which alter the microfibril orientation simultaneously alter the directionality of cell expansion (Green 1963 b, Richmond 1977). Recent evidence indicates that the inner 25% of the wall is the principal stress-bearing, and therefore growth regulating, region (Richmond 1977).

The location of the stress-bearing region on the inner side of the wall makes it more accessible to cytoplasmic regulation. This may be an important consideration for a submerged aquatic plant whose outer surface is exposed to a variable milieu. The contributions of metabolism to wall extension include wall synthesis and wall "loosening", although the nature of the loosening process has been the subject of much debate (e.g., Ray 1969, Green et al. 1977, Masuda 1978, Cleland and Rayle 1978, Stuart and Jones 1978). High resolution growth kinetics and in vitro cell wall extensibility measurements have been indispensable to the development of our views on the mechanism of wall loosening. The acid secretion theory of auxin-stimulated cell elongation in higher plants (see Cleland and Rayle 1978) prompted an investigation of the role of protons in regulating elongation in Nitella. There is now a considerable amount of evidence which suggests that wall acidification is an essential component of the extension process.

Fig. 1. Filament of *Nitella axillaris* Braun. Bar equals 3 cm.

II. The Control of Wall Extension

A. Growth Physics

In simplest terms the rate of cell elongation can be expressed as the product of turgor pressure and the yielding properties of the wall. Turgor pressure is maintained above a certain threshold level required to initiate extension, the apparent yield point, by solute accumulation in the vacuole (RAY *et al.* 1972). This relationship is summarized in the equation:

$$r = m \, (P - Y)$$

where r = growth rate, m = *in vivo* wall extensibility, P = turgor pressure and Y = apparent yield point (GREEN *et al.* 1971). Similar expressions have been described by LOCKHART (1965) and BOYER and WU (1978) in which turgor pressure is replaced by the parameters controlling it.

In studies on the physics of cell elongation, Green et al. (1971) substituted experimental values for three of the four unknowns in the above equation. Growth rate was measured by the displacement of resin bead markers and monitored by time-lapse cinemicrography. Values for turgor pressure were obtained directly by means of a micromanometer inserted into the vacuole (Green 1968). The apparent yield point was equated with the turgor pressure at which growth ceased. Shifts in turgor pressure were imposed by two different methods, either by varying the osmolarity of the external medium (Green et al. 1971) or by "sap dilution", a technique which involves drawing off a small volume of the vacuolar contents with a microcapillary tube (Stanton 1970). In both cases reductions in turgor pressure of as little as 0.2 atmospheres caused an immediate cessation of growth, implying that elongation normally proceeds with a very small margin of growth-promoting turgor pressure above the apparent yield point. When turgor was reduced by changing the osmolarity of the external medium, growth stopped immediately. Partial to full recovery of the initial growth rate followed after a period that depended on the magnitude of the turgor shift. Readjustment of the growth rate did not involve compensatory turgor changes but depended on a lowering of the yield threshold (Y). Growth recovery at reduced turgor (apparent yield point lowering) was inhibited by azide and was presumed to be metabolism-dependent. Growth rate regulation also occurred after a turgor step-up, and this was attributed to an increase in Y. Since Y raising was insensitive to azide, it was interpreted as a physical strain hardening of the wall.

Difficulties arise when attempts are made to attribute known physiological or mechanical properties to the m and Y terms. In addition to involving purely physical wall parameters, these terms also reflect regulatory processes of metabolic origin (Green and Cummins 1974). Another point to consider is that the observed cessation of growth during turgor step-downs may actually have been reductions in the growth rate below the resolving power of the photographic method. The significance of this possibility is that, so long as $(P - Y) > 0$, the growth rate can theoretically be increased by changes in m as well as by changes in Y.

Contrasting results were obtained when turgor pressure was reduced by the sap dilution method (Stanton 1970). No changes in either m or Y were observed. Instead, an osmoregulatory mechanism was stimulated which acted to restore the original turgor pressure. Growth resumed only after the original turgor pressure was attained. Thus, the cell possessses multiple pathways for the homeostatic regulation of steady state growth.

Biophysical studies of cellular growth were extended to higher plant tissues by Green and Cummins (1975). Again it was found that steady state growth was maintained by a stabilizing system acting to readjust growth rate after turgor shifts. Unlike Nitella, however, the relationship between growth rate and turgor was not the same when turgor was shifted upwards or downwards. Given the presence of a hysteresis in this relationship, the stabilizing system acting to reset steady state growth after a turgor change should not be interpreted as a simple physical change of the wall, but rather as some biochemical regulating process affecting wall properties.

B. *Physiology and Biochemistry of Cell Elongation*

1. Cell Wall Mechanical Properties

A distinct advantage of the Characean algae for cell wall studies is that it is possible to obtain intact sheets of primary wall or entire wall cylinders large enough for extensibility tests. Higher plant material is complicated by its multicellularity, with the attendant problems of heterogeneity of cell type and the presence of middle lamellae. Much of the early work on *Nitella* wall extensibility was performed by PRESTON and his colleagues (see reviews by CLELAND 1971 and PRESTON 1974). Wall extension studies have been of two general types depending on the directional nature of the applied load: uni-

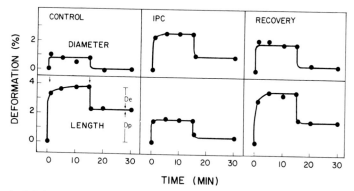

Fig. 2. Typical deformation responses of mercury-filled *Nitella* wall tubes to multiaxial stress. An internal pressure of five bars was applied at 0 minutes and lowered to 0.5 bars at 15 minutes. Elastic deformation, D_e, is recovered upon pressure lowering while plastic deformation, D_p, remains as a permanent set. The directional nature of wall deformation is seen for control, isopropyl *N*-phenylcarbamate (IPC) treated and drug-recovery cells. The walls had similar overall microfibrillar alignments but the control and drug-recovery cells had transverse inner walls while the IPC inner wall was nearly isotropic. The results indicate that the inner portion of the wall dictates most of the directional character of the deformation response. See section III.C. Data from RICHMOND *et al.* 1980.

axial vs. multiaxial. In either case, application of a load (stress) brings about a characteristic extension (strain). Upon application of a constant load, the material exhibits a slow yielding over long time periods; this time-dependent deformation is termed creep. The creep of plant cell walls declines exponentially with time and is roughly proportional to log time. Creep of this nature is typical of a viscoelastic material, one that has both elastic (reversible) and viscous (irreversible) components. Viscoelastic properties are characterized by the amount of creep for a defined stress and time interval. Upon removal of the load there is a partial or, in some cases, complete recovery of the length, which constitutes the elastic deformation, D_e. Elasticity is frequently expressed in terms of Young's modulus, the ratio of stress over strain. If recovery is incomplete, the material is not fully elastic and the remaining permanent deformation is referred to as the plastic deformation, D_p (see Fig. 2). Additional parameters can be obtained through stress relaxation studies carried out under conditions of constant strain rather than constant load. The decay of tension after the wall has been stretched to a predetermined length yields quantitative information which can be converted to energy units equivalent to the bonds limiting wall extension (Haughton et al. 1968, Preston 1974). These studies have shown that stress relaxation occurs in the matrix rather than the cellulose microfibrils, and that the matrix polymers have some degree of order.

Probine and Preston (1962) developed a procedure for uniaxial stress analysis using rectangular wall strips dried and glued to clamps in a constant load apparatus. Strips of equal dimensions were cut parallel and perpendicular to the longitudinal axis to allow a comparison of extensibility in the two directions. The dried strips were rehydrated and subjected to an applied load equal to the longitudinal force due to normal turgor pressure. Comparisons of the Young's modulus of strips extended longitudinally and transversely confirmed the mechanical anisotropy of the wall (see Part III.). However, given the irreversible nature of growth, plastic or viscoelastic wall properties must govern wall extension in vivo. Probine and Preston showed that the creep rate between 1 and 100 minutes of applied load correlated with elongation rates prior to cell excision. This correlation was confirmed by Métraux and Taiz (1978) using freshly isolated wall cylinders which had not undergone a drying step during preparation. Caution must be observed before concluding that there is either a quantitative or causal relationship between uniaxial creep and normal cell elongation. Quantitative correlations between creep and growth are not very meaningful given the difference in the deformation patterns (steady vs. declining) between the living cell and the excised wall. Values for creep rates reflect strain during an arbitrarily selected time interval. The discrepancy in the extension of living vs. nonliving cells demonstrates that the protoplast must act upon the wall to maintain a constant rate of extension. Stress-strain experiments have shown poor agreement between the yield stress for uniaxial creep and the "apparent yield stress" for growth as determined by Green (1968). It has also been found that stretching the wall uniaxially causes obvious structural modifications (Richmond et al. 1980). The sides of the wall neck in due to perpendicular contractile forces (Poisson ratio effects) and the cellulose microfibrils tend to realign in the direction of the applied force (see Fig. 3). These structural alterations account for the excessive amounts of extension obtained with uniaxial stress.

Multiaxial stress avoids the complications noted above by mimicking the nondirectional force of turgor pressure. Kamiya et al. (1963) were the first to employ the technique of mercury inflation for studying wall mechanical properties. Since they were primarily interested in osmoregulation, only walls from mature, nongrowing cells were analysed. Cells were excised, the cytoplasm gently squeezed out, and the cut end quickly dried and glued to a glass capillary. The capillary was then filled with mercury and connected to a syringe to which a pressure gauge was attached. To fill the wall with mercury, water was displaced with an initial pressure of 2–3 atmospheres. The surface tension of the hydrated wall was sufficient to block any mercury leakage at pressures up to 10 atmospheres. Length and volume changes were measured, and diameter readings were obtained indirectly by calculation. Consistent with the nongrowing status of the cells, little or no creep or plastic deformation was observed during a typical load on-load off cycle at pressures comparable to turgor pressure. Turgor-induced elongation was compared with uniaxial stress-induced extension and it was found that turgor (multiaxial stress) was only about $1/3$–$1/4$ as effective as uniaxial stress in causing wall

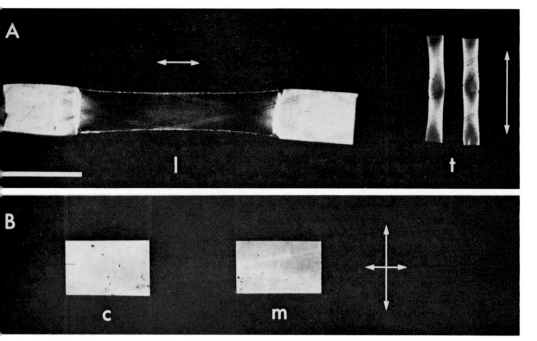

Fig. 3. Stress induced shape changes and rearrangements of microfibrils in isolated walls. All wall segments are viewed in polarized light, without compensation, at 45° to the crossed polars. The degree of wall brightness corresponds to net microfibrillar alignment and wall thickness. Microfibrils that are well aligned overall appear bright whereas a net random arrangement appears dark. Arrows give direction and relative magnitude of stresses. *A* Effects of uniaxially applied stresses on wall segments from an IPC treated internode. The longitudinally stressed segment (*l*) is a flattened cylinder. The transversely stressed segments (*t*) were loops that have been cut crosswise and rolled open. At the clamped ends of the longitudinal segment the shape is unchanged and the microfibrils remain transversely aligned. The stretched region, however, has been grossly distorted and the microfibrils dramatically realigned. With a first order red plate or a compensator it can be demonstrated that the slight brightness in the middle of the segment is due to a net longitudinal orientation of the microfibrils. The wider and darker regions of the transverse loops locate where glass hooks were positioned. These regions have not been significantly altered whereas the free wall regions show "necking" and a realignment of microfibrils into an improved transverse orientation. The striation lines on the walls reflect the position of cytoplasmic streaming null zones (GREEN and CHAPMAN 1955). Bar at lower left equals 1 mm. *B* Effects of multiaxial stress on microfibrillar arrangements. In this case microfibrillar arrangement are little altered. The coupled action of longitudinal and transverse stresses does not allow necking to occur. The flattened cylindrical segments were taken from a normal internode, *c* before and *m* after being pressurized at 5 bars. Data from RICHMOND *et al.* 1980.

extension, which can be explained in terms of the above-mentioned Poisson effects.

The mercury inflation technique was recently taken advantage of in our own investigation of the mechanical properties of growing *Nitella* cell walls (Métraux et al. 1980, Richmond et al. 1980). Two modifications of the Kamiya procedure were made: 1. length was determined as the distance between two resin bead markers placed on the wall (Green 1965), rather than the displacement of a single point; 2. diameter was measured directly using a Filar micrometer eyepiece. These modifications were essential to avoid artifacts caused by the elastic stretching of the glue between the wall and capillary. Upon application of a load equivalent to turgor pressure (5 atmospheres) the walls exhibited a viscoelastic extension which differed significantly from the extension elicited by uniaxial stress (see Fig. 2, control). The rate of multiaxial extension (viscoelastic strain during the first 15 minutes) was five to ten times less than the rate of uniaxial extension. The most rapid part of the extension occurred during the first five minutes, after which the strain rate declined considerably, making creep measurements difficult over short time periods. However, we have found that creep and plastic deformation, while being measures of different wall mechanical properties, are diagnostically equivalent in assessing the capacity of the wall for irreversible extension (Richmond et al. 1980). As noted previously, D_p represents the amount of irreversible extensions resulting from the initial load-on, load-off cycle. Probine and Preston (1962) ignored this initial expansion because of the uncertainties introduced by the drying procedure.

Before describing the results of the multiaxial D_p studies, an interesting feature of the extension should be mentioned. The initial length of the wall at zero stress is not the initial length of the wall at full turgor since the wall undergoes a viscoelastic contraction during cell excision. By marking the cell with resin beads it is possible to determine the percent shrinkage of the wall during excision. This amount is essentially recovered after fifteen minutes extension of the isolated wall at 5 atmospheres. When the load is released at that point a significant amount of plastic deformation remains after the elastic recovery is completed. Thus, multiaxial D_p does not represent irreversible deformation beyond the original extended length, but irreversible deformation beyond the contracted length of the wall. This seems to imply that during cell excision the wall undergones irreversible shrinkage, possibly due to structural rearrangements. Regardless of the origin of D_p, its developmental pattern parallels creep very closely.

A plot of the relative extension of a young, growing wall vs. the applied pressure reveals an abrupt yield stress at about 4.5 atmospheres, beyond which D_p increases dramatically and the elastic deformation decreases slightly (Fig. 4 *A*). This value is close to the *in vivo* apparent yield stress determined by Green (1968). In this regard multiaxial stress seems to better reflect *in vivo* wall properties than does uniaxial stress. In agreement with Kamiya et al., older, nongrowing walls exhibited no yield stress, and extension, in general, is much reduced. It should be pointed out that the "yield stress" at 4.5 atmospheres is not the stress at which extension begins, but the stress at which the rate is increased. If there is a primary yield stress which represents the minimum stress needed to initiate extension, it must occur between 0–2 atmospheres (Fig. 4 *A*). The 4.5 atmospheres value corresponds to a secondary yield stress. In the turgor step-down experiments of GREEN (1968), growth was observed to temporarily cease when the turgor was lowered below 4.8 atmospheres. CLELAND (1971) has suggested that the cessation of growth might have been caused by the viscoelastic contraction of the wall cancelling out the creep, while GREEN (1968) considers the elastic recovery phase to be completed within a few minutes (based on the measurements of KAMIYA *et al.*

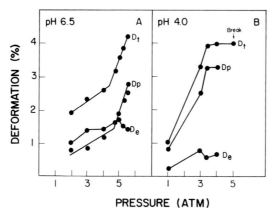

Fig. 4. Relation between the applied pressure and the deformation for multiaxially extended walls. *A* Wall taken from a young, growing cell and extended at pH 6.5. *B* Wall taken from a young, growing cell, extended at pH 4.0. Data from MÉTRAUX *et al.* 1980 and MÉTRAUX, unpublished.

1963). A short period of elastic recovery would make it unlikely that viscoelastic contraction could account for the prolonged growth cessation sometimes observed. Since there may actually be two yield points, it seems improbable that creep would be 100% inhibited during small turgor step-downs, unless the viscoelastic contraction of the wall cancelled out the low levels of creep as CLELAND suggests. The photographic method used to monitor growth may not have been sensitive enough to register the minute amounts of growth which may have persisted after the elastic recovery phase. At turgor pressures below 2 atmospheres no new steady state growth occurs (GREEN *et al.* 1971). This can be interpreted as evidence for a primary yield stress between 0–2 atmospheres (PROBINE and PRESTON 1962, MÉTRAUX *et al.* 1980).

The close correlation between creep rates and growth found in uniaxial studies (PROBINE and PRESTON 1962) has been cited as evidence for the significance of wall extensibility in regulating growth rate. It was therefore of interest to carry out similar studies using multiaxial stress, which better simulates the *in vivo* condition. Surprisingly, D_p correlated with the growth rates only for older cells having growth rates between 0–1% hour^{-1}. Younger cells with growth rates ranging from 1–3% hour^{-1} all had about the same D_p.

This result indicates that multiaxial wall extensibility as measured under standard conditions (neutral pH, low ionic strength) fails to account for the differences in the growth rates of young, rapidly growing cells, although it does correlate with the cessation of growth during maturation.

2. Effects of pH and Ions on Wall Mechanical Properties

Tagawa and Bonner (1957) first showed that the mechanical properties of *Avena* coleoptiles were influenced by inorganic ions. Potassium ions caused tissue (wall?) softening, while calcium appeared to cause stiffening. Early models proposed to explain the stiffening action of calcium on the wall involved the formation of divalent cross-links between pectin carboxyl groups (Bennett-Clark 1955). *Nitella* cell wall pectins are almost entirely non-esterified (Anderson and King 1961, Morikawa 1974) and could conceivably be cross-linked by divalent cations. Probine and Preston (1962) demonstrated that *Nitella* wall extensibility was dramatically enhanced by potassium ions and decreased by calcium ions. Wall loosening in response to potassium was accompanied by a loss of endogenous calcium from the wall, in accord with the divalent cross-link model for calcium action. Further studies on the effects of ions on *Nitella* wall extensibility were carried out by Morikawa, who monitored the orientation of the wall's carboxyl groups by the technique of polarized infrared spectroscopy (Morikawa and Senda 1964 a, b, Morikawa 1974). In uniaxial stress studies, acid pH was particularly effective in loosening the wall, and calcium antagonized the action of protons. However, since Mg^{2+} ions also enhanced wall extension, the simple model based on the disruption of divalent cation cross-links cannot be the entire explanation. Furthermore, protons differed from the other ions in their effects on the orientation of the carboxyl groups, which are generally aligned with their O–O groups perpendicular to the cell axis. Protons altered this orientation, as judged by a reduction in the dichroic ratio of the carboxylate group band, while other ions had no effect. Morikawa concluded that the bonds holding the carboxyl groups in place were nonionic in nature, possibly involving hydrogen bonds or solvation-like bonds with the OH groups in the wall (see below for further discussion).

The effects of pH and ions on extensibility were further characterized by Métraux and Taiz (1977, 1979) and by Métraux et al. (1980) using both uniaxial and multiaxial stress. Of the stimulatory ions, magnesium and ferrous ions were the most effective, while potassium was among the least effective. At saturating concentrations, however, Mg^{2+} and K^+ produced equal amounts of extension, suggesting that they may simply differ in their affinities for the same site. The threshold pH for acid-induced extensibility was first found to be around 4.8 using a 1 mM buffer, while later experiments with stronger buffers (10 mM) indicated that the actual pH threshold is about 5.3. According to Probine and Preston (1962) and Morikawa (1974) the loosening effects of protons and other ions are due to direct physical interactions with acid- and ion-labile bonds. Tepfer and Cleland (1979) reached a similar conclusion regarding the effects of acid pH on the extensibility of *Valonia* walls. In higher plants, however, acid-induced wall loosening is believed to be mediated by cell wall enzymes with acidic pH optima (MASUDA 1978, CLELAND and RAYLE 1978), although there is some evidence for direct physical interactions as well (JACCARD and PILET 1977). Boiling *Nitella* walls in either methanol or water for fifteen minutes had no effect on acid-induced extension in uniaxial stress studies. Boiling in water for 12 hours partially decreased the response, but also increased the extensibility at neutral pH (MÉTRAUX and TAIZ 1977). These findings make it highly unlikely that wall enzymes participate in acid-induced extension in *Nitella* and substantiate the previous

Table 1. *Effect of 1 mM Mes-Tris buffer at pH 4.0 on D_n, D_e, D_t. Walls extended at 5 atm. Percent increase in D_n, D_e, D_t calculated for 15 minute extension at pH 6.5, followed by 15 minute extension at pH 4.0. Average growth rate for the young cells: 1.3% hour^{-1} (± 0.4); average growth rate for old cells: 0% hour^{-1} (± 0). The percent change was calculated from the averages indicated. Creep at pH 6.5 measured as viscoelastic strain after 15 minutes. The solution was then changed to pH 4.0 and the total strain measured after an additional 15 minutes. The percent change was calculated from the average strains after each treatment (± SE, n = 5)*

	Young wall (± SE)			Old wall (± SE)		
	Rel. deformation or strain pH 6.5	Rel. deformation or total strain pH (6.5 + 4.0)	% change	Rel. deformation or strain pH 6.5	Rel. deformation or total strain pH (6.5 + 4.0)	% change
D_n	2.0 (± 0.5)	3.7 (± 0.8)	85	0.4 (± 0.04)	0.6 (± 0.03)	50
D_e	1.7 (± 0.1)	1.3 (± 0.04)	− 24	1.1 (± 0.13)	0.9 (± 0.08)	− 20
D_t	3.8 (± 0.4)	4.1 (± 1.2)	8	1.5 (± 0.08)	1.6 (± 0.13)	7
D_n/D_t	0.5 (± 0.04)	1.2 (± 0.3)	140	0.3 (± 0.04)	0.4 (± 0.03)	33
Creep	0.68 (± 0.2)	2.08 (± 0.4)	205	0.2 (± 0.03)	0.3 (± 0.04)	50

conclusions that direct physical interactions are involved. The effects of acid pH on extension are most dramatic under multiaxial stress conditions. Acid pH substantially increases creep and D_n, while D_e is slightly decreased (Table 1). Walls from old, nongrowing cells exhibit a greatly reduced response to acid.

Acid pH also alters stress-strain curves. As shown in Fig. 4 B, pH 4.0 apparently eliminates the secondary yield stress at 4.5 atmospheres. A limit of extension occurred at 4.0 atmospheres, beyond which the wall consistently broke. An alternative description of the acid effect would be that it greatly enhances extensibility (strain/stress) at stresses below the secondary yield stress. The distinction between the two interpretations may well be semantic.

We will now examine the evidence that protons regulate the growth rates of intact cells.

3. Control of Cell Elongation by Cell Wall pH Gradients

The observation that acid solutions can induce cell elongation was first made by STRUGGER (1932) on *Helianthus* hypocotyls, and it was further studied by BONNER (1934) in *Avena* coleoptiles. HAGER et al. (1971) and

Rayle and Cleland (1972) independently proposed the acid secretion theory of auxin-induced cell elongation. According to this model, auxin causes proton secretion into the cell wall, bringing about wall loosening and growth. The model is supported by the following types of evidence: 1. acid solutions enhance both growth and wall extension; 2. neutral buffers block auxin-stimulated growth; 3. auxin causes wall acidification to a degree sufficient to induce extension. It is not possible to study the effect of auxin on a putative acid secretion mechanism in *Nitella,* since this alga is not known to exhibit a rapid auxin growth response, although some long-term stimulations have been reported (Sandan and Ogura 1957). However, being both macroscopic and single-celled, *Nitella* internodes offer advantages in examining the significance of proton secretion in cellular growth.

It has long been observed that *Nitella* and ecorticate species of *Chara* frequently develop incrustations of calcium carbonate in distinct banded patterns along the length of the internode (Arens 1936, Hope and Walker 1975). Spear *et al.* (1969) showed that the banding pattern is a result of alternating zones of acid and alkaline pH in the cell wall. The appearance of these bands, visualized by floating cells in a dilute salt solution containing the pH indicator phenol red, is light-dependent, and it is apparently caused by the excretion of protons and hydroxyl ions (Fig. 5). Calcium carbonate precipitates in the alkaline zone when the concentrations of calcium (from the medium) and carbonate (from dissolved CO_2) reach their solubility product (Hope and Walker 1975). The biochemistry and biophysics of proton and hydroxyl transport in the *Characeae* have been extensively studied (*e.g.,* Kitasato 1968, Spanswick 1972, Lucas and Smith 1973, Lucas 1976, Lucas and Dainty 1977, Keifer and Spanswick 1979, Tazawa *et al.* 1979). Photosynthetic assimilation of CO_2 for HCO_3 uptake results in the production of OH^-, which is subsequently transported through the plasmalemma as a discrete efflux. It was recently demonstrated that the HCO_3^-/OH^- transport system is not an antiporter but functions through separate carriers (Lucas 1976). The partitioning of the OH^- exporting regions is not caused by specific localization of the OH^- carriers, but by a differential activation of the carriers (Lucas and Dainty 1977). Wall acidification is thought to arise from the activity of a light-stimulated electrogenic proton pump, probably an ATPase, located in the plasma membrane (Kitasato 1968, Spanswick 1972, Shimmen and Tazawa 1977). Scanning the cell surface with a miniature pH electrode indicates that the alkaline bands are about pH 9.5 and the adjacent acid bands are as low as pH 5.5 (Lucas and Smith 1973, Métraux *et al.* 1980). The presence of clearly demarcated regions of different pH values along a single internode cell provides an ideal system for testing the effect of proton gradients on growth.

In order to observe the pH banding pattern in *Nitella* the cells must be pre-incubated several hours in a dilute salt solution (K_b medium; Spear *et al.* 1969). Cells grown in autoclaved soil extract medium (Green 1968) normally show no pH banding patterns. Preincubated cells which have been marked with resin beads are plated on agar into which the K_b medium and phenol red indicator have been incorporated. Growth with acid or base bands can then be easily followed. Since the bands migrate every 1–4 hours, measurements are limited to the period during which the bands are stable. As shown in Table 2 elongation is almost entirely restricted to the acid bands. In addition, as the acid bands migrate to different locations, the regions of elongation shift correspondingly. Since cells grown only in autoclaved soil extract fail to

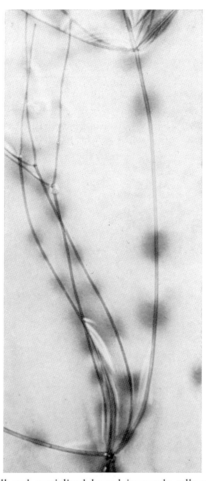

Fig. 5. Main internode cell and specialized lateral internode cells of *N. axillaris* producing alkaline bands (dark areas) and acid bands (alternating light areas) in the light. Cells are plated on agar containing 1.0 mM KCl, 0.1 mM each of $KHCO_3$, NaCl, $CaCl_2$, and $MgCl_2$, and 0.1 mM K salt of phenol red adjusted to pH 6.9 ($\times$ 3.7).

produce pH bands, it was of interest to compare their growth pattern with that of Kb-treated banding cells. In agreement with GREEN (1954) the growth of such cells was found to be uniform along the cell surface (MÉTRAUX 1979). Thus the localized pattern of growth correlates with the presence of localized regions of acid pH. To what extent acid pH contributes to the growth of non-banding cells remains to be determined. It is possible that the walls are

16*

uniformly acidic, the soil extract serving the repress the hydroxyl secretion process. Lucas and Smith (1973) regarded the entire cell surface as mildly acidic, with the alkaline banding superimposed whenever HCO_3^- is taken up. It is interesting to note that higher plant roots have recently been shown to generate acid and alkaline zones along their surface (Weisenseel et al. 1979) and that elongation is correlated with the acidic region (Evans and Mulkey 1980). These observations lend credence to the analogy Green (1980) has drawn between the *Nitella* cell wall and the epidermis of elongating organs.

Table 2. *Growth rates of acid and base regions of Nitella internodes plated on agar*

Time measured		Growth rate in:	
Cell	(hours)	acid band ($\%$ h^{-1})	base band ($\%$ h^{-1})
1	3.5	1.0	0
2	16	0.32	0.05
		0.33	
		0.26	
3	4	0.4	0.2
		0.25	
4	3	0.17	0.06
5	1.2	0.2	0
6	1.3	0.9	0
7	3	0.9	0.07
8	1	0.6	0
9	0.5	0.6	0
10	1.5	0.2	0
11	1.2	1.5	0.13
12	1	0.7	0.5
13	1	0.6	0
14	1.2	0.41	0
		0.26	
15	1.3	0.25	0
		0.4	

The growth rates were measured as long as the bands were stable. Growth rates were found to be lower in the Kb solution, than in soil extract. Similar results were found in cells of *Nitella clavata*, a closely related species.

Further support for the role of pH *in Nitella* wall extension was obtained in reversibility experiments. The nongrowing, alkaline regions can be induced to elongate by acidifying the region surrounding the band, and elongation can be prevented in the acid regions by submerging the band in alkaline buffer. Growth can be restored in an alkalinized acid band by adding an acid buffer (Fig. 6).

As previously noted, the pH of the wall surface in the acid band, as determined directly with a miniature pH electrode is about 5.5. This is very close to the threshold of acid induced extensibility (pH 5.3). The true wall pH is probably lower than the pH of the surface solution by about 0.5 pH units, due to the Donnan effects of the fixed negative charges (RUBERY and SHELDRAKE 1973). The concentration of negative charges in *Nitella* walls has been estimated at 0.6 N (DAINTY and HOPE 1959). As a result the wall acts as a sink for protons. MORIKAWA (1974) showed that *Nitella* walls took 300 hours to reach equilibrium with the medium when the walls were slowly titrated with HCl in the presence of 0.003 M $CaCl_2$ down to pH 3. The calculated maximum proton binding capacity of the cell wall was about 3 ueq/mg dry weight at pH 2.8. If the acidity in the acid region is due to

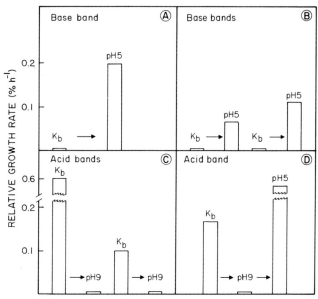

Fig. 6. Reversibility experiments on living *Nitella* cells producing acid and alkaline bands. Growth rates were measured between 1 and 4 hours. Cytoplasmic streaming continued undiminished throughout all treatments. *A* A nongrowing base band (cell in dilute salt Kb solution) is acidified with a buffered solution at pH 5, *B* two base bands (*Kb*) are treated with buffered pH 5 solution, *C* growing acid bands (*Kb*) are made basic with buffer solution at pH 9, *D* an acid band (*Kb*) is made basic (pH 9). The same region is then acidified at pH 5. Buffer solutions are 50 mM Mes-Tris in Kb solution.

a proton pump, transporting H^+ rather than a buffer, a pH gradient would develop across the wall due to the binding of protons. During the time the pH bands are stable in one place (at least several hours), the pH gradient across the wall could thus be considerable. That *Nitella* walls are capable of maintaining such gradients was shown experimentally by perfusing an empty wall tube with an unbuffered HCl solution at pH 4. The pH of the solution outside the wall was pH 9 for the first 15 minutes of perfusion, while the internal pH was still pH 4 (MÉTRAUX et al. 1980). Even in the absence of any other ions the outer surface took 35 minutes to equilibrate with the internal solution. The calculated proton binding capacity from this perfusion experiment was 0.18 ueq/mg dry weight, close to Morikawa's value.

In view of the wall's buffering capacity it is likely that the pH of the inner half of the wall is lower than the pH of the outer half, which is about 5.5. As will be developed in Part III, it is the inner portion of the wall which controls extensibility. Assuming protons are the main acidifying agent, the pH of the periplasmic space may be as low as 4.3 in the acid bands. This value was obtained by adding the estimated number of protons bound to the wall during transit to the concentration of protons on the outer surface (Métraux et al. 1980).

4. Wall Synthesis and Extension: ^{3}H-Glucose Incorporation

All *in vitro* extensibility tests of cell wall material carried out thus far, under a variety of conditions, agree on one point, that the extension rate decreases continuously with time in contrast to growth rates, which are approximately constant. Acid pH dramatically enhances the creep rate, but does not prevent its eventual decline. This constitutes strong evidence that an additional biochemical process is required to produce a constant rate of extension. Wall synthesis is a logical candidate, since new polymers are needed to maintain wall thickness. If wall extension and wall synthesis are tightly coupled in *Nitella,* we might expect that wall deposition would be restricted to the acid bands. To test this prediction, *Nitella* internode cells were allowed to produce pH bands in the presence of ^{3}H-glucose, while growth was monitored with the traveling microscope. Artifactual wall labeling due to incorporation by epiphytic microorganisms was avoided by injecting the label directly into the vacuole. At the end of the growth period, the acid and base bands were excised, cleaned, digested with α-amylase to remove starch, and counted. Surprisingly, radioactivity was not incorporated preferentially into the growing region. The ratio (counts per minute mm^2) of the acid band to the base band was approximately one (Fig. 7). If wall synthesis is equal in growing and nongrowing regions of the wall, slight transient differences in wall mass between the two zones are inevitable. However, during the time in which the bands remain stable the deviations would amount to no more than 4%, and would even out with time. Random ^{3}H-glucose incorporation was also observed in the case of cells without bands (diffuse growth) when the upper half of the cell was compared with the lower half (Fig. 7). Finally, the endogenous pH gradient in banding cells can be replaced by imposing an external pH gradient with buffered soil extract solutions. Cells are placed in a partitioned chamber, the upper half of the cell at pH 5, the lower half at pH 9. Again, ^{3}H-glucose incorporations was not significantly different in the two halves, despite differences in the growth rate (Fig. 7). The lack of correlation between growth and wall deposition demonstrates that general wall deposition alone does not induce wall extension. It has been suggested that there may be two types of wall synthesis, one which contributes to extension and one which does not (Ray 1969). If this were the case in *Nitella,* the growing region of the wall might be expected to be more extensible than the nongrowing region when tested *in vitro.* However, the acid and base bands have similar mechanical properties when extended at neutral pH (MÉTRAUX *et al.* 1980).

A lack of correlation between wall synthesis and growth has been noted before in the case of auxin stimulated growth in *Avena* coleoptiles and pea internode segments. Growth is enhanced by auxin within ten minutes whereas auxin stimulation of wall synthesis is not evident until one hour (EVANS and RAY 1969, RAY and ABDUL-BAKI 1968). In the absence of added sugar the effect of the hormone on wall synthesis is considerably reduced (RAY 1969). Nevertheless, despite a few reports of cases in which growth may occur with little wall synthesis (PRESTON 1974) we know of no demonstration of growth

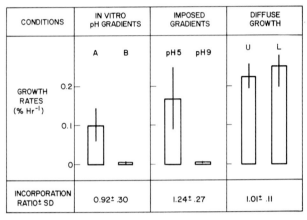

Fig. 7. Comparison of ³H-glucose incorporation (counts per minute per mm²) into the cell walls of the acid and base bands. Incorporation is expressed as the acid : base incorporation ratio in the presence of *in vivo* pH bands, imposed pH gradients or diffuse growth (no pH gradients). *A* Acid band, *B* base band, *U* upper (apical), and *L* lower (basal) end (n = 4). Data from MÉTRAUX 1979.

in the complete absence of wall synthesis. In *Nitella* the occurrence of active wall deposition in the acid bands suggests that both wall deposition and acidic wall solution pH are required for linear extension. *In vitro* multiaxial extension, even at acidic pH, proceeds in a nonlinear fashion. In fact, the possibility cannot be excluded that there are critical differences in the composition or structural arrangement of wall polymers in the acid and base regions, despite similar overall incorporation of label and mechanical properties. Experiments are underway to determine the nature of polymers deposited in acid and base regions. The site of insertion of new material within the wall might also be related to the effect of protons on wall deposition. A stimulation of intussusception was previously found in auxin-treated epidermal walls of *Avena* coleoptiles (RAY 1967). This may be related to auxin-induced proton secretion. Conceivably, acid pH could facilitate the penetration of polymers into the interior of the wall. Although speculative at this point, the issue is important, and *Nitella* may prove to be a useful system for this type of study.

5. Wall Composition and Structure: Implications for Acid-Enhanced Growth

Plant cell walls have been likened to fiber glass, the matrix polymers providing the continuous phase and the cellulose microfibrils acting as a reinforcing filler (Wainwright et al. 1976). In a cylindrical cells such as the *Nitella* internode, with transverse microfibrils, the stress in the longitudinal direction is borne principally by bonds in the matrix (Haughton et al. 1968). Until recently, however, our knowledge of matrix polymer interactions has been extremely limited. With the introduction of extraction procedures that rely on specific enzymes, in combination with methylation and hydrolysis to identify covalent bonds, it has been possible to construct a detailed molecular

Table 3. *Percent composition of neutral sugars in the major cell wall fractions of N. axillaris*

Sugar	Percent of Total		
	Pectin	Hemicellulose	Cellulose
Rhamnose	30.4	7.3	0.4
Fucose	6.3	4.2	0.8
Arabinose	13.5	4.6	1.3
Xylose	7.5	7.4	0.5
Mannose	4.3	18.8	5.7
Galactose	28.4	11.9	1.4
Glucose	9.5	45.7	89.9

Data from Gepstein and Taiz (unpublished). Polysaccharide fractions extracted by the procedure of Morikawa (1974). Sugars, obtained by hydrolysis with 2 N trifluoroacetic acid (121 °C), were analyzed by the method of Albersheim et al. 1967.

model of a dicotyledon cell wall (Keegstra et al. 1973). The essential feature of this model is that the matrix polymers form a covalently bonded network which is hydrogen-bonded by xyloglucan to the cellulose microfibrils. Wall creep occurs when xyloglucan detaches from cellulose, allowing the microfibrils to slide or rotate. Consistent with this hypothesis was the finding that a xyloglucan component of pea stem walls became solubilized during auxin- and acid-induced elongation (Labavitch and Ray 1974, Jacobs and Ray 1975). This model has recently been criticized on the grounds that a covalently bonded wall is too rigid a structure to account for the known extensibility properties of the wall (Preston 1979). The ability of xyloglucan to hydrogen bond to the surface of microfibrils was also questioned because of the highly branched structure of xyloglucan molecules. An alternative scheme was proposed which involved a variety of noncovalent interactions in the wall (Preston 1979).

The composition of *Nitella* walls is similar to that of higher plants. This resemblance in wall structure can be added to the criteria cited by systematists for classifying the Characean algae as a separate division or subphylum (*Charophyta*), more closely related to bryophytes than to the green algae (see Pickett-Heaps 1975). Our own analyses (Gepstein and Taiz, unpublished) indicate that *N. axillaris* walls are approximately 30% pectin, 24%

hemicellulose, 41% cellulose and 5% protein, similar to the values obtained by MORIKAWA (1974). An analysis of the neutral sugars released from the polysaccharide fractions by trifluoroacetic acid hydrolysis is shown in Table 3. The cellulose fraction contains largely glucose, with some mannose. The hemicellulose fraction is mainly glucose, with mannose and galactose next in order of abundance. The pectic fraction contains rhamnose, galactose, and arabinose. About 65% of the total wall uronic acids is in the pectic fraction (MORIKAWA 1974), and the pectic fraction, itself, is 67–75% galacturonic acid, almost all of which is nonesterified (MORIKAWA 1974, ANDERSON and KING 1961). For comparison, the pectic polysaccharide content of *Acer* pseudoplatanus walls is about 32%, and polygalacturonic acid is 13.4% of the total wall. The pectic fraction released by polygalacturonase is about 62% galacturonic acid, at least 70% of which is nonesterified (TALMADGE et al. 1973). The large amount of nonesterified polygalacturonic acid in *Nitella* walls accounts for the high concentration (0.6 N) of negative groups, which make the wall an excellent ion exchanger (DAINTY and HOPE 1959). During elongation of the internode from 4 to 70 mm, the total uronide content of the wall decreases, on a dry weight basis, from 29 to 16%, neutral sugars released by trifluoroacetic acid hydrolysis decline from 24 to 9%, while cellulose increases from 19 to 27% (MÉTRAUX 1981).

The *Nitella* wall protein differs in its amino acid composition from higher plant "extensin" in that it lacks hydroxyproline and is enriched in the acidic amino acids, asparatate and glutamate. The wall protein of monocotyledons is typically not rich in hydroxyproline (BURKE et al. 1974) and there appear to be two types of wall protein in runner beans (*Phaseolus coccineus*), one rich and one poor in hydroxyproline (SELVENDRAN 1975). The *Nitella* wall protein is extracted in the 24% KOH-soluble fraction along with the hemicelluloses (MORIKAWA 1974), as has been reported for some higher plant wall protein (MONRO et al. 1974). The report that *N. opaca* wall protein is rich in cysteic acid (THOMPSON and PRESTON 1967) has not been confirmed for *N. flexilis* walls (MORIKAWA 1974).

The ability of pH 5 buffers, ions and EDTA to cause wall loosening in *Nitella* indicates that noncovalent bonds play a significant role in maintaining wall structural integrity. In his thorough study of the orientation of wall hydroxyl and carboxyl groups using the technique of polarized infrared spectroscopy, MORIKAWA (1974) observed a correlation between the stimulatory effect of acid on extensibility and the alteration of carboxyl group orientation. Since ions, such as sodium and magnesium, did not alter the orientation, MORIKAWA concluded that protons primarily affected nonionic bonds, such as hydrogen bonds or solvation bonds with the hydroxyl groups of the wall. WUYTACKET and GILLET (1978) demonstrated that both exchangeable and non-exchangeable forms of calcium are present in *Nitella* walls. The non-exchangeable form, presumed to be chelated by COO anions, polysaccharide hydroxyl groups or protein amino groups, can be largely removed by acid. Polygalacturonic acid chains in gels are thought to occur as buckled ribbons which can pack tightly in parallel arrays in the presence of calcium ions (REES 1977). The calcium ions fit into the cavity between opposite kinks in the chains, like eggs in an egg box. Thus, instead of divalent cross-links between carboxyl groups, calcium is proposed to rigidify the wall through coordination spheres composed of the oxygen atoms from both hydroxyl and carboxyl groups (Preston 1979). The egg box model for wall pectins, together with Morikawa's conclusion that protons affect nonionic sites suggests a mechanism for acid-induced wall loosening based on the displacement of calcium from the relatively weak interactions with pectin hydroxyl groups in the coordination spheres. It remains to be seen whether cell wall pectins which have neutral sugars, can form gels in the same manner as polygalacturonic acid chains. The mechanism of ion-stimulated loosening is apparently different from acid-induced loosening, although, again, calcium displacement is thought to be involved (Morikawa 1974, Métraux and Taiz 1977, 1979). Thompson and Preston (1968) found no effect of proteases or the sulfhydril reagent dithiothreitol on *Nitella* wall extensibility, and concluded that the wall protein played no structural role. However, no evidence was given that the protease actually removed the wall protein.

As the cell approaches its mature length the elongation rate decreases and wall extensibility also declines (Probine and Preston 1962, Métraux *et al.* 1980). At the same time the walls become less sensitive to acid (Métraux *et al.* 1980). Green (1958 a) has shown that the microfibrils on the inner wall surface of mature internodes are random rather than transverse in orientation. This shift in deposition pattern by itself probably has little, if any, influence on elongation rate. As will be outlined below, elongating *Nitella* cells that are artificially induced to lay down random microfibrils begin to expand laterally. Given the likelihood that pectins are important for acid-stimulated extension, the loss of sensitivity to acid could be due to a decrease in the pectin content of the wall. Probine and Preston (1962), in fact, found that the percentage of water-soluble materials in the wall diminished with the decrease in growth rate in *N. opaca*. As noted above, during cell maturation the percent dry weight of wall uronides decline, while cellulose increases, consistent with a role for uronides in the acid-mediated loosening process. Water-soluble pectins have been implicated in auxin-induced elongation in *Avena* coleoptiles (Albersheim and Bonner 1959). An increase in the neutral arabinan-galactan pectic fraction has been correlated with active growth (Stoddart and Northcote 1967). It has recently been suggested that gibberellic acid-induced cell expansion is, in part, due to an increase in pectin synthesis (Fry 1980). In developmental studies we have found marked changes in the neutral sugar composition of the pectin and hemicellulose fractions as the internode cell matures (Gepstein and Taiz, unpublished). Speculations regarding the significance of these developmental changes must await a detailed linkage analysis of the wall polymers.

III. Directionality of Cell Expansion

The directionality of cell expansion in *Nitella* is dictated by the spatial organization of cellulose microfibrils in the inner portion of the wall and by the stress pattern on the wall. The arrangement of cellulose in the inner wall depends principally upon microfibrillar deployment during deposition. The organization of the middle and outer wall regions, on the other hand, is a passive product of the accumulated strain experienced by the microfibrils as they are displaced progressively outward. Wall organization and its relationship to growth have been studied *in vivo* and *in vitro* by a variety of optical and physical methods. Much of the work has involved comparisons of normal cells with chemically or physically perturbed ones.

A. Normal Growth

The highly axial growth found in *Nitella* internodes is achieved in the face of a distribution of stresses that would otherwise favor growth in girth. In a pressurized thin-walled cylindrical shell such as an internode, stresses on the wall relate mainly to wall curvature and are modified only by end effects (BYARS and SNYDER 1975). Since the end walls are partitions with essentially zero stress normal to them, their influence on the cylindrical side wall is minimal. Given these considerations, circumferential stress is twice longitudinal stress (CASTLE 1937) over nearly all of the exposed internodal wall. The cylindrical shape of *Nitella* internodes is maintained over a 3,000-fold increase in length (GREEN 1958 a). The five to one predominance in the relative rate of growth in length over that in width is also maintained over most of this extension (GREEN 1963 a).

The physical basis for axial growth in *Nitella* resides in the structural and, as a consequence, mechanical anisotropy of the wall. Structural anisotropy arises from the overall transverse orientation of cellulose microfibrils within the plane of the wall (GREEN 1958 a). This type of wall organization, termed tubular texture (FREY-WYSSLING 1976), is common to elongating plant cells that do not extend by tip growth (ROELOFSON 1959). As found for advanced composite materials (CORNSWEET 1970), the microfibrils greatly augment wall strength and confer directionality to wall mechanical properties.

Abundant evidence has accumulated to demonstrate the mechanical anisotropy of *Nitella* walls. Applying uniaxial stresses to sections of isolated walls, PROBINE and PRESTON (1962) found that the elastic modulus was about five times greater in the longitudinal than in the transverse direction for growing cells. We obtained similar results but used an entirely different method of testing (RICHMOND *et al.* 1980). Other parameters of wall mechanical strength, namely plastic deformation (D_p), creep, and extensibility (creep/stress), were also examined. Each exhibited an even greater degree of anisotropy than shown for elastic deformation (D_e) (RICHMOND *et al.* 1980, MÉTRAUX and TAIZ 1978). Since normal stress relationships are lost in uniaxially extended walls, wall anisotropy was further tested on isolated walls subjected to multiaxial stresses. This was achieved by pressurizing mercury-filled wall tubes (MÉTRAUX *et al.* 1980, KAMIYA *et al.* 1963). Again the wall was found to be highly anisotropic in its mechanical properties, especially for plastic deformation (Fig. 2, control) (RICHMOND *et al.* 1980). The anisotropic character of the wall is also seen in acid-induced extension, the transverse threshold being at a lower pH than the longitudinal one (see below). Cells with perturbed growth patterns provide additional evidence for the correlation between wall mechanical properties and the directionality of growth (see below).

The primary wall generates a complex three-dimensional organization of cellulose microfibrils during growth. A gradient in microfibrillar alignment between inner and outer wall surfaces forms as a consequence of a combination of synthetic and physical events. One facet of the wall synthesis processes is the deposition by apposition (Green 1958 b) of cellulose microfibrils in a scattered but predominantly transverse direction (Green 1958 a, 1960, Gertel and Green 1977). Subsequent to deposition the microfibrils are displaced progressively outward by continued deposition. Coincident with outward displacement they are strain realigned in a manner than reflects the strain pattern of the wall (Gertel and Green 1977). This pattern of deposition and passive realignment describes the essence of the multinet growth hypothesis (Roelofson and Houwink 1953). The hypothesis was first proposed for higher plants, but perhaps the best experimental evidence in support of it comes from work on *Nitella* by Green and co-workers. Polarized light microscopy of torn wall edges was used to determine the mean microfibril orientation at varying depths within the wall cross-section. Relative wall thicknesses along the wedge were deduced by interference microscopy. Green (1960) found a non-linear gradient in microfibril organization within the wall which varied with the strain history of the cell. In older cells that had experienced considerable longitudinal strain, the microfibrils in the outermost portion of the wall even exhibited a net axial alignment. These results were extended by the elegant experiments of Gertel and Green (1977). The growth pattern of *Nitella* internodes was physically perturbed to measured degrees by stretching or compressing the cells. A close correspondence was found between these perturbations and the passive reorganization experienced by the outer portions of the wall. On the other hand, microfibrillar alignment at the inner face (reflecting deposition) was consistently transverse for all physical perturbations examined except when surface growth was completely suppressed. Recently Erickson (1980) has provided a theoretical basis from which quantitative predictions may be made about the multinet growth hypothesis. The gradients in microfibril orientation found in walls of normally growing internodes (Green 1960, Gertel and Green 1977) appear to agree well with the theoretical curves of Erickson.

Cells of higher plants are not nearly as amenable to direct experimental manipulation and examination as are those of *Nitella*. Consequently, conclusions concerning the multinet growth hypothesis in higher plants are based mainly on apparent wall structures as visualized at the electron microscope level. Unlike *Nitella* the primary walls of many cells, in both algae and higher plants, have a cross-lamellar pattern, or ordered texture, of microfibrils within the wall (Roelofson 1965). Alternative proposals have been generated to account for these wall structures (Roland and Vian 1979). For example, the ordered subunit hypothesis (Roland *et al.* 1975, 1977, Roland and Vian 1979) proposes that the primary wall is organized into an ordered multi-ply construction (similar to plywood). The elongating *Nitella* wall, however, does not appear to correspond to any of the interpretations as proposed by

Roland and Vian (1979). They seem to allow for a substantially net transverse wall texture (predominantly transverse lamellae) only in the construction termed a primordial wall. This ontogenetically primordial wall is seen in the compound middle lamellae of meristematic cells. According to ordered subunit growth, anisotropic expansion is a response to directional stresses and is accomplished by selective loosening of bonds in the successive, ordered fibrillar layers. By proposing differential bond loosening the wall is considered to be more active in determining the directionality of expansion compared to the passive wall of the multinet growth hypothesis. The latter wall would be more homogeneous in bond loosening. The two hypotheses provide strongly contrasting views of the three-dimensional architecture of the primary wall, between an ordered criss-cross construction and a gradient in orientation of a more disperse texture. The ordered texture maintains itself and is not a product of reorganization during growth. The dynamics of microfibrillar strain realignment, on the other hand, imply rotations of microfibrils about their axes. Microfibril rotation has been challenged by Boyd and Foster (1975). They hypothesize intramicrofibrillar bending which leads to a trellis-like wall configuration. This pattern is felt to be a result of passive deformation of the wall in agreement with multinet growth.

B. Perturbed Growth

The study of perturbed growth is essential to understanding the parameters that determine the directionality of cell expansion. Alterations of growth in *Nitella* internodes have been achieved by both physical and chemical means. Physical perturbations of cell expansion do not generally affect the deposition of microfibrils (Gertel and Green 1977). This is perhaps not too unexpected since the mechanism for transverse microfibril deposition is normally maintained in the face of a large longitudinally biased strain anisotropy. Whatever mechanism guides the initial orientation of microfibrils, once established it appears to be relatively resistant to change by both endogenous and imposed physical forces.

The pattern of microfibril deposition can be significantly altered with anti-microtubule drugs; in turn, cell expansion is modified. Green (1962, 1963 a) found that initially short *Nitella* internodes will grow into spheres when treated with colchicine. The walls of these spherical cells are structurally isotropic as judged by polarized light microscopy. Green (1963 b) also demonstrated that the wall deposited during colchicine treatment is composed of randomly arranged cellulose microfibrils. This was determined by both polarized light microscopy (in combination with interference microscopy) and electron microscopy (replicas of the inner wall surface). These experiments foreshadowed the current view that microtubules act as causal agents in the alignment of microfibrils during deposition. In many cell types microtubules subjacent to the plasma membrane are found to run parallel to microfibrils at the inner wall surface (Newcomb 1969, Hepler and Palevitz 1974). Addition of colchicine removes these microtubules; subsequently deposited microfibrils are randomly organized within the plane of the wall (Palevitz and Hepler 1976, Hogetsu and Shibaoka 1978, Pickett-Heaps 1967).

Correlations have additionally been made between ordered cellulose deposition and rows of particles (presumably cellulose synthetase complexes) within the plasma membrane (BROWN and WILLISON 1977). Membrane particles of this type have been observed in both algae (BROWN and MONTEZINOS 1976, PENG and JAFFE 1976, GIDDINGS *et al.* 1980; see p. 160 f.) and in higher plants (MUELLER and BROWN 1980). It has been proposed that microtubules act in the positioning of these particles (HEPLER and FOSKET 1971, HEATH 1974). The causal chain in microfibril alignment then becomes: microtubules-membrane synthetases-microfibrils. Virtually nothing is known about the mechanisms which control either the orientation of microtubules or their organizing centers. For further discussions about the influence of microtubules on microfibril alignments, see articles by Gunning (p. 301 f.) and Hepler (p. 327 f.) in this volume.

The time-course of events during cell shaping of chemically perturbed *Nitella* internodes has been followed in detail by RICHMOND (1977). The anti-microtubule drugs colchicine, isopropyl *N*-phenylcarbamate (IPC) and trifluralin were used to alter the pattern of cell expansion. The results were essentially the same for each agent. The drugs act coordinately on microtubules and microfibrils in support of the concept of microtubule organization being linked to microfibril deposition. The major conclusions of this work concern the biophysics of cell shape generation and do not, however, depend upon a particular mechanism for microfibril deposition.

Changes in cell shape and cell wall texture were monitored with a high degree of spatial and temporal resolution by time-lapse cinemicrography in polarized light. The action of polarized light on the cylindrical cell wall of the living internode creates a banding pattern that is diagnostic of the net (overall) cellulose microfibril distribution (see cell b in Fig. 8). A quantitative value, termed the "birefringence index", is obtained simply as the ratio of the distance between the centers of the dark (isotropic) bands to the cell diameter. The index correlates directly with the overall degree of micro-fibrillar alignment. A value of 1.0 would indicate a perfectly transverse alignment whereas 0.0 would denote a random wall texture. The birefringence index is also inversely proportional to the net degree of microfibrillar scatter (mean angular dispersion) in the plane of the wall. In this sense it is similar to an expression derived by ERICKSON (1980) as part of a theoretical model of multinet growth.

The initial growth response of internodes to the addition of antimicrotubule drugs is a shift from axial to preponderantly circumferential expansion. The transition is complete after a relative increase of about 25% in cell surface area. For normal relative rates of growth, 1.5–2%/hour^{-1} in length, the old anisotropic growth pattern is fully replaced after about 15 hours. Preceding the growth response is a shift in the pattern of microfibril deposition from transverse to random (within the plane of the wall). This shift is deduced by the declining values in the birefringence index (Fig. 9).

A close correlation exists between the generation of a particular cell shape and the cell's initial proportions and attendant wall stress pattern. Cells initially of near isodiametric proportions become spherical as a result of their

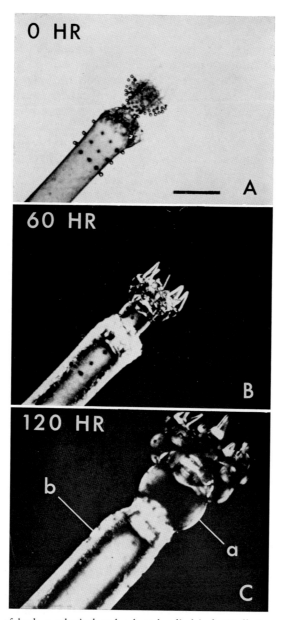

Fig. 8. Formation of both a spherical and a broad cylindrical *Nitella* internode in response to IPC. The lateral (leaf) cells have been removed at the node between cells *a* and *b*. Marking beads are placed on the cells to follow local deformation of cell surface. In B and C the internodes are viewed in polarized light (cells 45° to crossed polars). Cell *a* expands nearly isodiametrically and assumes a spherical shape. The dark "cross" in C signifies a net random array of microfibrils within the plane of the wall. Cell *b*, on the other hand, responds to the prevailing circumferential stress of a cylinder and expands mostly in girth. The lateral dark bands, with a bright stripe down the middle, characterize a scattered but predominantly transverse arrangement of birefringent microfibrils. Compare cell *b* to a flattened cylinder in Fig. 3. Expansion in both cells *a* and *b* is in accord with the passive yielding of an isotropic wall structure to the prevailing stress pattern on the cell surface. The bar in A equals 0.3 mm. Data from RICHMOND 1977.

uniform stress pattern. The combination of random microfibril deposition and isotropic expansion leads to a net random wall texture (see cell a in Fig. 8. Cells appreciably longer than wide, on the other hand, are transformed into broad cylinders as a consequence of random microfibril deposition in conjunction with the 2 : 1 diameter to length stress ratio on their walls. The circumferentially biased expansion of these cells serves to strain realign microfibrils in the transverse direction subsequent to their deposition, in accord with the multinet model. As a result, the overall transverse wall texture persists (see cell b in Fig. 8). In this case, overall transverse wall anisotropy does not correlate with axial extension. However, expansion does

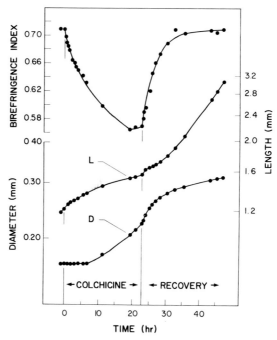

Fig. 9. Effect of colchicine addition and removal on a young *Nitella* internode. The growth response to colchicine treatment is an increase in the rate of growth in girth over that in length. Length (*L*) and diameter (*D*) are plotted on log scales; slopes on the curves express relative rates of growth. A new steady proportionality of growth is established at about 11 hours and after a relative surface area increase of about 20%. Subsequent to colchicine removal the normal growth pattern resumes within about 8 hours and after an increase of about 30% in relative surface area. Variations in birefringence index reveal shifts in both the pattern of microfibril deposition and net microfibrillar alignment. The drop in the index that follows colchicine addition indicates random microfibril deposition and a decrease in overall transverse wall organization. Upon drug removal transverse deposition returns and the original degree of organization is restored. Data from Richmond, unpublished.

correlate with the random (isotropic) microfibrillar texture of the inner portion of the wall. The growth response in all cells is consistent with the passive yielding of an isotropically reinforced wall to the cell's particular stress pattern.

The foregoing results indicate that the pattern of internodal growth may be controlled by only the inner portion of the cell wall. The wall maintains an almost constant thickness over the time course of transition in growth patterns. New wall is constantly being added by apposition (GREEN 1958 b) while old wall is being thinned by expansion. As a consequence of these wall dynamics, over a given time interval the percent increase in wall surface area is essentially equivalent to percent addition of new wall. Thus, the pattern of growth is totally transformed after a replacement of about the inner one-fourth of the wall. The influence of the inner wall on the directionality of expansion is further elucidated by the response of cells to recovery from drug treatment.

Both spherical and broad cylindrical internodes exhibit a remarkable ability to recover axial growth following the removal of anti-microtubule drugs. As determined by an abrupt increase in the birefringence index, transverse microfibril deposition rapidly resumes (Fig. 9). A stable pattern of elongation is usually reestablished after a surface area increase of less than 25%. Again the inner quarter of the wall appears to be instrumental in regulating the directional properties of growth. This concept is especially strengthened by the following results.

Spherical cells have no residual transverse wall texture to act as a basis for renewed longitudinal growth during the initial phase of recovery. Cells enlarging as broad cylinders, on the other hand, maintain a substantial overall transverse wall structure. If most of the wall influenced a cell's expansion pattern, then broad cylindrical internodes would be expected to regain longitudinal growth after a smaller increment of expansion than in spherical cells. The opposite was found to be the case. The results can be accounted for by considering the wall stresses in these cell types. Unlike a cylindrical cell, an initially spherical internode has no excess transverse stress to overcome before circumferential microfibrillar reinforcement can be translated into axial extension. Hence, due to wall stress factors, a spherical cell regains longitudinally directed growth sooner than does a broad cylinder. It is concluded that the pattern of internodal expansion in *Nitella* is principally a function of the microfibrillar arrangements of the inner portion of the wall and of the stress pattern on the wall. Further, this implies that the inner cell wall bears all of the wall stresses and is the primary site of wall metabolic activity.

It is considered likely that the general features of wall growth exhibited by *Nitella* internodes also exist in many other plant cells that enlarge by diffuse growth. With the inner wall dominant during expansion, a cell would probably be more responsive to growth stimuli and could more easily establish new patterns of expansion than if the entire wall controlled growth. Growth would not be constrained by the total legacy of the wall but would relate only to the more recent deposition. During normal development certain plant cells shift their growth patterns in a manner similar to that seen for drug-treated *Nitella* internodes. For example, the cylindrical cells comprising the net of the alga *Hydrodictyon africanum* become spherical as they mature (GREEN 1963 a). Only the initial and final states of cell shape and wall structure have been characterized, but it is apparent from polarized light analysis that the transition involves a shift from a transverse wall texture to an isotropic one. On the other hand, a form of broad cylinder formation is found in the latex vessels of *Euphorbia*. A final growth stage characterized by predominant widening follows vessel elongation (MOOR 1959). Wall deposition during lateral expansion consists of well-aligned microfibrils in a crossed-lamellar configuration. The arrangement produces a net random texture for these lamellae. As in *Nitella* broad cylinders, an overall transverse wall texture persists during vessel widening.

C. Directionality of Wall Mechanical Properties

It now becomes important to establish whether the factors which influence wall mechanical properties reside only in the inner part of isolated walls. Such comparisons of *in vitro* and *in vivo* expansion patterns provide a means of assessing the contributions of metabolism to the directional properties of the wall. Tests were carried out by comparing the directionality of mechanical properties of walls having similar overall microfibrillar orientations but with differing wall textures (RICHMOND et al. 1980). Walls were selected according to similar birefringence indices from normal (transverse inner wall, random to longitudinal outer wall), drug-treated (random inner wall, transverse to random outer wall), and drug-recovery cells. Plastic and elastic deformation and creep were measured by both uniaxial and multiaxial extension methods. Typical deformations in length and diameter of pressurized wall tubes are seen in Fig. 2. The normal wall is highly anisotropic in plastic deformation whereas the wall of an IPC-treated cell is more isotropic in its deformation behavior. Drug removal elicts a substantial but incomplete recovery of anisotropic behavior. IPC-recovery cells *in vivo*, likewise, do not usually fully attain normal growth anisotropies (RICHMOND 1977). The results indicate that the directional character of expansion is preserved to a large extent in the mechanical properties of isolated walls, and that most, if not all, of the deformation is determined by the inner wall.

The small but significant discrepancy between the directionalities of expansion in isolated and growing walls probably relates to the process of wall loosening. Isolated walls were routinely extended at pH 6.5 while *in vivo* the inner wall might well experience pH 4.3 (MÉTRAUX et al. 1980). When walls of IPC-treated cells were multiaxially stressed under acid conditions (pH 4.0–4.5), they became more isotropic than at neutral pH, in closer agreement with expectations from wall growth behavior (RICHMOND et al. 1980). Acidification might act to loosen matrix material in the middle portion of the wall to an extent that would render this region too weak to function in the expression of the wall's growth and mechanical characteristics. Through thinning and stretching the outer portion of the wall is thought to lose its integrity regardless of any action by hydrogen ions. The inner wall, having suffered little from stretching and thinning, would maintain a balance between wall loosening and bonding integrity of critical wall bonds. The yielding and directional properties of growth would, thus, reside exclusively in the inner portion of the cell wall.

Microfibrillar orientations in the inner wall also correlate with the thresholds for longitudinal and transverse acid-induced extension in isolated *Nitella* walls (RICHMOND *et al.* 1980). For normal walls the transverse threshold (pH 4.0, 10 mM buffer) (RICHMOND, MÉTRAUX, and TAIZ unpublished data) is appreciably lower than the longitudinal one (pH 5.3) (MÉTRAUX *et al.* 1980). Transversely oriented microfibrils apparently act to restrain the transverse acid effect. These microfibrils preferentially increase circumferential strength, and as a consequence, acid-induced loosening needs to be increased beyond the longitudinal threshold before an appreciable transverse stimulation is seen. Accordingly, a shift to equivalent thresholds would be expected to accompany a loss in preferred microfibrillar alignment. The difference in pH thresholds between longitudinal and transverse walls is substantially reduced for IPC walls (RICHMOND *et al.* 1980). The directionality of acid-induced extension, therefore, appears to be modulated by microfibrillar alignments and more specifically by those of the inner wall.

Acknowledgements

We thank Dr. PAUL B. GREEN of Stanford University for reviewing the manuscript. One of us (J.-P. MÉTRAUX) wishes to gratefully acknowledge the support of the Miller Institute for Basic Research in Science at the University of California, Berkeley. Portions of this research were funded by grants from the National Science Foundation to L. TAIZ.

References

ALBERSHEIM, P., NEVINS, D. J., ENGLISH, P. D., KARR, A., 1967: A method for the analysis of sugars in plant cell wall polysaccharides by gas-liquid chromatography. Carbohyd. Res. **5**, 340—345.

— BONNER, J., 1959: Metabolism and hormonal control of pectic substances. J. Biol. Chem. **234**, 3105—3108.

ANDERSON, D. M. W., KING, N. J., 1961: Polysaccharides of the *Characeae*. II. The carbohydrate content of *Nitella translucens*. Biochim. Biophys. Acta **52**, 441—449.

ARENS, K., 1936: Physiologisch polarisierter Massenaustausch und Photosynthese bei submersen Wasserpflanzen. II. Die Ca(HCO₃)₂-Assimilation. Jahrb. Wiss. Bot. **83**, 513.

BENNET-CLARK, T. A., 1955: A hypothesis on salt accumulation and the mode of action of auxin. In: The chemistry and mode of action of plant growth substances (WAIN, R. L., WIGHTMAN, F., eds.), pp. 284—294. London: Butterworths.

BONNER, J., 1934: The relation of hydrogen ions to the growth rate of the *Avena* coleoptile. Protoplasma **21**, 406—423.

— 1935: Zum Mechanismus der Zellstreckung auf Grund der Micellarlehre. Jahrb. Wiss. Botan. **82**, 377—412.

BOYD, J. D., FOSTER, R. C., 1975: Microfibrils in primary and secondary wall growth develop trellis configurations. Can. J. Bot. **53**, 2687—2701.

BOYER, J. S., WU, G., 1978: Auxin increases the hydraulic conductivity of auxin-sensitive hypocotyl tissue. Planta **139**, 227—237.

BROWN, R. M., MONTEZINOS, D., 1976: Cellulose microfibrils: visualization of the biosynthetic and orienting complexes of the plasma membrane. Proc. Nat. Acad. Sci. U.S.A. **73**, 143—147.

— WILLISON, J. H. M., 1977: Golgi apparatus and plasma membrane involvement in secretion and cell surface deposition with special emphasis on cellulose biogenesis. In: International cell biology 1976–1977 (BRINKLEY, B. R., PORTER, K. R., eds.), pp. 267—283. New York: The Rockefeller University Press.

BURKE, D., KAUFMAN, P. B., MCNEIL, M., ALBERSHEIM, P., 1974: The structure of plant cell walls. VI. A survey of the walls of suspension-cultured monocots. Plant Physiol. **54**, 109—115.

BYERS, E. F., SNYDER, R. D., 1975: Engineering mechanics of deformable bodies, pp. 161—166. New York: Intext Educational Publishers.

Castle, E. S., 1973: Membrane tensions and orientation of structure in the plant cell wall. J. Cell Comp. Physiol. 10, 113—121.
— 1955: The mode of growth of epidermal cells of the *Avena* coleoptile. Proc. Nat. Acad. Sci. U.S.A. 41, 197—199.
Cleland, R. E., 1971: Cell wall extension. Annu. Rev. Plant Physiol. 22, 197—222.
— 1973: Kinetics of hormone-induced H⁺ extrusion. Plant Physiol. 58, 210—213.
— Rayle, D. L., 1978: Auxin, H⁺-excretion and cell elongation. Bot. Mag. Tokyo Special Issue 1, 125—139.
Cornsweet, T. M., 1970: Advanced composite materials. Science 168, 433—438.
Dainty, J., Hope, A. B., 1959: Ionic relations of cells of *Chara australis*. Austr. J. Biol. Sci. 12, 395—411.
Erickson, R. O., 1980: Microfibrillar structure of growing plant cells walls. In: Lecture notes in biomathematics (Levin, S., ed.), Vol. 33: Mathematical modeling in biology and ecology (Getz, W. J., ed.), pp. 192—212. Berlin-Heidelberg-New York: Springer.
Evans, M. L., Ray, P. M., 1969: Timing of the auxin response in coleoptiles and its implications regarding auxin action. J. Gen. Physiol. 53, 1—20.
Mulkey, T. J., 1981: Geotropism in corn roots: evidence for its mediation by differential acid efflux. Science 212, 70—71.
Frey-Wyssling, A., 1976: The plant cell wall. In: Handbuch der Pflanzenanatomie (Zimmerman, M., Carlquist, S., Wulff, H. D., eds.), Vol. 4. Berlin: Gebrüder Borntraeger.
Fry, S. C., 1980: Gibberellin-controlled pectinic acid and protein secretion in growing cells. Phytochem. 19, 735—740.
Gertel, E. T., Green, P. B., 1977: Cell growth pattern and wall microfibrillar arrangement. Experiments with *Nitella*. Plant Physiol. 60, 247—254.
Giddings, T. H., Brower, D. L., Staehelin, L. A., 1980: Visualization of particle complexes in the plasma membrane of *Micrasterias denticulata* associated with the formation of cellulose fibrils in primary and secondary cell walls. J. Cell Biol. 84, 327—339.
Green, P. B., 1954: The spiral growth pattern of the cell wall in *Nitella axillaris*. Am. J. Bot. 41, 403—409.
— 1958 a: Structural characteristics of developing *Nitella* internodal cell walls. J. Biophys. Biochem. Cytol. 4, 505—516.
— 1958 b: Concerning the site of addition of new cell wall substances to the elongating *Nitella* cell wall. Am. J. Bot. 45, 111—116.
— 1960: Multinet growth in the cell wall of *Nitella*. J. Biophys. Biochem. Cytol. 7, 289—297.
— 1962: Mechanism for plant cellular morphogenesis. Science 138, 1404—1404.
— 1963 a: Cell walls and the geometry of plant growth. In: Meristems and differentiation. Brookhaven Symp. Biol., Vol. 16, pp. 203—217. Upton, N.Y.: Brookhaven National Laboratory.
— 1963 b: On mechanisms of elongation. In: Cytodifferentiation and macromolecular synthesis (Locke, M., ed.), pp. 203—234. New York: Academic Press.
— 1965: Anion exchange resin spheres as marking material for wet cell surfaces. Exp. Cell Res. 40, 195—196.
— 1968: Growth physics in *Nitella*: a method for continuous *in vivo* analysis of extensibility based on a micromanometer technique for turgor pressure. Plant Physiol. 43, 1169—1184.
— 1968: Cell morphogenesis. Ann. Rev. Plant Physiol. 20, 365—394.
— 1980: Organogenesis—a biophysical view. Ann. Rev. Plant Physiol. 31, 51—82.
— Chapman, G. B., 1955: On the development and structure of the cell wall in *Nitella*. Am. J. Bot. 42, 485—693.
— Erickson, R. O., Richmond, P. A., 1970: On the physical basis of cell morphogenesis. Ann. N.Y. Acad. Sci. 175, 712—731.
— — Buggy, J., 1971: Metabolic and physical control of cell elongation rate. *In vitro* studies in *Nitella*. Plant Physiol. 47, 423—430.
— Bauer, K., Cummins, W. R., 1977: Biophysical model for plant cell expansion: Auxin effects. In: Water relations in membrane transport in plants and animals (Jungreis, A. M., Hodges, T. K., Kleinzeller, A., Schultz, S. G., eds.), pp. 30—45. New York: Academic Press.

HAGER, A., MENZEL, H., KRAUSS, A., 1971: Versuche und Hypothese zur Primärwirkung des Auxins beim Streckungswachstum. Planta **100**, 47—75.

HAUGTON, P. M., SELLEN, D. B., PRESTON, R. D., 1968: Dynamic mechanical properties of the cell wall of *Nitella opaca*. J. Exp. Bot. **19**, 1—11.

HEARMON, R. F. S., 1961: Applied anisotropic elasticity. London: Oxford University Press.

HEATH, I. B., 1974: A unified hypothesis for the role of membrane bound enzyme complexes and microtubules in plant cell wall synthesis. J. Theor. Biol. **48**, 445—449.

HEPLER, P. K., FOSKET, D. E., 1971: The role of microtubules in vessel member differentiation in *Coleus*. Protoplasma **72**, 213—236.

— PALEVITZ, B. A., 1974: Microtubules and microfilaments. Annu. Rev. Plant Physiol. **25**, 309—362.

HOGETSU, T., SHIBAOKA, H., 1978: Effects of colchicine on cell shape and on microfibril arrangement in the cell wall of *Closterium acerosum*. Planta (Berl.) **140**, 15—18.

HOPE, A. B., WALKER, N. A., 1975: The physiology of giant algal cells. London-New York: Cambridge University Press.

JACCARD, M., PILET, P. E., 1977: Extensibility and rheology of collenchyma. II. Low-pH effect on the extenion of collocytes isolated from high- and low-growing material. Plant Cell Physiol. **18**, 883—891.

JACOBS, M., RAY, P. M., 1975: Promotion of xyloglucan metabolism by acid pH. Plant Physiol. **56**, 373—376.

— — 1976: Rapid auxin-induced decrease in the free-space pH and its relationship to auxin-induced growth in maize and pea. Plant Physiol. **58**, 203—209.

KAMIYA, N., TAZAWA, M., TAKATA, T., 1963: The relation of turgor pressure to cell volume in *Nitella* with special reference to the mechanical properties of the wall. Protoplasma **57**, 501—521.

KEEGSTRA, K., TALMADGE, K. W., BAUER, W. D., ALBERSHEIM, P., 1973: The structure of plant cell walls. III. A model of the walls of suspension-cultured sycamore cells based on the interconnections of the macromolecular components. Plant Physiol. **51**, 188—197.

KEIFER, D. W., SPANSWICK, R. M., 1979: Correlation of adenosine triphosphate levels in *Chara corallina* with the activity of the electrogenic pump. Plant Physiol. **64**, 165—168.

KITASATO, H., 1968: The influence of H^+ on the membrane potential and ion fluxes of *Nitella*. J. Gen. Physiol. **52**, 60—87.

LABAVITCH, J. M., RAY, P. M., 1974: Relationship between promotion of xyloglucan metabolism and induction of elongation by IAA. Plant Physiol. **54**, 499—502.

LOCKHART, J. A., 1965: Cell extension. In: Plant biochemistry (BONNER, J., VARNER, J. E., eds.), pp. 827—849. New York: Academic Press.

— 1967: Physical nature of irreversible deformation of plant cells. Plant Physiol. **42**, 1545—1552.

LUCAS, W. J., 1976: Plasmalemma transport of HCO_3^- and OH^- in *Chara corallina*. J. Exp. Bot. **27**, 19—31.

— DAINTY, J., 1977: Spatial distribution of functional OH^- carriers along a Characean internodal cell: Determined by the effect of cycochalasin B on HCO_3^- assimilation. J. Membrane Biol. **32**, 75—92.

— SMITH, F. A., 1973: The formation of alkaline and acid regions at the surface of *Chara corallina* cells. J. Exp. Bot. **24**, 15—28.

MASUDA, Y., 1978: Auxin-induced cell wall loosening. Bot. Mag. Tokyo Special Issue **1**, 103—123.

MÉTRAUX, J.-P., 1979: Studies on the regulation of cell expansion in *Nitella*. Ph.D. dissertation, University of California, Santa Cruz.

— 1981: Changes in cell wall composition of growing *Nitella* internodes. Plant Physiol. Supplement **67**, 127.

— RICHMOND, P. A., TAIZ, L., 1980: Control of cell elongation in *Nitella* by endogenous cell wall pH gradients. Multiaxial extensibility and growth studies. Plant Physiol. **65**, 204—210.

— Taiz, L., 1977: Cell wall extension in *Nitella* as influenced by acid and ions. Proc. Nat. Acad. Sci. U.S.A. **74**, 1565—1569.

— — 1978: Transverse viscoelastic extention in *Nitella*. I. Relationship to growth rate. Plant Physiol. **61**, 135—138.

— — 1979: Transverse viscoelastic extention in *Nitella*. II. Effect of acid and ions. Plant Physiol. **63**, 657—659.

Monro, J. A., Bailey, P. W., Penny, D., 1974: Cell wall hydroxyproline-polysaccharide associations in *Lupinus* hypocotyls. Phytochemistry **13**, 375—382.

Moor, H., 1959: Platin-Kohle-Abdruck-Technik angewandt auf den Feinbau der Milchröhren. J. Ultrastruct. Res. **2**, 393—422.

Morikawa, H., 1975: Studies on the ultrastructure of the plant cell wall in *Nitella*. Ph.D. dissertation, University of Kyoto, Japan.

— Senda, M., 1974 a: Effects of acids and ions on mechanical properties of *Nitella* cell walls. The 15th Annual Meeting of Japanese Society of Plant Physiologists, 6–8 April, Abstr., p. 107.

— — 1974 b: Nature of the bonds holding pectic substance in *Nitella* walls. Agr. Biol. Chem. **38**, 1955—1980.

Mueller, S. C., Brown, R. M., 1980: Evidence for an intramembrane component associated with a cellulose microfibril-synthesizing complex in higher plants. J. Cell Biol. **84**, 315—326.

Newcomb, E. H., 1969: Plant microtubules. Ann. Rev. Plant. Physiol. **20**, 253—288.

Palevitz, B. A., Hepler, P. K., 1976: Cellulose microfibril orientation and cell shaping in developing guard cells of *Allium*: the rate of microtubules and ion accumulation. Planta (Berl.) **132**, 71—93.

Peng, H. B., Jaffe, L. F., 1976: Cell wall formation in *Pelvetia* embryos. A freeze-fracture study. Planta (Berl.) **133**, 57—71.

Pickett-Heaps, J. D., 1967: The effects of colchicine on the ultrastructure of dividing plant cells, xylem wall differentiation and distribution of cytoplasmic microtubules. Devel. Biol. **15**, 206—236.

— 1975: Green algae. Sunderland, Mass.: Sinauer.

Preston, R. D., 1974: The physical biology of plant cell walls. London: Chapman and Hall.

— 1979: Polysaccharide conformation and cell wall function. Ann. Rev. Plant Physiol. **30**, 55—78.

Probine, M. C., Barber, N. F., 1966: The structure and plastic properties of the cell wall of *Nitella* in relation to extension and growth. Aust. J. Biol. Sci. **19**, 439—457.

— Preston, R. D., 1961: Cell growth and the structure and mechanical properties of the wall in internodal cells of *Nitella opaca*. I. Wall structure and growth. J. Exp. Bot. **12**, 261—282.

— — 1962: Cell growth and the structure and mechanical properties of the wall in internodal cells of *Nitella opaca*. Mechanical properties of the walls. J. Exp. Bot. **13**, 111—127.

Ray, P. M., 1967: Radioautographic study of cell wall deposition in growing plant cells. J. Cell. Biol. **35**, 659—674.

— 1969: The action of auxin on cell enlargement in plants. Dev. Biol. Suppl. **3**, 172—205.

— Abdul-Baki, A. A., 1968: Regulation of cell wall synthesis in response to auxin. In: Biochemistry and physiology of plant growth substances (Wightman, F., Setterfield, G., eds.), pp. 647—658. Ottawa: Runge Press.

— Green, P. B., Cleland, R. E., 1972: Role of turgor in plant cell growth. Nature **239**, 163—164.

Rayle, D. L., Cleland, R., 1970: Enhancement of wall loosening and elongation by acid solution. Plant Physiol. **46**, 250—253.

— — 1972: The *in vitro* acid-growth response: Relation to *in vivo* growth response and auxin action. Planta **104**, 282—296.

— Haughton, P. M., Cleland, R., 1970: An *in vitro* system that stimulates plant cell extension growth. Proc. Nat. Acad. Sci. U.S.A. **67**, 1814—1817.

REES, D. A., 1977: Polysaccharide shapes. Outline studies in botany series. London: Chapman and Hall; New York: Wiley.

RICHMOND, P. A., 1977: Control of plant cell morphogenesis by the cell wall: analysis in *Nitella*. Ph.D. thesis, University of Pennsylvania, Philadelphia.

— MÉTRAUX, J.-P., TAIZ, L., 1980: Cell expansion patterns and directionality of wall mechanical properties in *Nitella*. Plant Physiol. **65**, 211—217.

ROBINSON, D. G., 1977: Plant cell wall synthesis. Adv. Bot. Res. **5**, 89—151.

ROELOFSEN, P. A., 1965: Ultrastructure of the wall in growing cells and its relation to the direction of growth. Adv. Bot. Res. **2**, 69—149.

— HOUWINK, A. L., 1953: Architecture and growth of the primary cell wall in some plant hairs and in the *Phycomyces* sporangiophore. Acta Bot. Néerl. **2**, 218—225.

ROLAND, J. C., VIAN, B., 1979: The wall of the growing plant cell: its three-dimensional organization. Int. Rev. Cytol. **61**, 129—166.

— — REIS, D., 1975: Observations with cytochemistry and ultracryotomy on the fine structure of the expanding walls in actively elongating plant cells. J. Cell Sci. **19**, 239—259.

— — — 1977: Further observations on cell wall morphogenesis and polysaccharide arrangement during plant growth. Protoplasma **91**, 125—141.

RUBERY, P. H., SHELDRAKE, A. R., 1973: Effect of pH and surface charge on cell uptake of auxin. Nature New Biol. **244**, 285—288.

SANDAN, T., OGURA, T., 1957: Physiological studies on growth and morphogenesis of isolated plant cell cultured *in vitro*. III. The effects of pH, auxin and metabolic inhibitors. Bot. Mag. Tokyo **70**, 125—130.

SELVENDRAM, R. R., 1975: Cell wall glycoproteins and polysaccharides of parenchyma of *Phaseolus coccineus*. Phytochemistry **14**, 2175—2180.

SHIMMEN, T., TAZAWA, M., 1977: Control of membrane potential and excitability of *Chara* cells with ATP and Mg^{2+}. J. Membr. Biol. **37**, 167—192.

SPANSWICK, R. M., 1972: Evidence for an electrogenic ion pump in *Nitella translucens*. I. The effects of pH, K^+, Na^+, light and temperature on the membrane potential and resistance. Biochim. Biophys. Acta **288**, 73—89.

SPEAR, D. G., BARR, J. K., BARR, C. E., 1969: Localization of hydrogen ions and chloride fluxes in *Nitella*. J. Gen. Physiol. **54**, 397—414.

STANTON, F. W., 1970: A null method for the measurement of turgor pressure and the study of osmoregulation in growing *Nitella* cells. Masters Thesis, University of Pennsylvania, Philadelphia.

STODDART, R. W., NORTHCOTE, D. H., 1967: Metabolic relationships of the isolated fractions of the pectic substances of actively growing sycamore cells. Biochem. J. **105**, 45—59.

STRUGGER, S., 1932: Die Beeinflussung des Wachstums und des Geotropismus durch die Wasserstoffionen. Ber. dtsch. bot. Ges. **50**, 77—92.

STUART, D. A., JONES, R. L., 1978: The role of acidification in gibberellic acid- and fusicoccin-induced elongation growth of lettuce hypocotyl sections. Planta **142**, 135—142.

TAGAWA, T., BONNER, J., 1957: Mechanical properties of the *Avena* coleoptile as related to auxin and to ionic interactions. Plant Physiol. **32**, 207—312.

TALMADGE, K. W., KEEGSTRA, K., BAUER, W. D., ALBERSHEIM, P., 1973: The structure of plant cell walls. I. The macromolecular components of the walls of suspension-cultured sycamore cells with a detailed analysis of the pectic polysaccharides. Plant Physiol. **51**, 158—173.

TAZAWA, M., KAMIYA, N., 1965: Water relations of Characean internodal cell. Ann. Rep. Scient. Works, Fac. Sci. Osaka University **13**, 123—157.

— FUJII, S., KIKUYAMA, M., 1979: Demonstration of light-induced potential change in *Chara* cells lacking tonoplast. Plant and Cell Physiol. **20**, 271—280.

TEPFER, M., CLELAND, R. E., 1979: A comparison of acid-induced cell wall loosening in *Valonia ventricosa* and in oat coleoptiles. Plant Physiol. **63**, 898—902.

Thompson, E. W., Preston, R. D., 1967: Proteins in the cell walls of some green algae. Nature **213**, 684—685.

— — 1968: Evidence for a structural role of protein in algal cell walls. J. Exp. Bot. **19**, 690—697.

Wainwright, S. A., Biggs, W. D., Currey, J. D., Gosline, J. M., 1976: Mechanical design in organisms. London: Edward Arnold.

Wardrop, A., 1955: The mechanism of surface growth in parenchyma of *Avena* coleoptiles. Aust. J. Bot. **3**, 137—148.

Weisenseel, M. H., Dorn, A., Jaffe, L. F., 1979: Natural H^+ currents traverse growing roots and root hairs of barley (*Hordeum vulgare* L.). Plant Physiol. **64**, 512—518.

Wilson, K., 1957: Extension growth in primary cell walls with special reference to *Elodea canadensis*. Ann. Bot. **21**, 1—11.

Wuytacket, R., Gillet, C., 1978: Nature des liaisons de l'ion calcium dans la paroi de *Nitella flexilis*. Can. J. Bot. **56**, 1439—1443.

Yamagata, Y., Yamamoto, R., Masuda, Y., 1974: Auxin and hydrogen ion actions on light-grown pea epicotyl segments. II. Effect of hydrogen ions on extension of the isolated epidermis. Plant Cell Physiol. **15**, 833—841.

Morphogenesis and Polarity of Tubular Cells with Tip Growth*

A. Sievers and E. Schnepf

Botanical Institute, University of Bonn, Bonn, and University of Heidelberg, Heidelberg, Federal Republic of Germany

With 15 Figures

Contents

I. Introduction

Morphogenesis of a plant cell, as treated in this chapter, is, in general, morphogenesis of the cell wall with an emphasis on growth. The reason for this specialized treatment of cell morphogenesis is best explained through an examination of tip growth, the locally restricted enlargement of a cell

---

* Dedicated to Professor Dr. A. Frey-Wyssling on the occasion of his 80th birthday.

that usually has a tubular shape (Frey-Wyssling 1959). Tip growth is associated with polar, local secretion of wall material and represents a conspicuous example of cell polarity. Usually, cells with tip growth show unipolar growth. There are also, however, examples of a bipolar tip growth—phloem (Schoch-Bodmer and Huber 1951) and xylem (Bosshard 1952) fibers. Sometimes there are cells with multipolar growth, *i.e.*, branched systems of coenocytic fungi or algae, stellately armed parenchyma cells, and nonarticulated laticifers. Even the morphogenesis of cells with a complicated shape like the cells of *Micrasterias* (see p. 147 f.) can be regarded as multipolar growth.

This chapter does not treat polar growth processes of naked or transiently naked cells that form flagella or microvilli, or cytoplasmic tails of flagellates (*Ochromonas*, Brown and Bouck 1973; *Poterioochromonas*, Schnepf et al. 1975), or horns of diatoms (*Attheya*, Schnepf et al. 1980). It will also exclude intercalar or basal growth (*Antithamnion*, Kinzel 1956) of elongate cells. The morphogenesis of *Acetabularia* (see p. 119 f.) and *Nitella* (see p. 231 f.) is dealt with separately in this volume.

Because of its local restriction and high rate (in pollen tubes up to 600 μm × h^{-1}, Weisenseel et al. 1975), tip growth provides favorable models to study cell morphogenesis, cell polarity, and the functional and structural organization of cells. Free, tubular cell systems have the further advantage of being easily handled experimentally and directly observable. It should be mentioned here that it is possible to stop tip growth transiently without influencing cell polarity (*e.g.*, by treatment with low temperature or with some inhibitors) and that, in a few examples, it is possible to revert the polar axis without interrupting growth to a higher degree.

As will be shown, tip growth is only one manifestation of cell polarity; growth gradient is related to gradients of various reactions, to gradients in molecules, ions, and electric currents (see p. 379 f.), and to gradients in distribution and, occasionally, differentiation of cell organelles. It is our intention to identify these gradients and explain how they are interconnected and controlled by exogenous factors. Finally, we wish to show briefly the meaning of these cellular processes not only as they apply to cell morphogenesis but also as they apply to the organization of a multicellular plant body.

Problems of polarity have been reviewed by Bünning (1958), Bloch (1965), von Wettstein (1965), and more recently by Quatrano (1978) and Schnepf (1981). Surveys of the phenomena of tip growth have been given by Sievers (1964).

II. Course of Tip Growth and Its Manifestation

A. Initiation of Tip Growth and Polarity

Initiation of polarity is one of the first fundamental steps in cell morphogenesis. The problems of initiation of tip growth can be summarized in two main questions: How is the polar axis determined? How and when is growth initiated and directed?

In mature pollen grains and fungal spores, germination begins when the cells come into an appropriate environment; growth initiation is controlled by external factors. In many other cells, the effects of internal factors seem to be more conspicuous in initiating growth. One internal factor is cell differentiation. A rhizodermis cell of *Trianaea* undergoes several endo-mitotic cycles before it becomes a trichoblast and protrudes a root hair, whereas the adjacent atrichoblasts divide mitotically (TSCHERMAK-WOESS and HASITSCHKA 1953, CUTTER and FELDMAN 1970 a, b). The stage within the cell cycle is likewise important for tip growth initiation. In the normally growing caulonema of the moss *Funaria hygrometrica,* the third cell of a filament begins to form laterally a new growing tip that later develops into a side branch (SCHMIEDEL and SCHNEPF 1979 a). The local determina-tion is not a consequence of an apical dominance but of the developmental time. When the growth of the filament is retarded by cytochalasin B, the second cell begins to form a side branch at about the "right" time (Fig. 1), *i.e.,* when it would have been a third cell under normal conditions (SCHMIEDEL and SCHNEPF 1979 b).

There are only a few cells—*i.e.,* zygotes of the brown algae *Fucus* and *Pelvetia,* spores of the fungus *Botrytis,* the moss *Funaria,* the ferns *Dryopteris* and *Osmunda,* and the horsetail *Equisetum,* and protoplasts of coenocytic algae (ISHIZAWA *et al.* 1979)—whose polarity is not predetermined during the ontogenesis of the cell (for review, see WEISENSEEL 1979). Of all the external factors that polarize these cells, light is the most important. The main morphologic axis induced by light is oriented parallel to the largest gradient of absorption, which often is different from the direction of the incident light. In *Fucus* zygotes, unilateral light normally induces germina-tion at the shaded pole (JAFFE 1958; see Fig. 1 on p. 380). If spores of *Osmunda* and *Botrytis* are partially illuminated, they germinate at the shaded or the illuminated pole, respectively (JAFFE and ETZOLD 1962). In *Fucus* zygotes, the axis of polarity is really induced by light (JAFFE 1956, 1958, 1968). After illumination of the zygotes with plane-polarized light, about 50% of the cells produce two rhizoids in opposite directions; the single rhizoids of the other zygotes grow in either of these two directions. Thus, all rhizoids develop parallel to the electric vector. The induction of polarity is affected by blue light. It is assumed that carotenoids or flavins function as photoreceptors (HAUPT 1957, 1962, JAFFE and ETZOLD 1962, BENTRUP 1963, 1964).

Concentration gradients of chemical substances often induce polarity in *Fucus* zygotes. Zygotes lying close together mutually induce their polarity, forming their basal poles away from or toward each other (JAFFE and NEUSCHELER 1969). If they are exposed to a laminar flow, they germinate downstream, the direction in which a diffusible substance would be trans-ported (BENTRUP and JAFFE 1968). Gradients of H^+, K^+ (see Fig. 1 on p. 380) and Ca^{2+} are also effective in inducing polarity of the zygotes (BENTRUP *et al.* 1967, ROBINSON and JAFFE 1976). *Fucus* zygotes, *Equisetum* spores (BENTRUP 1968 a, b), and *Pelvetia* zygotes (PENG and JAFFE 1976) respond to applied electric fields by germinating in a parallel direction to the field. Steady electric fields and forced calcium entry caused by the ionophore A 23187 can be used to control the point at which the rhizoids emerge from *Funaria* spores (CHEN and JAFFE 1979).

Among the internal factors that influence the polar outgrowth of cells, and thus the position of a new polar axis, is the nucleus. In a second cell of a *Funaria* caulonema, the nucleus lies near the side wall about 90 μm from the apical cross

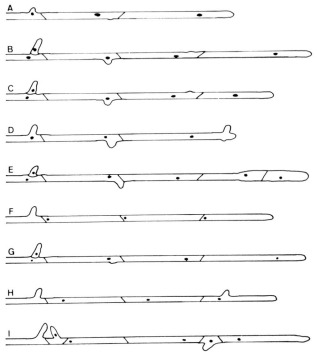

Fig. 1. Polar growth and development in the *Funaria* caulonema: *A* developing tip, tip cell and third cell in preprophase; *B* normal development, about 7 hours later than A, the same situation in the cell cycle; *C* after treatment with cytochalasin B, growth is slowed down but the development of the subapical cells is unaffected; *D* after treatment with colchicine, elongation is inhibited, the development but not the initiation of the side branches is stopped; *E* after treatment with the ionophore A 23 187, enlargement of cell diameter; *F* interphase cells after centrifugation (similar effects after treatment with D₂O), the nuclei are translocated in proximal direction; *G* after treatment as in F, recovery; *H* after treatment as in F, recovery inhibited by colchicine; *I* like G, but the tip cell and the third cell have been in preprophase when the filament was centrifuged or treated with D₂O: reversal of polarity.
From SCHNEPF 1981.

wall. Side branch formation begins 40 to 50 μm apically from the nucleus, obliquely opposite to it; the side wall bulges out here (Fig. 1). Then the nucleus moves toward the cell protrusion, divides, and one of the daughter nuclei migrates into the outgrowth that is then cut off from the mother cell (SCHMIEDEL and SCHNEPF 1979 a). In vertically placed Petri dishes, the nucleus of horizontally growing filaments tends to follow gravity and lie on the "bottom" of the cell. Under these conditions, most side branches arise negatively gravitropically. The nucleus seems to function as a statolith (SCHMIEDEL and SCHNEPF 1979 a). The orientation of the cross wall is also influenced by gravity in this situation.

If the nucleus is displaced experimentally (Fig. 1), for instance by centrifugation or by the application of heavy water, it moves back into its

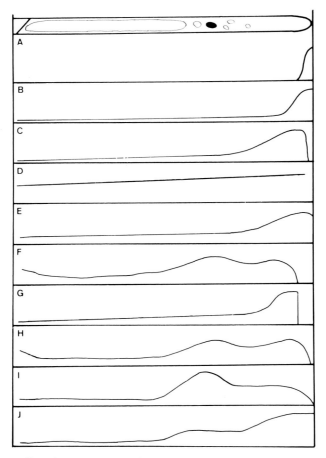

Fig. 2. Polar gradient in a *Funaria* caulonema tip cell in arbitrary units (the values are based on rough calculations rather than on exact measurements). Above: Tip cell, viewed schematically, to show the position of the gradients—nucleus black, vacuoles open circles. *A–J* gradients, *A* growth; *B* exocytosis of Golgi vesicles; *C* distribution of dictyosomes; *D* number of Golgi cisternae per dictyosome; *E* distribution of coated regions and coated pits in the plasmalemma; *F* distribution of plastids; *G* starch content; *H* distribution of mitochondria; *I* distribution of microtubules; *J* intensity of fluorescence after staining with chlorotetracycline. From SCHNEPF 1981.

normal position. When, however, the remigration fails, the new, abnormal position of the nucleus determines the site of side branch formation, *i.e.*, it controls the establishment of the new polar axis. The growth direction of side branches is determined also by unknown internal factors that lead to a rather constant angle between side and main branch (about 70° in the *Funaria* caulonema).

As in germinating spores or pollen grains, the tip of the developing side branch (or hypha or tube, respectively) ruptures the existing cell wall.

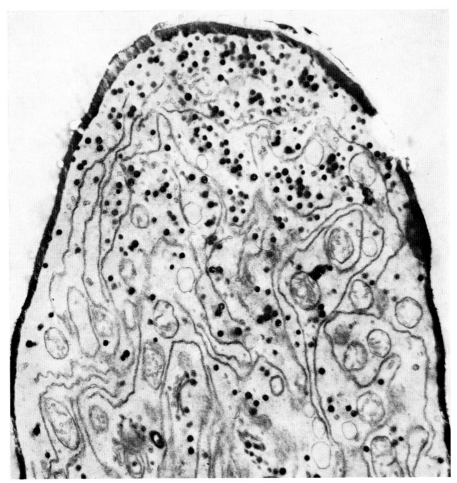

Fig. 3. Longitudinal section through the tip of a growing root hair of *Zea mays*. In the apical zone, dark Golgi vesicles are accumulated. Fixation: KMnO₄ (×16,200). From Sievers 1963 a.

Local secretion of hydrolytic enzymes seems to be involved (Knox and Heslop-Harrison 1970, 1971, Auvity *et al.* 1974, Larpent-Gourgaud and Aumaître 1976, Mullins 1979, Fèvre 1979), *i.e.*, tip growth initiation is accompanied by cell wall lysis. Inhibition of RNA-synthesis affects side branch formation in moss protonema (Aumaître and Larpent-Gourgaud 1971). In the case of the watermold *Achlya*, the hormone antheridiol is effective in producing antheridial branches in male hyphae. A sharp rise in cellulase activity has been reported at the time when antheridial side branches are initiated (THOMAS and MULLINS 1967, 1969). Inhibition of either transcription or translation represses the release of cellulase and the production of branches (KANE et al. 1973). It has been proposed that a prerequisite for the formation of side branches is a localized softening of the hyphal wall caused by cellulase.

Other early expressions of polarity in *Fucus* zygotes are the local incorporation of fucoidin into the cell wall at the rhizoid pole (NOVOTNY and

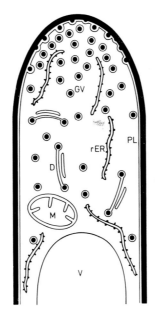

Fig. 4. Schematic drawing of the polar arrangement of cell organelles in a tubular cell with tip growth. In the apical zone (limited to the curved portion of the tip), accumulation and exocytosis of Golgi vesicles occur; the following subapical zone contains all cell organelles except vacuoles; the basal zone is characterized by the central vacuole. *GV* Golgi vesicle; *PL* plasmalemma; *M* mitochondrion; *rER* rough endoplasmic reticulum; *V* vacuole.

FORMAN 1974), the accumulation of vesicles in this region (NUCCITELLI 1978, see also QUATRANO 1972), and the local secretion of mucilage (SCHRÖTER 1978).

B. Morphological Gradients

Most tubular cells with tip growth show a polar arrangement of cell organelles that becomes especially visible in the distribution of the Golgi apparatus elements. The apical zone, which is limited to the curved portion of the apex, is characterized by accumulation and exocytosis of secretory Golgi vesicles (Figs. 2–4). This is shown in root hairs (SIEVERS 1963 a, b; NEWCOMB and BONNETT 1965), pollen tubes (SASSEN 1964, ROSEN et al. 1964), *Funaria* caulonema cells (SCHMIEDEL and SCHNEPF 1980), *Chara* rhizoids (SIEVERS 1965, 1967 a), and fungal hyphae (GIRBARDT 1969, GROVE *et al.* 1970). In the case of the hyphae of different fungal species, GIRBARDT demonstrated that the apical body (Spitzenkörper, BRUNSWIK 1924) mainly consists of vesicles. The population of vesicles in the tip of the growing *Chara* rhizoid can also be made visible as an apical body by dark-field light microscopy (Figs. 5 *a–c*). The secretory vesicles have a diameter of 0.1 to 0.3 μm. Often a second species of vesicles with a diameter of 40 nm is found (in pollen tubes, LARSON 1965; coated vesicles in rhizoids, SIEVERS

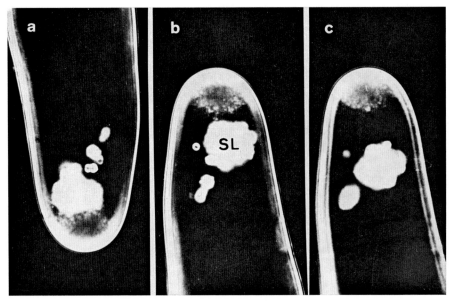

Figs. 5 *a–c*. Dark-field micrographs of the tip of a living *Chara* rhizoid demonstrating the apical body (Spitzenkörper) containing Golgi vesicles, some of which can be seen as light spots. *a* normal vertical exposure; *b* and *c* inverted vertical exposure of the rhizoid (*b* 30 seconds, *c* 80 seconds after inversion). *SL* statoliths. Photographs by Eckart Bartnik (×1,200)

1967 a, and in root hairs, BONNETT and NEWCOMB 1966), the function of which is still unknown. In addition to the vesicles, only cisternae of the endoplasmic reticulum (ER) occur in the apical zone.

The subapical zone often has a length of some μm. In the *Chara* rhizoid, the total length of the apical and the subapical zone is constant—300 μm. The subapical zone contains active dictyosomes, Golgi vesicles, rough ER, mitochondria, and—excepting fungal cells—plastids. In some cases, the apical part of the subapical zone is free of plastids. Sometimes lipid droplets, lysosomes, and multivesicular bodies are described.

In addition to these cell organelles, the main criterion of the basal zone is the occurrence of the central vacuole and cytoplasmic streaming. In pollen tubes and in root hairs, cyclosis reaches to the apical body; in caulonema tip cells of *Funaria,* cytoplasmic streaming does not occur. Two populations of microfilaments occur in root hairs—bundles of microfilaments throughout the cytoplasm and single microfilaments near the plasmalemma, specifically associated with microtubules (SEAGULL and HEATH 1979; see also p. 166 f.). Both microfilament populations are oriented parallel to the direction of and are involved in cytoplasmic streaming (SEAGULL and HEATH 1980 b). Microtubules are found near the plasmalemma in root hairs (NEWCOMB and BONNETT 1965, SEAGULL and HEATH 1980 a) and in *Funaria* caulonema tip cells (SCHMIEDEL and SCHNEPF 1980). In these tip cells, the microtubules also occur in the interior of the cells where they are associated with the nucleus, mitochondria, and plastids. Both microfilaments and microtubules occur in the apical body of hyphal tip cells in fungi (HOWARD 1981). The nucleus is located in the subapical zone or in the apical part of the basal zone (more than one nucleus in some fungal hyphae, GROVE *et al.* 1970). In the *Chara* rhizoid, the basal part of the apical zone is marked by special vacuoles acting as statoliths; furthermore, the nucleus is located at the basal end of the subapical zone (SIEVERS 1967 a).

The polarity of cells with tip growth becomes manifest not only in polar growth and polar distribution of cell constituents (Fig. 2) but also, in a few cases, in a polar differentiation of organelles. In caulonema tip cells of *Funaria,* the cell apex contains chloroamyloplasts with many large starch grains and only a few thylakoids, short or even globular mitochondria, and dictyosomes with up to nine Golgi cisternae. In the basal region of the tip cells, the plastids have less and smaller starch grains but more thylakoids, mitochondria are elongated, and the dictyosomes consist of only four to six cisternae (Fig. 2). If the apical chloroamyloplasts are displaced, they become similar to the other plastids of the cell (SCHMIEDEL and SCHNEPF 1980). A similar gradient in starch content was found in plastids of fern protonema (WADA and O'BRIEN 1975, OOTAKI and FURUYA 1969, for *Acetabularia* see PUISEUX-DAO and DAZY 1970).

The secretory vesicles are transported from dictyosomes in the subapical and basal zones to the apical zone. In the living *Chara* rhizoid, the individual vesicles can be observed by means of dark-field light microscopy. They show a saltatory movement with a netto flux to the apical cell wall where they sometimes are reflected until they disappear in it (HEJNOWICZ and SIEVERS, unpublished). *In statu nascendi,* the secretory vesicles often develop a central dark sphere that during exocytosis has the same contrast as the cell wall (SIEVERS 1963 a, b). The membrane thickness of the vesicles increases from 5 to 6 nm at the secretion side of the dictyosomes to 8 nm during exocytosis, thus demonstrating a structural assimilation to the apical plasmalemma (SIEVERS 1967 a).

The secretory vesicles deliver membranes and export matrix substances of the growing cell wall (DASHEK and ROSEN 1966). Polymerization and methylation of the wall polysaccharides are reported to occur in the vesicles and, after extrusion, in the cell wall. In root hairs, the apical cell wall is a typical primary wall with randomly oriented cellulose microfibrils; in the subapical and the basal part of the cell wall, two layers of cellulose microfibrils are present—an outer layer similar to the apical cell wall and an inner layer of parallel, mainly axially oriented microfibrils (Frey-Wyssling and Mühlethaler 1949, Newcomb and Bonnett 1965).

C. Physiological Gradients

Structural polarity may lead to polar transport of matter and to electric potentials (Bünning 1952) and is expressed also in cytochemical gradients (Turian 1979). The gradient in growth is, *per definitionem,* the characteristic feature of cells with tip growth. The restriction of elongation to the tip does not, however, exclude thickening of the wall in more basal parts of the cell (Bopp and Fell 1976).

The occurrence of an apical zone of Golgi vesicles seems to be related to growth rate. It is found in root hairs (Sievers 1963 a, b, growth rate about 80 μm h^{-1}: Moerz 1977), pollen tubes (Rosen and Gawlik 1966, growth rate more than 600 μm h^{-1}; Weisenseel et al. 1975, Reiss and Herth 1979 b), rhizoids (Sievers 1965, growth rate up to 180 μm h^{-1}), fungal hypha (*Lagenisma:* Schnepf et al. 1978, growth rate about 60 μm h^{-1}: Schnepf and Drebes 1977), and *Funaria* caulonema cells (Schmiedel and Schnepf 1980, growth rate nearly 50 μm h^{-1}) but is lacking in the more slowly elongating (4 to 10 μm h^{-1}) chloronematic side branches of the *Funaria* protonema (Schmiedel and Schnepf 1979 a) and in fern protonemata (DeMaggio and Stetler 1977, growth rate 2 to 10 μm h^{-1}). After inhibition of tip growth the apical vesicle zone usually disappears in fungal hyphae (Girbardt 1969, Howard and Aist 1977), pollen tubes (Reiss and Herth 1979 a) and *Funaria* caulonema cells (Schmiedel and Schnepf 1980).

The apical accumulation of vesicles results from a directed vesicle transport necessary for the polar exocytosis of wall material produced by the Golgi apparatus. The Golgi vesicles mainly or exclusively secrete matrix substances (Van Der Woude et al. 1971). The cellulose is believed to be synthesized by enzymes localized within the plasmalemma (Giddings et al. 1980, Mueller and Brown 1980, see p. 160 f.). Therefore, it can be assumed that these enzymes are concentrated or at least are more active in the apical portion than in the basal parts of tip growing cells. Cellulose formation, hence, also depends on polar transport of precursors.

The polarity of the transport could be based on a polar loading and un-loading of the cells (compare the observation of Schumacher, 1936, on polar fluoresceine transport in plant hairs and his interpretation of the phenomena, 1967); it is a characteristic feature for many multicellular, tubular systems (*e.g.,* Bopp and Knoop 1974).

Cytoplasmic streaming could also be involved in polar transport. Many but not all cells with tip growth reveal rapid cytoplasmic streaming, usually in the form of a fountain or counter-fountain movement. However, in *Chara* rhizoids, cyclosis only occurs in the basal zone, which contains the central vacuole. The apical and subapical zones with a length of 300 μm do not show cytoplasmic streaming where only the movement of individual Golgi vesicles can be observed (*cf.,* Section II. B.). This stream seems to be driven by the actin myosin system (cf., FRANKE et al. 1972). Experiments with the inhibitor, cytochalasin B, indicate that in pollen tubes and other cells with tip growth, the streaming is involved in the polar transport of wall material (HERTH et al. 1972, NOVOTNY and FORMAN 1974), especially of wall material included in Golgi vesicles (cf., MOLLENHAUER and MORRÉ 1976).

The apical vesicle zone seems to be excluded from the bulk cytoplasmic streaming. The mature Golgi vesicles have to leave the stream here, otherwise they would be retransferred into the basal part of the cell. They accumulate in the apex before they discharge their contents. Specific receptors may be important for membrane-membrane recognition. As shown by the occurrence of other patterns of cytoplasmic streaming (for instance in Characeae), and by high growth rates despite the lack of a noticeable streaming in Funaria the function of streaming in polar transport is not yet fully clear. As mentioned above, growth is slowed down in Funaria by cytochalasin B (up to 20 µg ml^{-1}) but not inhibited completely. Besides contractile mechanisms, electrophoretic ones are also assumed to be involved in polar transport (QUATRANO 1978).

The secretion of wall material by Golgi vesicles is related to an incorporation of vesicle membranes into plasmalemma. For growing pollen tubes, MORRÉ and VAN DER WOUDE (1974) calculated that the amount of membranes supplied by exocytosis equals the increase in plasma membrane area (cf., MORRÉ, KARTENBECK, and FRANKE 1979). The data of SCHMIEDEL and SCHNEPF (1980), however, indicate that in the Funaria tip cells (which differ in wall thickness and vesicle diameter from pollen tubes), the incorporation rate is five to ten times higher than the plasmalemma growth rate. It is interesting to note in this connection that in the pollen tube tips, coated pits are rather seldom (REISS, personal communication), whereas they are more frequent in tip cells of Funaria caulonema (Fig. 2). Coated pits are believed to play a prominent role in membrane internalization and recycling (HEUSER and EVANS 1980); our observations are in line with this concept (see also Section III. B. 3.).

Cells with tip growth are also characterized by polar electric currents (WEISENSEEL et al. 1975, WEISENSEEL and JAFFE 1976, see also p. 379 f.) and by polar distribution of ions such as calcium. As shown by proton microprobe analysis (BOSCH et al. 1980) and by ^{45}Ca autoradiography (JAFFE et al. 1975), the calcium content of pollen tubes is highest in the very tip and decreases toward the base. Another approach to analyse calcium gradients in tip growing cells is their visualization by fluoresence with chlorotetracycline (REISS and HERTH 1978, 1979 b; Fig. 2). The intensity of the fluorescence gradient seems to vary with the growth rate. Chlorotetracycline is believed to complex with Ca^{2+} and cytoplasmic membranes (CHANDLER and WILLIAMS 1978). Hence, this method traces only a certain portion of Ca within the cell, whereas the microprobe analysis and autoradiography do not discriminate between functional and storage calcium.

Experiments with gemmae of Achlya germinating in a calcium free medium explain the importance of the compartmentation of calcium. The resting gemma does not show a remarkable fluorescence in contrast to the growing tips of new hyphae, though the latter had not taken up any calcium from the medium. Again the fluorescence gradient depends on the growth rate (STADLER 1980). The gradient visualized with chlorotetracycline reflects the polar distribution of calcium rather than the distribution of membranes, as shown by treatment with fluorescamine, which stains membranes (POCCIA et al. 1979). Calcium probably even participates in organizing the morphological gradients (WEISENSEEL et al. 1975, CHEN and JAFFE 1979, REISS and HERTH 1979 a, b).

D. Termination of Tip Growth

Tip growth is per definitionem locally restricted; by unknown mechanisms, the cells are able to grow with a more or less constant, specific width. In the Funaria caulonema, the width can be increased by about 50% through the ionophore A 23187 without changing the cell volume considerably (SCHMIEDEL and SCHNEPF 1980; Fig. 1). High external Ca^{2+} concentrations increase the pollen tube width (MORRÉ and VAN DER WOUDE 1974). It is, therefore, probable that calcium ions are involved in width control.

The longitudinal termination of tip growth is difficult to understand. Generally, the cells within a filament have a specific length; root hair growth is limited. Obviously, the cells can measure their developmental time or their length (or volume). Root hairs reach their specific length independent of the growth rate. Developmental "time" must not necessarily equal "physical" time but is the consequence of consecutive series of metabolic steps in which "time" is implicated in "production rate".

There are examples where the cell can "measure" length, i.e., during the outgrowth of flagella (ROSENBAUM et al. 1969) or microvilli. In these cases, the surface-to-volume ratio seems to be the determining factor (SITTE 1978). It must be mentioned here that in the Chara rhizoid and in the Funaria caulonema, the nucleus of the tip cell maintains a constant distance from the growing apex (300 µm and 120 µm, respectively); it migrates at the same rate as the cell elongates (SIEVERS 1967 a, SCHMIEDEL and SCHNEPF 1980). Because of these constant distances and because of a constant rhythm in mitoses and cytokineses in the caulonema, a constant length of the nongrowing daughter cells is achieved. Experiments in which the nucleus is dislocated or its movement is inhibited also affect regular growth (for details see Section II. C.). Electric currents, diffusion processes, and the cytoskeleton could be involved in the recognition of cellular distances.

Besides cells that grow up to a specific, constant length, there are those that stop elongation when they have reached their destination—the egg cell in the case of a pollen tube, the host cell in the case of an infectious hypha of a parasitic fungus, etc. In these cases, tip growth is terminated in unknown but surely different ways in which cell-to-cell recognition reactions are dominating.

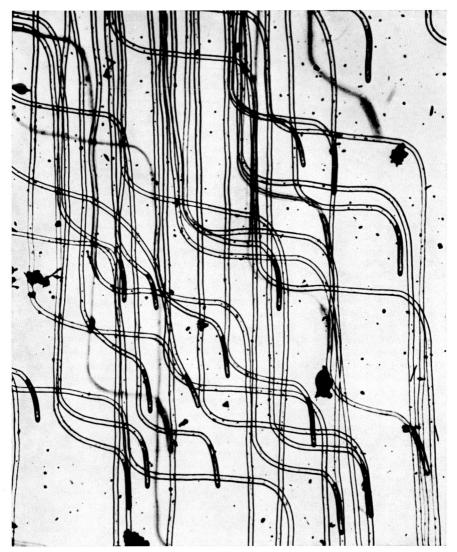

Fig. 6. Cluster of living *Chara* rhizoids demonstrating that each cell responds to turning from the vertical to the horizontal exposure by a positive ortho-gravitropic curvature. Rhizoids with weak graviresponse contain less statoliths (×35).

III. Factors Controlling Tip Growth and Polarity

A. Exogenous Factors

1. Gravity

With regard to gravity as a stimulus that orients the direction of tip growth, the rhizoids of the Characeae are well-known objects (SIEVERS and VOLKMANN 1979). A cluster of *Chara* rhizoids (Fig. 6) shows that each cell responds to turning from the vertical to the horizontal exposure by a

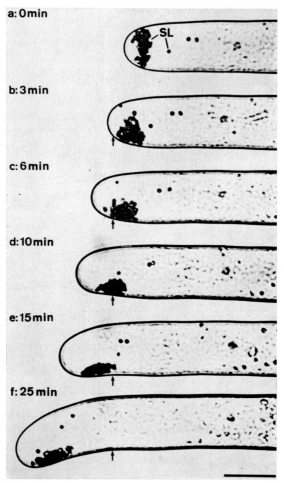

Fig. 7. A series of time-lapse photographs illustrating the graviresponse of a *Chara* rhizoid after turning from the vertical to the horizontal exposure. Note sedimentation of the statolith complex followed by downward bending of the rhizoid tip. The arrow indicates the same point on the cell wall in each micrograph. Note the slight indentation on the cell wall at the lower flank in d, which indicates a local inhibition of cell wall growth by the presence of statoliths. Later it disappears (f). *SL* statoliths ($\times 470$). From SIEVERS and SCHRÖTER 1971.

positive orthogravitropic curvature. At room remperature, this graviresponse ends after 2 to 3 hours. In vertical exposure, 50 to 60 special vacuoles functioning as statoliths are distributed over the cross section of the cell, 10 to 20 μm basal to the rhizoid tip (Fig. 5 *a–c*).

After application of cytochalasin B (30 μg ml^{-1}, 1 minute), the statoliths sediment to the plasmalemma of the apical zone where they inhibit the longitudinal growth of the cell (HEJNOWICZ and SIEVERS 1981). Twenty to thirty minutes later, the statoliths are relocated to their normal position; simultaneously the longitudinal growth of the cell starts again. From this

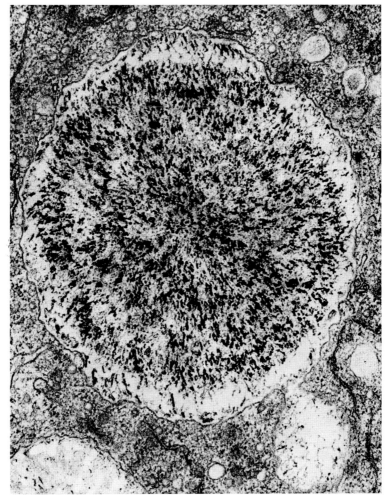

Fig. 8. A statolith of the *Chara* rhizoid is included in a special compartment that contains crystallites of $BaSO_4$ in preferentially radial arrangement. Fixation: OsO_4 ($\times$44,500). From SIEVERS 1967 a.

observation it must be concluded that microfilaments are involved in controlling the position of statoliths.

A series of time-lapse photographs of a rhizoid in horizontal exposure (Fig. 7) illustrates the beginning of the graviresponse in detail. The statoliths sediment within a few minutes onto the physically lower apical flank. Initially, the cell continues to grow straight. After a latent period of 10 minutes, however, the cell tip shows an asymmetrical lateral outline; 15 minutes later, a clear curvature can be observed. After completion of the graviresponse, the statoliths are redistributed normally (SIEVERS and SCHRÖTER 1971). If the statoliths are translocated from their apical site into the base by centrifugation (6 minutes, 400 to 450 g; BUDER 1961), graviresponse does not take place and growth remains unchanged. Statoliths first reappear in the cell tip several hours later, and the ability to perform a graviresponse returns simultaneously. This confirms that the statoliths bring about the perception of gravity.

The statolith vacuole is approximately 2 μm in diameter (Fig. 8). It is probably formed from the endoplasmic reticulum (SCHRÖTER et al. 1973). By electron microprobe analysis, selected area electron diffraction, and

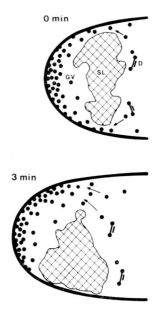

Fig. 9. Scheme of the displacement of statoliths and Golgi vesicles in the apex of the *Chara* rhizoid at the beginning of graviresponse. At the lower flank, exocytosis of Golgi vesicles is inhibited by the position of statoliths (3 minutes). *D* dictyosome, *GV* Golgi vesicle, *SL* complex of statoliths. From SIEVERS and SCHRÖTER 1971.

atomic absorption spectrophotometry, it was proved that the statolith compartments are filled with mainly radially arranged crystallites of $BaSO_4$ (SCHRÖTER et al. 1975, STEUDLE et al. 1978).

Together with the sedimentation of the statoliths, a further disturbance of the radial symmetric distribution of the cell organelles occurs in the apical zone of the cell. More Golgi vesicles and an increase of their exocytosis into the plasmalemma were observed in the physically upper apical cell flank than in the lower one (Fig. 9; SIEVERS 1967 b). This asymmetry leads to differential growth of the apical flanks and thus results in gravitropic curvature. The symmetry of the cell growth is reestablished by the renewed symmetrical distribution of the statoliths at the end of the graviresponse. In the *Chara* rhizoid, therefore, statolith distribution and local cell wall growth marked by exocytosis of Golgi vesicles are coupled by a very simple feedback principle.

It has been proved that statoliths inhibit local cell wall growth when they are sedimented so that they lie near the plasmalemma:

(1) The slight indentation of the cell wall at the lower apical flank of the rhizoid (Fig. 7 *d*, arrow) is a clear indication of a local inhibition of cell wall growth by the presence of statoliths. The inhibition occurs exactly where the statoliths sedimented first. Later it disappears (Fig. 7 *f*).

(2) By placing resin spheres on a rhizoid tip, the growth of both flanks can be measured exactly (HEJNOWICZ *et al.* 1977). After horizontal exposi-

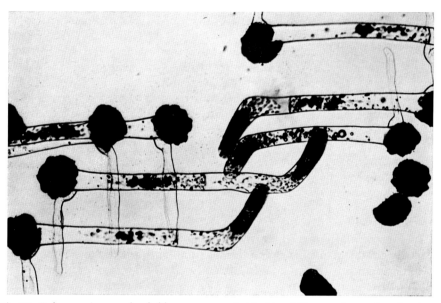

Fig. 10. Polarotropic growth of chloronemata and rhizoids of *Dryopteris austriaca* in plane-polarized red light. After 2 days, the filter was turned 60°; the photograph was taken 1 day later. The chloronemata grow perpendicular and the rhizoids, parallel to the E vector. Photograph by Liane Caspers (×300).

tion, the extreme tip and the physically upper, apical flank continue growing, while the growth of the lower flank is immediately inhibited by the sedimented statoliths. If the statoliths happen to be displaced somewhat basipetally, the cell wall area that is now free starts growing again. This growth stops when the statoliths are relocated to the apical area in the course of gravitropic curvature (SIEVERS *et al.* 1979).

(3) If statoliths in a vertically growing rhizoid are forced to sediment to the apical plasmalemma where maximal growth occurs, longitudinal growth is inhibited. This growth inhibition is irreversible if an irreversible sedimentation is caused by turgor reduction (SIEVERS and SCHRÖTER 1971); however, a reversible growth inhibition was observed if a reversible sedimentation is caused by a dose of cytochalasin B, which does not stop cytoplasmic streaming in the basal zone (HEJNOWICZ and SIEVERS 1981).

2. Light

In morphogenesis of plants, light is the most effective external factor. In day light the spores of the fern *Dryopteris filix-mas* develop to typical prothallia. However, in darkness and in red light, filamentous chloronemata develop (for review see DENNISON 1979). Their apical cells are typical tubular cells with tip growth (ETZOLD 1965) and a polar arrangement of cell organelles (FALK and STEINER 1968). The growth rate is very low (5 μm × h⁻¹ at room temperature, HEYDER 1975). This may be the reason for the fact, that an apical body is lacking, though exocytosis of Golgi vesicles occurs in the very tip (HEYDER 1975; see Section II. C.).

The apical zone of the chloronema shows a positively phototropic response to unilateral red light by bending toward the light (MOHR 1956, ETZOLD 1965). If the chloronema is illuminated by plane-polarized red light of equal intensity at two opposite flanks (to compensate the phototropic effects), the direction of tip growth is perpendicular to the electric vector of light. This can also be demonstrated if the spores germinate at the bottom of an overturned Petri dish containing nutrient agar and are illuminated from above. The young rhizoids normally grow in a direction parallel to the electric vector. If the polarizing filter is turned 60°, the direction of growth of the chloronema tip cell changes by 60° (Fig. 10). This bending response is called polarotropism. The same photo- and polarotropic responses are shown in the chloronema of the liverwort *Sphaerocarpus donnellii* (STEINER 1967 b) and in short filaments of the green alga *Mougeotia* (NEUSCHELER-WIRTH 1970).

The photo- and polarotropically responding tip cells are well investigated with respect to the identification, localization, and orientation of the photoreceptor pigments (ETZOLD 1965, STEINER 1967 a, b, NEUSCHELER-WIRTH 1970). Blue light receptors and phytochrome are involved in the polarotropism of *Dryopteris* and *Mougeotia;* however, in *Sphaerocarpus*, only blue light effects are observed in spite of the fact that phytochrome is present. The photoreceptor dipoles are oriented parallel or perpendicular to the plasmalemma of the apical zone. Maximum growth occurs where light absorption is maximal. The links in the stimulus-response chain between the photoreceptor pigments and the growth mechanisms of the cell wall are still unknown.

Phototropic responses, by tip growth, are also found in the protonema of the moss *Physcomitrium* (NEBEL 1968, 1969), in coenocytic giant algae (ISHIZAWA and WADA 1979 a, b), in the sporangiophore of *Pilobolus* (PAGE 1968, PAGE and CURRY 1966), and during phase I of the *Phycomyces* sporangiophore development.

3. Bending Mechanisms

Both the gravitropic *Chara* rhizoid and the phototropic tip cell of the fern chloronema respond to the outer stimuli—gravity and light, respectively—by bending of the apical zones. In the rhizoid, bending is performed by differential growth of the opposite cell wall flanks in the apical zone

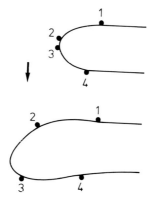

Fig. 11. Position of resin spheres attached to the tip of a *Chara* rhizoid before (above) and 25 minutes after (below) the beginning of the gravitropic bending. Bending is performed by bowing, *i.e.*, by differential growth of the opposite cell wall flanks in the apical zone. The growth center stays at the same point of the cell wall. (For further information about the growth behavior of the two opposite flanks, see SIEVERS *et al.* 1979.) *Arrow:* direction of the gravity vector.

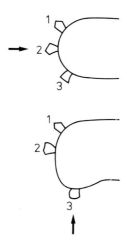

Fig. 12. Position of starch grains attached to the cell wall of a chloronema tip cell of *Dryopteris* before (above) and after (below) the beginning of the phototropic bending. Bending is performed by bulging, *i.e.*, by displacement of the growth center from the tip to the apical flank. *Arrows:* direction of light. After ETZOLD 1965.

(Fig. 11; SIEVERS *et al.* 1979). This is a special case of differential intercalar growth located basal to the very tip. It is caused by a local inhibition of cell wall growth at that area where statoliths are sedimented (see Section III. A. 1). In the case of the phototropic chloronema tip cell, an alternative mechanism is realized (Fig. 12; ETZOLD 1965) in which bending is achieved by "having rapid growth on the prospective concave side, provided the gradient on that side can be modified to produce a new center of growth (point of maximum rate) somewhat proximal to the original tip" (Green et al. 1970).

The bending mechanism of the *Chara* rhizoid is called "bowing" growth, that of the chloronema tip cell, "bulging" growth. The main difference between bowing and bulging concerns the behavior of the growth center. It stays at the same point of the cell wall in bowing growth, while it is displaced in bulging growth (Hejnowicz and Sievers 1971).

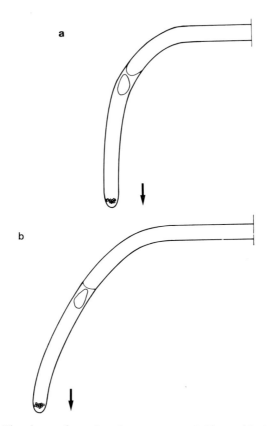

Figs. 13 *a* and *b*. The shape of gravitropic curvatures of *Chara* rhizoids is a bow. It is caused by bowing growth: *a* a rhizoid containing all statoliths during the response period; *b* a rhizoid the statoliths of which are first translocated to the basal zone by centrifugation; during the following horizontal exposure the single statoliths are retranslocated to the tip with different transport rates. Thus, the greater the number of statoliths, the stronger the graviresponse. *Arrows:* direction of the gravity vector.

Bowing and bulging growth result in different shapes of curvature; bowing produces a bow (Fig. 13 *a*, *b*) the radius of which depends on the length of the growing zone, however, a relatively sharp angle results by bulging (Fig. 14). These differences are especially evident if gravity or light stimulate the cells in such a manner that the new growth direction lies perpendicular to the preceding one. In the *Chara* rhizoid, the radius of the bow is short at the beginning of the response, later on it becomes infinitesimal. Thus, the normal shape of the bow is a hyperbola (Fig. 13 *a*). The bow radius is much longer in those rhizoids in which only a few statoliths are present, thus transducing the gravistimulus less effectively. If after translocation of the statoliths into the basal zone (see Section III. A.) the number of statoliths in the apex increases, the radius slowly becomes shorter at the beginning of the response period (Fig. 13 *b*). Thus, the greater the number of statoliths, the stronger the gravitropic response (BUDER 1961).

In the chloronema tip cell of *Struthiopteris filicastrum*, after turning the polarizing filter 90° HARTMANN *et al.* (1965) observed bulging not only

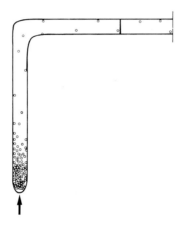

Fig. 14. The shape of the phototropic curvature of a chloronema tip cell (*Dryopteris*) is a relatively sharp angle. It is caused by bulging growth. *Arrow:* direction of light. After MOHR 1956.

in the apical zone but also in the basal one. Occasionally, only in the basal zone, one or two new maxima of growth were initiated, whereas the growth of the former apex was irreversibly inhibited. Thus, shifting of the growth center is a clear criterion of phototropic bending by bulging. Finally, shifting of the growth center results in branching of the cell.

In comparison to the gravitropic bending by bowing, the causal analysis of bending by bulging, including shifting of the growth center to basal surface points, has yet to be analyzed. In the *Chara* rhizoid, branching only occurs at the border between the apical and the subapical zone after irreversible inhibition of the growth center (SIEVERS and SCHRÖTER 1971). In the rhizoid, only branching is a clear case of bulging growth.

4. Magnetic Field

The first case of a magnetotropic response of tubular cells with tip growth is reported in pollen tubes of *Lilium longiflorum* (SPERBER *et al.* 1981). If pollen grains germinate and their tubes grow in the homogeneous magnetic field (14 Tesla) of a horizontal Bitter magnet for 3 hours, the tubes are oriented parallel to the magnetic field with equal tendency to grow toward the north or toward the south pole. This magnetotropic response becomes weaker with decreasing field strengths. In inhomogeneous fields, the pollen tubes grow preferentially toward the region of decreasing field strength. The authors speculate that the magnetic fields may act on the plasmalemma where they might influence the localization of membrane proteins necessary for exocytosis of vesicles.

5. Chemical Factors

In comparison with chemotaxis in bacteria and in unicellular eukaryotes (for reviews, see Macnab 1979, Bean 1979, Poff and Whitaker 1979), chemotropism in cells with tip growth, as well as in multicellular organs, is less analysed. The reason for this lack may be that the concentration gradient of the chemical substance that seems to cause the orientation of growth often only promotes growth.

The germ hyphae of the watermold *Achlya* are positively chemotropic to casein hydrosylate and amino acids (Musgrave *et al.* 1977). It is assumed that the fungus detects the amino acids, *per se*, if sensing the gradient. *Achlya* is one of the best studied fungi with respect to the hormonal mechanisms involved in the regulation of its sexual reproduction. The hormone antheridiol (C-29 steroid) has been extracted from female hyphae. It is responsible for the chemotropic growth of the antheridial hyphae to the oogonial initials. Furthermore, it is also effective in producing antheridial branches in male strains in concentrations as low as 5 to 10 pg ml^{-1}. Male *Achlya* strains seem to remove antheridiol added to the culture medium, and it cannot be extracted back from the hyphae (see Mascarenhas 1978).

If plus and minus hyphae of *Mucor* grow in close proximity with each other, zygophores are produced. The zygophores of opposite mating type are mutually attracted to each other (zygotropism, Burgeff 1924). The chemotropic substances are volatile and act through air (Plempel 1960, 1962, Plempel and Dawid 1961). Their chemical nature is still unknown. The induction of zygophore formation is caused by trisporic acids (Mascarenhas 1978).

Molisch (1889) was the first to demonstrate that pistil tissue contains some substances functioning as a chemotropic factor for pollen tube growth. It appears safe to conclude that the canal cells and the cells of the stigma surface secrete substances of chemotropic activity. In pollen tubes of *Antirrhinum*, Ca^{2+} is chemotropically active (Mascarenhas and Machlis 1962) but is chemotropically inactive in *Lilium* (Rosen 1971) and in *Oenothera* (Glenk *et al.* 1971). Chemotropically active substances, which direct the pollen tubes in the gynoecium, are probably mixtures of different inorganic and organic substances such as K^+, amino acids, peptides, amines, and sugars (Schildknecht and Benoni, 1963).

The germ hyphae of some parasitic fungi (*e.g.*, *Puccinia*) grow through the stomata into the leaves of their host plants. This phenomenon often is interpreted as a positive hydrotropism. Whether a gradient of humidity is the orienting factor remains to be clarified. Chemo- and hydrotropism have been reviewed by Ziegler (1962 a, b).

B. Endogenous Factors

1. Growth Substances

There are a few reports which indicate that growth substances participate in controlling tip growth. Auxin can suppress the polarization of moss spores; the germinating spore enlarges but does not grow locally to form a tubular cell (HEITZ 1940). It was assumed that a gradient in auxin is necessary to establish polar growth and that external auxin destroys the auxin gradient (VON WETTSTEIN 1953). It was possible to reproduce these results largely (SCHNEPF, unpublished results). High auxin concentrations (5 × 10^{-5} M) also disturb regular tip growth in the caulonema and cause malformations that resemble those after colchicine application (see Section III. A. 4.). In general, growth substances in more "physiological" concentrations seem to be rather ineffective in influencing tip growth and polarity (LARPENT-GOURGAUD 1969). In pollen tube growth, *in vitro* various growth hormones stimulate elongation to a certain extent (IWANAMI 1980); in *Campsis*, the endogenous level of auxin is believed to be regulated by an IAA oxidase system (SHARMA and MALIK 1978).

2. Turgor and Tensional Stress

Plant cell growth needs turgor pressure (see p. 234). In *Tradescantia* the growth of root hairs stops if their turgor is reduced by a medium containing 0.05 M glucose; however, the growing root produces new root hairs (SCHRÖTER and SIEVERS 1971). The formation of root hairs is irreversibly inhibited if the glucose concentration is higher than 0.22 M. If the growth of root hairs is stopped by turgor reduction, the cell wall substances are still secreted for some time so that apical wall thickenings are formed (see p. 168 f.). Three hours after the beginning of turgor reduction, the polar organization of the formerly growing root hairs is destroyed; their cell organelles are distributed at random. Particularly, the number of the dictyosome cisternae is increased (SCHRÖTER and SIEVERS 1971). This may be a sign of an inhibition of membrane flow via Golgi vesicles between dictyosomes and plasmalemma. If the turgor reducing medium is replaced by the original one so that the root hairs regain their normal turgor, 75% of the hairs start growing again by bulging in the subapical zone (SCHRÖTER and SIEVERS 1971).

Subplasmolytical concentrations of Ca^{2+} inhibit the formation of side branches in the *Funaria* caulonema but do not affect their initiation. Side branch cells are cut off from the mother cell, though their tip growth is suppressed completely; even the local bulging of the main filament or a local wall thickening at this site do not occur (SCHMIEDEL and SCHNEPF 1979 b).

In growing *Chara* rhizoids (about 30 µm in diameter), the displacement of resin spheres attached to the surface of the apical cell wall (Fig. 15) was measured to calculate the distribution of elemental extension rates in meridional and latitudinal direction (HEJNOWICZ *et al.* 1977). The meridional extension is limited to the very apical region extending from the growth center at the very tip to a relatively small distance from the tip (maximum

15 μm). The latitudinal extension covers a longer region than the meridional one. This is the obvious consequence of the increase of the rhizoid diameter. In the extreme apical region, the two extension rates are equal; this means there is isotropy of extension. Behind this region, up to about 12 μm, the meridional extension rate seems to predominate slightly, then the relation changes rapidly, and soon the latitudinal extension rate occurs alone.

In addition to the distribution of extension rates, the tensional stress (force per unit area) in meridional and latitudinal direction was calculated (Hejnowicz et al. 1977). Tensional stress in the wall of a cell subjected to internal pressure (turgor) is a function of the pressure and cell geometry. The meridional and the latitudinal stresses are equal in the extreme apical

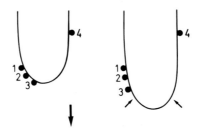

Fig. 15. Position of resin spheres attached to the tip of a *Chara* rhizoid growing in normal vertical direction (right photograph taken after 6 minutes). The maximum meridional extension is limited to the extreme apical region between the two small arrows. *Large arrow: direction of the gravity vector.*

region as is the case with the extension rates. In the cylindrical region, the latitudinal stress is twice as high as the meridional. The change from this anisotropy to the apical isotropy is achieved by means of drastically minimising the latitudinal stress. There is also some minimization of the meridional stress in the apical region. Both minima are not zero. The maximum and the isotropy of extension rates occurs where there is the minimum and the isotropy of stresses.

Hejnowicz et al. (1977) concluded from the coincidence of maximal extension rates and minimal stress at the tip of the growing rhizoid that there is a feedback between the extension pattern and the shape of the cell through the stress pattern with the cooperation of the turgor. This feedback produces a tendency toward the stabilization of the shape. The biological mechanisms controlling the shape of a growing cell and linked to the genetic information need information about the actual shape. The latter can be provided by means of measuring stresses in the cell wall and/or plasmalemma (Hejnowicz et al. 1977). To be fully informed about the shape and dimension, the cell should be able to measure (a) the ratio of stresses in the orthogonal directions in the cell wall or in the plasmalemma, (b) the absolute value of stress at least at one point, and (c) the turgor pressure.

3. Ions and Electric Currents

Polar movements of ions and electric currents play important roles not only in initiation of cell polarization (ROBINSON and JAFFE 1975, CHEN and JAFFE 1979) but also in tip growth (BENTRUP 1968 a, WEISENSEEL *et al.* 1975, WEISENSEEL and JAFFE 1976); their effects on cell morphogenesis are treated separately by WEISENSEEL and KICHERER on p. 379 f. Therefore, only some experiments on the effects of ionophores on tip growing cells are mentioned here. It must be added that a sufficient supply of ions is a prerequisite for growth; if *Achlya* hyphae grow in the absence of calcium, growth slows down and, finally, the tips of the hyphae burst; calcium is needed to form a rigid cell wall at the apex (STADLER 1980).

In *Lilium* pollen tubes, the calcium ionophore A 23187 (5×10^{-5} M) stops tip growth completely after 20 minutes, whereas cytoplasmic streaming remains unaffected for a longer time (HERTH 1978, REISS and HERTH 1979 a). Simultaneously, the apical vesicle zone is reduced and the zonation of the other organelles is no longer visible. The contents of the Golgi vesicles are discharged irregularly, not preferentially in the tip, and are not integrated into the pollen tube wall. Tip growth as well as oriented exocytosis are affected. It is unknown whether the equilibration of Ca^{2+} directly causes these variations or whether changes in turgor also are involved. In spite of the severe effects that also concern the structure of dictyosomes and mitochondria, the polarity of the cells is maintained.

After transfer into an ionophore free medium, the pollen tubes resume growth. The calcium gradient, at least that gradient which is visualized with chlorotetracycline, is destroyed by the ionophore (REISS and HERTH 1978). Apparently, it is necessary for tip growth but it seems to be a consequence rather than a cause of polar organization. As was already mentioned, in the *Funaria* caulonema, the ionophore A 23187 slows down growth and causes the formation of wider and shorter cells but does not inhibit polar growth completely (SCHMIEDEL and SCHNEPF 1980), (Fig. 1).

The broad-range cationophore X-537 A (5×10^{-5} M) also inhibits pollen tube growth after 30 minutes but at least 60 minutes elapse before it stops cytoplasmic streaming. The zonation of the cytoplasm in the tip region remains rather normal (REISS and HERTH 1980). This is an unexpected effect for, in general, the apical vesicle zone disappears if tip growth is inhibited (HOWARD and AIST 1977, SCHMIEDEL and SCHNEPF 1980). The cell wall at the very tip is thickened considerably. At the same time, the plasmalemma has coated pits here in high number; they are seldom in untreated, growing tips (see Section II. C.). As with A 23187, tip growth, but not cell polarity, is affected.

4. Organelles

Tip growth is an active process and depends on a normal metabolism, *i.e.*, on suitable environmental conditions and on fully active organelles of the cell, mitochondria to produce ATP, dictyosomes to synthesize wall material, etc. Certain organelles also play a role in defining time, site, and direction of polar growth (see Section II. A.). In addition, cellular structures determine the polar axis of growing polarized cells and/or are "effectors" of polar growth in that they participate in polar transport, in the distribution of synthetases within the plasmalemma, etc.

Certainly, microtubules are involved in such processes. As in *Micrasterias* (see p. 166 f.) and in hyphae of basidiomycetes (Girbardt 1968, Raudaskoski and Koltin 1973, Raudaskoski 1980), they seem to participate in moving the nucleus. In the *Funaria* caulonema tip cells stop polar elongation when they are treated with colchicine. The nucleus drifts somewhat backward then. The growth processes first are not interrupted completely, for the cells protrude subapically one or more new growing tips. These tips, however, also soon cease to grow. Similar observations have been made by Yamasaki (1954) and Nakazawa (1959) in fern protonema. Obviously, true elongation in *Funaria* caulonema cells is possible only if the nucleus can keep pace with the growing apex. When this movement is stopped, elongation is interrupted and growth dislocated (Schmiedel and Schnepf 1980), (Fig. 1). Since colchicine affects microtubules, these organelles can be regarded as effectors for polar growth. Colchicine also inhibits the chloroplasts to move into a developing side branch; the initiation of the latter is not influenced by this drug. However, the cell does not elongate, presumably because mitosis also cannot take place (Schmiedel and Schnepf 1979 b). Yet, there are no indications that microtubules participate directly in polar vesicle transport in cells with tip growth (Franke et al. 1972), perhaps with the exception of *Fusarium* hyphae in which the apical accumulation of vesicles and the polar zonation of organelles are disturbed by colchicine (Howard and Aist 1977). Similar effects in *Funaria* (Schmiedel and Schnepf 1980) seem to be unspecific. Microfilaments also seem to participate in moving the nucleus. Treatment with cytochalasin B dislocates the nucleus in apical direction; it does not inhibit growth completely but reduces its rate (Fig. 1).

If tip cells of the *Funaria* caulonema are centrifuged in the right way (Schmiedel and Schnepf 1980), (Fig. 1), the polar zonation is destroyed completely and the nucleus is dislocated on the basal cross wall. Growth is interrupted then. Within one hour, most cells begin to reconstitute the normal distribution of organelles and the nucleus returns to its normal position with a migration rate of about 40 μm h^{-1}, which corresponds to the rate in growing cells. Colchicine retards the migration of the nucleus and of other organelles considerably. If it comes to rest somewhere within the cell, a new tip is formed here after some hours. In long tip cells, *i.e.*, in cells in preprophase, the displaced nucleus does not remigrate; it divides in the basal position. The newly formed short basal cell produces a new tip that often grows with reversed polarity, *i.e.*, toward the center of the protonema. In some cases, a new, reversed tip is formed before mitosis (Schmiedel and Schnepf 1980; *cf.*, Ootaki 1963 for fern protonema cells).

Similar morphogenetic abnormalities can be found in *Funaria* after treatment with heavy water (80 to 90%; Fig. 1). Growth stops, the zonation of organelles breaks down, and the nucleus moves toward the basal cross wall. After several hours, the polar organization is restored and the cells continue to grow in the normal direction. Again, cells in preprophase react in another way: the displaced nucleus cannot be retransferred; the cells directly develop a new tip at their basal end, which grows with reversed polarity, or they first undergo "intercalar" divisions as they usually do under suboptimal conditions. In some cases, basal "growth" is transient and becomes manifest only by a local thickening of the cell wall, which is not continued when the nucleus remigrates (SCHMIEDEL and SCHNEPF 1979 b, 1980).

These results seem to indicate that the site of growth and the direction of the polar axis are determined by the position of the nucleus within the cell. So far, the nucleus directs the position of, e.g., synthetases in the plasmalemma and the movement of secretory vesicles. On the other hand, the plasmalemma or, perhaps more precisely, the ectoplasm (the plasmalemma with some underlying cytoskeletal elements, etc.) determines the position of the nucleus. It is, together with the cell wall, the only part of the cell that is not displaced during centrifugation. The ectoplasm, therefore, seems to be polarly organized, perhaps by the unequal distribution of channels or pumps for ions in the plasma membrane, and directs the migration and the position of the nucleus. The retransfer of the nucleus requires a moving system with contractile elements and the transmission of the mechanical force on the nuclear envelope. Microtubules seem to be a part of this system. Indeed, the interphase nucleus is closely associated with many microtubules. In preprophase the moving system does not function. One possible explanation could be that the nuclear envelope is modified then; another explanation could be that the microtubular system is being reorganized (SCHMIEDEL and SCHNEPF 1979 b, 1980).

In several filamentous plant systems such as staminal hairs of *Tradescantia*, a preprophase band of microtubules appears before mitosis (BUSBY and GUNNING 1980; see also p. 306 f.). Probably it is involved in the positioning of the nucleus in cells that have no elaborated interphase transport machinery for the nucleus; it does not occur in the moss protonemata, in contrast to moss leaflets that lack noncortical microtubules in interphase (SCHNEPF 1973).

Anti-microtubule drugs also affect growth and morphogenesis in filamentous fern gametophytes (YAMASAKI 1954, NAKAZAWA 1959, MILLER and STEPHANI 1971) and in the fungus, *Trichophyton* (DALL'OLIO and VANNINI 1979), but do not influence tip growth of pollen tubes (FRANKE et al. 1972), root hairs (MOERZ 1977), and hyphae of the oomycete, *Lagenisma* (SCHNEPF and HEINZMANN 1980). The latter are unicellular systems that do not divide and, thus, do not require a strict coordination between growth and position of the nucleus, in contrast to the former, in which cells of uniform size have to be formed (SCHNEPF and HEINZMANN 1980). Influences of the position of the nucleus and its division on the structure of apical cells and the development of septa in the basidiomycete, *Schizophyllum commune*, have been described by RAUDASKOSKI (1980).

19*

IV. Conclusion

Tip growth of tubular cells is a process in which the relations between structure and function as well as between the single organelles and the cell as an integral whole are perhaps more transparent than in other examples of cell morphogenesis. It not only provides convincing observations that Golgi vesicles participate in cell wall formation (Sievers 1963 a, b, 1964) and good examples to explain the response of a cell to external factors like gravity or light but also demonstrates that a cell has to be regarded as a *system* with integrated elements (organelles, etc.) functioning by integrated processes. As is life itself, the cell is characterized by interdependence rather than by dependence, by feedback mechanisms rather than by cause and effect, and it should be considered as a network of relationships. In that respect, questions about the cause of polarity may be misleading. Polarity is one of the meshes of this network and, in contrast to Bünning (1958), not an irreversibly fixed structural characteristic of the cell. Feedback mechanisms also become obvious when tip growth is inhibited. Secretion of wall material in root hairs proceeds after plasmolysis to a certain extent (Schröter and Sievers 1971), but generally the production of wall material is reduced considerably in such cases.

Similar relationships seem to exist in the morphogenesis of the multicellular plant; it is the consequence of the morphogenesis of its single parts, the cells. But cell morphogenesis, to the contrary, is controlled by the organism which, like the cell, acts as an integral whole. Thus, the development of its parts is dominated by mutual inhibition and promotion. As shown by coenocytic plants, *e.g.*, *Caulerpa*, a subdivision of the organism into cells is not necessary to reach a relatively high level of organization, but cellular organization has proved its superiority in evolution.

References

Aumaître, M.-P., Larpent-Gourgaud, M., 1971: Action de quelques inhibiteurs de synthèses nucléiques et protéiniques sur la croissance et la ramification du protonéma de *Ceratodon purpureus* (Hedw.) Brid. cultivé *in vitro* à l'obscurité. C. R. Acad. Sci. Paris **272**, 1503—1506.

Auvity, M., Fèvre, M., Larpent-Gourgaud, M., Larpent, J. P., 1974: Mécanismes de la ramification des systèmes filamenteux chez les végétaux inférieurs. C. R. Soc. Biol. **168**, 1344—1349.

Bean, B., 1979: Chemotaxis in unicellular eukaryotes. In: Encyclopedia of plant physiology, New Series, Vol. 7 (Haupt, W., Feinleib, M. E., eds.), pp. 335—354. Berlin-Heidelberg-New York: Springer.

Bentrup, F. W., 1963: Vergleichende Untersuchungen zur Polaritätsinduktion durch das Licht an der *Equisetum*-Spore und der *Fucus*-Zygote. Planta **59**, 472—491.

— 1964: Zur Frage eines Photoinaktivierungs-Effektes bei der Polaritätsinduktion in *Equisetum*sporen und *Fucus*zygoten. Planta **63**, 356—365.

— 1968 a: Die Morphogenese pflanzlicher Zellen im elektrischen Feld. Z. Pflanzenphysiol. **59**, 303—339.

— 1968 b: Zur Funktion der Zellmembran bei der Cytomorphogenese. Ber. Dtsch. Bot. Ges. **81**, 311—314.

— Jaffe, L. F., 1968: Analyzing the "group effect": rheotropic responses of developing *Fucus* eggs. Protoplasma **65**, 25—35.

BENTRUP, F. W., SANDAN, T., JAFFE, L. F., 1967: Induction of polarity in *Fucus* eggs by potassium ion gradients. Protoplasma **64**, 254—266.

BLOCH, R., 1965: Polarity and gradients in plants: a survey. In: Handbuch der Pflanzenphysiologie, XV/1 (LANG, A., ed.), pp. 234—274. Berlin-Heidelberg-New York: Springer.

BONNETT, jr., H. T., NEWCOMB, E. H., 1966: Coated vesicles and other cytoplasmic components of growing root hairs of radish. Protoplasma **62**, 59—75.

BOPP, M., FELL, J., 1976: Manifestation der Cytokinin-abhängigen Morphogenese bei der Induktion von Moosknospen. Z. Pflanzenphysiol. **79**, 81—87.

— KNOOP, B., 1974: Régulation de la différenciation chez le protonéma des Mousses. Bull. Soc. Bot. Fr. **171**, 145—152.

BOSCH, F., EL GORESY, A., HERTH, W., MARTIN, B., NOBILING, R., POVH, B., REISS, H. D., TRAXEL, K., 1980: The Heidelberg proton microprobe. Nuclear Science Applications **1**, 1—39.

BOSSHARD, H. H., 1952: Elektronenmikroskopische Untersuchungen im Holz von *Fraxinus excelsior* L. Ber. Schweiz. Bot. Ges. **62**, 482—508.

BROWN, D. L., BOUCK, G. B., 1973: Microtubule biogenesis and cell shape in *Ochromonas*. II. The role of nucleating sites in shape development. J. Cell Biol. **56**, 360—378.

BRUNSWIK, H., 1924: Untersuchungen über die Geschlechts- und Kernverhältnisse bei der Hymenomyzetengattung *Coprinus*. Bot. Abhandlungen **5**, 1—152.

BUDER, J., 1961: Der Geotropismus der Charazeenrhizoide. Ber. Dtsch. Bot. Ges. **74**, (14)—(23).

BÜNNING, E., 1952: Morphogenesis in plants. Surv. Biol. Prog. **2**, 105—140.

— 1958: Polarität und inäquale Teilung des pflanzlichen Protoplasten. Protoplasmatologia VIII, 9 a. Wien: Springer.

BURGEFF, H., 1924: Untersuchungen über Sexualität und Parasitismus bei Mucorineen. Bot. Abhandlungen **4**, 1—135.

BUSBY, C. H., GUNNING, B., 1980: Observations on pre-prophase bands of microtubules in uniseriate hairs, stomatal complexes of sugarcane, and *Cyperus* root meristems. Europ. J. Cell Biol. **21**, 214—223.

CHANDLER, D., WILLIAMS, J. A., 1978: Intracellular divalent cation release in pancreatic acinar cells during stimulus-secretion coupling. II. Subcellular localization of the fluorescent probe chlorotetracycline. J. Cell Biol. **76**, 386—399.

CHEN, T.-H., JAFFE, L. F., 1979: Forced calcium entry and polarized growth of *Funaria* spores. Planta **144**, 401—406.

CUTTER, E. G., FELDMAN, L. J., 1970 a: Trichoblasts in *Hydrocharis:* I. Origin, differentiation, dimensions and growth. Amer. J. Bot. **57**, 190—201.

— — 1970 b: Trichoblasts in *Hydrocharis:* II. Nucleic acids, proteins and a consideration of cell growth in relation to endopolyploidy. Amer. J. Bot. **57**, 202—211.

DALL'OLIO, G., VANNINI, G. L., 1979: Coumarin-induced disturbances of morphological development and cell wall formation in *Trichophyton mentagrophytes*. Cytobiol. **18**, 390—397.

DASHEK, W. V., ROSEN, W. G., 1966: Electron microscopical localization of chemical components in the growth zone of lily pollen tubes. Protoplasma **61**, 192—204.

DEMAGGIO, A. E., STETLER, D. A., 1977: Protonemal organization and growth in the moss *Dawsonia superba:* ultrastructural characteristics. Amer. J. Bot. **64**, 449—454.

DENNISON, D. S., 1979: Phototropism. In: Encyclopedia of plant physiology, New Series, Vol. 7 (HAUPT, W., FEINLEIB, M. E., eds.), pp. 506—566. Berlin-Heidelberg-New York: Springer.

ETZOLD, H., 1965: Der Polaritropismus und Phototropismus der Chloronemen von *Dryopteris filix-mas* (L.) Schott. Planta **64**, 254—280.

FALK, H., STEINER, A. M., 1968: Phytochrome-mediated polarotropism: An electron microscopical study. Naturwiss. **55**, 500.

FÈVRE, M., 1979: Glucanases, glucan synthases and wall growth in *Saprolegnia monoica*. In: Fungal walls and hyphal growth (BURNETT, TRINCE, eds.), pp. 225—263. Cambridge: University Press.

Franke, W. W., Herth, W., Van Der Woude, W. J., Morré, D. J., 1972: Tubular and filamentous structures in pollen tubes: possible involvement as guide elements in protoplasmic streaming and vectorial migration of secretory vesicles. Planta **105**, 317—341.

Frey-Wyssling, A., 1959: Die pflanzliche Zellwand. Berlin-Göttingen-Heidelberg: Springer.

— Mühlethaler, K., 1949: Über den Feinbau der Zellwand von Wurzelhaaren. Mikroskopie (Wien) **4**, 257—320.

Giddings, jr., T. H., Brower, D. L., Staehelin, L. A., 1980: Visualization of particle complexes in the plasma membrane of *Micrasterias denticulata* associated with the formation of cellulose fibrils in primary and secondary cell walls. J. Cell Biol. **84**, 327—339.

Girbardt, M., 1968: Ultrastructure and dynamics of the moving nucleus. In: Aspects of cell motility. 22. Symp. Soc. Exp. Biol., Oxford (Miller, P. L., ed.), pp. 249—259. Cambridge: University Press.

— 1969: Die Ultrastruktur der Apikalregion von Pilzhyphen. Protoplasma **67**, 413—441.

Glenk, H. O., Wagner, W., Schimmer, O., 1971: Can Ca^{++} ions act as a chemotropic factor in *Oenothera* fertilization? In: Pollen: development and physiology (Heslop-Harrison, J., ed.), pp. 255—261. London: Butterworths.

Green, P. B., Erickson, P. O., Richmond, P. A., 1970: On the physical basis of cell morphogenesis. Ann. N.Y. Acad. Sci. **175**, 712—731.

Grove, S. N., Bracker, C. E., Morré, D. J., 1970: An ultrastructural basis for hyphal tip growth in *Pythium ultimum*. Amer. J. Bot. **57**, 245—266.

Hartmann, K. M., Menzel, H., Mohr, H., 1965: Ein Beitrag zur Theorie der polarotropischen und phototropischen Krümmung. Planta **64**, 363—375.

Haupt, W., 1957: Die Induktion der Polarität bei der Spore von *Equisetum*. Planta **49**, 61—90.

— 1962: Die Entstehung der Polarität in pflanzlichen Keimzellen, insbesondere die Induktion durch Licht. Ergeb. Biol. **25**, 1—32.

Heitz, E., 1940: Die Polarität keimender Moossporen. Verh. schweiz. naturf. Ges. (Locarno) **1940**, 168—170.

Hejnowicz, Z., Heinemann, B., Sievers, A., 1977: Tip growth: patterns of growth rate and stress in the *Chara* rhizoid. Z. Pflanzenphysiol. **81**, 409—424.

— Sievers, A., 1971: Mathematical model of geotropically bending *Chara* rhizoids. Z. Pflanzenphysiol. **66**, 34—48.

— — 1981: Regulation of the position of statoliths in *Chara* rhizoids. Protoplasma **108**, 117—137.

Herth, W., 1978: Ionophore A 23 187 stops tip growth, but not cytoplasmic streaming, in pollen tubes of *Lilium longiflorum*. Protoplasma **96**, 275—282.

— Franke, W. W., Van Der Woude, W. J., 1972: Cytochalasin stops tip growth in plants. Naturwiss. **59**, 38—39.

Heuser, J., Evans, L., 1980: Three-dimensional visualization of coated vesicle formation in fibroblasts. J. Cell Biol. **84**, 560—583.

Heyder, L., 1975: Feinstrukturelle Untersuchungen zum Spitzenwachstum von Farnchloronemen und -rhizoiden. Diplomarbeit, Botanisches Institut der Universität Bonn.

Howard, R. J., 1981: Ultrastructural analysis of hyphal tip cell growth in fungi: Spitzenkörper, cytoskeleton and endomembranes after freeze-substitution. J. Cell Sci. **48**, 89—103.

— Aist, J. R., 1977: Effects of MBC on hyphal tip organization, growth, and mitosis of *Fusarium acuminatum*, and their antagonism by D_2O. Protoplasma **92**, 195—210.

Ishizawa, K., Enomoto, S., Wada, S., 1979: Germination and photo-induction of polarity in the spherical cells regenerated from protoplasma fragments of *Boergesenia forbesii*. Bot. Mag. Tokyo **92**, 173—186.

— Wada, S., 1979 a: Growth and phototropic bending in *Boergesenia* rhizoid. Plant and Cell Physiol. **20**, 973—982.

— — 1979 b: Action spectrum of negative phototropism in *Boergesenia forbesii*. Plant and Cell Physiol. **20**, 983—987.

Iwanami, Y., 1980: Stimulation of pollen tube growth *in vitro* by dicarboxylic acids. Protoplasma **102**, 111—115.

JAFFE, L. F., 1956: Effect of polarized light on polarity of *Fucus*. Science **123**, 1081—1082.

— 1958: Tropistic response of zygotes of the *Fucaceae* to polarized light. Exp. Cell Res. **15**, 282—299.

— 1968: Localization in the developing *Fucus* egg and the general role of localizing currents. Adv. Morphogen. **7**, 295—328.

— ETZOLD, H., 1962: Orientation and locus of tropic photoreceptor molecules in spores of *Botrytis* and *Osmunda*. J. Cell Biol. **13**, 13—31.

— NEUSCHELER, W., 1969: On the mutual polarization of nearby pairs of Fucaceous eggs. Dev. Biol. **19**, 549—565.

JAFFE, L. A., WEISENSEEL, M. H., JAFFE, L. F., 1975: Calcium accumulations within the growing tips of pollen tubes. J. Cell Biol. **67**, 488—492.

KANE, B. E., REISKIND, J. B., MULLINS, J. T., 1973: Hormonal control of sexual morphogenesis in *Achlya:* dependence on protein and ribonucleic acid synthesis. Science **180**, 1192—1193.

KINZEL, H., 1956: Untersuchungen über Bau und Chemismus der Zellwände von *Antithamnion cruciatum* (Ag.) Näg. Protoplasma **46**, 445—474.

KNOX, R. B., HESLOP-HARRISON, J., 1970: Pollen-wall proteins: localization and enzymic activity. J. Cell Sci. **6**, 1—27.

— — 1971: Pollen-wall proteins: electron-microscopic localization of acid phosphatase in the intine of *Crocus vernus*. J. Cell Sci. **8**, 727—733.

LARPENT-GOURGAUD, M., 1969: Déterminisme de la ramification et du bourgeonnement chez le protonéma de *Bryales*. Ann. Sci. Nat., Bot. Biol. Veget. 12. Ser., **10**, 1—102.

— AUMAÎTRE, M.-P., 1976: Protéines, enzymes et morphogenèse comparées de deux espèces de *Bryales*. Phyton (Argent.) **34**, 45—49.

LARSON, D. A., 1965: Fine-structural changes in the cytoplasm of germinating pollen. Amer. J. Bot. **52**, 139—154.

MACNAR, R. M., 1979: Chemotaxis in bacteria. In: Encyclopedia of plant physiology, New Series, Vol. 7 (HAUPT, W., FEINLEIB, M. E., eds.), pp. 310—334. Berlin-Heidelberg-New York: Springer.

MASCARENHAS, J. P., 1978: Sexual chemotaxis and chemotropism in plants. In: Receptors and recognition, Series B, Vol. 5; Taxis and behavior, elementary sensory systems in biology (HAZELBAUER, G., ed.), pp. 169—203. London: Chapman and Hall.

— MACHLIS, L., 1962: The pollen-tube chemotropic factor from *Antirrhinum majus:* Bioassay, extraction, and partial purification. Amer. J. Bot. **49**, 482—489.

MILLER, J. H., STEPHANI, M. C., 1971: Effects of colchicine and light on cell form in fern gametophytes. Implications for a mechanism of light-induced cell elongation. Physiol. Plant. **24**, 264—271.

MOERZ, G., 1977: Wachstum der Wurzelhaare. Wiss. Arbeit zum Staatsexamen, Univ. Heidelberg.

MOHR, H., 1956: Die Abhängigkeit des Protonemawachstums und der Protonemapolarität bei Farnen vom Licht. Planta **47**, 127—158.

MOLISCH, H., 1889: Über die Ursachen der Wachstumsrichtungen bei Pollenschläuchen. Sitz.-Ber. math.-nat. Kl. Akad. Wiss. Wien (Anz. Akad. Wissensch.) **28**, 11—13.

MOLLENHAUER, H. H., MORRÉ, D. J., 1976: Cytochalasin B, but not colchicine, inhibits migration of secretory vesicles in root tips of maize. Protoplasma **87**, 39—48.

MORRÉ, D. J., KARTENBECK, J., FRANKE, W. W., 1979: Membrane flow and interconversions among endomembranes. Biochim. Biophys. Acta **559**, 71—152.

— VAN DER WOUDE, W. J., 1974: Origin and growth of cell surface components. In: Macromolecules regulating growth and development. 30. Symp. Soc. develop. Biol., pp. 81—111. New York-San Francisco-London: Academic Press.

MUELLER, S. C., BROWN, jr., R. M., 1980: Evidence for an intramembrane component associated with a cellulose microfibril-synthesizing complex in higher plants. J. Cell Biol. **84**, 315—326.

MULLINS, J. T., 1979: A freeze-fracture study of hormone-induced branching in the fungus *Achlya*. Tissue Cell **11**, 585—595.

Musgrave, A., Loes, E., Scheffer, R., Oehlers, E., 1977: Chemotropism of *Achlya bisexualis* germ hyphae to casein hydrolysate and amino acids. J. Gen. Microbiology 101, 65—70.

Nakazawa, S., 1959: Morphogenesis of the fern protonemata. I. Polar susceptibility to colchicine in *Dryopteris varia*. Phyton (Argent.) 12, 59—64.

Nebel, B. J., 1968: Action spectra for photogrowth and phototropism in protonemata of the moss *Physcomitrium turbinatum*. Planta 81, 287—302.

— 1969: Responses of moss protonemata to red and far-red polarized light: evidence for disc-shaped phytochrome photoreceptors. Planta 87, 170—179.

Neuscheler-Wirth, H., 1970: Photomorphogenese und Phototropismus bei *Mougeotia*. Z. Pflanzenphysiol. 63, 238—260.

Newcomb, E. H., Bonnett, jr., H. T., 1965: Cytoplasmic microtubule and wall microfibril orientation in root hairs of radish. J. Cell Biol. 27, 575—589.

Novotny, A. M., Forman, M., 1974: The relationship between changes in cell wall composition and the establishment of polarity in *Fucus* embryos. Develop. Biol. 40, 162—173.

Nuccitelli, R., 1978: Oöplasmic segregation and secretion in the *Pelvetia* egg is accompanied by a membrane-generated electrical current. Develop. Biol. 62, 13—33.

Ootaki, T., 1963: Modification of the developmental axis by centrifugation in *Pteris vittata*. Cytologia 28, 21—29.

— Furuya, M., 1969: Experimentally induced apical dominance in protonemata of *Pteris vittata*. Embryologia 10, 284—296.

Page, R. M., 1968: Phototropism in fungi. In: Photophysiology (Giese, A. C., ed.), Vol. III, pp. 65—90. New York: Academic Press.

— Curry, G. M., 1966: Studies on phototropism of young sporangiophores of *Pilobolus kleinii*. Photochem. Photobiol. 5, 31—40.

Peng, H. B., Jaffe, L. F., 1976: Polarization of fucoid eggs by steady electrical fields. Dev. Biol. 53, 277—284.

Plempel, M., 1960: Die zygotropische Reaktion bei Mucorineen. I. Mitteilung. Planta 55, 254—258.

— 1962: Die zygotropische Reaktion der Mucorineen. III. Mitteilung. Planta 58, 509—520.

— Dawid, W., 1961: Die zygotropische Reaktion bei Mucorineen. II. Mitteilung. Planta 56, 438—446.

Poff, K. L., Whitaker, B. D., 1979: Movement of slime molds. In: Encyclopedia of plant physiology, New Series, Vol. 7 (Haupt, W., Feinleib, M. E., eds.), pp. 355—382. Berlin-Heidelberg-New York: Springer.

Poccia, D. L., Palevitz, B. A., Campisi, J., Lyman, H., 1979: Fluorescence staining of living cells with fluorescamine. Protoplasma 98, 91—113.

Puiseux-Dao, S., Dazy, A.-C., 1970: Plastid structure and the evolution of plastids in *Acetabularia*. In: Biology of *Acetabularia* (Brachet, J., Bonotto, S., eds.), pp. 111—122. New York-London: Academic Press.

Quatrano, R. S., 1972: An ultrastructural study of the determined site of rhizoid formation in *Fucus* zygotes. Exp. Cell Res. 70, 1—12.

— 1978: Development of cell polarity. Ann. Rev. Plant Physiol. 29, 487—510.

Raudaskoski, M., 1980: Griseofulvin-induced alterations in site of dividing nuclei and structure of septa in a dikaryon of *Schizophyllum commune*. Protoplasma 103, 323—331.

— Koltin, Y., 1973: Ultrastructural aspects of a mutant of *Schizophyllum commune* with continuous nuclear migration. J. Bact. 116, 981—988.

Reiss, H.-D., Herth, W., 1978: Visualization of the Ca^{2+}-gradient in growing pollen tubes of *Lilium longiflorum* with chlorotetracycline fluorescence. Protoplasma 97, 373—377.

— — 1979 a: Calcium ionophore A 23 187 affects localized wall secretion in the tip region of pollen tubes of *Lilium longiflorum*. Planta 145, 225—232.

— — 1979 b: Calcium gradients in tip growing plant cells visualized by chlorotetracycline fluorescence. Planta 146, 615—621.

— — 1980: Effects of the broad-range ionophore X 537 A on pollen tubes of *Lilium longiflorum*. Planta 147, 295—301.

ROBINSON, K. R., JAFFE, L. F., 1975: Polarizing fucoid eggs drive a calcium current through themselves. Science **187**, 70—72.

— — 1976: Calcium gradients and egg polarity. J. Cell Biol. **70**, 37 a.

ROSEN, W. G., 1968: Ultrastructure and physiology of pollen. Ann. Rev. Plant Physiol. **19**, 435—462.

— 1971: Pistil-pollen interactions in *Lilium*. In: Pollen: Development and physiology (HESLOP-HARRISON, J., ed.), pp. 239—254. London: Butterworths.

— GAWLIK, S. R., 1966: Fine structure of lily pollen tubes following various fixation and staining procedures. Protoplasma **61**, 181—191.

— — DASHEK, W. V., SIEGESMUND, K. A., 1964: Fine structure and cytochemistry of *Lilium* pollen tubes. Amer. J. Bot. **51**, 61—71.

ROSENBAUM, J. L., MOULDER, J. E., RINGO, D. L., 1969: Flagellar elongation and shortening in *Chlamydomonas*. The use of cycloheximide and colchicine to study the synthesis and assembly of flagellar proteins. J. Cell Biol. **41**, 600—619.

SASSEN, M. M. A., 1964: Fine structure of *Petunia* pollen grain and pollen tube. Acta Bot. Neerl. **13**, 175—181.

SCHILDKNECHT, H., BENONI, H., 1963: Über die Chemie der Anziehung von Pollenschläuchen durch die Samenanlagen von Oenotheren. Z. Naturforsch. **18 b**, 45—54.

SCHMIEDEL, G., SCHNEPF, E., 1979 a: Side branch formation and orientation in the caulonema of the moss, *Funaria hygrometrica*: normal development and fine structure. Protoplasma **100**, 367—383.

— — 1979 b: Side branch formation and orientation in the caulonema of the moss, *Funaria hygrometrica*: experiments with inhibitors and with centrifugation. Protoplasma **101**, 47—59.

— — 1980: Polarity and growth of caulonema tip cells of the moss, *Funaria hygrometrica*. Planta **147**, 405—413.

SCHNEPF, E., 1973: Mikrotubulus-Anordnung und -Umordnung, Wandbildung und Zellmorphogenese in jungen *Sphagnum*-Blättchen. Protoplasma **78**, 145—173.

— 1981: Polarity and gradients in tip growing plant cells. In: International cell biology 1980/1981 (SCHWEIGER, H. G., ed.), pp. 483—488. Berlin-Heidelberg-New York: Springer.

— DEICHGRÄBER, G., DREBES, G., 1978: Development and ultrastructure of the marine, parasitic oomycete, *Lagenisma coscinodisci* Drebes (*Lagenidiales*). The infection. Arch. Microbiol. **116**, 133—139.

— — — 1980: Morphogenetic processes in *Attheya decora* (*Bacillariophyceae, Biddulphiineae*). Plant Syst. Evol. **135**, 265—277.

— DREBES, G., 1977: Über die Entwicklung des marinen parasitischen Phycomyceten *Lagenisma coscinodisci* (*Lagenidiales*). Helgol. wiss. Meeresunters. **29**, 291—301

— HEINZMANN, J., 1980: Nuclear movement, tip growth and colchicine effects in *Lagenisma coscinodisci* Drebes (*Oomycetes, Lagenidiales*). Biochem. Physiol. Pfl. **175**, 67—76.

— RÖDERER, G., HERTH, W., 1975: The formation of the fibrils in the lorica of *Poteriochromonas stipitata*: tip growth, kinetics, site, orientation. Planta **125**, 45—62.

SCHOCH-BODMER, H., HUBER, P., 1951: Das Spitzenwachstum der Bastfasern bei *Linum usitatissimum* und *L. perenne*. Ber. Schweiz. Bot. Ges. **61**, 377—404.

SCHRÖTER, K., 1978: Asymmetrical jelly secretion of zygotes of *Pelvetia* and *Fucus*: an early polarization event. Planta **140**, 69—73.

— LÄUCHLI, A., SIEVERS, A., 1975: Mikroanalytische Identifikation von Bariumsulfat-Kristallen in den Statolithen der Rhizoide von *Chara fragilis* Desv. Planta **122**, 213—225.

— RODRIGUEZ-GARCIA, M. I., SIEVERS, A., 1973: Die Rolle des endoplasmatischen Retikulums bei der Genese der *Chara*-Statolithen. Protoplasma **76**, 435—442.

— SIEVERS, A., 1971: Wirkung der Turgorreduktion auf den Golgi-Apparat und die Bildung der Zellwand bei Wurzelhaaren. Protoplasma **72**, 203—211.

SCHUMACHER, W., 1936: Untersuchungen über die Wanderung des Fluoresceins in den Haaren von *Cucurbita Pepo*. Jb. wiss. Bot. **82**, 507—533.

— 1967: Der Transport von Fluorescein in Haarzellen. In: Handbuch der Pflanzenphysiologie, XIII (SCHUMACHER, W., ed.), pp. 17—19. Berlin-Heidelberg-New York: Springer.

SEAGULL, R. W., HEATH, I. B., 1979: The effects of tannic acid on the *in vivo* preservation of microfilaments. Europ. J. Cell Biol. **20**, 184—188.
— — 1980 a: The organization of cortical microtubule arrays in the radish root hair. Protoplasma **103**, 205—229.
— — 1980 b: The differential effects of cytochalasin B on microfilament populations and cytoplasmic streaming. Protoplasma **103**, 231—240.
SHARMA, R., MALIK, C. P., 1978: Effect of light on pollen germination, tube elongation and IAA-oxidase in *Campsis grandiflora*. Biochem. Physiol. Pfl. **173**, 451—455.
SIEVERS, A., 1963 a: Beteiligung des Golgi-Apparates bei der Bildung der Zellwand von Wurzelhaaren. Protoplasma **56**, 188—192.
— 1963 b: Über die Feinstruktur des Plasmas wachsender Wurzelhaare. Z. Naturforsch. **18 b**, 830—836.
— 1964: Zur Feinstrukturanalyse pflanzlicher Zellen mit Spitzenwachstum. Ber. Dtsch. Bot. Ges. **77**, 388—390.
— 1965: Elektronenmikroskopische Untersuchungen zur geotropischen Reaktion. I. Über Besonderheiten im Feinbau der Rhizoide von *Chara foetida*. Z. Pflanzenphysiol. **53**, 193—213.
— 1967 a: Elektronenmikroskopische Untersuchungen zur geotropischen Reaktion. II. Die polare Organisation des normal wachsenden Rhizoids von *Chara foetida*. Protoplasma **64**, 225—253.
— 1967 b: Elektronenmikroskopische Untersuchungen zur geotropischen Reaktion. III. Die transversale Polarisierung der Rhizoidspitze von *Chara foetida* nach 5 bis 10 Minuten Horizontallage. Z. Pflanzenphysiol. **57**, 462—473.
— HEINEMANN, B., RODRIGUEZ-GARCIA, M. I., 1979: Nachweis des subapikalen differentiellen Flankenwachstums im *Chara*-Rhizoid während der Graviresponse. Z. Pflanzenphysiol. **91**, 435—442.
— SCHRÖTER, K., 1971: Versuch einer Kausalanalyse der geotropischen Reaktionskette im *Chara*-Rhizoid. Planta **96**, 339—353.
— VOLKMANN, D., 1979: Gravitropism in single cells. In: Encyclopedia of plant physiology, New Series, Vol. 7 (HAUPT, W., FEINLEIB, M. E., eds.), pp. 567—572. Berlin-Heidelberg-New York: Springer.
SITTE, P., 1978: Die lebende Zelle als System, Systemelement und Übersystem. Nova Acta Leopoldina **47**, Nr. 226, 195—216.
SPERBER, D., DRANSFELD, K., MARET, G., WEISENSEEL, M. H., 1981: Oriented growth of pollen tubes in strong magnetic fields. Naturwiss. **68**, 40—41.
STADLER, U., 1980: Untersuchungen zum Spitzenwachstum bei *Achlya* (Oomycetes). Wiss. Arbeit zum Staatsexamen, Univ. Heidelberg.
STEINER, A. M., 1967 a: Dose-response curves for polarotropism in germlings of a fern and a liverwisc. Naturwiss. **54**, 497.
— 1967 b: Action spectra for polarotropism in germlings of a fern and a liverworth. Naturwiss. **54**, 497—498.
STEUDLE, E., LÄUCHLI, A., SIEVERS, A., 1978: X-ray microanalysis of barium and calcium in plant material: significance for the analysis of statoliths. Z. Naturforsch. **33 c**, 444—446.
THOMAS, D. DES S., MULLINS, J. T., 1967: Role of enzymatic wall-softening in plant morphogenesis: hormonal induction in *Achlya*. Science **156**, 84—85.
— — 1969: Cellulase induction and wall extension in the water mold *Achlya ambisexualis*. Physiol. Plant. **22**, 347—353.
TSCHERMAK-WOESS, E., HASITSCHKA, G., 1953: Über Musterbildung in der Rhizodermis und Exodermis bei einigen Angiospermen und einer Polypodiacee. Österr. Bot. Z. **100**, 646—651.
TURIAN, G., 1979: Cytochemical gradients and mitochondrial exclusion in the apices of vegetative hyphae. Experientia **35**, 1164—1166.
VANDERWOUDE, W. J., MORRÉ, D. J., BRACKER, C. E., 1971: Isolation and characterization of secretory vesicles in germinated pollen of *Lilium longiflorum*. J. Cell Sci. **8**, 331—351.

WADA, M., O'BRIEN, T. P., 1975: Observations on the structure of the protonema of *Adiantum capillus-veneris* L. undergoing cell division following white-light irradiation. Planta **126**, 213—227.

WEISENSEEL, M. H., 1979: Induction of polarity. In: Encyclopedia of plant physiology, New Series, Vol. 7 (HAUPT, W., FEINLEIB, M. E., eds.), pp. 485—505. Berlin-Heidelberg-New York: Springer.

— JAFFE, L. F., 1976: The major growth current through lily pollen tubes enters as K$^+$ and leaves as H$^+$. Planta **133**, 1—7.

— NUCCITELLI, R., JAFFE, L. F., 1975: Large electrical currents traverse growing pollen tubes. J. Cell Biol. **66**, 556—567.

WETTSTEIN, D. VON, 1953: Beeinflussung der Polarität und undifferenzierte Gewebebildung aus Moossporen. Z. Bot. **41**, 199—226.

— 1965: Die Induktion und experimentelle Beeinflussung der Polarität bei Pflanzen. In: Handbuch der Pflanzenphysiologie, XV/1 (LANG, A., ed.), pp. 275—330. Berlin-Heidelberg-New York: Springer.

YAMASAKI, N., 1954: Über den Einfluß von Colchicin auf Farnpflanzen. I. Die jungen Prothallien von *Polystichum craspedosorum* Diels. Cytologia **19**, 249—254.

ZIEGLER, H., 1962 a: Chemotropismus. In: Handbuch der Pflanzenphysiologie, XVII/2 (BÜNNING, E., ed.), pp. 396—431. Berlin-Göttingen-Heidelberg: Springer.

— 1962 b: Hydrotropismus. In: Handbuch der Pflanzenphysiologie, XVII/2 (BÜNNING, E., ed.), pp. 432—450. Berlin-Göttingen-Heidelberg: Springer.

Microtubules and Cytomorphogenesis in a Developing Organ: The Root Primordium of *Azolla pinnata*

B. E. S. GUNNING

Department of Developmental Biology, Research School of Biological Sciences,
Australian National University, Canberra City, A.C.T., Australia

With 31 Figures

Contents

I. Introduction

It is a major theme of this book that cytomorphogenetic events participate in all aspects of growth and development. No cell can grow, change shape, assume or maintain or alter a state of polarity, differentiate, or divide without concomitant control and adjustment of its cytoskeleton. Such events occur in sequence during the life of an individual cell, and, in multicellular tissues and organs, are co-ordinated so that the population of cells develops harmoniously and produces a defined overall organization.

This chapter is an attempt to survey several facets of cytomorphogenesis, not analysing them as disparate phenomena so much as treating them as sub-processes of larger developmental systems. To highlight their partipation in the larger scene the account will be focused on a single multicellular object, composed of many thousands of cells of several different types, all developing in an integrated fashion.

The organ on which the chapter is centred is the root primordium of *Azolla pinnata* R. Br. (Figs. 2 and 3). Reference to other material will be made where relevant but comprehensive reviews will not be presented. The root of *Azolla* is unusually small compared with most roots of flowering plants. Its development is highly ordered and accordingly advantageous for electron microscope studies (Gunning, Hughes, and Hardham 1978). Having become familiar with its features, it is possible not only to be retrospective and deduce the past history of any given cell, but also to be predictive and state with confidence what the future course of development of that cell would have been, had not the root been fixed, embedded and sectioned.

The processes that will be considered are: control of the site and plane of cell division, control of cell shaping, and certain subcellular aspects of differentiation. As will be seen, all of these processes involve microtubules in the cell cortex. In order to perform them the cells in the root must be able to develop particular types of microtubule array in particular places and at particular times—a capability that emerges as being of central importance in cytomorphogenesis.

II. The Root of *Azolla pinnata*

Azolla pinnata R. Br. is a water fern, widespread in Africa and the Pacific region and economically important in S.E. Asia and China for its symbiotic nitrogen-fixation properties (Lumpkin and Plucknett 1980). The frond axis terminates in an apical cell and bears alternate leaves in dorsal and ventral ranks. Lateral axes develop alternately at regular intervals. One root is associated with each branch. Root anatomy in this and other members of the genus has been described in detail many times (Gunning, Hughes, and Hardham 1978 and references therein) and only relevant features are given here.

Ignoring its enclosing sheath and 2-layered root cap, the root proper has at its extreme apex a cell shaped approximately like an octant of a sphere (Fig. 3). This, the apical cell, divides sequentially at three cutting faces (Fig. 1). At each division cycle it reproduces itself together with a flattened derivative which tapers towards the central axis of the root and has a curved outer face. Three derivatives, or merophytes, overlapping one another as a consequence of the sequential division process, occupy one complete turn of a helix. The mature root consists of about 18 such turns, *i.e.*, about 50–55 cell divisions at the apical cell. No doubt the actual number of merophytes varies according to growth conditions but at any event the meristematic activity of the apical cell is limited and root growth is determinate.

The merophytes expand and become partitioned internally by a closely programmed sequence of formative divisions which introduce tangential-longitudinal (periclinal) and radial-longitudinal (anticlinal) walls in a set order (Figs. 1 and 3). Transverse sections cut near the apex show 3-fold

---

Fig. 1. Diagrammatic representation of the production of three files of merophytes by sequential divisions in the apical cell of *Azolla pinnata*. The sequence of formative divisions that gives rise to the population of initial cells is shown in merophytes 4–12. The final outcome of the proliferative (transverse) divisions for each cell type is shown at the top. Merophyte 6 shows the first example of the "sextant" wall (from the apex to the mid point of the first tangential wall) referred to in the caption to Fig. 21.

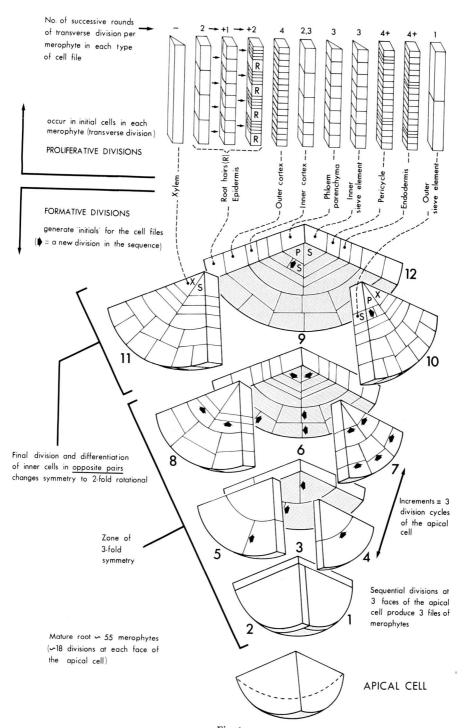

No. of successive rounds
of transverse division per
merophyte in each type
of cell file

occur in initial cells in each
merophyte (transverse division)
PROLIFERATIVE DIVISIONS

FORMATIVE DIVISIONS

generate 'initials' for the cell files
(◆ = a new division in the sequence)

Final division and differentiation
of inner cells in opposite pairs
changes symmetry to 2-fold rotational

Zone of
3-fold
symmetry

Mature root ∿ 55 merophytes
(∿18 divisions at each face of
the apical cell)

Increments ≡ 3
division cycles
of the apical
cell

Sequential divisions at
3 faces of the apical
cell produce 3 files of
merophytes

APICAL CELL

Fig. 1

symmetry, as would be expected if each merophyte undergoes the same set of events. However, the final division occurs in just two cells lying opposite one another in two of the three files of merophytes. This, and the eventual differentiation of the central cells in opposite *pairs*, overrides the underlying 3-fold symmetry of the basic construction and creates a 2-fold rotational pattern.

The development of the cell pattern by the formative division sequence is perhaps not truly morphogenetic in that the external form of the root may not be markedly influenced, but the internal partitioning (into 18 or 19 cells per merophyte) permits differentiation of a range of cell types with their own developmental programmes and ultimate functions. The merophytes themselves are aligned in three longitudinal files (sometimes slightly spiralled) and their internal partitioning walls are also longitudinally in register so that longitudinal files of cells are produced (usually 56 per root). With the exception of xylem precursors, which differentiate directly, each cell in each merophyte undergoes a set programme of proliferative transverse divisions (Figs. 1 and 3) and the products then complete their differentiation. The mature root consists of root hairs, epidermal cells, outer and inner cortex, endodermis, pericycle, outer and inner sieve elements, phloem parenchyma and xylem. The lineages of all members of the total population of about 9,000 cells can thus be mapped.

Each cell file exhibits a developmental gradient from young cells near the apex towards more mature regions at the base of the root. There is, however, no steady state. Progression of differentiation along the files towards the apex does not occur at the same rate as the addition of new initial cells to the files so that, as the root ages, the sites at which developmental events can first be detected occur closer and closer to the apical cell. This applies to formative and transverse divisions, xylem differentiation, formation of root hairs, and even subcellular events such as differentiation of plastids (WHATLEY and GUNNING 1981). Further, the successive merophytes are not fully equivalent. Those near the apex have fewer plasmodesmata connecting them to their neighbours than those formed earlier (GUNNING 1978). Progressive diminution in the frequency of plasmodesmata correlates with reduction in the capacity for cell-to-cell communication, as assessed by measurements of electrical coupling, and could well underlie the determinate nature of development in this root (SCOTT and GUNNING, unpublished).

It is a very general phenomenon of plant morphogenesis that the shaping

---

Figs. 2 and 3. Fig. 2 is a low magnification view of a longitudinal section of a root of *Azolla pinnata*. Note that there is an apical zone where the girth of the root increases rapidly, while the stele has virtually parallel sides from its inception. The increase in overall girth occurs in the extra-stelar cell files. The bulbous cells on the surface of the root are trichoblasts. ×100.

Fig. 3 shows the zone of formative divisions in the apical region (*AC* apical cell) and the early proliferative divisions (vertical arrows at first appearance). The outer tangential walls of the endodermis, delimiting the stele, are shown where they first appear (horizontal arrows). The cell files are labelled: *DG* dermatogen, *OC* outer cortex, *IC* inner cortex, *E* endodermis, *P* pericycle, *IS* inner stele. ×1,400.

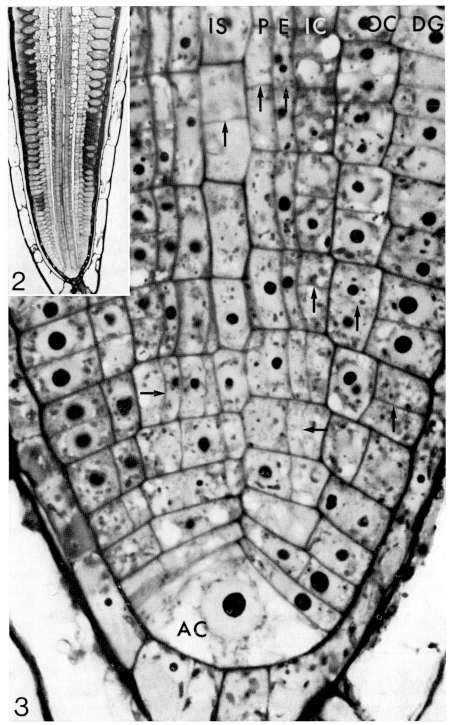

Figs. 2 and 3

of the final structure is a composite of the shaping of individual cells, the individual cells having arisen by the insertion of cell walls in particular sites and planes in pre-existing cells. This does not mean that individual cells have full independence, indeed the existence of global controls that harmonize and integrate shaping within the total population of cells has often been emphasized, from de Bary's famous aphorism "the plant makes cells, not cells the plant" up to the most recent reviews (Green 1980). Nor does it mean that the phases of cell division and cell shaping are separated either in time or in space. They are, in fact, related in various ways.

In the apical cell of *Azolla,* directed cell expansion is closely coupled to mitosis, so that the cell more or less regains a set volume after each division. The expansion that led to division at a face of the apical cell continues smoothly in the young merophyte, at first without further division, but later accompanied by internal partitioning that reduces the average cell volume. The axes of expansion then alter subtly and locally. Some cells merely elongate; others elongate *and* increase in girth. In aggregate these anisotropies generate the shape of the root tip. Most cells then divide in a new—transverse—plane, the successive mitoses at first approximately restoring cell volume to a standard size for each cell type, but eventually giving way to expansion without division. Although all cell files elongate to the same extent, the relationship between elongation and the number of rounds of transverse division varies according to cell type (Fig. 3). For instance, xylem elements end up at least 32-fold longer than their neighbouring pericycle cells.

The majority of plant organs exhibit varied interrelationships of cell division and cell expansion, comparable to the above (Green 1976). Two other extreme possibilities are internal partitioning without growth, and expansion of the plant body as a coenocyte without any preparatory internal partitioning. These rarities aside, the root of *Azolla* illustrates how regulation of the site and plane of division sets the scene for the subsequent patterned cell differentiation and locally-directed expansion processes that produce an organ. The following sections look more closely at certain cytomorphogenetic events that underlie the spatial control of division, shaping and differentiation.

III. The Site and Plane of Division

In vacuolated cells the formation of a specifically-positioned fenestrated raft of cytoplasm—the phragmosome—before the nucleus enters prophase shows that the cell anticipates its future site and plane of division by establishing a particular form of spatially-organized cytoskeleton (Sinnott and Bloch 1941, Venverloo *et al.* 1980). Coincidence between the position of a phragmosome and a subsequently-formed phragmoplast is not so readily seen

---

Figs. 4–6. Pre-prophase bands of microtubules in *A. pinnata*. Fig. 4 shows a dense mass of band microtubules in longitudinal view in a thick section (0.25–0.5 μm) viewed at 1000 kV. ×60,000. Figs. 5 and 6 show the cell plate and the corresponding pre-prophase band at late telophase in the asymmetrical division that divides the epidermis precursor cell (*E*) from a trichoblast (*R*). ×12,000.

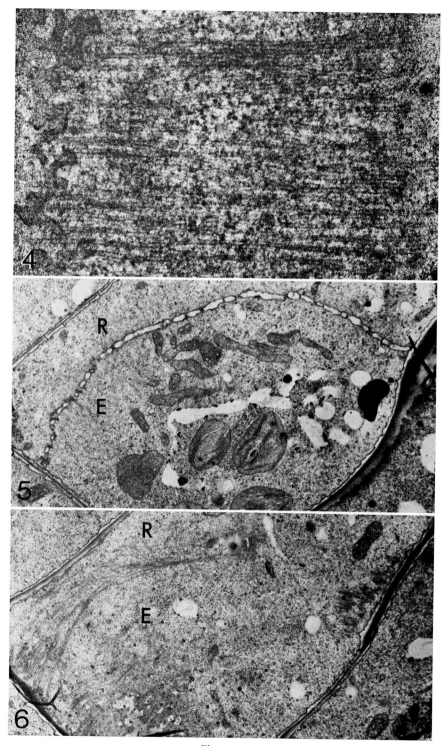

Figs. 4–6

when, as in most meristematic cells, the vacuolar component is much smaller. Yet anticipatory cytoplasmic organization can be detected. ÔTA, in 1961, centrifuged dividing cells in stamen hairs and concluded that "a specially differentiated pattern is imprinted [before mitosis] in the equatorial region of the cytoplasmic cortex", and that in telophase this region recognizes and exerts a force upon the margin of the cell plate (ÔTA 1961). We now know that in a wide variety of cells the "specially differentiated pattern" includes a girdle of microtubules (Fig. 4) that is formed before prophase and transitorily encircles the cell in the cytoplasmic cortex at the site where, in late telophase, the edge of the cell plate will fuse with the parental walls (see also p. 327 f.).

PICKETT-HEAPS and NORTHCOTE (1966) called this girdle of microtubules the pre-prophase band. The intervening years have seen many additional observations but not very much additional understanding. The band occurs in roots, shoots, leaves and hairs in flowering plants and in various organs in cryptogams, including bryophytes (see GUNNING, HARDHAM, and HUGHES 1978 a, BUSBY and GUNNING 1980, VENVERLOO et al. 1980, GALATIS and MITRAKOS 1979, 1980), yet it is clearly not an essential part of the morphogenetic machinery of plants in general. It apparently does not occur in the branching division of moss protonemata (SCHMIEDEL and SCHNEPF 1979), nor has it been seen in any alga (PICKETT-HEAPS 1975). It has not been seen in suspension cultures or callus tissue, nor when multinucleate endosperm cellularizes, nor preceding division of a microspore or sperm nucleus (see GUNNING, HARDHAM, and HUGHES 1978 a).

However, the pre-prophase band accurately predicts the line of fusion of cell plate and parental walls when it *is* present (BUSBY and GUNNING 1980). This is a genuine prediction phenomenon: the band disappears at prophase, yet its site somehow remains "imprinted", to use Ôta's word. Further, it is ubiquitous in the cell divisions that generate a root of *Azolla pinnata*. All formative and proliferative divisions are preceded by development of a pre-prophase band, be they anticlinal, periclinal, transverse, symmetrical or asymmetrical (GUNNING, HARDHAM, and HUGHES 1978 a). The fidelity of the prediction is extraordinary. Figs. 5 and 6 illustrate a pre-prophase band

---

Figs. 7 and 8. Cortical microtubule viewed at two tilt angles, + and − 30°. At the left hand side the microtubule tapers (at arrow) to a few protofilaments which extend off the edge of the micrograph. Discrete bridges to the plasma membrane are not readily distinguished but the space between the tubule and the membrane is very uniform. Note how a discontinuity between two microtubules (at arrowhead, Fig. 8) can virtually vanish if the two ends are superimposed in the direction of viewing (Fig. 7).

Fig. 9. Overlapping microtubules bridged to each other and to the plasma membrane. Note that the ends can be abrupt (upward arrow) or gradual (downward arrows).

Fig. 10. One member of a set of pictures taken at a range of tilt angles showing discrete bridges (arrows) that were visible in all views.

Fig. 11. A microtubule with an abruptly capped end (arrowhead) near the corner of a cell. Figs. 7–11 all *A. pinnata,* ×80,000.

Fig. 12. Glancing section of the cell cortex and wall of a developing sieve element in *A. pinnata* showing microtubules parallel to microfibrils of wall material. ×38,000.

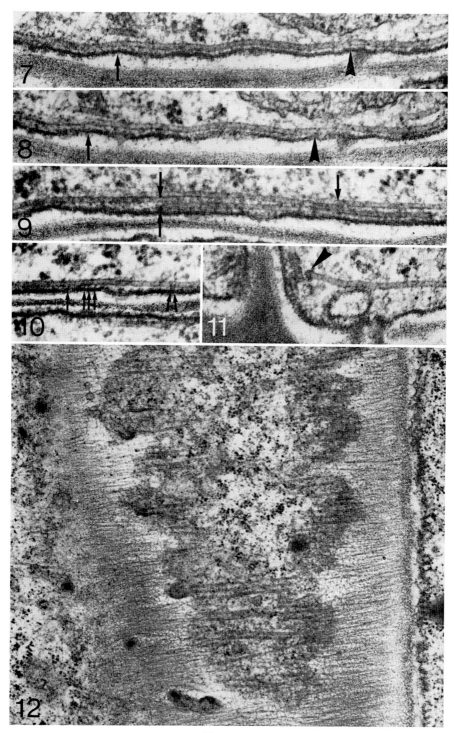

Figs. 7–12

for a highly asymmetrical division where extreme subtleties of wall placement are anticipated. In *Azolla* roots and in a number of other anatomical locations it can be seen that the new wall becomes placed at least approximately along the mid-line of the pre-prophase band site, which is some 1.5–3 µm in width. Each daughter cell thus receives about half of the site that was established at pre-prophase—a feature to be mentioned in another context in Section V.

The microtubules of the pre-prophase band cannot do more than participate as morphogenetic tools in regulating the site and plane of division, the ultimate specifications presumably being a combination of internal genetic instructions and external signals (PICKETT-HEAPS and NORTHCOTE 1966, PICKETT-HEAPS 1974). More, their absence from certain divisions which obviously are precisely placed [*e.g.,* at the apex of *Chara* (PICKETT-HEAPS 1975)] could be taken to indicate that they are not even essential participants. They do not themselves adjust the positioning of the cell plate, for they have long disappeared by telophase, and in any case placement of the cell plate is more sensitive to cytochalasin than to colchicine (PALEVITZ and HEPLER 1974). An involvement in establishing the site to which the edge of the cell plate is brought and at which it fuses with the parental walls seems the most likely role, although the nature of the involvement is debatable. It could be that the cortical site becomes specialized or perhaps stabilized by deposition of a particular set of wall microfibrils under the influence of the pre-prophase band microtubules. It has been observed that slight thickening of the wall occurs under the band (PACKARD and STACK 1976, GALATIS and MITRAKOS 1979, 1980).

An alternative argument, which perhaps does less than justice to the economy and efficiency with which cells in general seem to operate, states that the pre-prophase band microtubules have no specific role except as a reservoir of tubulin (PICKETT-HEAPS 1974). On this view, it is suggested that they form where they do because the cortical site has, amongst other properties, a role as a microtubule organizing centre (see also HEPLER and PALEVITZ 1974). However, the orientation of the microtubules in the pre-prophase band gives every impression of being very highly controlled, making it hard to accept that they are mere by-products. Work on *Azolla* root tip cells supports the idea that the pre-prophase band site might possess microtubule generating capacities which are important after division (see Section V).

IV. Cell Shaping

To impute a role for pre-prophase band microtubules in depositing a specific band of cell wall material, as above, is merely to extend a long-standing hypothesis on the role of cortical microtubules during interphase. This role has been discussed fully on p. 327 f. and it suffices here to reiterate that the many observations of congruence in orientation of cortical microtubules and currently deposited microfibrils in the wall have given rise to the idea that the microtubules somehow guide oriented apposition of wall material. The microtubules neither synthesize cellulose nor influence the quantity

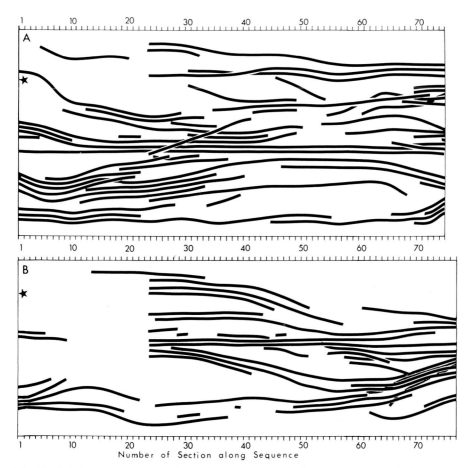

Fig. 13. Serial section reconstructions of interphase cortical arrays of microtubules against longitudinal walls in two back-to-back cells of *A. pinnata*. The axis of cell elongation is vertical on the page and the microtubules are thus predominantly transverse. The view is from the centre of a cell looking outwards at array A, then on through the cell wall at array B in the neighbouring cell (the asterisks are reference marks by which the reconstructions can be superimposed). Note the numerous short microtubules and the lines of terminations around section 23 in both arrays. Scale marker 1 μm. (Reproduced by permission, from HARDHAM and GUNNING 1978.)

of synthesis (GRIMM *et al.* 1976); what they offer is a means of imposing spatial control on cell wall formation, thereby helping to determine the shape that is assumed by the expanding cell (MARCHANT 1979).

The transverse cortical microtubules of longitudinal walls of *Azolla* root tip cells are oriented in the same way as the wall microfibrils, as in other systems (Fig. 12). However, they are considerably shorter than the cell wall against which they lie. The same applies to microtubules in root hairs in radish (SEAGULL and HEATH 1980), interphase arrays in *Zea* and *Impatiens* (HARDHAM and GUNNING 1978) and *Phleum* (GUNNING 1979), and to microtubules in pre-prophase bands as well as those at xylem thickenings (HARDHAM and Gunning 1978). In the majority of reconstructions derived from serial sections it was found that the ends of these short microtubules do not lie in any particular pattern against the face of the cell, but occasionally clusters or lines of endings were seen (Fig. 13). Microtubules *can* be very long, as in heliozoan axopodia, and the observed organization of the cortical arrays in plant cells poses a number of questions in relation to their supposed function.

The point has been made, both for the predominantly transverse arrays in elongating cells of *Azolla* roots (Hardham and Gunning 1978) and the predominantly longitudinal arrays in regions of root hairs that are undergoing secondary wall deposition (Seagull and Heath 1980), that short microtubules could not provide effective "guide tracks" for cellulose deposition. Possible mechanisms by which the array *as a whole* could control directionality of wall orientation are discussed on p. 341 f.). Perhaps the short microtubules found on the faces of the cells are depolymerizing, having been released or broken from the sites at which they were initiated (see later). Serial sectioning of *Azolla* cells shows that a substantial proportion of the cortical microtubules have one end close to the *edge* of a cell face, the other end lying somewhere against the *face* of the cell, amongst the short free microtubules. Both tapered and abrupt ends can occur (Figs. 7–9 and 11) and the microtubules may be bridged to each other and to the plasma membrane (Figs. 9 and 10; see also p. 166 and Fig. 8 C on p. 169).

Cortical microtubules are associated with localized developments of the cell wall at three levels of organization. The first is within cells, as in xylem elements and at stomatal pores (see p. 327 f.). The second level is concerned with shaping at the level of the whole cell. In elongating cells, predominantly transverse microtubule arrays lie against longitudinal walls which possess initially transverse wall microfibrils. This is well shown by *Azolla* (Figs. 12 and 13). Other aspects of microtubules and cell elongation are mentioned in the next section. At the third level of organization there is evidence that microtubule orientation is important in both the initiation of organs and their subsequent overall shaping.

Production of a radially symmetrical organ from a parental organ with apico-basal polarity demands that certain cells undergo a polarity shift. Detailed examination of one such system (Hardham, Green, and Lang 1980) has revealed that morphological alterations are preceded by a shift in the orientation of the microtubules, coupled with changes in their abundance. There was consistent parallelism of microtubules and microfibrils, the latter observed by a combination of polarizing and electron microscopy. The indications that crucial cytoskeletal changes occur at an extremely early stage of the overall morphogenetic sequence are very strong. They occur in those cells at the periphery of the future meristem which require a polarity shift in order to generate its radial symmetry.

Moving from the initiation to the shaping of an organ, the root of *Azolla pinnata* provides another striking example of controls that operate *via* the cytoskeleton in localized cells. The overall shape of the root is cylindrical, but the outer and cortical layers of cells narrow where the root tapers towards the apical cell. Within this tapered outline the stele remains virtually constant

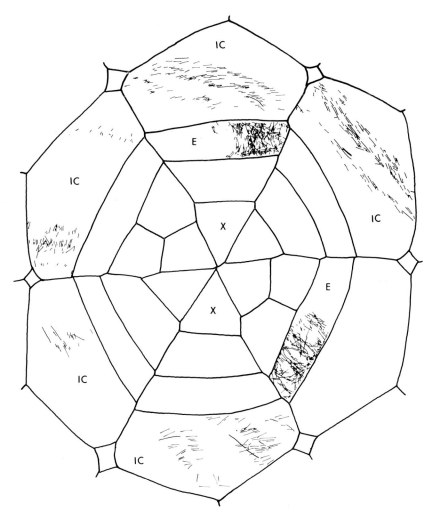

Fig. 14. Microtubule orientations against transverse walls in inner cortex and endodermis cells, mapped from serial transverse sections (not all adjacent) cut at a level in the root of *A. pinnata* where it is increasing in girth by radial expansion of the cells external to the stele. In the inner cortex (*IC*) the microtubules are predominantly tangential. In the endodermis (*E*) they are much more dense and predominantly radial. The cells left blank did not have transverse walls in the sequence of sections. X = xylem.

in diameter (Fig. 2). Two components of shaping can thus be distinguished: elongation of all cells, and radial expansion of cells lying external to the endodermis. The transverse orientation of microtubules at the side walls of longitudinally extending cells has already been mentioned; what is very interesting to find is a marked contrast in microtubule orientation between the transverse faces of the endodermis cells and the corresponding faces of their neighbouring inner cortex cells (Fig. 14). The former do *not* expand radially and have numerous predominantly radially-oriented microtubules;

the latter *do* expand radially and their microtubules, as in elongating cells, are oriented normal to the axis of expansion. Many of the radial microtubules of the transverse faces of the endodermis cells terminate near the tangential edges of those faces; many of the tangential microtubules of the corresponding faces of the inner cortex cells terminate near the radial edges. The contrast between the cell types is observable only in that small zone of the root where its girth increases. Thereafter the oft-described random orientation against the transverse walls pertains (Busby and Gunning, unpublished).

It would seem that oriented arrays of cortical microtubules can be called into play on selected faces of *all* cells (*e.g.*, in the elongation zone of roots), on selected faces of *selected* cells (*e.g.*, at sites of organ initiation and girth increases as described above), and on *selected parts* of selected faces of selected cells (*e.g.*, at sites of thickenings on longitudinal walls of developing xylem elements). The microtubular component of the cytoskeleton clearly serves many purposes. Taken together with what is known about preprophase bands, it is evident that cells, tissues and organs must possess systems that enable them to deploy their microtubules accurately. If we accept that microtubules themselves are important in cytomorphogenesis, then it follows that the microtubule-generating systems are equally vital as links in the overall morphogenetic process.

V. Development of Microtubule Arrays

It has almost become dogma that in the cell microtubules do not form except at specialized nucleating sites, widely known as "microtubule organizing centres" (MTOCs). Some MTOCs are impaired if protein (Stearns and Brown 1979) or nucleic acid (Heidemann *et al.* 1977) is digested from them; some are apparently arranged in the form of a template and others are much less organized (Tucker 1979). The microtubules that emerge from them have an intrinsic molecular polarity based on the sequence of α- and β-tubulin subunits along the protofilaments. They also have one end, designated "+", at which addition of subunits exceeds loss during the growth of the tubule, and the other end, designated "—", at which loss exceeds addition, unless reactions at that end are blocked by attachment to an MTOC (see Kirschner 1980). The + and — polarity may or may not be related to the intrinsic molecular polarity, or to the conformation with respect to an MTOC, or to the direction in which mechanochemical forces might be generated along the microtubule. The local concentration of free tubulin and other co-factors influences the kinetics of addition and loss. Theoretical considerations of these parameters have yielded the conclusions that microtubules are unlikely to form spontaneously without a nucleator (MTOC) being present, and that if free microtubules *do* arise by fragmentation or release from an MTOC, then, unless tubulin is present in very high concentrations, they are likely to depolymerize (Kirschner 1980).

These conclusions are relevant to the arrays of cortical microtubules in plant cells. Are there nucleating sites for these arrays and, if so, are they spatially restricted, and do they help to establish particular orientations? Further, recognizing that free microtubules do seem to be present, what is their origin and life-span? General questions are raised about the stability of the system, especially in relation to the postulated role in cell shaping.

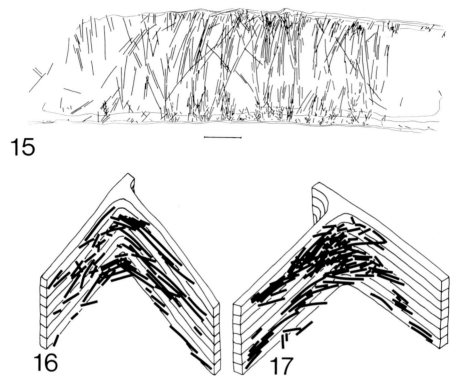

Figs. 15–17. Microtubules in the cortex of cells in root tips of *Cyperus eragrostis*. Fig. 15 shows a longitudinal face in an elongating cell, with the positions of the microtubules traced from five adjacent serial sections. Note the non-randomness at the edges of the cell face. Many microtubules radiate from loose clusters dispersed along the cell edges. If these clusters represent sites of origin, then the population of microtubules on the cell face consists of two interdigitating sets. Scale marker 1 μm. Figs. 16 and 17 show portions of cell edges analyzed by serial sectioning in the transverse plane (7 sections). In these cases it is known that the cells had very recently divided. As compared with the areas of cell face remote from the edges, the corners contain large numbers of clustered microtubules. It is difficult and in places impossible to track microtubules reliably in the planes of section used for Figs. 14–16 and no attempts have been made to join up microtubules that passed from section to section. Therefore no significance is to be attached to the numerous apparent ends of microtubules in these diagrams.

Some of the attributes of the microtubule generating system of the plant cell cortex have already been specified. It must be susceptible to localized spatial control at intercellular *and* intracellular levels, and also to quantitative regulation. Temporal regulation also occurs, as witnessed by the cyclic progression from the interphase array to the pre-prophase band to the spindle

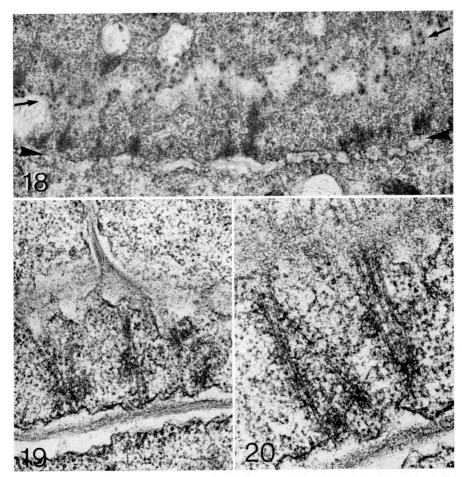

Figs. 18–20. Putative nucleating sites for cortical microtubules, seen at post-cytokinesis in cells of *A. pinnata* roots. Fig. 18 shows a longitudinal edge in a recently divided cell, with the new cell plate (between arrows) cut obliquely close to its juncture with the parental wall (between arrowheads). Dense clusters of microtubules and associated material are formed at intervals in the angle between the two. ×19,000. Figs. 19 and 20 illustrate similar clusters at higher magnification (×45,000 and ×66,000 respectively). All high voltage electron micrographs of thick sections.

Fig. 21. Longitudinal section of successive merophytes in *A. pinnata*. The mid line of the root is at the left. The micrograph demonstrates that a given cell edge can repeatedly form nucleating sites (arrowheads) in successive cell cycles. In merophyte *A* they are seen near the juncture of a newly laid down sextant wall (in grazing section, star) and a longitudinal parental wall. In merophyte *B*, which is equivalent but older by an increment equivalent to three cycles of the apical cell (see Fig. 1), the same cell edge is present but is one cell cycle older (in this case the sextant wall is just out of the plane of section) and still carries nucleating sites although the cell has now divided in a different plane (new wall arrowed). In merophyte *C* the sextant is again seen in grazing section (star) and where it joins the basal parental wall there are nucleating sites for a longitudinal pre-prophase band predicting the site of the arrowed wall in merophyte *D*, in which nucleating sites are once again present along the edge made by the sextant (star) and longitudinal walls. ×7,500.

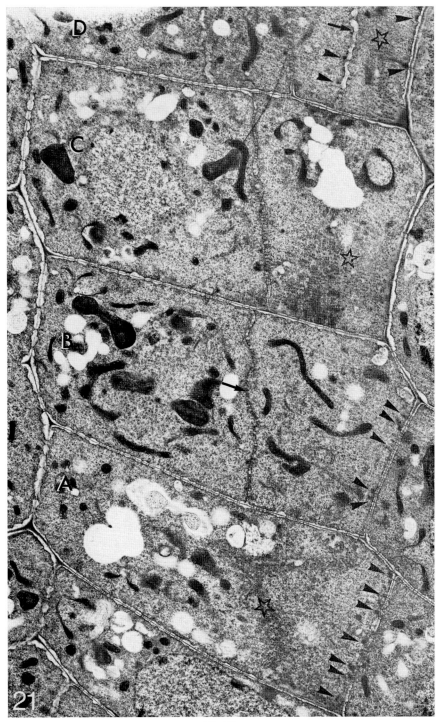

Fig. 21

to the phragmoplast and back to the interphase array. Other attributes emerge from detailed examination of the microtubule population during cell expansion.

There is a phase of cell elongation in which the rate of interpolation of microtubules against the longitudinal walls is strongly correlated with the rate of cell growth. During this phase the number of microtubules per unit length of longitudinal cell surface can be maintained at a set level in the face of several- or even many-fold cell elongation (Hardham and Gunning 1979, Withers 1980). Different cell types can maintain different densities, and it is even possible for different faces of a single cell type to maintain slight [Azolla endodermis (Hardham and Gunning 1979)] or marked [cereal root epidermis (Withers 1980)] differences. Complex interactions between the effects of growth regulators on cell elongation and the frequency of microtubules have also been detected (Schnepf and Deichgräber 1979).

It has also emerged from counts of transverse microtubules during cell elongation that their frequency is correlated with the length of the cell and not with its surface area or volume. The inference is that at this time the generating system is linear and in the axis of cell elongation (Withers 1980), at least for the microtubules that lie against the longitudinal walls. There is one precedent for a linear MTOC, in the form of a rhizoplast from which microtubules emerge at right angles (Brown and Bouck 1973), but no comparable structure has been seen in cells of higher plants. Approximately linear arrangements of microtubule ends have nevertheless been detected in two locations in Azolla. One location was the face of a cell wall (Fig. 13) and while the observation remains most intriguing, it must be stated that it was made only once in numerous reconstructions (Hardham and Gunning 1978). The other location occurs in polyhedral cells, where microtubule ends are consistently seen along the edges of the polyhedra, for example the longitudinal edges in the case of elongating cells in root apices.

Serial sectioning shows that many microtubules terminate at the edges of the polyhedra, both in established interphase arrays during cell expansion, and during the immediate post-cytokinesis phase when the interphase arrays are being reinstated after having been removed during mitosis (Figs. 18–20). The microtubules fan out from clusters that lie scattered along the edge of the cell. The clusters are quite conspicuous in Azolla (Gunning, Hardham, and Hughes 1978 b, Gunning and Hardham 1979, Gunning 1980) and certain other ferns (Busby, unpublished). In higher plants in which they have been sought they seem much more elusive but serial sectioning does give indications that they can occur, again along the edges of the cell (Figs. 15–17, and Galatis 1980).

It is not known whether all the clusters that have been observed in different taxa are functionally equivalent. Those in Azolla pinnata have been examined in sufficient detail to allow them to be plausibly equated with "nucleating sites" for microtubules—a term that in our current state of ignorance seems preferable to "MTOC" (the organizational abilities implicit in "MTOC" have not been established in relation to the development of interphase cortical arrays). The putative nucleating sites develop around the

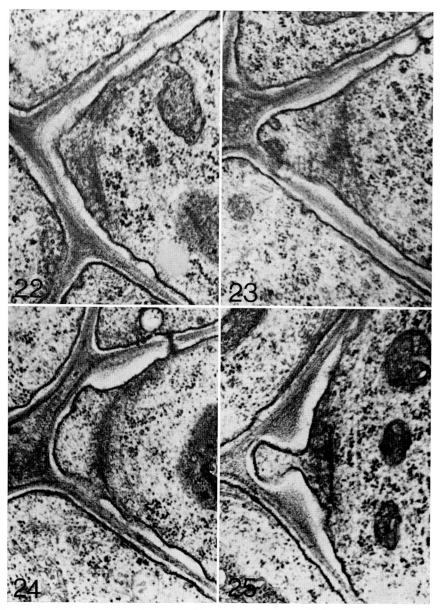

Figs. 22–25. Four successive early stages of formation of xylem wall thickenings and associated clusters of microtubules in *A. pinnata*. The sections are transverse to the root axis and depict various corners of the triangular pre-xylem cells (see cells marked X in Fig. 14). The micrographs shown here were selected from complete sets of serial sections and each one represents the middle of the microtubule cluster (for corresponding longitudinal plane of section see Fig. 30). Tubules initially appear to terminate in the angle of the two walls (Fig. 22). Later, a bundle of microtubules forms but its radius of curvature is then too large to allow it to penetrate into the extreme angle (Fig. 24). Initial wall deposition follows the contour of the bundle where it is in contact with the plasma membrane (Fig. 24), thus leaving a gap in the angle (Fig. 25). All ×44,000.

edges of new cell plates, at the line of fusion with the parental walls. They also develop at cell edges that were generated in previous cell cycles, showing that the capacity to form them persists at this location in the cell (Fig. 21). They are more obvious early in interphase than later. The edges along which they develop generally lie roughly at right angles to the final predominant orientation of the microtubule array. In *Azolla* they have been seen at all types of edges—transverse radial, transverse tangential, and longitudinal. Similar clusters of microtubules have been seen where pre-prophase bands intersect the edges of polyhedral cells, especially if the orientation of the band differs markedly from the orientation of the interphase array that it supplants (Fig. 21) (Gunning 1980).

There is also evidence for the nucleation of microtubules in dense clusters at localized sites where wall thickenings will be deposited in developing xylem elements. In *Azolla pinnata* it is obvious that these thickenings have their inception along the edges of the xylem precursor cells (indeed many types of wall thickening first appear at cell edges, *e.g.*, Galatis and Mitrakos 1980). The anticipatory clusters of microtubules are seen in transverse and longitudinal sections in Figs. 22–25, 30 and 31 respectively. Later in development the familiar association between wall thickenings and microtubules (see p. 329 f.) is established. Colchicine treatments disrupt the orderly deposition of wall thickenings (Figs. 27 and 28) but when an *Azolla* root recovers from a temporary treatment it re-establishes patterned formation of microtubule clusters along the edges of its pre-xylem cells (Fig. 29) (Hardham and Gunning 1980). In accord with the idea that these clusters are sites of production of microtubules, the drug IPC has been observed to disrupt the development of xylem thickenings (p. 330 f.). IPC is held to damage certain MTOCs (Coss and Pickett-Heaps 1974) and it is of interest in relation to the possible role of microtubule nucleating sites in the development of pre-prophase bands that it also leads to abnormal placement of new cell walls in *Azolla* (Wick and Busby, unpublished).

The nucleating sites (if such they are) appear to participate in the development of all categories of cortical array in cells of *Azolla* roots. In geometrical terms the set of edges has many of the attributes listed above as being implicit in what is known about the formation of cortical microtubules. Thus the cell

---

Figs. 26–31. Developing xylem in *A. pinnata*. Fig. 26 shows the two files of xylem elements (arrowheads) in an intact cleared root viewed by Nomarski optics ($\times 340$). Note also the root hair divisions (arrows), as seen in Figs. 5 and 6. Fig. 27 illustrates a similar preparation from a plant that had been treated with colchicine ($\times 200$). Fig. 28 is the same preparation at higher magnification ($\times 570$) and shows the loss of the pattern of wall thickenings in the xylem. Fig. 29 shows recovery of wall patterning (at bracket) after a short period in colchicine which disrupted wall deposition (xylem above bracket). $\times 530$. Figs. 30 and 31 are longitudinal sections illustrating the clusters of microtubules that precede the development of wall thickenings, forming at intervals along the edges of the cells. In stage of development they are probably comparable to Figs. 22–24. Fig. 30 is from a control and Fig. 31 is from a "recovery treatment" comparable to younger cells than those marked by the bracket in Fig. 29. Note that the inter-cluster spacing is very similar in the two cases. Both $\times 32,000$.

(Figs. 27–30 reproduced by permission, from Hardham and Gunning 1980.)

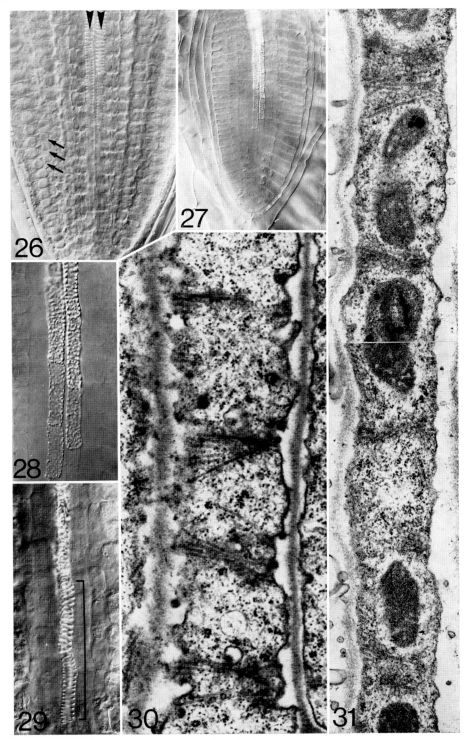

Figs. 26–31

can "activate" selected edges or selected regions of edges and thereby at least partially regulate the deployment of its microtubules. Activation of transverse radial edges could give rise to tangential arrays like those of the transverse walls of the inner cortex in *Azolla* roots, while activation of transverse tangential edges could generate the radial arrays of the neighbouring endodermal walls (Fig. 14). The known ability to produce arrays of different densities on different faces of a given cell suggests that individual edges, or sub-sets of edges circumscribing individual faces, can be controlled independently. Pre-prophase bands provide examples of even more localized activation. The microtubules that precede xylem thickenings not only lie in evenly spaced clusters along the cell surface, but the spacing between the clusters is constant irrespective of the length of the cells, which are shorter as the root ages (Hardham and Gunning 1980). Further, the spacing is the same after recovery from colchicine disruption as it is before treatment (Figs. 30 and 31).

In biochemical or biophysical terms it is not known how nucleation is stimulated at particular sites, how fine details of microtubule orientation are controlled, how microtubules with both ends apparently free against the faces of the cells arise, or how quantitative controls such as that relating polymerization of tubulin to cell elongation are imposed. Moreover there are numerous cells which do not conform to the polyhedral geometry of those in root tips of *Azolla:* how, for instance, does a cylindrical cell in a *Tradescantia* stamen hair generate its circumferentially oriented interphase and preprophase arrays? There are no edges at right angles to these systems (Busby and Gunning 1980). Where are the short microtubules of cylindrical root hairs (Seagull and Heath 1980) formed? It may well emerge that the system for which there now is evidence in the fern *Azolla* is far from universal, although observations on *Cyperus* (Figs. 15–17) and recent work on *Graptopetalum* meristematic cells (Hardham, Green, and Lang 1980) and *Vigna* stomatal guard cells (Galatis and Mitrakos 1980) provides a measure of corroboration based on angiosperm material.

There remains another fundamental question. What is the basis of the concentration of microtubule nucleating activity at the edges of the cell? While recognizing that it may be irrelevant, the edges of the cell do have a developmental history that involves a distinctive previous exposure to microtubules. First, each cell plate meets and fuses with the parental walls approximately along the mid line of the former pre-prophase band site in the cell cortex (Gunning, Hardham, and Hughes 1978 a, Busby and Gunning 1980). Thus the edges of the new wall (in both of the daughter cells) are at half former pre-prophase band sites. Second, the phragmoplast is considered to carry a microtubule nucleating capacity. Usually the development of the phragmoplast microtubules proceeds centrifugally as the cell plate extends towards the parental walls. Eventually it is concentrated at the *edges* of the plate. Given that the great majority of walls (though not the outer walls of root hairs or stamen hairs!) are formed as cell plates, there are therefore at least two stages at which a microtubule nucleating capacity could be conferred on the cell edges—at pre-prophase and at late telophase. In either case the conferment could be considered as the functional equivalent of other modes of equipping daughter cells with MTOCs, such as duplication of kinetochores, centrioles, basal bodies, or spindle plaques.

VI. Concluding Comments

A detailed knowledge of the anatomy and kinetics of root primordium development in *Azolla* has provided glimpses of what could be an important cytomorphogenetic system. Cortical microtubules participate as morphogenetic tools in determining the site and plane of cell division and in specific shaping processes that occur within cells and at the level of whole cells and at the level of multicellular aggregates. A possible source for these arrays of cortical microtubules has been described in morphological terms. The development of the microtubules is subject to precise temporal and spatial regulation. The regulatory mechanisms, of which virtually nothing is known, must be of great morphogenetic significance.

It is appropriate, however, to close by noting that although microtubules have been given emphasis in this chapter, they are but one component of the cytoskeleton. Other components must play their part. For example, the cell plate is guided to the pre-prophase band site long after the microtubules of the band have disappeared. The unicellular alga *Micrasterias* can undergo much of its elaborate shape development in the presence of colchicine (TIPPIT and PICKETT-HEAPS 1974; see p. 181 f.). Key events of directed cell enlargement by which lateral organs are initiated can also commence in the presence of colchicine (FOARD *et al.* 1965, CHARLTON 1977). These few examples serve as reminders that microtubules may indeed be important, but not all-important.

References

BROWN, D. L., BOUCK, G. B., 1973: Microtubule biogenesis and cell shape in *Ochromonas*. II. The role of nucleating sites in shape development. J. Cell Biol. **56**, 360—378.

BUSBY, C. H., GUNNING, B. E. S., 1980: Observations on pre-prophase bands of microtubules in uniseriate hairs, stomatal complexes of sugar cane, and *Cyperus* root meristems. Europ. J. Cell. Biol. **21**, 214—223.

CHARLTON, W. A., 1977: Evaluation of sequence and rate of lateral root initiation in *Pontederia cordata* L. by means of colchicine inhibition of cell division. Bot. Gaz. **138**, 71—79.

COSS, R. A., PICKETT-HEAPS, J. D., 1974: The effects of isopropyl N-phenyl-carbamate on the green alga *Oedogonium cardiacum*. I. Cell division. J. Cell Biol. **63**, 84—98.

FOARD, D. E., HABER, A. H., FISHMAN, T. N., 1965: Initiation of lateral root primordia without completion of mitosis and without cytokinesis in uniseriate pericycle. Amer. J. Bot. **52**, 580—590.

GALATIS, B., 1980: Microtubules and guard cell morphogenesis in *Zea mays* L. J. Cell Sci. **45**, 211—244.

GALATIS, B., MITRAKOS, K., 1979: On the differential divisions and preprophase microtubule bands involved in the development of stomata of *Vigna sinensis*. J. Cell Sci. **37**, 11—37.

— — 1980: The ultrastructural cytology of the differentiating guard cells of *Vigna sinensis*. Amer. J. Bot. **67**, 1243—1261.

GREEN, P. B., 1976: Growth and cell pattern formation on an axis: critique of concepts, terminology, and modes of study. Bot. Gaz. **137**, 187—202.

— 1980: Organogenesis—a biophysical view. Ann. Rev. Plant Physiol. **31**, 51—82.

GRIMM, I., SACHS, H., ROBINSON, D. G., 1976: Structure, synthesis and orientation of micro-fibrils. II. The effects of colchicine on the wall of *Oocystis solitaria*. Cytobiologie **14**, 61—74.

GUNNING, B. E. S., 1978: Age-related and origin-related control of the number of plasmo-desmata in cell walls of developing *Azolla* roots. Planta **143**, 181—190.

— 1979: Nature and development of microtubule arrays in cells of higher plants. In: Proc. 37th Ann. Meeting Electron Microscopy Soc. Amer. (BAILEY, G. W., ed.), pp. 172—175. Baton Rouge: Claitor's Pub. Div.

— 1980: Spatial and temporal regulation of nucleating sites for arrays of cortical micro-tubules in root tip cells of the water fern *Azolla pinnata*. Europ. J. Cell Biol. **23**, 53—65.

— HARDHAM, A. R., 1979: Microtubules and morphogenesis in plants. Endeavour N.S. **3**, 112—117.

— — HUGHES, J. E., 1978 a: Pre-prophase bands of microtubules in all categories of formative and proliferative cell division in *Azolla* roots. Planta **143**, 145—160.

— — — 1978 b: Evidence for initiation of microtubules in discrete regions of the cell cortex in *Azolla* root-tip cells, and an hypothesis on the development of cortical arrays of microtubules. Planta **143**, 161—179.

— HUGHES, J. E., HARDHAM, A. R., 1978: Formative and proliferative cell divisions, cell differentiation, and developmental changes in the meristem of *Azolla* roots. Planta **143**, 121—144.

HARDHAM, A. R., GREEN, P. B., LANG, J. M., 1980: Reorganization of cortical microtubules and cellulose deposition during leaf formation in *Graptopetalum paraguayense*. Planta **149**, 181—195.

— GUNNING, B. E. S., 1978: Structure of cortical microtubule arrays in plant cells. J. Cell Biol. **77**, 14—34.

— — 1979: Interpolation of microtubules into cortical arrays during cell elongation and differentiation in roots of *Azolla pinnata*. J. Cell Sci. **37**, 411—442.

— — 1980: Some effects of colchicine on microtubules and cell division in roots of *Azolla pinnata*. Protoplasma **102**, 31—51.

HEIDEMANN, S. R., SANDER, G., KIRSCHNER, M., 1977: Evidence for a functional role of RNA in centrioles. Cell **10**, 337—350.

HEPLER, P. K., PALEVITZ, B. A., 1974: Microtubules and microfilaments. A. Rev. Pl. Physiol. **25**, 309—362.

KIRSCHNER, M. W., 1980: Implications of treadmilling for the stability and polarity of actin and tubulin polymers *in vivo*. J. Cell Biol. **86**, 330—334.

LUMPKIN, T. A., PLUCKNETT, D. L., 1980: *Azolla*: botany, physiology, and use as a green manure. Econ. Bot. **34**, 111—153.

MARCHANT, H. J., 1979: Microtubules, cell wall deposition and the determination of plant cell shape. Nature **278**, 167—168.

ÔTA, T., 1961: The role of cytoplasm in cytokinesis of plant cells. Cytologia **26**, 428—447.

PACKARD, M. J., STACK, S. M., 1976: The pre-prophase band: possible involvement in the formation of the cell wall. J. Cell Sci. **22**, 403—411.

PALEVITZ, B. A., HEPLER, P. K., 1974: The control of the plane of division during stomatal differentiation in *Allium*. I. Spindle reorientation. Chromosoma **46**, 297—326.

PICKETT-HEAPS, J. D., 1974: Plant microtubules. In: Dynamic aspects of plant ultrastructure (ROBARDS, A. W., ed.). London-New York: McGraw-Hill.

— 1975: Green algae. Sunderland, Mass.: Sinauer Assoc.

— NORTHCOTE, D. H., 1966: Organization of microtubules and endoplasmic reticulum during mitosis and cytokinesis in wheat meristems. J. Cell Sci. **1**, 109—120.

SCHMIEDEL, G., SCHNEPF, E., 1979: Side branch formation and orientation in the caulonema of the moss, *Funaria hygrometrica*. Protoplasma **100**, 367—384.

SCHNEPF, E., DEICHGRÄBER, G., 1979: Elongation growth of setae of *Pellia* (*Bryophyta*): fine structural analysis. Zeit. f. Pflanzenphysiol. **94**, 283—297.

SEAGULL, R. W., HEATH, I. B., 1980: The organisation of cortical microtubule arrays in the radish root hair. Protoplasma 103, 205—230.

SINNOTT, E. W., BLOCH, R., 1941: Division in vacuolate plant cells. Amer. J. Bot. 28, 225—232.

STEARNS, M. E., BROWN, D. L., 1979: Purification of cytoplasmic tubulin and microtubule organizing center proteins functioning in microtubule initiation from the alga Polytomella. Proc. Natl. Acad. Sci. U.S.A. 76, 5745—5749.

TIPPIT, D. H., PICKETT-HEAPS, J. D., 1974: Experimental investigations into morphogenesis in Micrasterias. Protoplasma 81, 271—296.

TUCKER, J. B., 1979: Spatial organisation of microtubules. In: Microtubules (ROBERTS, K., HYAMS, J. S., eds.), pp. 315—357. London: Academic Press.

VENVERLOO, C. J., HORENKAMP, P. H., WEEDA, A. J., LIBBENGA, K. R., 1980: Cell division in Nautilocalyx explants. Z. Pflanzenphys. 100, 161—174.

WHATLEY, J. M., GUNNING, B. E. S., 1981: Chloroplast development in Azolla roots. New Phytol. (in press).

WITHERS, G. R. A., 1980: Microtubule distribution in the root tips of barley, wheat, oats and rye-grass. Ph.D. Thesis, Monash University.

Morphogenesis of Tracheary Elements and Guard Cells

P. K. Hepler

Botany Department, University of Massachusetts, Amherst, Massachusetts, U.S.A.

With 11 Figures

Contents

I. Introduction

Plant cell shape is determined by the cell wall. The pattern and orientation of the reinforcing cellulose microfibrils of the wall define the manner and places in which a cell will strain in response to the internally generated turgor stress (Green 1980). To understand morphogenesis one must explain the mechanism of cell wall formation and in particular how unique patterns of cellulose reinforcements arise. Two cell types that possess specific wall patterns related to cell shape and function, and that have received considerable attention in studies of cytomorphogenesis are tracheary elements and stomatal guard cells (Hepler and Palevitz 1974). Despite their obvious differences in overall shape and function they possess remarkable similarities in the structure of their secondary wall and in the apparent manner by which the wall develops. It is the purpose of this chapter to consider the mechanism of secondary wall formation in differentiating tracheary elements and guard cells and to focus especially on those structures and processes at the cell wall-cytoplasmic interface that seems to be responsible for controlling the location of wall formation and the orientation of cellulose deposition.

II. Tracheary Elements

A. Formation of the Secondary Wall

During differentiation of primary xylem, bands of secondary wall are deposited onto the primary wall into patterns often described as annular, helical, reticulate or scalariform (Esau 1953) (Fig. 1). Electron microscopy and polarized light microscopy both show that the individual bands comprising the secondary thickenings are composed of cellulose microfibrils that are oriented parallel to one another and parallel to the major axis of the band (Hepler et al. 1970). The cytoplasmic factors underlying the formation of the wall banding pattern have been sought in studies of primary xylem during its differentiation in growing meristems and also in xylem that is regenerating in response to wounding. Sinnott and Bloch (1945) many years ago observed that strands of cytoplasm appeared in differentiating wound vessel elements of *Coleus* and became organized into a reticulate pattern identical to that of the subsequently thickened secondary wall. Ultrastructural examination of differentiating cells, while showing an enrichment of cytoplasm (Hepler and Newcomb 1963) have failed to uncover a specific cytoplasmic pattern (Cronshaw and Bouck 1965, Esau et al. 1966 a, Pickett-Heaps and Northcote 1966 c, Wooding and Northcote 1964). Nevertheless, with the introduction of glutaraldehyde as a superior fixative cytoplasmic specializations structurally associated with the developing secondary wall thickening have been discovered; these are the cytoplasmic microtubules that occur next to the plasmalemma, grouped specifically over the bands of secondary wall, and oriented parallel to the cellulose microfibrils (Hepler and Newcomb 1964) (Fig. 3). These observations have been repeated in several laboratories and have provided the basis for new thought and experimental direction in our quest to understand cell wall formation and cytomorphogenesis (Cronshaw and Bouck 1965, Esau et al. 1966 b, Gossen-de Roo 1973 b, Maitra and De 1971, Pickett-Heaps 1967 a, Wooding and Northcote 1964).

Microtubules are composed of 13 protofilament rows of tubulin protein dimers that are bonded laterally to form a linear, rigid structure circular in cross section and 24–26 µm in diameter (Hepler and Palevitz 1974). They are grouped over the secondary thickening, but not usually in a strictly regular pattern (Hepler and Fosket 1971). Some occur close to the plasmalemma and electron microscopic observations have revealed fine interconnections or cross-bridges projecting laterally between the tubule surface and the cell membrane (Brower and Hepler 1976). Other tubules are found deeper in the cytoplasm and although they remain localized over the wall band they appear less intimately associated with the cell surface (Fig. 3).

Initial studies from tangential sections led to the conclusion that microtubules were very long, extending entirely around the cells as "hoops" around a barrel (Ledbetter and Porter 1963). The serial section analyses by Hardam and Gunning (1978, 1979) have shown both for differentiating tracheary elements and cells with primary walls that in reality microtubules are much shorter than the cell circumference; in tracheary elements they are 4–8 µm long and exist in extensively overlapping arrays (1979; see also p. 311 f.).

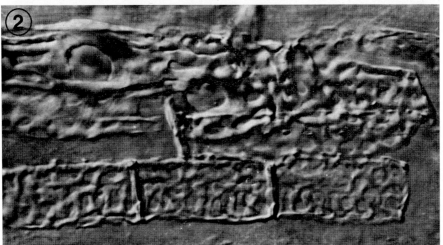

Fig. 1. Wound vessel element of *Coleus* examined by differential interference contrast microscopy. The secondary wall bands are deposited in a reticulate pattern. From HEPLER and FOSKET 1971. ($\times$1,400.)

Fig. 2. Wound vessel element of *Coleus* cultured in the presence of colchicine. The secondary wall has become smeared over the primary wall and the normal reticulate pattern is lost. From HEPLER and FOSKET 1971. ($\times$1,400.)

The evidence that microtubules participate in the generation of wall pattern and the orientation of cellulose derives from several investigations. PICKETT-HEAPS (1967 a) and more recently HARDHAM and GUNNING (1979) have convincingly demonstrated in differentiating xylem initials in root meristems that groups of microtubules aggregate along the cell cortex in a pattern that precedes and anticipates the position of the future secondary wall thickening. The report of HARDHAM and GUNNING (1979) in addition shows that prior to differentiation there is an increase in the number of microtubules and that during early stages of wall deposition the average length of individual microtubules increases from 5 to 7 µm. During normal development, microtubules thus acquire orientations and patterns of distribution that are parallel to, but precede the formation and structure of the secondary wall.

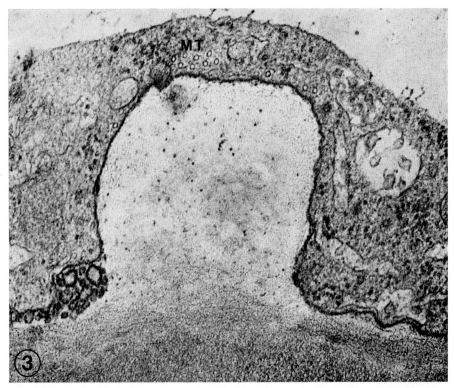

Fig. 3. Electron micrograph from a transverse section of secondary wall thickening. Microtubules (*MT*), circular in cross section, are clustered over the secondary wall. From HEPLER and FOSKET 1971. (×60,000.)

Removal of microtubules with various drugs provides further information on their role in wall formation (Fig. 2). Differentiating cells cultured in colchicine form highly aberrant secondary thickenings (HARDHAM and GUNNING 1980, HEPLER and FOSKET 1971, PICKETT-HEAPS 1967 a). This is especially clear in a wound vessel element of *Coleus* in which the normally discrete bands of secondary wall, during successive deposition, become smeared out over the primary wall surface (HEPLER and FOSKET 1971) (Figs. 2 and 4). If a cell receives colchicine at the inception of wall deposition then the pattern of secondary bands is totally lost. Isopropyl N-phenyl carbamate (IPC), a drug that disrupts microtubules in several different plants also has been

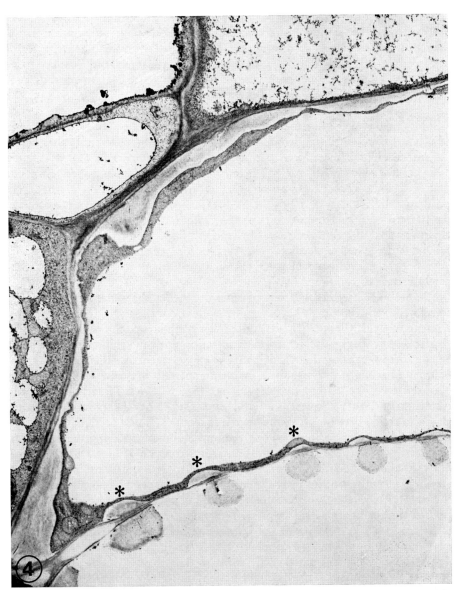

Fig. 4. Ultrastructural view of a wound vessel element of *Coleus* cultured in colchicine. The secondary wall is smeared unevenly over the surface of the primary wall except along the bottom of the micrograph. In this instance a neighboring differentiated cell has caused the thickenings to form in discrete places (∗). From HEPLER and FOSKET 1971. (×6,000.)

shown to exert a pronounced effect upon xylem secondary wall pattern in root meristem of *Triticum* (BROWER and HEPLER 1976).

Examination of the cellulose microfibrillar orientation in these drug-induced aberrant wall thickenings reveals that it too is greatly altered. Interestingly, though, the microfibrils are not randomly oriented but instead occur in swirls of organized material (BROWER and HEPLER 1976, HEPLER and FOSKET 1971). Individual bundles, however, lie at acute angles to one another, thus producing an overall disordering of wall structure.

Microtubules participate in determining wall pattern and the orientation of cellulose microfibrils, but they are not solely responsible. Wall thickenings in adjacent differentiating xylem elements often occur opposite each other across the intervening primary wall. In wound vessel elements of *Coleus* treated with colchicine, wall thickenings in a neighboring xylem member, even one that is dead, control the position of aberrant wall deposition in the drug-cultured cell (HEPLER and FOSKET 1971) (Fig. 4). Under these circumstances the pattern of wall banding mirrors the differentiated cell. Similarly, in the determination of cellulose microfibril orientation, factors other than microtubules emerge as regulators. The regions or patches of ordered material suggest that microfibrils of a given alignment influence the orientation of newly deposited cellulose.

Microtubules are the only cytoplasmic organelles that have been primarily correlated with the time and place of secondary wall deposition in tracheary elements, nevertheless, a variety of other cellular components appear to possess certain degrees of structural association with the thickenings and must be considered as potential participants in morphogenesis. Elements of endoplasmic reticulum, for example, are interspersed between the thickenings (GOOSEN-DE ROO 1973 a, PICKETT-HEAPS 1967 a, 1968, PICKETT-HEAPS and NORTHCOTE 1966 c, SRIVASTAVA and SINGH 1972 b). Dictyosome cisternae may also reside between the bands but frequently vesicles, presumably derived from the dictyosome appear among the microtubules overlying the secondary wall (ESAU *et al.* 1966 a, HARDHAM and GUNNING 1979, HEPLER *et al.* 1972, MAITRA and DE 1971, PICKETT-HEAPS 1968). It seems certain that these membrane compartments provide enzymes, precursors and even pre-formed components to the growing secondary wall, but there is no evidence that they control the place of wall deposition. An important unknown in the process of localized wall deposition is the role of the plasmalemma itself. Might this membrane contain localized domains of vesicle or microtubule binding factors that ultimately control pattern? I return to this provocative question and to possible mechanisms of microtubule action in cellulose orientation in more detail towards the end of the chapter.

III. Stomatal Guard Cells

A. General Consideration

For studies on cell morphogenesis differentiating guard cells are excellent and offer key advantages over tracheary elements. Most importantly, guard cells reside on the epidermis and can be examined directly by light microscopy.

The use of paradermal slices from young onion seedlings has permitted us to observe the division of the guard mother cell into two guard cells and certain aspects of the subsequent differentiation (PALEVITZ and HEPLER 1974 a, b). Largely for this reason, it has been possible to acquire detailed information on the dynamics of the normal morphogenesis process and to study its perturbation through the action of selected agents.

Formation of the stomatal complex involves first a sequence of asymmetric and/or highly oriented cell divisions that give rise to the undifferentiated guard cells. Subsequently, through deposition of a patterned secondary wall, the guard cells acquire their particular shape and ability to function in the control of gas exchange (RASCHKE 1979). Different development plans and degrees of complexity exist among stomatal apparatuses of various species, but for the present discussion two types will be considered: 1. the simple stomatal apparatus as observed primarily in *Allium* and *Pisum,* and 2. the complex system as typified by grasses, *Avena, Hordeum, Phleum,* and *Zea.*

B. Formation of the Undifferentiated Guard Cell

The early steps in the formation of the undifferentiated guard cell are an important part of cell morphogenesis; they define the placement of the cross walls and thus the initial size and shape of the cells that will ultimately produce the complex. In simple complexes, as observed in onion, a nuclear migration followed by an asymmetric cytokinesis in an elongated epidermal cell yields two unequal cells, the smaller of which is the guard mother cell. The guard mother cell then divides longitudinally producing the pair of guard cells and these, through subsequent deposition of a specially thickened wall, acquire their typical kidney shape (PALEVITZ and HEPLER 1974 a, 1976) (Fig. 8).

Grasses form their stomatal complexes in serial order along defined triplet rows of cells. The central row, through asymmetric division, yields a repeating sequence of guard mother cell and intervening cell (ZEIGER 1971) (Fig. 5). The two flanking rows of lateral cells, also through nuclear migration and asymmetric divisions, produce small lens-shaped subsidiary cells that are positioned to each side of the guard mother cell (ZEIGER 1971) (Fig. 5). Concomitantly the guard mother cell divides symmetrically, but longitudinally to produce the pair of guard cells that elongates and differentiates into the distinctive dumbbell or "bone" shape (PALEVITZ 1980 b).

Within these divisions there are several important but poorly understood events. Nuclear migration and asymmetric divisions are abundantly evident, and it is through them that the initial position and shape of the stomatal complex is determined. Some evidence showing the sensitivity of nuclear position to colchicine suggests that microtubules are involved, and similar kinds of experiments with cytochalasin B implicate acto-myosin microfilaments (PALEVITZ and HEPLER 1974 b). The data are not compelling and furthermore direct examination of nuclei in the process of migration has failed to uncover a noticeable array of microtubules, microfilaments, or any particular distribution of other structures or organelles that might participate in causing motion (HEPLER and ZEIGER, unpublished observations).

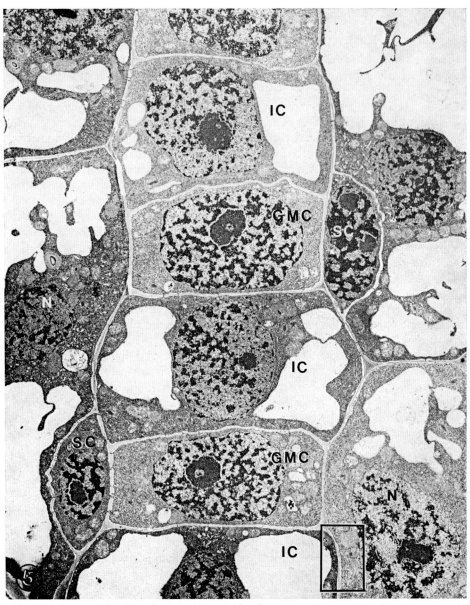

Fig. 5. Low magnification electron micrograph of a stomatal row of barley. Two guard mother cells (*GMC*) are shown separated by intervening cell (*IC*). Each guard mother cell at this stage of development is flanked by a subsidiary cell (*SC*). The nuclei marked (*N*) are in the process of migrating to the sides of the guard mother cells in order to form the additional subsidiary cells. The area outlined in the box contains a preprophase band of microtubules and is shown at higher magnification in Fig. 6. From HEPLER and ZEIGER, unpublished observations. (×2,500.)

Fig. 6. Numerous microtubules make up the preprophase band. In this example, the nucleus is still migrating into position. During cytokinesis the new cross wall will fuse with the parent wall at a point that bisects the region of the preprophase band. From HEPLER and ZEIGER, unpublished observations. (×50,000.)

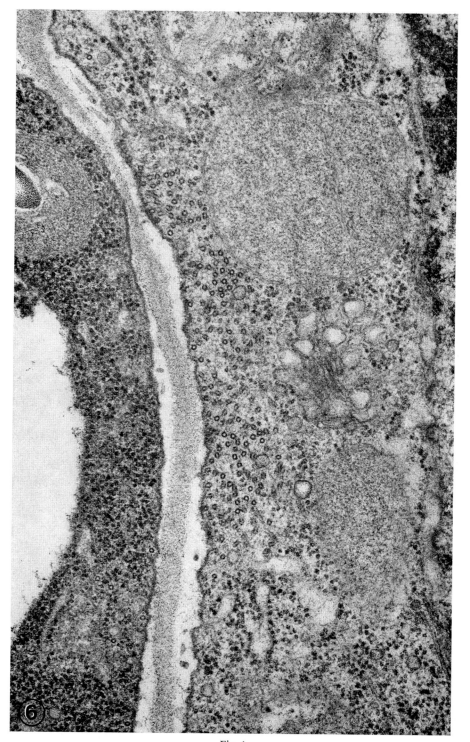

Fig. 6

Even more perplexing to understand in terms of their regulation are asymmetric division. The process is especially striking in the formation of subsidiary cells in grasses. The cell plate firstly arises in a position that is markedly displaced towards the cell pole adjacent to the guard mother cell. Subsequently, during centrifugal growth it curves towards and fuses with the parent wall across from the intervening cell just to each end of the guard mother cell (Fig. 5). The crescent-shaped cell plate thus cuts off a small lens-shaped cell.

Finally, among the early events in stomatal morphogenesis, I draw attention to the symmetrical, but longitudinal division of the guard mother cell into the guard cell pair. Because of the longitudinal orientation it was expected that the spindle apparatus, at the onset of mitosis, would form transversely across the cell. In onion the results show, however, that the spindle originally arises at some angle diverging up to 90° away from the anticipated position (Palevitz and Hepler 1974 a) (Figs. 7 a–c). This erroneous orientation is maintained well into anaphase since chromosomes often move diagonally into opposite corners of the cell. But during late anaphase, early telophase a dramatic reorientation occurs; the entire spindle apparatus including chromosomes and phragmoplast rotates and comes to rest in a position that insures an equal, longitudinal portioning of the guard mother cell (Palevitz and Hepler 1974 a) (Figs. 7 a–c).

What factors regulate and drive the rotating spindle apparatus? Drug studies reveal that the motion requires energy and that it is dependent upon the structural integrity of the spindle-phragmoplast. Colchicine, vinblastine and IPC, all of which either destroy or modify microtubules, inhibit reorientation, but they also block mitosis and cytokinesis (Palevitz and Hepler 1974 b). Cytochalasine B and phalloidin, however, appear to selectively inhibit spindle rotation and do not stop the preceding mitosis or the subsequent completion of cytokinesis or even the formation of the thickened secondary wall (Palevitz 1980 a). While considerable controversy has surrounded the mechanism of action of cytochalasin B and the efficacy of its use, recent results show that the drug has a pronounced inhibitory effect on actin filament polymerization (Lin et al. 1980). Phalloidin is also a microfilament antagonist and causes its effects through stabilization of actin filaments (Wieland 1977). That two anti-microfilament drugs specifically inhibit spindle rotation provides strong circumstantial evidence in favor of force generation through an acto-myosin system. Presumably these microfilaments reside at the edges of the cell and create shear force between the phragmoplast-spindle and the plasmalemma. Unfortunately, to date the presumptive microfilaments have not been observed; additional studies are needed to solve this intriguing morphogenetic movement.

Throughout the early events of stomatal complex formation one is struck

Figs. 7 a–c. A time lapse sequence of a guard mother cell undergoing spindle reorientation in onion. The spindle axis is over 45° off from the expected orientation (a). Through rotation of the nuclei, spindle and phragmoplast (b), the cell plate becomes longitudinally aligned (c). From Palevitz and Hepler 1974 a. (×3,000.)

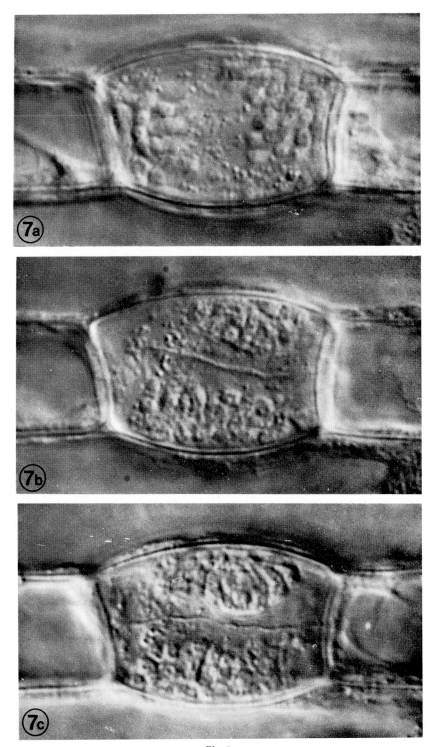

Fig. 7

by the number of instances in which the cell plate is positioned along precisely determined planes. Some of these divisions are highly asymmetrical with curved cell plates, while others are symmetric with straight longitudinal partitions. In one example (*i.e.,* onion guard mother cells) a novel motile process has been uncovered that participates in cell plate alignment (Palevitz and Hepler 1974 a). However, while we can speculate on the nature of the motile macromolecules, and their location in the cell, we still do not know how they became positioned. Several years ago the discovery of the preprophase band of microtubules by Pickett-Heaps and Northcote (1966 a, b) provided striking new evidence of a cytoplasmic differentiation that might be involved in regulating the plane of division (Fig. 6). The preprophase band consists of microtubules that are localized along the cytoplasmic cortex at the place where the cross wall will fuse with the parent wall during cytokinesis (Gunning *et al.* 1978 a, Pickett-Heaps and Northcote 1966 a, b, Pickett-Heaps 1969) (Fig. 6). These bands are seen in a great many cell types undergoing symmetrical and asymmetrical division. In *Azolla* roots an exhaustive investigation reveals that they occur without exception in all cells at the place where the new wall and parent wall fuse (Gunning *et al.* 1978 a). The perplexing aspect of these observations is the fact that the microtubules disappear by prophase long before cell plate formation begins (Pickett-Heaps and Northcote 1966 a). Nevertheless, it seems possible that the preprophase band indicates a specialization of the cell cortex, in particular the plasmalemma, at that site. In a theoretical discussion of the mechanism of spindle rotation in onion guard mother cells the idea seemed attractive to us that the preprophase band site might contain specific binding or organizing factors to which the centrifugally growing phragmoplast microtubules would preferentially attach (Palevitz and Hepler 1974 a). A slight modification of the above idea suggests that the preprophase site might itself initiate a small but important increment of centripetal phragmoplast growth. [Note, for example, the presence of wall stubs that arise despite caffeine inhibition of cell plate formation in dividing stomatal cells of wheat (Pickett-Heaps 1969)]. A cortically-positioned and membrane-anchored portion of phragmoplast could then interact laterally with the outward growing cell plate and either through microtubule cross-bridge formation or bridge-generated shearing forces bring the misaligned cell plate into correct orientation. Many aspects of the determination of the plane of cytokinesis lack explanation, but it seems evident that the plasmalemma plays a key role in controlling its spatial location. It is important now to examine the structure of the cell membrane at the preprophase band site to see if discernible specialization exists.

C. Formation of the Secondary Wall

The secondary wall thickenings of guard cells, like those of tracheary elements, are composed of highly oriented cellulose microfibrils embedded in a matrix of encrusting substances. In a guard cell pair of simple stomatal complexes, the thickening arises along both sides of the common wall that will subsequently delimit the pore (Palevitz 1980 b, Palevitz and Hepler 1976, Singh and Srivastava 1973). Initially the thickenings resemble a single pair

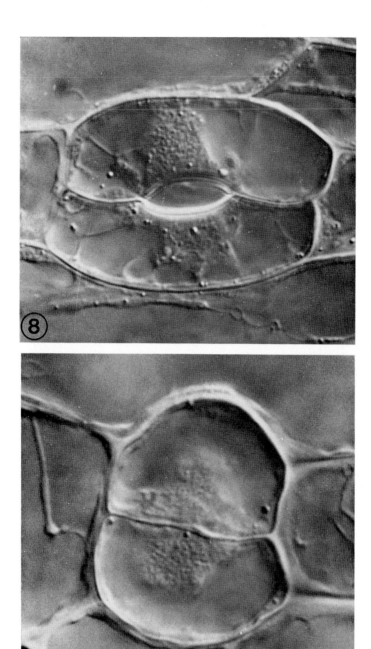

Fig. 8. Normal mature guard cells of onion. The thickened wall in the region of the pore and the radial cellulose reinforcement gives each cell its kidney shape. From PALEVITZ and HEPLER 1974 a. ($\times$2,000.)

Fig. 9. Guard cells allowed to differentiate in the presence of colchicine. The cells become swollen and polarized light microscopy reveals that the cellulose pattern is randomized. From PALEVITZ and HEPLER 1976. ($\times$2,000.)

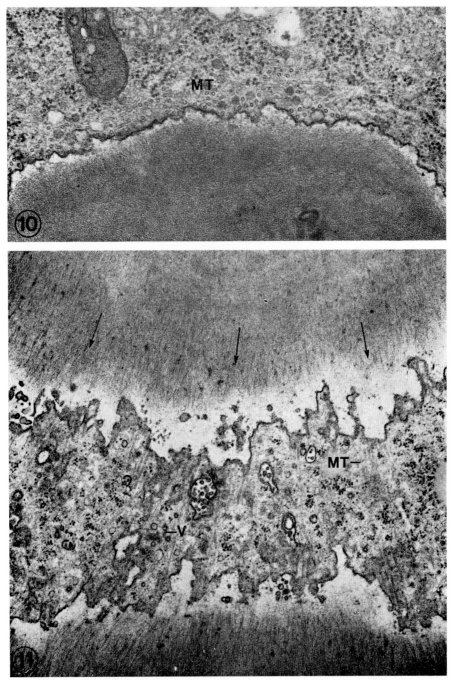

Fig. 10. Electron micrograph from a transverse section of the thickened wall of an onion guard cell. Microtubules (*MT*) are clustered over the thickening. From PALEVITZ and HEPLER 1976. (×34,000.)

Fig. 11. An electron micrograph from a tangential section of a differentiating onion guard cell shows the cortical cytoplasmic layer beneath the thickened wall. Microtubules (*MT*) are parallel to the microfibrils of the wall (arrows). Vesicles (*V*) are also observed among the microtubules. From PALEVITZ and HEPLER 1976. (×25,000.)

of young wall bands in two adjacent differentiating tracheary elements. How-ever, as development continues, the wall, while greatly enlarged at the mid-point of the cells, spreads in decreasing thickness towards the ends. Polarized light analysis of maturing cells reveals that the cellulose microfibrils, which lie parallel to one another in the central highly thickened portions of the wall, fan outward from the pore site to the sides and ends of the cell (Palevitz and Hepler 1976). During the process of wall formation the cell becomes kidney shaped (Fig. 8).

Differentiation in guard cells of grasses begins almost identically to that in onion (Kaufman et al. 1970, Palevitz 1980 b, Srivastava and Singh 1972 a). At an intermediate stage of development the walls are composed of radially oriented cellulose microfibrils and the cells are kidney shaped. This is only a transition phase and is soon followed by an elongation of the cell and constriction of the distal side wall. Concomitantly new cellulose micro-fibrils are deposited in a near axial direction to form an elongated pad of thickened wall. Cellulose microfibrils fan out over the bulbous ends of the cell, which together with the thickened, straight mid-section give the complex its "bone" shaped appearance (Palevitz 1980 b).

In either simple or complex stomatal apparatuses the structure of the wall defines the appearance of the cell and dictates the changes in shape that are permitted. Thus, once the wall has been formed, stomatal pore opening and closing can be explained on the basis of passive strain deformation in response to internally generated turgor stress (Raschke 1979).

Investigation of the cytoplasm factors responsible for generating these unique secondary wall patterns again reveals microtubules as the most likely candidate (Figs. 10 and 11). Microtubules cluster at the place of wall deposi-tion prior to thickening. They increase in number as wall deposition begins. They are parallel to the underlying cellulose microfibrils and fan outward in an identical manner (Kaufman et al. 1970, Palevitz 1980 b, Palevitz and Hepler 1976, Singh and Srivastava 1973, Srivastava and Singh 1972 a) (Fig. 11). In grasses in which there is a transition from radial cellulose align-ment to one that is near axial, the microtubules show the exact same pattern (Palevitz 1980 b). Finally, application of drugs (colchicine and IPC) that destroy or alter microtubules markedly modify secondary wall deposition. This is especially clear with colchicine which causes thickenings of abnormal shape to form and within which the cellulose microfibrils are randomized (Palevitz and Hepler 1976) (Fig. 9).

IV. How Do Microtubules Control Cellulose Orientation?

That microtubules participate in defining the pattern and orientation of cellulose in primary and secondary wall formation is firmly established in numerous, but not all, examples of higher plants (Hepler and Palevitz 1974) and algae (Quader et al. 1978). How microtubules achieve their effects is not known, but some suggestions have been made. The original idea that they control cytoplasmic streaming which in turn orients the cellulose (Ledbetter and Porter 1963) seems unlikely for several reasons. Streaming is not evident in young differentiated guard cells. More compelling are the studies on *Nitella* internode cells undergoing ordered primary wall formation which reveal that the microtubules are located next to the wall in the stationary ectoplasm and thus physically separated from the mobile endoplasm (PICKETT-HEAPS 1967 b). Furthermore, the cytoplasmic stream and the microtubules possess markedly different orientation; the streaming occurs in a longitudinal steep helical pattern while microtubules are oriented transversely, almost at right angles to the direction of streaming (HEPLER and PALEVITZ 1974, PICKETT-HEAPS 1967 b).

A modification of this idea suggests that microtubules cause a localized flow of vesicles or precursor material in the cortical cytoplasm (NORTHCOTE 1969, PICKETT-HEAPS 1968). As the vesicles fuse with the plasmalemma they would carry with them a directional component of force which might be transmitted to the membrane components themselves and thus align elongated macromolecules or growing cellulose chains. Golgi and coated vesicles, and other granules are frequently observed among the cortical microtubules overlying bands of thickened wall in tracheary elements (MAITRA and DE 1971) and guard cells (PALEVITZ and HEPLER 1976) (Fig. 11), but directed motion of these particles has not been detected. However, we must not hastily dismiss this theory. One is reminded of the linear motion of particles and states in kinetochore microtubule bundles of dividing *Haemanthus* endosperm observed with differential interference contrast microscopy several years ago by ALLEN and co-workers (1969). Only with the highest resolution optical systems and most ideal specimens are these motions demonstrated. Thus a similar motion of vesicles might occur locally in the cytoplasmic cortex that has not been detected.

A third idea suggests that microtubules directly generate a shear oriented motion of macromolecules within the plane of the plasmalemma and that this flow ultimately orients cellulose microfibrils (BROWER and HEPLER 1976, HEATH 1974, HEPLER and FOSKET 1971, HEPLER and PALEVITZ 1974). The observation of cross-bridges between microtubules and the plasmalemma led to the genesis of this idea (HEPLER and FOSKET 1971), since it seemed that these links might be analogous to dynein arms of cilia and flagella that are known to cause sliding between adjacent doublet tubules in axonemes. Even more pertinent to the argument are the recent observations of DENTLER et al. (1980) showing the presence of a dynein-ATPase bridge that binds the doublet microtubules to the plasma membrane in cilia. Because ciliary beating can be arrested by a photoactivated agent that cross-links membrane proteins it appears that axonemal bending requires, in part, a dynein interaction with mobile proteins in the plasmalemma. When these results are coupled to the earlier investigations of BLOODGOOD (1977), revealing rapid, linear saltations of attached particles along the outer flagellum surface, the concept of a microtubule-membrane, dynein-ATPase, capable of generating flow within and through the membrane gains considerable credence (DENTLER 1980). It is alternatively possible that force is actually generated by a membrane-based but microtubule-oriented actomyosin system. However, in this instance it is worth noting that cytochalasin B and phalloidin, which inhibit spindle rotation in onion guard mother cells, exerts no influence on the subsequent oriented deposition of cellulose microfibrils in the thickened secondary wall (PALEVITZ 1980 a). Regardless of the mechanism of force production, the theory suggests that macromolecular complexes, conceivably the cellulose synthetases themselves (HEATH 1974, HEPLER and PALEVITZ 1974), are moved in a direction parallel to the microtubules. Any long, asymmetric molecule such as a cellulose microfibril, even though it might have no direct attachment to the microtubule itself, because of the oriented shear forces would become aligned parallel to the microtubules.

There is no experimental finding from studies of plants to support this model, but there is evidence from investigations of animal cells that show the cooperative involvement of the motile macromolecules in the movement of lectin binding sites (EDELMAN 1976). Glycoproteins thus are moved laterally within the plane of the plasmalemma by a system of microtubules and microfilaments, and it seems plausible that a similar process occurs in plants and might be linked to the ordered deposition of cellulose.

Even if we explain how microtubules control the oriented deposition of cellulose we are left with vexing questions of how the microtubules themselves become aligned. Recent studies by GUNNING et al. (1978 b) on *Azolla* roots have begun to unravel this problem as it relates to primary wall formation. They find aster-like clusters of microtubules that are associated with small vesicles and embedded in dense amorphous material along the edges of cells where two walls meet and suggest that there are microtubule organizing centers. Microtubules appear to grow out from these focal points and become situated along the faces of the wall. Similar aggregates or organizing centers have been noted in differentiating guard cells of *Phleum* (PALEVITZ 1978). In both examples the question of how the microtubules are properly aligned is not answered, although the presence of discrete organizing centers provides an important new basis for experimentation and observation.

V. Membranes and Morphogenesis

Microtubules, since their discovery in the plant cell cortex 18 years ago, have been the single most important evidence of cytoplasmic structural specialization coupled to cell wall formation. However, they stand to one side of the cell membrane while the cellulose microfibrils, which they influence, are on the other. As the interface between microtubules and cell wall, the plasmalemma becomes the most important but least understood intermediary in the process of wall formation. Recent freeze-fracture studies have revealed ordered rosettes of particles on the cell membrane overlying secondary wall in the desmid *Micrasterias* (KIERMAYER and SLEYTR 1979, GIDDINGS et al. 1980; see p. 160 f.) and in corn and mung bean cells (MUELLER and BROWN 1980). Similar rosettes occur in *Vicia* guard cell protoplasts (SCHNABL et al. 1980). These are exciting new findings and have led to a model in which the rosettes are viewed as synthetases that are spinning out a growing cellulose chain (GIDDINGS et al. 1980).

Notwithstanding the importance and elegance of the current freeze-fracture observations, we are just beginning to elucidate structural differentiation within the plasmalemma as it relates to wall formation. We have alluded to the possible localization of microtubule organizing or stabilizing factors such as the preprophase band site that may reside in defined regions of the membrane and may control the placement of the cross wall during cytokinesis. In this instance the plasmalemma must respond to signals from neighbouring cells in order to correctly position the postulated morphogenetic determinants. Here then is one example of the array of information that the plasmalemma of a cell in a complex tissue receives and processes into decisions having positional components. The signals (ion fluxes, hormones, turgor stress), the morphogenetic determinants (synthetases, microtubules, microfilaments) that respond to these signals, and the mechanisms that cause a specific spatial distribution of determinants (membrane flow) appear to be parts of the system that create pattern. Further characterization of these parts and in particular how they work together may bring us closer to an understanding of morpho-genesis.

Together with the cell membrane we are increasingly coming to realize that motility may play a central role in morphogenesis. The most plausible theories of cellulose orientation, for example, involve directed flow of macromolecules within the plane of the membrane. We also view the cell membrane as the surface against which shearing forces act to cause spindle rotation in guard mother cells of *Allium* (Palevitz and Hepler 1974 a) and possibly even nuclear migration in many different species. Microtubules are closely positioned, even cross-bridged to the plasmalemma and it seems reasonable to expect that actomyosin filaments will also be found here. The plasmalemma, in turn, could reasonably be expected to regulate the formation and/or activation of these motile macromolecules through the local control of calcium ion concentration.

If the cell membrane serves such crucial functions in morphogenesis how can it be investigated when it is covered by a thickened wall. For certain investigations protoplasts could be used and the recent study applying fluorescently labelled antibodies to map out microtubule distributions (Lloyd et al. 1980) dramatically emphasizes the efficacy of this approach. But even if the wall is present it should be possible to capitalize upon its specific pattern (e.g., in tracheary elements) and use it as an indicator for the position of selectively differentiated regions on the membrane. Through comparison of the alteration in membrane structure with concomitant changes in microtubules and cellulose microfibrils during development of normal and experimentally perturbed systems, important clues may be learned about the function of the plasmalemma in morphogenesis.

Acknowledgements

I thank Dr. Barry A. Palevitz for sharing with me his recent work on stomatal differentiation in grasses. I also thank Mr. Dale A. Callaham for technical assistance in the preparation of this chapter. This work has been supported by grants from the National Science Foundation and the National Institutes of Health (R 01-GM 25120).

References

ALLEN, R. D., BAJER, A., LA FOUNTAIN, J., 1969: Poleward migration of particles or states in spindle fiber filaments during mitosis in *Haemanthus*. J. Cell Biol. **43**, 4 a.

BLOODGOOD, R. A., 1977: Motility occurring in association with the surface of the *Chlamydomonas* flagellum. J. Cell Biol. **75**, 983—989.

BROWER, D. L., HEPLER, P. K., 1976: Microtubules and secondary wall deposition in xylem: the effects of isopropyl N-phenylcarbamate. Protoplasma **87**, 91—111.

CRONSHAW, J., BOUCK, G. B., 1965: The fine structure of differentiating xylem elements. J. Cell Biol. **24**, 415—431.

DENTLER, W. L., 1981: Microtubule-membrane interactions in cilia and flagella. Inter. Rev. Cytol. **72**, 1—47.

— PRATT, M. M., STEPHENS, R. E., 1980: Microtubule-membrane interactions in cilia. II. Photochemical cross-linking of bridge structures and the identification of a membrane-associated dynein-like ATPase. J. Cell Biol. **84**, 381—403.

EDELMAN, G., 1976: Surface modulation in cell recognition and cell growth. Science **192**, 218—226.

ESAU, K., 1953: Plant anatomy, p. 735. New York: J. Wiley.

— CHEADLE, V. I., GILL, R. H., 1966 a: Cytology of differentiating tracheary elements. I. Organelles and membrane systems. Amer. J. Bot. **53**, 756—764.

— — — 1966 b: Cytology of differentiating tracheary elements. II. Structures associated with cell surfaces. Amer. J. Bot. **53**, 765—771.

GIDDINGS, T. H., jr., BROWER, D. L., STAEHELIN, L. A., 1980: Visualization of particle complexes in the plasma membrane of *Micrasterias denticulata* associated with the formation of cellulose fibrils in primary and secondary cell walls. J. Cell Biol. **84**, 327—339.

GOOSEN-DE ROO, L., 1973 a: The relationship between cell organelles and cell wall thickenings in primary tracheary elements of the cucumber. II. Quantitative aspects. Acta Bot. Neerl. **22**, 301—320.

— 1973 b: The fine structure of the protoplast in primary tracheary elements of the cucumber after plasmolysis. Acta Bot. Neerl. **22**, 467—485.

GREEN, P. B., 1980: Organogenesis—a biophysical view. Ann. Rev. Plant Physiol. **31**, 51—82.

GUNNING, B. E. S., HARDHAM, A. R., HUGHES, J. E., 1978 a: Preprophase bands of microtubules in all categories of formative and proliferative cell division in *Azolla* roots. Planta **143**, 145—160.

— — — 1978 b: Evidence for initiation of microtubules in discrete regions of the cell cortex in *Azolla* root-tip cells, and an hypothesis on the development of cortical arrays of microtubules. Planta **143**, 161—179.

HARDHAM, A. R., GUNNING, B. E. S., 1978: Structure of cortical microtubule arrays in plant cells. J. Cell Biol. **77**, 14—34.

— — 1979: Interpolation of microtubules into cortical arrays during cell elongation and differentiation in roots of *Azolla pinnata*. J. Cell Sci. **37**, 411—442.

— — 1980: Some effects of colchicine on microtubules and cell division in roots of *Azolla pinnata*. Protoplasma **102**, 31—51.

HEATH, I. B., 1974: A unified hypothesis for the role of membrane bound enzyme complexes and microtubules in plant cell wall synthesis. J. Theor. Biol. **48**, 445—449.

HEPLER, P. K., FOSKET, D. E., 1971: The role of microtubules in vessel member differentiation in *Coleus*. Protoplasma **72**, 213—236.

— — NEWCOMB, E. H., 1970: Lignification during secondary wall formation in *Coleus*: an electron microscopic study. Amer. J. Bot. **57**, 85—96.

— NEWCOMB, E. H., 1963: The fine structure of young tracheary xylem elements arising by redifferentiation of parenchyma in wounded *Coleus* stem. J. Exp. Bot. **14**, 496—503.

— — 1964: Microtubules and fibrils in the cytoplasm of *Coleus* cells undergoing secondary wall deposition. J. Cell Biol. **20**, 529—533.

Hepler, P. K., Palevitz, B. A., 1974: Microtubules and microfilaments. Ann. Rev. Plant Physiol. **25**, 309—362.

— Rice, R. M., Terranova, W. A., 1972: Cytochemical localization of peroxidase activity in wound vessel members of *Coleus*. Can. J. Bot. **50**, 977—983.

Kaufman, P. B., Petering, L. B., Yocum, C. S., Baic, D., 1970: Ultrastructural studies on stomatal development in internodes of *Avena sativa*. Amer. J. Bot. **57**, 33—49.

Kiermayer, O., Sleytr, U. B., 1979: Hexagonally ordered "rosettes" of particles in the plasma membrane of *Micrasterias denticulata* Bréb. and their significance for microfibril formation and orientation. Protoplasma **101**, 133—138.

Ledbetter, M., Porter, K. R., 1963: A "microtubule" in plant cell fine structure. J. Cell Biol. **19**, 239—250.

Lin, D. C., Tobin, K. D., Grumet, M., Lin, S., 1980: Cytochalasins inhibit nuclei-induced actin polymerization by blocking filament elongation. J. Cell Biol. **84**, 455—460.

Lloyd, C. W., Slabas, A. R., Powell, A. J., Lowe, S. B., 1980: Microtubules, protoplasts and plant cell shape. Planta **147**, 500—506.

Maitra, S. C., De, D. N., 1971: Role of microtubules in secondary thickening of differentiating xylem element. J. Ultrastruct. Res. **34**, 15—22.

Mueller, S. C., Brown, R. M., jr., 1980: Evidence for an intramembrane component associated with a cellulose microfibril-synthesizing complex in higher plants. J. Cell Biol. **84**, 315—326.

Northcote, D. H., 1969: Fine structure of cytoplasm in relation to synthesis and secretion in plant cells. Proc. Roy. Soc. **B 173**, 21—30.

Palevitz, B. A., 1978: Cortical microtubules in plant cells: a high voltage electron microscope (HVEM) study. J. Cell Biol. **79**, 278 a.

— 1980 a: Comparative effects of phalloidin and cytochalasin B on motility and morphogenesis in *Allium*. Can. J. Bot. **58**, 773—785.

— 1980 b: The structure and development of stomatal cells. In: Stomatal physiology (Jarvis, P. G., Mansfield, T. A., eds.). Soc. Exp. Biol. Sem. Ser. Cambridge University Press, Cambridge, U.K.

— Hepler, P. K., 1974 a: The control of the plane of division during stomatal differentiation in *Allium*. I. Spindle reorientation. Chromosoma **46**, 297—326.

— — 1974 b: The control of the plane of division during stomatal differentiation in *Allium*. II. Drug studies. Chromosoma **46**, 327—341.

— — 1976: Cellulose microfibril orientation and cell shaping in developing guard cells of *Allium*: the role of microtubules and ion accumulation. Planta **132**, 71—93.

Pickett-Heaps, J. D., 1967 a: The effects of colchicine on the ultrastructure of dividing plant cells, xylem wall differentiation, and distribution of cytoplasmic microtubules. Dev. Biol. **15**, 206—236.

— 1967 b: Ultrastructure and differentiation in *Chara* sp. I. Vegetative cells. Aust. J. Biol. Sci. **20**, 539—551.

— 1968: Xylem wall deposition. Radioautographic investigations using lignin precursors. Protoplasma **65**, 181—205.

— 1969: Preprophase microtubule bands in some abnormal mitotic cells of wheat. J. Cell Sci. **4**, 397—420.

— Northcote, D. H., 1966 a: Organization of microtubules and endoplasmic reticulum during mitosis and cytokinesis in wheat meristems. J. Cell Sci. **1**, 109—120.

— — 1966 b: Cell division in the formation of the stomatal complex of the young leaves of wheat. J. Cell Sci. **1**, 121—128.

— — 1966 c: Relationship of cellular organelles to the formation and development of the plant cell wall. J. Exp. Bot. **17**, 20—26.

Quader, H., Wagenbreth, I., Robinson, D. G., 1978: Structure, synthesis and orientation of microfibrils. V. On the recovery of *Oocystis solitaria* from microtubule inhibitor treatment. Cytobiologie **18**, 39—51.

RASCHKE, K., 1979: Movements of stomata. In: Physiology of movements (HAUPT, W., FEINLEIB, M. E., eds.), pp. 383—441. (Encycl. Plant Physiol., New Series, Vol. 7.) Berlin-Heidelberg-New York: Springer.

SCHNABL, H., VIENKEN, J., ZIMMERMAN, U., 1980: Regular arrays of intramembranous particles in the plasmalemma of guard cell and mesophyll cell protoplasts of *Vicia faba*. Planta **148**, 231—237.

SINGH, A. P., SRIVASTAVA, L, M., 1973: The fine structure of pea stomata. Protoplasma **76**, 61—82.

SINNOTT, E. W., BLOCH, R., 1945: The cytoplasmic basis of intracellular patterns in vascular differentiation. Amer. J. Bot. **32**, 151—157.

SRIVASTAVA, L. M., SINGH, A. P., 1972 a: Stomatal structure in corn leaves. J. Ultrastruct. Res. **39**, 345—363.

— — 1972 b: Certain aspects of xylem differentiation in corn. Can. J. Bot. **50**, 1795—1804.

WIELAND, T., 1977: Modification of actin by phallotoxins. Naturwiss. **64**, 303—309.

WOODING, F. B. P., NORTHCOTE, D. H., 1964: The development of the secondary wall of the xylem in *Acer pseudoplatanus*. J. Cell Biol. **23**, 327—337.

ZEIGER, E., 1971: Cell kinetics, development of stomata and some effects of colchicine in barley. Planta **99**, 89—111.

Sporogenesis and Pollen Grain Formation

Brigitte Buchen and A. Sievers

Botanical Institute, University of Bonn, Bonn, Federal Republic of Germany

With 9 Figures

Contents

Abbreviations

CW callosic spore mother cell wall, *D* dictyosome, *ER* endoplasmic reticulum, *GV* Golgi vesicle, *N* nucleus, *NE* nexine, PA-TCH-SP periodic acid-thiocarbohydrazide-silver proteinate treatment for the localization of acidic and neutral polysaccharides (Thiéry 1967), *PM* plasma membrane, *S* septum, *SE* sexine, *SMC* spore mother cell, *SP* sporoderm, *W* wart.

I. Introduction

Pollen grains and the spores of nonseed vascular plants [1] develop from diploid somatic cells in a largely identical way. Ultimately, the cells are characterized by a highly ordered cell wall, the so-called sporoderm (Figs. 1 and 2). Both the chemical and the principal ultrastructural features of sporoderms show little variation in different taxa, although the sculpture, number, and thickness of the individual wall layers display species-specificity. Fig. 3 shows a schematic cross section of a sporoderm and the surface view of a tectum, using in part the terminology proposed by Erdtman (1952).

---

[1] Referred to collectively as "spores" in this chapter.

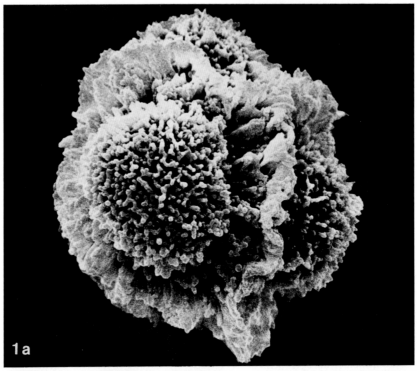

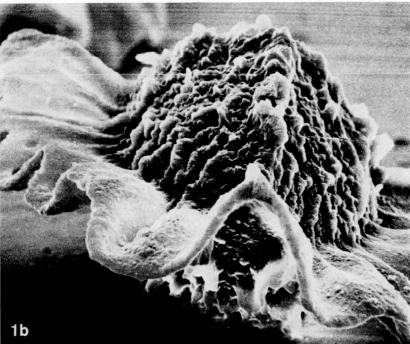

Figs. 1 *a* and *b*. Scanning electron micrographs of young *Selaginella caulescens* megaspores. Sculpture of (*a*) the tetrad and (*b*) a single spore (a ×1,750, b ×2,790).

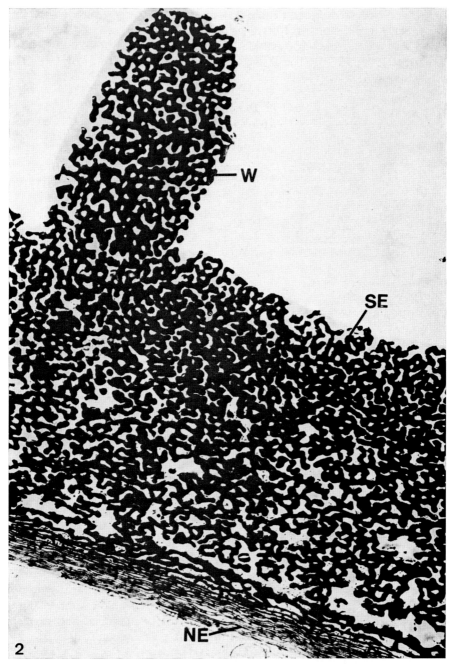

Fig. 2. Cross-section of the sporoderm of a growing *Selaginella helvetica* megaspore with lamellated nexine (*NE*) and spongy sexine (*SE*). After conventional fixation and contrasting (glutaraldehyde/OsO$_4$), the sporoderm appears electron dense. *W* wart (×7,150).

The intine is the innermost layer of the sporoderm (Fig. 3). This layer consists of polysaccharides (Table 1)—namely pectins, hemicelluloses, cellulose (Sitte 1953, Roland 1971, Hara et al. 1977), and callose (Waterkeyn 1964, Albertini and Souvré 1978), together with proteins (Heslop-Harrison 1975, 1979). The intine has, therefore, the features of a primary cell wall.

Table 1. *Development of spores*

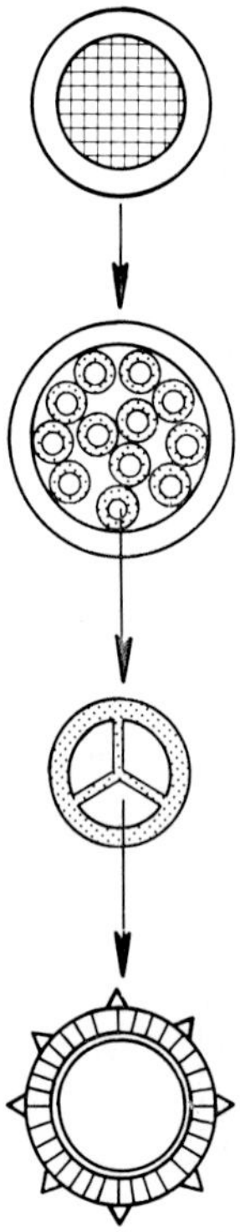

Cell types in the sporangium	Cell wall	
	Types	Chemical compounds
cells of sporo-genous tissue	primary wall	pectins, hemicelluloses, cellulose
spore mother cells (meiocytes)	spore mother cell wall	outer part: as above inner part: callose (plus acidic polysaccharides)
meiospores and tetrads	as above, plus septa plus primexine early exine	pectins, hemicelluloses, callose matrix: pectins, hemicelluloses, cellulose pattern: lipoproteins (?) protosporopollenin

free spores	sporoderm:
	  exine
	  intine
		sporopollenin
		pectins, hemicelluloses, cellulose, callose

The substance characteristic of the exine (Figs. 1–2, and 3) is sporopollenin, a highly polymeric lipid consisting of carotins and carotinoid esters (Shaw 1971, Southworth 1974). It is resistant to microbial and chemical degradation. Recent analyses demonstrate that proteins and polysaccharides are also located in the exine (Heslop-Harrison 1975, Pettitt 1979, Vithanage and Knox 1979). The surface of the exine often is covered by "Pollenkitt" consisting of lipoids and proteins.

The aim of this article is to summarize the results concerning the morphogenesis of sporoderms beginning with the premeiotic development of the sporogenous tissue and ending with the formation of the intine (Table 1). First, we shall deal with the cytoplasmic structures involved in synthesis, secretion, and deposition of the different cell wall materials. Second, we shall consider the influence of the surrounding tissue on sporoderm development. In this context, one should ask whether the haploid spore itself or the diploid cells of the sporangium control the morphogenetic processes (see also reviews by HESLOP-HARRISON 1971 a–c, ECHLIN 1971 a, b, DICKINSON 1976 a).

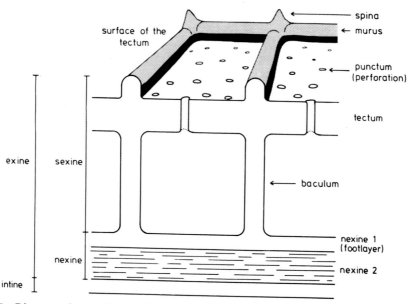

Fig. 3. Diagram of sporoderm architecture representing a surface view of the tectum and a cross-section.

II. Development of Spores

A. Spore Mother Cells Preceding Meiosis

The sporangia are differentiated into parietal layers, tapetal cells, and central sporogenous tissue. The ultrastructural organization of the sporogenous cells, including the presence of plasmodesmata in their walls, shows no special features when compared with other meristematic cells (HORNER and LERSTEN 1971, GENEVÈS 1974, HORNER 1977, BUCHEN and SIEVERS 1978 a, DICKINSON and POTTER 1978).

Before the beginning of meiosis, the spore mother cell (SMC) produces a characteristic inner wall layer containing callose (special mother cell wall, FITTING 1900, BEER 1911; Fig. 4). Callose (MANGIN 1890) mainly consists of β-1,3-glucan (KESSLER 1958) and may also contain β-1,4-glucosidic linkages (HERTH et al. 1974, VITHANAGE et al. 1980). Callose is found in the walls of nearly all micro- and mega-SMCs (see, however, Riccardia, HORNER et al. 1966; Lophocolea, OLTMANN 1974; Podocarpus, VASIL and ALDRICH 1970;

some ferns, WATERKEYN and BIENFAIT 1971 a, SHEFFIELD and BELL 1979; *Oenothera-mega*-SMC, JALOUZOT 1970; *Gasteria-mega*-SMC, WILLEMSE and FRANSSEN-VERHEIJEN 1978). For histochemical identification of callose, the aniline blue induced fluoroscence is used (ESCHRICH and CURRIER 1964, REYNOLDS and DASHEK 1976, SMITH and McCULLY 1978, WATERKEYN 1981). At present, however, on the ultrastructural level, no distinct test exists for proving callose. Callose is transparent in the electron microscope, as is shown, *e.g.*, in sieve tubes (CRONSHAW and ANDERSON 1969, CATESSON 1973), in pollen tubes (CRESTI and WENT 1976), and in SMCs (HESLOP-HARRISON 1966, ECHLIN and GODWIN 1968 b, CHRISTENSEN *et al.* 1972). In some species, however, callose of the SMC-wall is layered by or associated with other polysaccharides (PETTITT 1971 b, WATERKEYN and BIENFAIT 1972, ALBERTINI and SOUVRÉ 1978, BUCHEN and SIEVERS 1978 a, Table 1; Fig. 4).

The beginning and time course of callose deposition vary in different plants. Normally, its production starts in the prophase of meiosis I, and the SMC-wall is completed before meiosis II begins (HESLOP-HARRISON 1966). In most SMCs, callose is uniformly deposited over the whole cell surface. The situation is more complicated in the case of many monocytoledons and during *mega*-sporogenesis of vascular plants (RODKIEWICZ 1970, CHRISTENSEN *et al.* 1972, CHRISTENSEN and HORNER 1974, RODKIEWICZ and BEDNARA 1976, NOHER DE HALAC and HARTE 1977, KAPIL and TIWARI 1978 b, RUSSELL 1979, SCOLES and EVANS 1979, KAPIL and BHATNAGAR 1981).

In *Sorghum* anthers, the sporogenous tissue consists of a single layer of cells, each of which has contact with the tapetum (CHRISTENSEN and HORNER 1974). In this case, callose formation starts at the inner periclinal walls of the sporogenous cells where a greater amount of callose is produced than later on at the radial and outer periclinal walls. This polarity in callose deposition is important for the morphogenetic steps of spore development discussed later (CHRISTENSEN and HORNER 1974). A similar polarization was observed during the monoaxial development of, *e.g.*, *Oenothera*- (NOHER DE HALAC and HARTE 1977) and *Zea-mega*spores (RUSSELL 1979), the mega-sporocytes of which also produce asymmetrical callose layers. The functional *mega*spore originates at that place where the SMC-wall has the lowest amount of callose.

The question arises: how are the substances of the SMC-wall secreted? Further, is the polar deposition of callose an effect of the structural polarity of the callose producing cells? Concerning the formation of callose, two different mechanisms can be postulated as with the formation of all cell wall components. First, synthesis of callose may occur inside the cell. As a consequence it must be secreted. Second, callose may be formed outside or at the outer face of the plasmalemma. Vacuoles of unknown origin (ESCHRICH 1964) and Golgi vesicles (HESLOP-HARRISON 1966, GABARA 1971, ECHLIN and GODWIN 1968 b, p. 175, however, see p. 177) are thought to be likely candidates for the synthesis and export of callose (see also, in the case of pollen tube development, CRESTI *et al.* 1977). However, since an increase in dictyosome activity at the time of the appearance of the callose and the elec-tron transparency of both callose and vesicle content are the only grounds for

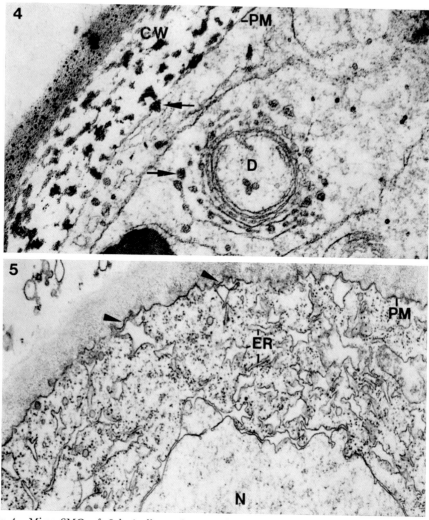

Fig. 4. Micro-SMC of *Selaginella caulescens* after treatment with PA-TCH-SP (periodic acid-thiocarbohydrazide-silver proteinate). The outer wall layer (primary wall) and the particles in the callose layer of the SMC wall (*CW*) react positively. The particles containing pectins and hemicelluloses are secreted by Golgi vesicles (arrows). *D* dictyosome; *PM* plasma membrane (×36,000).

Fig. 5. *Mega*-SMC of *Selaginella caulescens* during callose formation. Numerous rER cisternae (ER) in the peripheral cytoplasm are located near to the plasma membrane (arrow heads). *N* nucleus; *PM* plasma membrane (×30,000).

this assumption, we should conclude that at present there exists no proof for the function of the Golgi apparatus in callose formation (see also CHRISTENSEN and HORNER 1974). In *Selaginella*, only distinct particles in the callose layer are secreted via Golgi vesicles (BUCHEN and SIEVERS 1978 a; Fig. 4). They contain pectins and hemicelluloses.

During deposition of callose, a remarkable increase of ER is often observed in SMCs (Fig. 5). It is interesting to see that at this stage of development both the ER and the plasma membrane are difficult to preserve during the fixation procedure (SKVARLA and LARSON 1966, ANGOLD 1967, ECHLIN and GODWIN 1968 b; cf., Fig. 15 in DICKINSON and POTTER 1978, the peripheral cytoplasm of the *Lilium-mega*-SMC shows dilatations of the ER cisternae). This may be interpreted as an indication of a high transport activity of both membrane types. In the *mega*-SMC of *Zea*, ER cisternae are located mainly at that pole where callose is deposited (RUSSELL 1979). Participation of the ER and the plasmalemma in the formation of callose is also assumed for sieve tubes (CATESSON 1973) and pollen tubes (CRESTI and WENT 1976). The specific role of the ER in this process, however, is not yet clear. A secretion of callose by ER derived vesicles and a transport of enzymes or precursor molecules to the plasmalemma have been discussed. However, clear proof of the fusion of an ER membrane with the plasmalemma is still lacking (CATESSON 1973, BUCHEN and SIEVERS 1978 a; Fig. 5). Since in all cases callose has been identified only outside the plasma membrane and neither in the ground cytoplasm nor in other intracellular compartments, it can be concluded that the plasma membrane itself regulates the polarity of the cell and thereby also regulates the asymmetrical callose formation.

B. Formation of Septa After Meiosis

In Pteridophyta and Angiospermae, during the prophase of meiosis, the plasmodesmata develop into large channels up to 1.5 μm in diameter (WEILING 1965, LUGARDON 1968, RISUENO et al. 1969, WHELAN 1974, SHEFFIELD and BELL 1979). It has been assumed that these plasmatic connections are a precondition for the synchronization of the cellular activities during meiosis (HESLOP-HARRISON 1966, 1971 a). Before meiosis II, the channels are plugged by additional callose. Thus, the meiocytes are isolated during the completion of their karyokinesis.

During meiosis, the cytoplasmic elements are partly degraded, and subsequently the haploid cells are reorganized (DICKINSON and HESLOP-HARRISON 1977, DICKINSON and POTTER 1978, PETTITT 1978). The karyokinesis is followed by cytokinesis, which produces the four haploid spores. Formation and spatial orientation of the septa are important in spore morphogenesis. By incomplete septation in *Eleocharis*, e.g., only one viable spore develops from the tetrad (DUNBAR 1973). Normally, in monocotyledons the spores are formed by successive cell divisions. After meiosis I, a dyad originates, and by the following division after meiosis II the tetrad is formed. On the other hand, dicotyledons divide simultaneously by septation after meiosis II (HESLOP-HARRISON 1971 c). In each case, the septa originate in the equatorial planes of the phragmoplasts. The type of cytokinesis establishes the configuration of spores in the tetrad and the axes of symmetry. Hence, the final geometry of the spore, including the arrangement of pores, is determined by the orientation of the spindle microtubules (DRAHOWZAL 1936, HESLOP-HARRISON 1971 a, c). It is not clear whether during meiosis a preprophase

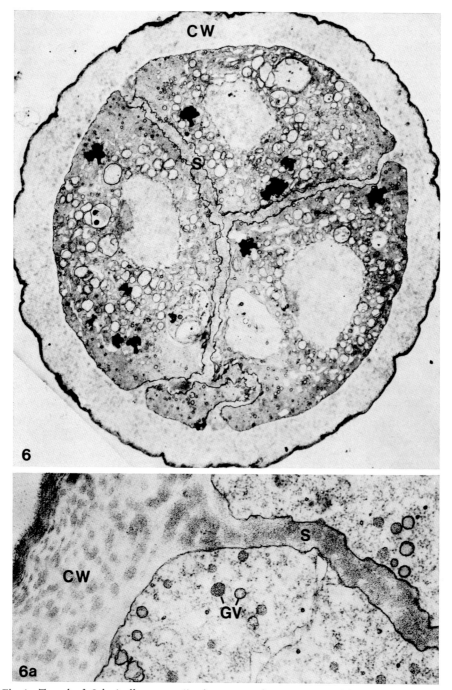

Fig. 6. Tetrad of *Selaginella martensii* microspores after treatment with PA-TCH-SP. The septa (S) react positively. CW SMC wall (×6,750); *a* detail of the identical reaction of the particles in the SMC wall (CW), of the septa (S), and of the Golgi vesicle content (GV) to the test for acidic and neutral polysaccharides (*PA-TCH-SP*) (×30,000).

band of microtubules exists that, as in cytokinesis after mitosis, indicates the future fusion zone of daughter and mother cell walls (PICKETT-HEAPS and NORTHCOTE 1966, BURGESS 1970, BROWN and LEMMON 1980 a).

Like cell plates between diploid nuclei (reviews: SCHNEPF 1969, FREY-WYSSLING 1976), the septa originate centrifugally or centripetally in the phragmoplast, characterized by parallel arrays of microtubules and ER tubules. The septa are composed of pectins, hemicelluloses (GENEVÈS 1972, AUDRAN 1974, ALBERTINI and SOUVRÉ 1978, BUCHEN and SIEVERS 1978 a; Fig. 6) and callose (HESLOP-HARRISON 1971 a, ROWLEY and DAHL 1977, BIDDLE 1979; Table 1).

The secretion of the material of the septa has rarely been studied in detail. Vesicles of various origins (WILLEMSE 1971, DUNBAR 1973, AUDRAN 1974, GENEVÈS 1974) are reported to form the septa. In cases with septa similar to primary walls, the function of the Golgi apparatus is clearly established (Fig. 6 a; BUCHEN and SIEVERS 1978 a); however, it is not clear whether this holds for septa consisting only of callose (ECHLIN and GODWIN 1968 b, GABARA 1971). In Selaginella, the number and diameter of the Golgi vesicles and of the Golgi cisternae are nearly identical with those from dictyosomes in the sporogenous tissue during rapid cell wall formation. However, they are different from the inactive ones in the phase of callose deposition (see p. 355). The Golgi vesicles contain acidic polysaccharides; callose is also present in the septa.

The causal relationship between the arrangement of microtubules and ER with the process of septum formation has not been elucidated. Treatments of Lilium meiocytes with colchicine inhibits the formation of the phragmoplast microtubules. During meiosis I, the cells are especially sensitive and show anomalies in cell cleavage (HESLOP-HARRISON 1971 c). The incompletely divided or undivided cells of such syncytia are still capable of forming the exine (see also Eleocharis, DUNBAR 1973). This shows that secretion and deposition of sporopollenin does not depend on the microtubule system. The sites of apertures, however, and thus the polar organization of the spores, are disturbed.

C. Formation of the Sporoderm in the Tetrad

Sporoderm formation begins as soon as the spores are isolated by their special callosic walls (Fig. 7). In the case of megaspore development in Zea, a thin osmiophilic layer around the functional megaspore (RUSSELL 1979) may be a relict of sporoderm formation as seen in microsporogenesis.

The exine, or at least the preceding "primexine" (HESLOP-HARRISON 1963) with its species-specific pattern, originates within the SMC-wall. The early extrusion of sporoderm substances and the expression of its characteristic pattern is determined by the spore itself (BEER 1911, SIEVERS and BUCHEN 1970, DICKINSON 1971, HESLOP-HARRISON 1971 a). The callose is considered to be a physiological barrier for various substances (KNOX and HESLOP-HARRISON 1970, SOUTHWORTH 1971). However, the concept of callose as a molecular filter allowing only small molecules to pass through (HESLOP-HARRISON and MACKENZIE 1967, RODKIEWICZ 1970) is probably too simple (MASCARENHAS 1975) as a selective passage of substances is demonstrated that cannot be explained only by molecular weight.

It is confirmed that the steps in sporoderm formation shown for *Silene* (HESLOP-HARRISON 1963, scheme, HESLOP-HARRISON 1971 a) are also valid for further Spermatophyta (*Zea*, SKVARLA and LARSON 1966; *Endymion*,

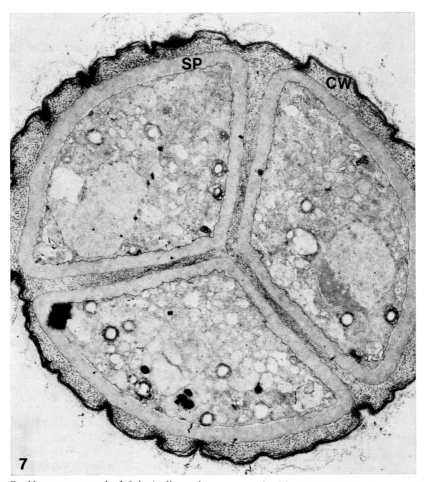

Fig. 7. *Megaspore tetrad of Selaginella caulescens* treated with PA-TCH-SP. The sporoderm (*SP*) is deposited between the SMC wall (*CW*) and the plasma membrane. The sporoderm reacts negatively to the polysaccharide test (its low contrast results from the treatment with periodic acid during the PA-TCH-SP procedure) (×8,000).

ANGOLD 1967; *Helleborus*, ECHLIN and GODWIN 1968 b; *Lilium*, HESLOP-HARRISON 1968; *Pinus*, DICKINSON 1971; *Citrus*, HORNER and LERSTEN 1971; *Eleocharis*, DUNBAR 1973; *Sorghum*, CHRISTENSEN et al. 1972; *Ribes*, GENEVÈS 1974; *Helianthus*, HORNER and PEARSON 1978). Divergent results are reported for lower plants (PETTITT 1971 b, DENIZOT 1976, BROWN and LEMMON 1980 b). A first step in the developmental process is the production of a fibrous layer (primexine matrix) between the SMC-wall and the plasma membrane. The matrix consists of acidic and neutral polysaccharides (WATERKEYN and BIENFAIT 1970, GENEVÈS 1974, Table 1) and is secreted by Golgi vesicles (HESLOP-HARRISON 1971 a, WILLEMSE 1971, ROLAND-HEYDACKER 1979). HESLOP-HARRISON (1968) and DICKINSON (1970) suppose that the matrix contains cellulose.

Modifications of this primexine formation are reported in *Onagraceae* and *Asteraceae*, which form "rod-shaped units" or "tufts" of mucopolysaccharides and proteins (glycocalyx, ROWLEY 1973, 1976, ROWLEY and DAHL 1977). Following detachment from the plasma membrane by the production of additional tufts, a honeycomb-like pattern is constructed in *Artemisia*. The tufts are connected by an electron dense fibrillar network of neutral polysaccharides. The consequence of irregularities in this pattern formed by locally increased numbers of tufts is the production of wart-like structures and surface sculptures. In *Canna*, the glycocalyx gives rise to a three-dimensional network that reflects the shape of the exine (ROWLEY and SKVARLA 1975).

In the region of the future germination pore, the primexine is thin or absent (see, however, *Epilobium*, ROWLEY 1976). There is evidence that the secretion of the matrix in this special area is blocked by the apposition of ER cisternae to the plasma membrane (HESLOP-HARRISON 1963, ECHLIN and GODWIN 1968 b, CHRISTENSEN and HORNER 1974, HORNER and PEARSON 1978). Similarly, exocytosis of Golgi vesicles is probably also prevented by the close association of ER and plasmalemma during formation of the pores in angiosperm sieve tubes (ESAU et al. 1962, NORTHCOTE and WOODING 1966, CATESSON 1973). Special vacuoles containing crystallites of $BaSO_4$ inhibit a local exocytosis of Golgi vesicles in growing *Chara* rhizoids, too (SIEVERS and SCHRÖTER 1971). Experiments with centrifugation of *Lilium* pollen during the early tetrad phase can alter the distribution of pores (HESLOP-HARRISON 1971 c).

Subsequently, patterns are moulded in the primexine matrix. Observations on the first initiation of species-specific patterning are divers.

In *Pinus*, the primexine matrix forms a surface relief reported to be shaped by a spatial differential exocytosis of Golgi vesicles (DICKINSON 1971, WILLEMSE 1971). According to DICKINSON (1971, 1976 a, b), the Golgi vesicles first contact the plasma membrane. Thereafter, both the plasma membrane and the distal parts of the vesicle membranes become "diffuse and more electron opaque" (DICKINSON 1976 a). The proximal parts of the vesicle membranes, however, are said to fuse and to form finally a new plasmalemma. The opinion that the limiting membranes of adjacent Golgi vesicles may break and portions fuse to form a new membrane is speculative. The assumption that a cell replaces its original plasma membrane by a new internal membrane formed in this way cannot be considered as proved. According to WILLEMSE (1971), the typical relief of the *Pinus* primexine is achieved by local incorporation of Golgi vesicles alternating with deposition of callose.

The regular polygonal exine pattern of *Ipomoea* (GODWIN et al. 1967) demonstrates the cooperation of the callose envelope and the primexine matrix during patterning (WATERKEYN and BIENFAIT 1970). The callose layer thereby serves as a template for the primexine matrix, which builds a hexagonal network. The center of each mesh is occupied by a callose knob. In the corners and at the adjacent margins of the primexine-hexagons are holes in which the first exine elements originate: the probacula and the spines. In *Eleocharis* both local intrusions of callose and plasma membrane evaginations into the primexine matrix are assumed to play a role in exine patterning (DUNBAR 1973).

After conventional procedures of fixation and contrasting, numerous species show electron-dense globuli or radial columns (probacula) in the primexine matrix. The ends of the probacula often fuse laterally to build a protectum and/or a footlayer (Fig. 3, HESLOP-HARRISON 1963, ECHLIN and GODWIN 1968 b, HORNER and PEARSON 1978). Early in development, the probacula are not yet resistant to acetolysis, the standard test for sporopollenin (DICKINSON 1976 b). In *Lilium* they become resistant as soon as the complete future exine ornament is formed (HESLOP-HARRISON 1971 c). In *Pinus* the deposition of an electron-dense layer begins at the interfaces between primexine matrix and callose. This layer is acetolysis-resistant and becomes the surface of the exine (DICKINSON 1971). The primexine, therefore, already contains sporopollenin or precursors ("protosporopollenin", HESLOP-HARRISON 1968).

The formation of the probacula and the specific lipid accumulation are central problems in sporogenesis. The chemical nature of the probacula is not yet clearly known and probably changes during development (HESLOP-HARRISON 1971 a, DICKINSON 1976 b). At the beginning, the elements show a medium electron contrast that increases with time (ECHLIN and GODWIN 1968 b, HESLOP-HARRISON 1968, CHRISTENSEN and HORNER 1974). The probacula of *Ribes* and *Mahonia* stay electron transparent after treatments specific for polysaccharides, whereas the surrounding matrix fibrils react positively (GENEVÈS 1974, ROLAND-HEYDACKER 1979). The increasing affinity to osmium suggests an increasing content in lipids. Possibly, these substances are protosporopollenin, as indicated by the acetolysis resistance. This idea is strengthened by investigations with fluorescence microscopy (WATERKEYN and BIENFAIT 1971 b) that show sporopollenin-specific fluorescence with primuline already at the stage of probacula in *Lilium*. It remains to be clarified, however, to what extent protosporopollenin differs from sporopollenin in the mature exine. Differences in their contrasting and staining behavior hint that they are not identical substances (DICKINSON 1976 b). But it is true that lipophilic material is secreted by the spore still enclosed in the SMC-wall (MEPHAM 1970, HESLOP-HARRISON 1971 a, DICKINSON 1976 b, BUCHEN and SIEVERS 1978 b; Figs. 7 and 8).

The question now is how the position of the probacula is determined and how the lipophilic substances penetrate them. The spatial arrangement of the bacula is supposed to be determined by ER (HESLOP-HARRISON 1963, SKVARLA and LARSON 1966, GABARA 1974, DICKINSON and POTTER 1976), by microtubuli (ECHLIN and GODWIN 1968 b, DICKINSON and POTTER 1976), or by mitochondria (Vazart 1970). In *Lilium,* "irregular protrusions arise from the plasma membrane" (Dickinson 1970). They form groups of tubuli each of which is surrounded by a membrane. Later in development, the tubular groups appear as radial lamellae in the probacula. The origin of these lamellae is not clear (Dickinson 1976 b). In the region of the probacula, accumulations of ribosomes are found directly below the plasmalemma. Golgi vesicles are also supposed to play a role in the formation of the muri. In other plants, however, no cytoplasmic structures could be related to the arrangement of probacula (Dunbar 1973, Christensen and Horner 1974, Horner and Pearson 1978). These examples show that a high degree of uncertainty exists concerning the participation of cytoplasmic structures in determining the arrangement of probacula.

In order to clarify the problem, *Lilium* pollen in different developmental stages was centrifuged or incubated with colchicine (Heslop-Harrison 1971 c). The geometrical pattern was altered by treatments during the early development, *i.e.,* before the probacula become impregnated with sporopollenin. However, since the intracellular order of the organelles was completely disarranged by centrifugation, it remains open as to which subcellular structure is involved in probacula orientation. The influence of colchicine demonstrates that microtubules play a role in the process, but it is not known in which precise way.

The lipophilic substances accumulate either at the sculptured boundaries between the primexine-matrix and the SMC-wall (*e.g., Pinus*), in the matrix papillae (*Lophocolea*), or in the probacula (*e.g., Lilium*). In some liverworts, the first exine elements are seen at the inner face of the SMC-wall but directly outside the plasma membrane (Denizot 1976). They become isolated from the plasma membrane by the successive deposition of polysaccharides (callose). Thus, the whole centripetal exine formation appears to be similar to that of the nexine in angiosperms (Brown and Lemmon 1980 b; see also, Southworth 1974).

How are lipid substances extruded? The secretion of plant lipids (oils, cutin, suberin, etc.) probably follows eccrine routes (Schnepf 1969). The monomeric precursors may pass through the plasma membrane and polymerize in the extracellular space. Occasionally, sporoderm lipids are said to be secreted by Golgi- (Horner et al. 1966, Skvarla and Larson 1966, Dunbar 1973) or ER-vesicles (Dickinson 1976 b). The number of ER cisternae often is increased during lipid secretion, thus indicating that the ER participates in lipid biosynthesis. However, the data do not support a granulocrine secretion of lipids.

Most investigators do not see any association of subcellular structures and lipid secretion. However, there are many examples in which an obvious increase in thickness and contrast of the plasma membrane has been observed (Echlin and Godwin 1969, Dickinson 1970, Robert 1971; see Figs. 39–43 in Christensen and Horner 1974). In *Selaginella* the plasma membrane is asymmetrical from the onset of the lipid secretion. The outer electron dense leaflet is at least twice as thick as the inner one (Fig. 9). Thus, a functional correlation may exist between the asymmetry of the plasma

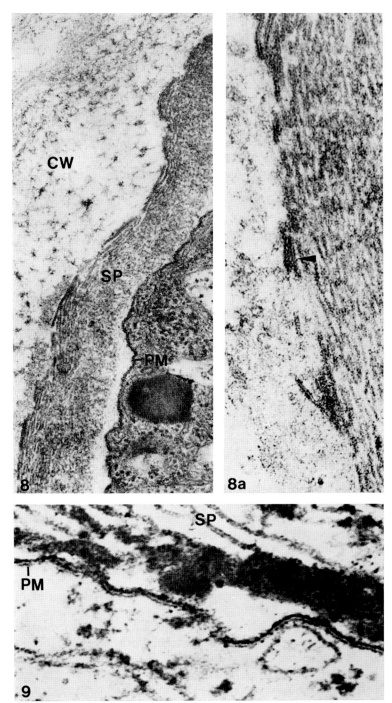

Fig. 8. Periphery of a young tetrad-*mega*spore (*Selaginella caulescens*) with early developmental stages of sporoderm (*SP*). *CW* SMC wall; *PM* plasma membrane (×72,000). *a* Detail of a young sporoderm. Tripartite lamellae (arrow head) (×168,000).

Fig. 9. Detail of the peripheral cytoplasm of a growing *mega*spore (*Selaginella caulescens*) after dissolution of the SMC wall. During the secretion of sporoderm lipids, the plasma membrane (*PM*) is asymmetrical. *SP* sporoderm (×168,000).

membrane and lipid secretion (Buchen and Sievers 1978 b). Similar phenomena have been observed during the differentiation of oil bodies in *Marchantia* (Galatis *et al.* 1978).

The molecular extrusion is followed by polymerization. Various models have been formulated as to how the accumulation of sporopollenin can occur in highly ordered patterns:

Model 1. The lipids appear as characteristic tripartite lamellae arranged parallel to the plasma membrane, which are, in cross sections, similar to unit membranes (Fig. 8 a; "white lines": Rowley and Southworth 1967, Dickinson 1970, 1976 b, Atkinson *et al.* 1972, Rowley and Dahl 1977, Brown and Lemmon 1980 b). The thickness, including the electron translucent central zone is 7 to 10 nm (Fig. 8 *a*). Possibly these lamellae are special surfaces on which the polymerization takes place. This is indicated by the increasing thickness and contrast of the outer surface of the lamellae (see, *e.g.*, Figs. 39 and 40 in Christensen and Horner 1974). Sometimes such tripartite lamellae occur within the sexine only during the early phase of lipid secretion.

The origin and chemical nature of the lamellae are unclear. It is quite clear, however, that they are not biomembranes *sensu stricto*. Such unit membrane-like structures can also be found in other organisms, for example the cell wall of the gram-negative bacteria is tripartite. It is often thought that the plasma membrane participates in the formation of the tripartite lamellae (Rowley and Dunbar 1967, Denizot 1976, Dickinson 1976 b) although other theories have been proposed (*e.g.*, extracellular de novo synthesis: Rowley and Dunbar 1967). They are regarded as lipoproteins (Heslop-Harrison 1971 a). The outer surfaces react positively for proteins and mucopolysaccharides (Rowley and Dahl 1977) and have affinity for silver as lipid droplets have (Buchen and Sievers 1978 b). The inner central space shows no reaction in these cytochemical methods.

According to Rowley and Dahl (1977), the lamellae could originate from the agglutination of two filaments of polyanions. Bridging by basic molecules could effect the parallel orientation of these filaments, whereas their fusion would be inhibited by the negative charges. Even after destruction of sporopollenin, the filaments are still visible (Sengupta and Rowley 1974). The alterations of the filaments in lamellae and their increasing contrast might be understandable as a consequence of further association of different molecules.

Model 2. A second mode of ordered lipid polymerization might involve proteins and/or glycoproteins in cooperation with physical forces at the interfaces of callose and primexine matrix. At these boundaries, physicochemical interactions may well exist that are different from those between molecules of a more uniform matrix. The lipid monomers might accumulate earlier and polymerize to a greater extent at these sites (low permeability of callose?). Thus, a pattern of spore wall is initiated and then preserved with a continuous addition of sporopollenin monomers. Such a step-by-step polymerization is supported by data of *Pinus* (Dickinson 1971, Willemse 1971) and *Eleocharis* (Dunbar 1973). Dickinson (1976 b) assumes that in *Pinus* the polymers are deposited on special membranes derived from Golgi vesicles (*cf.*, p. 360). In *Eleocharis,* the sculptured exine develops only around the whole tetrad, that is in regions where the primexine matrix has been established beneath the callose wall (DUNBAR 1973). Lipids are also extruded into the callose-free septa but form only globuli lacking a specific pattern. Relevant also is a *Glycine* mutant with coenocytic microspores that are occasionally connected by small sporoderm bridges (ALBERTSEN and PALMER 1979). These originate where individual SMCs have been attached to each other by callose bars. This is a further indication of the role of the callose cover in development of primexine and exine.

Model 3. A type of lipid polymerization similar to theory 2 can be deduced from observations in *Epilobium* (ROWLEY 1973), *Canna* (ROWLEY and SKVARLA 1975), *Cosmos* (DICKINSON and POTTER 1976), and *Artemisia* (ROWLEY and DAHL 1977). The lipids accumulate in the gaps or free spaces in the primexine matrix. The *Epilobium* primexine consists of rod-shaped mucopolysaccharides. Lipidic material polymerizes at the surfaces of these rods. Thus, the exine is constructed with uniformly large channels consisting of the rod-like mucopolysaccharides surrounded by sporopollenin. The ramified system of bacula is formed by filling the cavities between them. In *Artemisia* the protectum tufts are impregnated with lipids (ROWLEY and DAHL 1977). Wherever fibrils of neutral polysaccharides have delimited the tufts in the primexine, microcapillaries remain in the exine. In this way, a discontinuous deposition of sporopollenin and a pattern is achieved. Even in *Sorghum,* small knobs and discontinuities in the primexine matrix can be recognized (see Figs. 19 and 20 in CHRISTENSEN *et al.* 1972). Such weak irregularities may already suffice to direct local accumulation of lipid precursors and preserve channels in the exine. The primexine matrix in the region of the germination pore is secreted somewhat later than in the other cell surface. It is rather thin at its margin and thicker and more irregularly shaped in the middle region. At the margin, only a thin layer of lipids is deposited, but the center of the pore (operculum) acquires a pattern similar to the rest of the sexine. As pointed out (see p. 360), *Ipomoea* produces in the exine a replica of its primexine matrix.

It should be emphasized that all these ideas of how an ordered polymerization of sporopollenin precursors is achieved are speculative. It is by no means clear in which way proteins, polysaccharides, and lipids cooperate or how this complex system may be regulated. The above mentioned models may give indications of possible mechanisms but also show that different reaction pathways may be involved that do not exclude each other. As with many other patternings, space and time are decisive factors.

D. Growth of the Sporoderm in the Free Spore Phase

The lysis of callose surrounding the tetrad spores begins at different developmental stages: in *Sorghum* when a distinct exine pattern is not yet visible (CHRISTENSEN *et al.* 1972); in *Tradescantia,* however, after the bulk of sporopollenin is deposited (MEPHAM 1970). In general, only the pattern of the sexine and nexine I (Fig. 3) with small lipid content is developed when callose is digested (HESLOP-HARRISON 1971 a, DICKINSON 1976 b). For the further development of the sexine, monomers from the locular fluid are required. In *Selaginella mega*spores, for instance, the total volume of the exine increases by a factor of 1.5×10^4 from the tetrad to the mature stage (BUCHEN and SIEVERS 1978 b); despite a large increase in *Lilium* spore volume, the thickness of the exine remains constant (WILLEMSE and REZNICKOVA 1980). The substances needed for this kind of growth cannot be produced by the spore cytoplasm alone.

Even though the period of nexine formation and the number and structure of the nexine layers vary, some general construction principles can be described. The nexine originates by deposition of sporopollenin onto lamellae that usually run parallel to the cell surface (*cf.*, p. 364; ECHLIN and GODWIN 1969, HESLOP-HARRISON 1971 a, DICKINSON 1976 a, HORNER and PEARSON 1978). At the beginning of the lipid accumulation, the light central zone of the lamellae is visible but often becomes masked by additional lipid deposition (HESLOP-HARRISON 1968, CHRISTENSEN and HORNER 1974, ROWLEY and DAHL 1977). The lamellae then appear homogeneous, and by fusing they form a network around the spore (SIEVERS and BUCHEN 1970). In *Sorghum* a nonlamellar amorphous inner exine layer is produced, except at the region of the pores (CHRISTENSEN *et al.* 1972); a situation similar to the innermost part of the nexine of *Podocarpus* (VASIL and ALDRICH 1970).

During this period, the central vacuole of the spore develops. The turgor pressure effects the increase in spore volume. Thus, the shape of the cell changes and the nexine lamellae are pressed against each other. The materials for the nexine and the intine are synthesized by the spore cytoplasm (HESLOP-HARRISON 1971 a, 1975). The sexine thickening is made by incorporation of lipid monomers into the already existing elements. This polymerization may be a process in which the cytoplasm is not directly involved (SIEVERS and BUCHEN 1970, BROWN and LEMMON 1980 b). The exine structures are also stretched by the increase in cell volume, but the pattern is preserved by continuous enlargement of the single elements.

The intine develops between plasma membrane and nexine. The chemical properties of the intine are similar to a normal primary cell wall (Table 1). The Golgi apparatus participates in its secretion (HESLOP-HARRISON 1975), but the different types of vesicles cannot, as yet, be correlated to the secretion of distinct wall substances. The low number of microtubules found in the spore cytoplasm during intine formation corresponds to the low degree of order in the distribution of cellulose fibrils (HESLOP-HARRISON 1971 a).

Near the aperture, the intine is usually thicker than in other regions (in *Artemisia*, about tenfold, ROWLEY and DAHL 1977). Tubular structures are remarkable elements that may occur in the whole intine (*Canna*) but are often accumulated near the apertures (HESLOP-HARRISON 1975). They originate as protrusions of the plasma membrane resembling microvilli. During further wall growth, they become detached from the plasmalemma with the membranes being degraded (ROWLEY and DAHL 1977, HESLOP-HARRISON 1979). Proteins have been demonstrated in these tubular structures (see Figs. 7 and 8 in HESLOP-HARRISON 1975).

The factors that cause the asymmetrical thickening of the intine and the accumulation of proteins near the germination pores are still unknown. Participation of cytoplasmic structures in the establishment of this polarity has not been demonstrated. In other cases, microtubules are involved in the development of local cell wall thickenings (SCHNEPF 1973).

III. The Role of the Tapetum in Sporogenesis

The tapetum is considered to have a nutritive function for the maturing, heterotrophic spores (ECHLIN 1971 a; concerning *mega*sporogenesis of higher plants and the integumentary tapetum, see KAPIL and TIWARI 1978 a).

During the development of SMCs, the tapetum differentiates into a secretion tissue (HOEFERT 1971, DICKINSON and BELL 1976, STEER 1977 b). The secretory function of the tapetum has been documented (SAUTER 1970, MASCARENHAS 1975), but the role of the different organelles in these processes is not yet clarified (STEER 1977 a, b). Before meiosis, the cellular contacts between SMCs and tapetum mediated by plasmodesmata are interrupted. As a consequence, the nutrients can be transported only via the apoplast with subsequent passage through the spore plasmalemma. After callose dissolution, the transport of substances to the spore cell is facilitated by channels and holes in the maturing exine (ROWLEY 1976).

The question arises as to the significance of the tapetum secretory products for sporogenesis. Besides soluble substances in the loculus, there are lipophilic electron dense particles (Ubisch bodies or orbicules, ECHLIN and GODWIN 1968 a, ECHLIN 1971 a) outside the plasma membrane of the tapetum cells. They appear at the beginning of sporoderm formation, preferably at the inner tangential walls. In *Poaceae, Cyperaceae,* and *Asteraceae,* they form a complete orbicular wall (peritapetal wall) around the loculus. This wall is acetolysis resistant, indicating sporopollenin; its ultrastructure is similar to the sporoderm (HESLOP-HARRISON 1969, ECHLIN 1971 a, WATERKEYN and BIENFAIT 1971 b, CHRISTENSEN et al. 1972, DUNBAR 1973). The function of the orbicules is still unknown. It is clear that the precursors first polymerize in the extraplasmic space (ECHLIN 1971 a, DICKINSON 1976 b). The suggestion that these polymers contribute to the exine growth probably is derived from artefacts. They do not have direct influence on the sporoderm development and neither mould nor thicken the exine.

Despite the lack of precise knowledge about the soluble secretory products of the tapetum, one can conclude from data obtained with *Sorghum* (CHRISTENSEN and HORNER 1974), *Oenothera* and *Zea* (NOHER DE HALAC and HARTE 1977, RUSSELL 1979, see p. 354) that a physiological gradient arises in the spore cells by the uptake of these substances. In *Sorghum* an area of the pollen grain wall remains in contact with the tapetum until the accumulation of reserve material begins (CHRISTENSEN and HORNER 1974). The germination pores develop in these pollen sides facing the tapetum. It cannot be excluded that the formerly established gradient in supply is maintained and results in a sequence of "polar events" (*e.g.,* thickening of the intine, storage of reserve material). In the *mega*spore development, a similar gradient possibly determines the position and, thereby, the morphogenesis of the functional *mega*spore (NOHER DE HALAC and HARTE 1977). Plasmodesmata in the cell wall between the functional *mega*spore and the tapetum as well as callose barriers around the nonfunctional ones are the structural elements that are the basis of this assumption. The functional *mega*spore originates where the supply with nutrients is greater. In heterosporous ferns in which only one *mega*spore tetrad develops from many sporogenous cells, spatial and physiological correlations of the tapetum and of this viable SMC/tetrad have not been observed (PETTITT 1971 a).

Furthermore, the sporogenesis is influenced to a great extent by the secretion of callase from the tapetum. Callase liberates the spores from the tetrad by degradation of callose (MEPHAM and LANE 1969, ECHLIN 1971 a). STIEGLITZ (1977) found two enzymes with different molecular weights in *Lilium* whose activity maxima coincide exactly with the time of callose dissolution. Exact timing of the secretion of the callase and the activation of the enzyme are critical for the ordered development of spores. Alterations in any of these steps often result in lethality (see IV.).

IV. Sterility of Pollen and Sporogenesis

Pollen abortion can begin at various phases of spore development. In a genic male sterile (gms) mutant of *Glycine*, sporogenesis is identical to that of fertile spores except that cytokinesis does not occur (ALBERTSEN and PALMER 1979). The resulting coenocytic tetrad with four nuclei forms a species specific pattern of the exine. The number of the pores and their regular arrangement, however, may be disturbed. This disturbance has often been described as a consequence of a false or omitted cell division (see p. 356).

In a *Zea*-gms-mutant (CHENG et al. 1979) and in cytoplasmic male sterile (cms) mutants of other species, pollen sterility is caused by a non- or malfunctioning tapetum (review: LASER and LERSTEN 1972; NANDA and GUPTA 1974, HORNER 1977, SCOLES and EVANS 1979). Fertile pollen is produced only if the correct development of the tapetum is coordinated in time with spore development. A premature dissolution of callose can depend on hypertrophy of the tapetum cells or a noncoordinated time course of callase activity (*Petunia*, IZHAR and FRANKEL 1971; *Sorghum*, WARMKE and OVERMAN 1972; *Allium*, NANDA and GUPTA 1974; *Zea*, LEE et al. 1979). In case the tapetum degenerates too early, callose may not be digested (*Capsicum*, HORNER and ROGERS 1974) or nutrients necessary for the growth of the spores cannot be synthesized (*Helianthus*, HORNER 1977; *Secale*, SCOLES and EVANS 1979). In *Helianthus*-cms-plants, the degeneration of the tapetum starts during meiosis. The tetrad spores develop a primexine and a rudimentary exine (as in *Capsicum*) before they abort. This suggests that the tapetum does not influence initiation and moulding of the sporoderm. During growth and maturation of the pollen grains, however, callose has to be removed to allow entry of the nutrients to the spores. After dissolution of callose, the spores still degenerate if the tapetum does not supply the essential substances (*Helianthus*, HORNER 1977; *Secale*, SCOLES and EVANS 1979).

Intermediate wheatgrass plants derived from wheat $\times$ wheatgrass hybrids (YOUNG *et al.* 1979) produce sterile pollen. The sterile pollen usually exhibits a malformed exine in which channels do not form a continuous system through the sporoderm or in which the operculum in the centre of the aperture is continuous with the exine. Assuming that specific transport routes in the sporoderm for nutrients are a condition for normal development, the sterility may be caused by such malformations of the exine. It is significant that in a hybrid of wheatgrass, the orbicular wall around the tapetum is thick and exine-like, whereas the microspore exine consists only of scattered orbicules (YOUNG *et al.* 1979). Thus, both spore and tapetum synthesize sporopollenin and are able to form a structurally similar cell wall. This strengthens the assumption that in the tapetum and in the spore cells the lipidic material is synthesized, secreted and polymerized by the same basic mechanisms. In addition, this also demonstrates that in normal development the genetic information for the construction of an exine is expressed by the spore cell and not by the tapetum.

From the foregoing we may make these conclusions:

1. The tapetum strongly influences sporogenesis by the secretion of callase.

2. The pattern formation of the sporoderm is independent of the tapetum.

3. Growth and maturation of the spore depend on nutrients supplied by the tapetum.

4. Channels or holes in the exine are usually a prerequisite for growth and maturation of the spores.

5. Both the tapetum and the spore itself have the same capacity to form structurally similar cell walls of sporopollenin.

V. Control of Sporogenesis

The shape of the spores and the pattern of the sporoderm are influenced by the general organization of the sporangium. This is clearly shown in *Poaceae* in which the asymmetrical callose formation already points to the polarity in the sporangium and in SMCs (see pp. 354 and 367). After successive cytokinesis, the septa of the tetrad run perpendicularly to the surface of the tapetum, thus ordering the spores into a ring-like plane. Their germination pores lie in a constant spatial relation to the cleavage-axes (DOVER 1972) and to the inner surface of the tapetum (CHRISTENSEN and HORNER 1974, SCOLES and EVANS 1979). The tetrads of dicotyledons normally have a tetrahedral configuration of the spores (HESLOP-HARRISON 1971 c). Although in these cases the pores are also arranged with a constant pattern involving both position and number, it is not possible for each spore to have a pore in contact with the tapetum because of both the geometry of the tetrad and the organization of the sporangium (see p. 354 one layered sporogenous tissue).

The primary cause of the early initiation of polarity, thus of the symmetry of spores, is unknown. This is also the case for other examples of polarity.

The participation of ER cisternae in the actual arrangement of the pores and of the spindle microtubuli in the orientation of the cell plates only transfer the problem of causality "one stage further back in the sequence of events" (Dover 1972).

An interesting hypothesis suggests the development of polarity resulting from physical stress (Lintilhac 1974 a, b). Accordingly, cell plates originate in areas of the cell with zero stress and perpendicular to the zones with greatest stress. In *Zea* the functional *mega*spore and the future embryo sack are reported to develop in such zero stress areas. The zero stress zones have been deduced from the position of the cell plates in the nucellus (Russell 1979). Investigations on cells with tip growth have shown that wall growth is maximal at the point of lowest stress (Hejnowicz *et al.* 1977). Different stress situations could influence the geometry of the sporangium and the symmetry of the tetrad spores and could be responsible for the asymmetrical wall formations (callose) and the orientation of the spindle microtubules.

Certainly, the pattern of the exine is determined very early in sporogenesis. In *Lilium*, experimental inhibition of cell cleavage in late prophase by colchicine shows that the patterning is independent of nuclear behavior. However, because of the missing spindle poles, the pores are either not initiated or appear with random orientation (Heslop-Harrison 1971 c).

According to Dover (1972), in *Triticum* the position of the spindle and, in correlation with this, the position of the pores depend on "cytoplasmic determinants" that, he suggests, determine SMC polarity during the premeiotic interphase. If colchicine is applied at this stage, poreless monads develop. However, if the polarity is induced just at the beginning of prophase, such that the axis of the spindle apparatus is determined, then after colchicine application, the spindle formation is completely suppressed and monads with four randomly oriented pores develop. These experiments demonstrate that prevention of the development of the spindle apparatus at this stage no longer influences the number of pores but does influence their position.

Thus, the genetic information for wall patterning is already transcribed in the premeiotic nucleus and transmitted to the cytoplasm (Dover 1972, Young *et al.* 1979). Later on, the physiological state of the nucleus is of no importance for the sporoderm formation. This seems to be supported by investigations with *Lilium* hybrids that have an incomplete set of chromosomes which form a normal exine (Rogers and Harris 1969). Stronger support of this idea is provided by experiments with centrifugation. The resulting cell fragments—cells with fragments of nuclei and cells with increased numbers of nuclei—form an exine with a normal pattern (Heslop-Harrison 1971 c). The amount of spore cytoplasm seems to influence the thickness of the exine (Rogers and Harris 1969, Dover 1972). Even if the genetic information is transferred early into the spore cytoplasm, it remains to be expressed by the cytoplasmic structures. If these are damaged or impeded in their functions, then this information is translated in an imperfect or anomalous fashion.

Acknowledgements

We are grateful to Professor P. R. Bell, London, and to Professor J. R. Rowley, Stockholm, for critical reading of the manuscript. Help in translation by Drs. G. E. F. Scherer and T. J. Buckhout, Bonn, is greatly appreciated.

References

Albertini, L., Souvré, A., 1978: Les polysaccharides des microsporocytes et du tapis chez le *Rhoeo discolor* Hance. Étude cytochimique et autoradiographique (glucose ³H). Bull. Soc. bot. Fr. **125**, 45—50.

Albertsen, M. C., Palmer, R. G., 1979: A comparative light- and electron-microscopic study of microsporogenesis in male sterile (MS₁) and male fertile soybeans [*Glycine max*. (L.) Merr.]. Amer. J. Bot. **66**, 253—265.

Angold, R. E., 1967: The ontogeny and fine structure of the pollen grain of *Endymion non-scriptus*. Rev. Palaeobotan. Palynol. **3**, 205—212.

Atkinson, jr., A. W., Gunning, B. E. S., John, P. C. L., 1972: Sporopollenin in the cell wall of *Chlorella* and other algae: Ultrastructure, chemistry, and incorporation of ¹⁴C-acetate, studied in synchronous cultures. Planta **107**, 1—32.

Audran, J.-C., 1974: Aspects ultrastructuraux de l'individualisation des microspores du *Ceratozamia mexicana* (Cyadacées). C. R. Acad. Sci. Paris D **278**, 1023—1027.

Beer, R., 1911: Studies in spore development. Ann. Bot. **25**, 199—214.

Biddle, J. A., 1979: Anther and pollen development in garden pea and cultivated lentil. Can. J. Bot. **57**, 1883—1900.

Brown, R. C., Lemmon, B. E., 1980 a: Ultrastructure of sporogenesis in a moss, *Ditrichum pallidum*. I. Meiotic prophase. The Bryologist **83**, 137—152.

— — 1980 b: Ultrastructure of sporogenesis in a moss, *Ditrichum pallidum*. III. Spore wall formation. Amer. J. Bot. **67**, 918—934.

Buchen, B., Sievers, A., 1978 a: Megasporogenese von *Selaginella*. I. Ultrastrukturelle und cytochemische Untersuchungen zur Sekretion von Polysacchariden. Protoplasma (Wien) **96**, 293—317.

— — 1978 b: Megasporogenese von *Selaginella*. II. Ultrastrukturelle und cytochemische Untersuchungen zur Sekretion von Lipiden. Protoplasma (Wien) **96**, 319—328.

Burgess, J., 1970: Microtubules and cell division in the microspore of *Dactylorchis fuschii*. Protoplasma (Wien) **69**, 253—264.

Catesson, A.-M., 1973: Observations cytochimiques sur les tubes criblés de chelques Angiospermes. J. Microscopie (Paris) **16**, 95—104.

Cheng, P. G., Greyson, R. I., Walden, D. B., 1979: Comparison of anther development in genic male-sterile (ms 10) and in male-fertile corn (*Zea mays*) from light microscopy and scanning electron microscopy. Can. J. Bot. **57**, 578—596.

Christensen, J. E., Horner, jr., H. T., 1974: Pollen pore development and its spatial orientation during microsporogenesis in the grass *Sorghum bicolor*. Amer. J. Bot. **61**, 604—623.

— — Lersten, N. R., 1972: Pollen wall and tapetal orbicular wall development in *Sorghum bicolor* (*Gramineae*). Amer. J. Bot. **59**, 43—58.

Cresti, M., van Went, J. L., 1976: Callose deposition and plug formation in *Petunia* pollen tubes *in situ*. Planta **133**, 35—40.

— Pacini, E., Ciampolini, F., Sarfatti, G., 1977: Germination and early tube development *in vitro* of *Lycopersicum peruvianum* pollen: ultrastructural features. Planta **136**, 239—247.

Cronshaw, J., Anderson, R., 1969: Sieve plate pores of *Nicotiana*. J. Ultrastruct. Res. **27**, 134—148.

Denizot, J., 1976: Remarques sur l'édification des différentes couches de la paroi sporale à exine lamellaire de quelques *Marchantiales* et *Sphaerocarpales*. In: The evolutionary significance of the exine (Ferguson, I. K., Muller, J., eds.), pp. 201—210. London: Academic Press.

DICKINSON, H. D., 1970: Ultrastructural aspects of primexine formation in the microspore tetrad of *Lilium longiflorum*. Cytobiologie 1, 437—449.

— 1971: Development of pollen in *Pinus*. In: Sporopollenin (BROOKS, J., GRANT, P. R., MUIR, M., VAN GIJZEL, P., SHAW, G., eds.), pp. 31—65. London: Academic Press.

— 1976 a: Common factors in exine deposition. In: The evolutionary significance of the exine (FERGUSON, I. K., MULLER, J., eds.), pp. 67—89. London: Academic Press.

— 1976 b: The deposition of acetolysis-resistant polymers during the formation of pollen. Pollen et Spores 18, 321—334.

— BELL, P. R., 1976: Development of the tapetum in *Pinus banksiana* preceding sporogenesis. Ann. Bot. 40, 103—113.

— POTTER, U., 1976: The development of patterning in the alveolar sexine of *Cosmos bipannatus*. New Phytol. 76, 543—550.

— — 1978: Cytoplasmic changes accompanying the female meiosis in *Lilium longiflorum* Thunb. J. Cell Sci. 29, 147—169.

— HESLOP-HARRISON, J., 1977: Ribosomes, membranes and organelles during meiosis in angiosperms. Phil. Trans. R. Soc. Lond. B 277, 327—342.

DOVER, G. A., 1972: The organization and polarity of pollen mother cells of *Triticum aestivum*. J. Cell Sci. 11, 699—711.

DUNBAR, A., 1973: Pollen development in the *Eleocharis palustris* group (*Cyperaceae*). I. Ultrastructure and ontogeny. Bot. Notiser 126, 197—254.

DRAHOWZAL, G., 1936: Beiträge zur Morphologie und Entwicklungsgeschichte der Pollenkörner. Österr. Bot. Z. 85, 241—269.

ECHLIN, P., 1971 a: The role of the tapetum during microsporogenesis of angiosperms. In: Pollen: Development and physiology (HESLOP-HARRISON, J., ed.), pp. 41—61. London: Butterworths.

— 1971 b: Production of sporopollenin by the tapetum. In: Sporopollenin (BROOKS, J., GRANT, P. R., MUIR, M., VAN GIJZEL, P., SHAW, G., eds.), pp. 220—247. London: Academic Press.

— GODWIN, H., 1968 a: The ultrastructure and ontogeny of pollen in *Helleborus foetidus* L. I. The development of the tapetum and Ubisch bodies. J. Cell Sci. 3, 161—174.

— — 1968 b: The ultrastructure and ontogeny of pollen in *Helleborus foetidus* L. II. Pollen grain development through the callose special wall stage. J. Cell Sci. 3. 175—186.

— — 1969: The ultrastructure and ontogeny of pollen in *Helleborus foetidus* L. III. The formation of the pollen grain wall. J. Cell Sci. 5, 459—479.

ERDTMAN, G., 1952: Pollen morphology and plant taxonomy. Angiosperms. Stockholm: Almquist & Wiksell.

ESAU, K., CHEADLE, V. I., RISLEY, E. B., 1962: Development of sieve-plate pores. Bot. Gaz. 123, 233—243.

ESCHRICH, W., 1964: Die Callosesynthese bei Pollenmutterzellen von *Cucurbita ficifolia*. In: Pollen physiology and fertilization (LINSKENS, H. F., ed.). pp. 48—51. Amsterdam: North-Holland.

— CURRIER, H. B., 1964: Identification of callose by its diachrome and fluorochrome reactions. Stain Technol. 39, 303—307.

FITTING, H., 1900: Bau und Entwickelungsgeschichte der Makrosporen von *Isoëtes* und *Selaginella* und ihre Bedeutung für die Kenntniss des Wachsthums pflanzlicher Zellmembranen. Bot. Z. 58, 107—163.

FREY-WYSSLING, A., 1976: The plant cell wall. Berlin-Stuttgart: Gebrüder Borntraeger.

GABARA, B., 1971: Cytokinesis in pollen mother cells. II. *Magnolia soulangeana* Soul. Biochem. Physiol. Pflanzen (BPP) 162, 450—458.

— 1974: A possible role for the endoplasmic reticulum in exine formation. Grana palynol. 14, 16—22.

GALATIS, B., KATSAROS, C., APOSTOLAKOS, P., 1978: Ultrastructural studies on the oil bodies of *Marchantia paleacea* Bert. II. Advanced stages of oil-body cell differentiation: synthesis of lipophilic material. Can. J. Bot. 56, 2268—2285.

Genevès, L., 1972: Aspects ultrastructuraux de la formation de l'enveloppe des spores, chez l'*Hypnum rusciforme* (Hypnacées). C. R. Acad. Sci. Paris D 275, 197—200.

— 1974: Distribution des polysaccharides pariétaux, en relation avec l'évolution de diverses cytomembranes, dans les méiocytes staminaux du *Ribes rubrum* (Grossulariacées), et principalement, au cours du cloisonnement. J. Microscopie (Paris) 19, 65—88.

Godwin, H., Echlin, P., Chapman, B., 1967: The development of the pollen grain wall in *Ipomoea purpurea* (L.) Roth. Rev. Palaeobotan. Palynol. 3, 181—195.

Hara, A., Yamashita, H., Kobayashi, A. 1977: Isolation of a polysaccharide from the inner cell wall, intine, of pollen of *Cryptomeria japonica*. Plant and Cell Physiol. 18, 381—386.

Hejnowicz, Z., Heinemann, B., Sievers, A., 1977: Tip growth: Patterns of growth rate and stress in the *Chara* rhizoid. Z. Pflanzenphysiol. 81, 409—424.

Herth, W., Franke, W. W., Bittiger, H., Kuppel, A., Keilich, G., 1974: Alkali-resistant fibrils of β-1,3- and β-1,4-glucans: structural polysaccharides in the pollen tube wall of *Lilium longiflorum*. Cytobiologie 9, 344—367.

Heslop-Harrison, J., 1963: An ultrastructural study of pollen wall ontogeny in *Silene pendula*. Grana palynol. 4, 7—24.

— 1966: Cytoplasmatische Verbindungen während der Sporenbildung bei höheren Pflanzen. Endeavour 25, 65—72.

— 1968: Wall development within the microspore tetrad of *Lilium longiflorum*. Can. J. Bot. 46, 1185—1192.

— 1969: An acetolysis-resistant membrane investing tapetum and sporogenous tissue in the anthers of certain *Compositae*. Can. J. Bot. 47, 541—542.

— 1971 a: The pollen wall: Structure and development. In: Pollen: Development and physiology (Heslop-Harrison, J., ed.), pp. 75—98. London: Butterworths.

— 1971 b: Sporopollenin in the biological context. In: Sporopollenin (Brooks, J., Grant, P. R., Muir, M., van Gijzel, P., Shaw, G., eds.), pp. 1—30. London: Academic Press.

— 1971 c: Wall pattern formation in angiosperm microsporogenesis. Symp. Soc. Exp. Biol. 25, 277—300.

— 1975: The physiology of the pollen grain surface. Proc. R. Soc. London B 190, 275—299.

— 1979: Aspects of the structure, cytochemistry and germination of the pollen of rye (*Secale cereale* L.). Ann. Bot. 44, Suppl. 1, 1—47.

— Mackenzie, A., 1967: Autoradiography of soluble 2-¹⁴C thymidine derivates during meiosis and microsporogenesis in *Lilium* anthers. J. Cell Sci. 2, 387—400.

Hoefert, L. L., 1971: Ultrastructure of tapetal cell ontogeny in *Beta*. Protoplasma (Wien) 73, 397—406.

Horner, jr., H. T., 1977: A comparative light- and electron-microscopic study of microsporogenesis in male-fertile and cytoplasmic male-sterile sunflower (*Helianthus annuus*). Amer. J. Bot. 64, 745—759.

— Lersten N. R., 1971: Microsporogenesis in *Citrus limon* (*Rutaceae*). Amer. J. Bot. 58, 72—79.

— — Bowen, C. C., 1966: Spore development in the liverwort *Riccardia pinguis*. Amer. J. Bot. 53, 1048—1064.

— Rogers, M. A., 1974: A comparative light and electron microscopic study of microsporogenesis in male-fertile and cytoplasmic male-sterile pepper (*Capsicum annuum*). Can. J. Bot. 52, 435—441.

— Pearson, C. B., 1978: Pollen wall and aperture development in *Helianthus annuus* (*Compositae: Heliantheae*). Amer. J. Bot. 65, 293—309.

Izhar, S., Frankel, R., 1971: Mechanisms of male sterility in *Petunia*: The relationship between pH, callase activity in the anthers, and the breakdown of the microsporogenesis. Theor. Appl. Genet. 41, 104—108.

Jalouzot, M. F., 1970: Mise en evidence de parois callosiques au cours de la mégasporogenèse et de l'oogenèse d'*Oenothera biennis*. C. R. Acad. Sci. Paris D 270, 317—319.

Kapil, R. N., Bhatnagar, A. K., 1981: Ultrastructure and biology of female gametophyte in flowering plants. Int. Rev. Cytol. 70, 291—341.

Kapil, R. N., Tiwari, S. C., 1978 a: The integumentary tapetum. Bot. Rev. (N.Y.) **44**, 457—490.

— — 1978 b: Plant embryological investigations and fluorescence microscopy: an assessment of integration. Int. Rev. Cytol. **53**, 291—331.

Kessler, G., 1958: Zur Charakterisierung der Siebröhrencallose. Ber. Schweiz. Bot. Ges. **68**, 5—43.

Knox, R. B., Heslop-Harrison, J., 1970: Direct demonstration of the lower permeability of the angiosperm meiotic tetrad. Z. Pflanzenphysiol. **62**, 451—459.

Laser, K. D., Lersten, N. R., 1972: Anatomy and cytology of microsporogenesis in cytoplasmic male sterile angiosperms. Bot. Rev. (N. Y.) **38**, 425—454.

Lee, S. L. J., Gracen, V. E., Earle, E. D., 1979: The cytology of pollen abortion in c-cytoplasmic male-sterile corn anthers. Amer. J. Bot. **66**, 656—667.

Lintilhac, P. M., 1974 a: Differentiation, organogenesis, and the tectonics of cell wall orientation. II. Separation of stresses in a two-dimensional model. Amer. J. Bot. **61**, 135—141.

— 1974 b: Differentiation, organogenesis, and the tectonics of cell wall orientation. III. Theoretical considerations of cell wall mechanics. Amer. J. Bot. **61**, 230—238.

Lugardon, B., 1968: Sur l'existence de liaisons protoplasmiques entre les cellules-mères des microspores de Ptéridophytes au cours de la prophase hétérotypique. C. R. Acad. Sci. Paris **D 267**, 593—596.

Mangin, L., 1890: Sur la callose nouvelle substance fondamentale existant dans la membrane. C. R. Acad. Sci. Paris **110**, 644—647.

Mascarenhas, J. P., 1975: The biochemistry of angiosperm pollen development. Bot. Rev. (N.Y.) **41**, 259—314.

Mepham, R. H., 1970: Development of the pollen grain wall: Further work with *Tradescantia bracteata*. Protoplasma (Wien) **71**, 39—54.

— Lane, G. R., 1969: Formation and development of the tapetal periplasmodium in *Tradescantia bracteata*. Protoplasma (Wien) **68**, 175—192.

Nanda, K., Gupta, S. C., 1974: Malfunctioning tapetum and callose wall behavior in *Allium cepa* microsporangia. Beitr. Biol. Pflanzen **50**, 465—472.

Noher de Halac, I., Harte, C., 1977: Different patterns of callose wall formation during megasporogenesis in two species of *Oenothera* (*Onagraceae*). Plant Syst. Evol. **127**, 23—38.

Northcote, D. H., Wooding, F. B. P., 1966: Development of sieve tubes in *Acer pseudoplatanus*. Proc. R. Soc. Lond. **B 163**, 524—537.

Oltmann, O., 1974: Licht- und elektronenmikroskopische Untersuchungen zur Sporogenese von *Lophocolea heterophylla* (Schrad.) Dum. I. Die Entwicklung von der Sporenmutterzelle bis zum Tetradenstadium. Pollen et Spores **16**, 5—25.

Pettitt, J. M., 1971 a: Developmental mechanisms in heterospory. I. Megasporocyte degeneration in *Selaginella*. Bot. J. Linn. Soc. **64**, 237—246.

— 1971 b: Some ultrastructural aspects of sporoderm formation in Pteridophytes. In: Pollen and spore morphology. Plant taxonomy. Pteridophyta (Erdtman, G., Sorsa, P., eds.), pp. 227—251. Copenhagen: Munksgaard.

— 1978: Regression and elimination of cytoplasmic organelles during meiosis in *Lycopodium*. Grana palynol. **17**, 99—105.

— 1979: Developmental mechanisms in heterospory: Cytochemical demonstration of spore-wall enzymes associated with β-lectins, polysaccharides and lipids in water ferns. J. Cell Sci. **38**, 61—82.

Pickett-Heaps, J. D., Northcote, D. H., 1966: Organization of microtubules and endoplasmic reticulum during mitosis and cytokinesis in wheat meristems. J. Cell Sci. **1**, 109—120.

Reynolds, J. D., Dashek, W. V., 1976: Cytochemical analysis of callose localization in *Lilium longiflorum* pollen tubes. Ann. Bot. **40**, 409—416.

Risueno, M. C., Giménez-Martín, G., Lopez-Saez, J. F., R.-García, M. I., 1969: Connexions between meiocytes in plants. Cytologia **34**, 262—272.

Robert, D., 1971: Nouvelle contribution à l'étude de l'origine des parois microsporales chez le *Selaginella kraussiana* A. Br. C. R. Acad. Sci. Paris **D 272**, 385—388.

RODKIEWICZ, B., 1970: Callose in cell walls during megasporogenesis in angiosperms. Planta 93, 39—47.

RODKIEWICZ, B., BEDNARA, J., 1976: Cell wall ingrowths and callose distribution in megasporogenesis in some *Orchidaceae*. Phytomorphology 26, 276—281.

ROGERS, C. M., HARRIS, B. D., 1969: Pollen exine deposition: a clue to its control. Amer. J. Bot. 56, 1209—1211.

ROLAND, F., 1971: Characterization and extraction of the polysaccharides of the intine and of the generative cell wall in the pollen grains of some *Ranunculaceae*. Grana palynol. 11, 101—106.

ROLAND-HEYDACKER, F., 1979: Aspects ultrastructuraux de l'ontogénie du pollen et du tapis chez *Mahonia aquifolium* Nuff. *Berberidaceae*. Pollen et Spores 21, 259—278.

ROWLEY, J. R., 1973: Formation of pollen exine bacules and microchannels on a glycocalyx. Grana palynol. 13, 129—138.

— 1976: Dynamic changes in pollen wall morphology. In: The evolutionary significance of the exine (FERGUSON, I. K., MULLER, J., eds.), pp. 39—65. London: Academic Press.

— DAHL, A. O., 1977: Pollen development in *Artemisia vulgaris* with special reference to glycocalyx material. Pollen et Spores 19, 169—284.

— DUNBAR, A., 1967: Sources of membranes for exine formation. Svensk Bot. Tidskrift 61, 49—64.

— SOUTHWORTH, D., 1967: Deposition of sporopollenin on lamellae of unit membrane dimensions. Nature 213, 703—704.

— SKVARLA, J. J., 1975: The glycocalyx and initiation of exine spinules on microspores of *Canna*. Amer. J. Bot. 62, 479—485.

RUSSELL, S. D., 1979: Fine structure of megagametophyte development in *Zea mays*. Can. J. Bot. 57, 1093—1110.

SAUTER, J. J., 1969: Autoradiographische Untersuchungen zur RNS- und Proteinsynthase in Pollenmutterzellen, jungen Pollen und Tapetumzellen während der Mikrosporogenese von *Paeonia tenuifolia* L. Z. Pflanzenphysiol. 61, 1—19.

SCHNEPF, E., 1969: Sekretion und Exkretion bei Pflanzen. (Protoplasmatologia VIII/8.) Wien-New York: Springer.

— 1973: Mikrotubulus-Anordnung und -Umordnung, Wandbildung und Zellmorphogenese in jungen *Sphagnum*-Blättchen. Protoplasma (Wien) 78, 145—173.

SCOLES, G. J., EVANS, L. E., 1979: Pollen development in male-fertile and cytoplasmic male-sterile rye. Can. J. Bot. 57, 2782—2790.

SENGUPTA, S., ROWLEY, J. R., 1974: Re-exposure of tapes at high temperature and pressure in the *Lycopodium clavatum* spore exine. Grana palynol. 14, 143—151.

SHAW, G., 1971: The chemistry of sporopollenin. In: Sporopollenin (BROOKS, J., GRANT, P. R., MUIR, M., VAN GIJZEL, P., SHAW, G., eds.), pp. 305—348. London: Academic Press.

SHEFFIELD, E., BELL, P. R., 1979: Ultrastructural aspects of sporogenesis in a fern, *Pteridium aquilinum* (L.) Kuhn. Ann. Bot. 44, 393—405.

SIEVERS, A., BUCHEN, B., 1970: Über den Feinbau der wachsenden Megaspore von *Selaginella*. Protoplasma (Wien) 71, 267—279.

— SCHRÖTER, K., 1971: Versuch einer Kausalanalyse der geotropischen Reaktionskette im *Chara*-Rhizoid. Planta 96, 339—353.

SITTE, P., 1953: Untersuchungen zur submikroskopischen Morphologie der Pollen- und Sporenmembranen. Mikroskopie (Wien) 8, 290—299.

SKVARLA, J. J., LARSON, D. A., 1966: Fine structural studies of *Zea mays* pollen. I. Cell membranes and exine ontogeny. Amer. J. Bot. 53, 1112—1125.

SMITH, M. M., MCCULLY, M. E., 1978: A critical evaluation of the specificity of aniline blue induced fluorescence. Protoplasma (Wien) 95, 229—254.

SOUTHWORTH, D., 1971: Incorporation of radioactive precursors into developing pollen walls. In: Pollen: Development and physiology (HESLOP-HARRISON, J., ed.), pp. 115—120. London: Butterworths.

— 1974: Solubility of pollen exines. Amer. J. Bot. 61, 36—44.

Steer, M. W., 1977 a: Differentiation of the tapetum in *Avena*. I. The cell surface. J. Cell Sci. **25**, 125—138.

— 1977 b: Differentiation of the tapetum in *Avena*. II. The endoplasmic reticulum and Golgi apparatus. J. Cell Sci. **28**, 71—86.

Stieglitz, H., 1977: Role of β-1,3-glucanase in postmeiotic microspore release. Develop. Biol. **57**, 87—97.

Thiéry, J.-P., 1967: Mise en evidence des polysaccharides sur coupes fines en microscopie électronique. J. Microscopie (Paris) **6**, 987—1018.

Vasil, I. K., Aldrich, H. C., 1970: A histochemical and ultrastructural study of the ontogeny and differentiation of pollen in *Podocarpus macrophyllus* D. Don. Protoplasma (Wien) **71**, 1—37.

Vazart, B., 1970: Morphogenèse du sporoderme et participation des mitochondries à la mise en place de la primexine dans le pollen de *Linum usitatissimum* L. C. R. Acad. Sci. Paris D **270**, 3210—3212.

Vithanage, H. J. M. V., Knox, R. B., 1979: Pollen development and quantitative cytochemistry of exine and intine enzymes in sunflower, *Helianthus annuus* L. Ann. Bot. **44**, 95—106.

— Gleeson, P. A., Clarke, A. E., 1980: The nature of callose produced during self pollination in *Secale cereale*. Planta **148**, 498—509.

Warmke, H. E., Overman, M. A., 1972: Cytoplasmic male sterility in *Sorghum*. I. Callose behavior in fertile and sterile anthers. J. Hered. **63**, 103—108.

Waterkeyn, L., 1964: Callose microsporocytaire et callose pollinique. In: Pollen physiology and fertilization (Linskens, H. F., ed.), pp. 52—58. Amsterdam: North-Holland Publ. Comp.

— 1981: Cytochemical localization and function of the 3-linked glucan callose in the developing cotton fibre cell wall. Protoplasma (Wien) **106**, 49—67.

— Bienfait, A., 1970: On a possible function of the callosic special wall in *Ipomoea purpurea* (L.) Roth. Grana palynol. **10**, 13—20.

— — 1971 a: Morphologie et nature des parois sporocytaires chez les Ptéridophytes. La Cellule **69**, 7—25.

— — 1971 b: Primuline induced fluorescence of the first exine elements and Ubisch bodies in *Ipomoea* and *Lilium*. In: Sporopollenin (Brooks, J., Grant, P. R., Muir, M., van Gijzel, P., Shaw, G., eds.), pp. 108—127. London: Academic Press.

— — 1972: Sur une localisation particulière de la callose dans la paroi microsporocytaire et l'endospore mégasporale des Sélaginelles. C. R. Acad. Sci. Paris D **274**, 2489—2491.

Weiling, F., 1965: Zur Feinstruktur der Plasmodesmen und Plasmakanäle bei Pollenmutterzellen. Planta **64**, 97—118.

Whelan, E. D. P., 1974: Discontinuities in the callose wall, intermeiocyte connections, and cytomixis in angiosperm meiocytes. Can. J. Bot. **52**, 1219—1224.

Willemse, M. T. M., 1971: Morphological and fluorescence microscopical investigation on sporopollenin formation at *Pinus silvestris* and *Gasteria verrucosa*. In: Sporopollenin (Brooks, J., Grant, P. R., Muir, M., van Gijzel, P., Shaw, G., eds.), pp. 68—107. Academic Press.

— Franssen-Verheijen, M. A. W., 1978: Cell organelles changes during megasporogenesis and megagametogenesis in *Gasteria verrucosa* (Mill.) Haw. Bull. Soc. Bot. Fr. **125**, 187—191.

— Reznickova, S. A., 1980: Formation of pollen in the anther of *Lilium*. I. Development of the pollen wall. Acta Bot. Neerl. **29**, 127—140.

Young, B. A., Schulz-Schaeffer, J., Carroll, T. W., 1979: Anther and pollen development in male-sterile intermediate wheatgrass plants derived from wheat × wheatgrass hybrids. Can. J. Bot. **57**, 602—618.

III. General Aspects of Cytomorphogenesis in Plants

Ionic Currents as Control Mechanism in Cytomorphogenesis

M. H. Weisenseel and Rosalinde M. Kicherer

Botanical Institute, Technical University of Karlsruhe, Karlsruhe, and
Institute for Botany and Pharmaceutical Biology, University of Erlangen, Erlangen
Federal Republic of Germany

With 8 Figures

Contents

I. Introduction: The Problem of Spatial Development

As plant cells develop into multicellular organisms, three-dimensional patterns are formed without the cells possessing genetically laid down plans for these patterns. Therefore, a cell needs additional information for its proper spatial development to indicate, for instance, in which direction it should grow or in which plane it should divide. This information generally comes from external physical factors such as light, gravity, or pressure or from chemical factors such as ion or hormone gradients. Some examples of the effects of such factors on the morphogenesis of plant cells are illustrated in Figs. 1 and 2. When, for instance, zygotes of the brown algae *Fucus* and *Pelvetia*, which are practically nonpolar cells, are exposed to light from one side or only partly illuminated or exposed to a K^+ gradient they grow on the side turned away from the light or at the shaded side or on the side with the higher K^+ concentration, respectively (BENTRUP 1971, BENTRUP et al. 1967, JAFFE 1968). When the tip-growing tubes of the yellow-green alga *Vaucheria* are locally irradiated with blue light, an additional growth zone is formed on this spot, which leads to branching within a few hours (KICHERER and WEISENSEEL, unpublished).

All the morphogenetically effective physical and chemical factors only function as signals and do not supply the cell with the necessary energy for the subsequent reactions. This can be concluded from their very short duration of action and the low intensity necessary to give rise to a certain

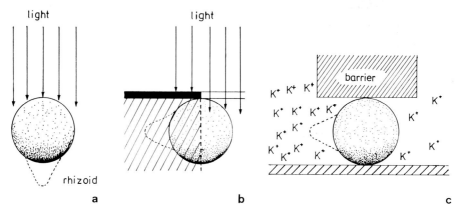

Fig. 1. The effect of external factors such as light and ionic concentration gradients on the cytomorphogenesis of *Fucus* zygotes: *a* Unilateral light induces the rhizoid to grow away from the light source. *b* Half-side illumination of a zygote results in germination of a rhizoid at the shaded side. This demonstrates that growth occurs at the darkest place within the cell, which in unilateral light is the side turned away from the light source. *c* In a gradient of K^+ ions, the zygotes grow toward the side of higher K^+ concentration if the average K^+ concentration is 10 mM or higher. According to BENTRUP 1971.

spatial development in the cell some time later. For instance, the spores of the horsetail *Equisetum* require only a few seconds of irradiation to induce a localized germination several hours later (HAUPT 1957). If information for cytomorphogenesis, especially for the spatial differentiation, comes from outside the cell and if short and weak signals are sufficient, two questions arise: First, where are these signals perceived in the cell? Second, how are the impressions left behind by the signal transduced and amplified by the cell?

To deal with these questions, we shall proceed from a hypothesis that is supported by known results and seems to be verifiable by the methods available at present. We assume that each signal is perceived by specific receptors localized in the cell membrane and that the perception leads to an initially unstable, asymmetrical distribution of molecules, particularly membrane pumps and membrane channels, along the cell membrane. The asymmetry in the ionic conductivity of the membrane thus produced then leads to a transcellular ionic current. Ions flow passively into the cell at

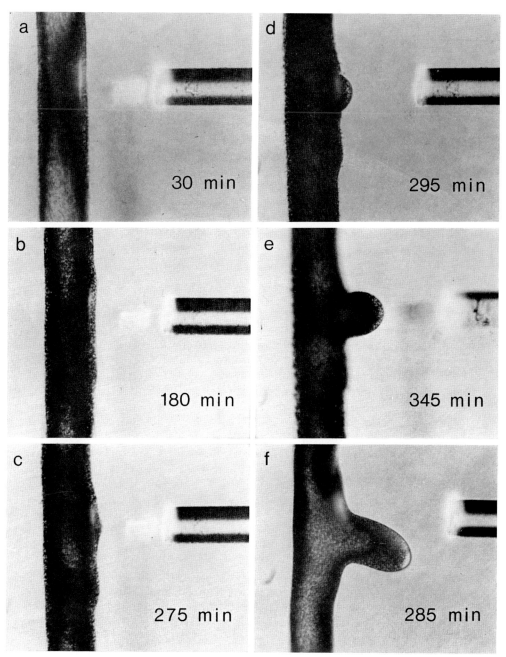

Fig. 2. The effect of local irradiation on the cytomorphogenesis of *Vaucheria* tubes. Continuous irradiation of a tube with blue light from a single 50 μm thick glass fiber (on the right in each frame) results in aggregation of chloroplasts and, within 3 to 5 hours, results in the establishment of a new outgrowth at the irradiated site: *a–f* typical stages of development after the beginning of irradiation; *a–e* in the same alga; *f* a later stage in a different tube.

one point and are actively transported out of the cell at another point. The electric fields associated with this current then amplify the asymmetric distribution of pumps and channels and, consequently, the current and produce concentration gradients of mobile ions.

II. Self-Generated Ionic Currents in Plant Cells and Tissues

A. First Current Measurements in Fucus zygotes

That there are bioelectric phenomena has been known since Galvanis' and Voltas' experiments about 200 years ago (cf., Geddes and Hoff 1971). Most investigations in this field, however, were concerned with the electric membrane potential in nerve and muscle cells. They have hardly dealt with the long-term bioelectric potentials that occur in single cells and in tissues, nor have they dealt with the ionic currents associated with them. The first indications of transcellular ionic currents relating to pattern formation in plant cells were discovered by Jaffe (1966) in brown algae. He observed that the zygotes and embryos of Fucus generate currents of about 100 pA with surface density of $1 \mu A \ cm^{-2}$ that enter at the rhizoid pole and leave the cell at the thallus pole.

As these zygotes are only about 70 to 80 μm in diameter, the resistance of conventional microelectrodes proved to be too high to measure the current in single cells directly. Therefore, the transcellular currents could only be obtained indirectly from the potential drop developing across several hundred zygotes in series by growing them in the same direction in a loosely fitting glass tube filled with sea water and applying unilateral illumination. At about the time the first cells in the capillary began to germinate, an electric potential appeared between the ends of the tube showing that a current was flowing through the medium. No potential developed along control tubes in which the zygotes were randomly orientated due to light from all sides. This elegant method had disadvantages—it could only be employed with cells, that can be grown in series, and it allowed no spatial and temporal resolution of the current pattern of individual cells. Jaffe and Nuccitelli (1974) therefore developed a new method to measure transcellular currents in single cells.

B. A New Method for Measuring Transcellular Currents

Most of the self-generated currents (described in this section) that traverse plant cells during their development were measured by this new method. This method combines three advantages that make it especially suitable for identifying ionic currents in growing cells: (1) it is more sensitive than measurements with static microelectrodes by a factor of 10^2 to 10^3, so even potential differences as small as nanovolts can be measured with accuracy; (2) it has a spatial resolution that permits it to identify membrane areas that have different properties and lie only a few μm apart within a single cell; (3) the currents are measured extracellularly, i.e., in the medium that surrounds the cell, so the cell is neither damaged nor is it disturbed in its normal development.

The "sensor" of the measuring instrument is an electrode consisting of a glass micropipette filled with metal whose tip bears a ball of platinum-black 10 to 30 μm in diameter. This ball is electroplated, and its size is kept small to yield high spatial resolution. On the shaft of the micropipette, several mm above the tip, lies a further layer of platinum-black that functions

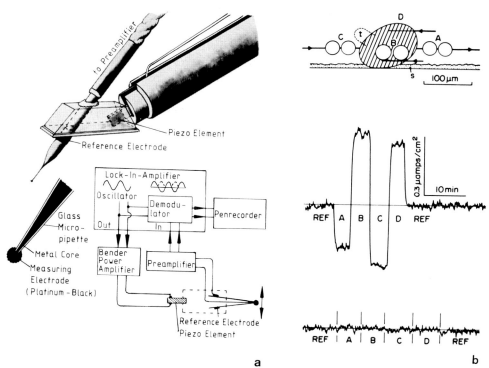

Fig. 3. Vibrating electrode and typical recording of transcellular currents: *a* Illustration of the vibrating electrode and schematic representation of the measuring circuit (according to Jaffe and Nuccitelli 1974). *b* Representative measurement of the endogenous current traversing an ungerminated pollen grain of *Lilium*. The downward deflections of the recorder (middle) indicate current flux from left to right in the diagram on top. *A*, *B*, *C*, and *D* are the various measuring positions around the cell, and *REF* means reference position. The vibrating electrode is symbolized by two circles; *t* indicates the pollen tube that emerged near position C about 2 hours later; *s* indicates layer of silicon grease to fasten the cell. The recording at the bottom shows a measurement from a dead pollen grain after fixation in chloraldehyde. From Weisenseel *et al.* 1975.

as a coaxial reference electrode. The measuring electrode and the reference electrode are connected through a preamplifier to a lock-in amplifier. The compound electrode is cemented into a small plexiglass piece that is fixed to a piezoelectric element (Fig. 3 *a*).

This electrode is vibrated by applying a sinusoidal alternating current of several hundred Hertz to the piezoelectric element. If a current flows through the medium at the measuring position, the ohmic drop in potential caused by this current is detected by the electrode by means of capacitative coupling, converted to a sinusoidal alternating current by the vibration, amplified and measured by the lock-in amplifier. From this voltage (the effective voltage, Veff), the current density at the site (J_a) can be calculated by means of the following equation: $J_a = \dfrac{Veff \cdot 2.83}{d \cdot \rho \cdot W}$, where d is the amplitude of the electrode's vibration, ρ is the resistivity of the medium, and W is the efficiency factor of the electrode. (The efficiency factor is determined by calibrating against a known current source and generally has a value of about 0.9.) Several measurements at various points around an object with different vibrational directions allow the direction of the current to be determined, and from this information, a picture of the current pattern around the object can be built up. Fig. 3 b shows a typical example of a measurement from a growing lily pollen grain and the conclusions drawn about the gross current pattern.

The main reason for the high sensitivity of this method lies in an optimal relationship between unknown signal voltage and noise voltages. These noise voltages (V_N) are mostly caused by random movements of charges and are defined by the equation: $V_N = \sqrt{4 \, k \, T \, Z \, B}$, whereby k is the Boltzmann constant; T, the absolute temperature; Z, the impedance of the electrode; and B, the band width of the measuring frequencies. As $Z = \sqrt{(X_R)^2 + (X_C)^2}$, *i.e.*, is a function of the ohmic resistance X_R and the capacitative resistance X_C, and $X_C = 1/\omega \, C$ is a function of the angular velocity (ω) and the capacity (C) of the electrode tip, the value for Z can be greatly reduced by a metallic electrode and by increasing the frequency. Furthermore, the width of the frequency band can be greatly limited through measurements within a narrow range of frequencies. Thus, it is possible to limit the background interferences almost to the theoretical minimum and, therefore, measure very small signals reliably.

As the vibrating electrode measures at a certain distance from the object, the current density on the surface of the object must be determined by calculation. With freely accessible objects, this current density is obtained by determining the current density at at least three different distances at right angles to the surface. From this the current density on the surface can be calculated. For instance, in a lily pollen grain that has not yet germinated, the following relationship was found: $J_s \approx J_a \cdot (\frac{r}{a})^{-3}$, whereby J_s is the current density at the surface; J_a, the current density at the measuring point; a, the distance of the midpoint of vibration from the centre of the sink; and r, the radius of the current sink (r was approximately equal to the radius of the pollen grain in this case). For the tip of root hairs and *Vaucheria* tubes, the following relationship applied: $J_s \approx J_a \cdot (\frac{r}{a})^{-2}$ (r was approximately 10 to 20 µm in this case; *cf.*, Fig. 5 a). For the barley root, a cylindrical current sink, the following relationship was valid: $J_s \approx J_a \cdot (\frac{r}{a})^{-1}$ (r was equivalent to the radius of the root). In the case of a spherical cell, therefore, the current falls with the cube of the distance; at the tip of an elongated cell, it falls with the square of the distance; and at a cylindrical organ it falls with the simple distance from the surface. The determination of the surface density of the current becomes more complicated in objects that are not freely accessible, as, for example, in a growing pollen tube fastened to an ion-permeable membrane (compare Weisenseel et al. 1975).

C. Transcellular Ionic Currents

In all plant cells investigated so far, ionic currents have been found that precede pattern formation and accompany local growth. These currents always enter at the future or actual growth pole and leave the cell at nongrowing regions. (It is convention to equate the direction of current flow with the movement of positive charge. The measured current densities can be converted into ion flows whereby $1 \mu A \, cm^{-2}$ corresponds to $10 \, pmol$ $cm^{-2} \, s^{-1}$ for univalent ions.)

In *Pelvetia zygotes*, a current begins to flow into the cell on one side of the cell (in cells illuminated unilaterally it is the side averted from the light) within an hour after fertilization, *i.e.*, 8 to 10 hours before the visible germination of a rhizoid begins. This current spreads within the cell and leaves it at a relatively extended area of the future thallus pole. The current returns through the medium to the future rhizoid pole. A rhizoid always grows at the point at which a stable current flows into the zygote. After visible growth begins, a current of about $1 \mu A \, cm^{-2}$ enters the rhizoid and leaves the thallus part of the cell. This current has an absolute value of about $100 \, pA$ per cell, rises in the light with an increasing growth rate, and falls as the growth rate decreases in the dark (NUCCITELLI and JAFFE 1974, NUCCITELLI 1978, JAFFE and NUCCITELLI 1977) (Fig. 4 *a*).

In *lily pollen*, a current begins to flow into one area of the pollen grain and leaves the opposite side of the cell about 1/2 hour after sowing the pollen grains in a liquid medium. This current increases in the next 1 to 3 hours and reaches a density of approximately 3 to $5 \mu A \, cm^{-2}$. Later a pollen tube grows out of the region where the current enters. After the pollen tube has sprouted, the current enters all along the growing tube, except next to the pollen grain, and exits over the whole surface of the pollen grain (WEISENSEEL *et al.* 1975) (Fig. 4 *b*). A growing pollen grain is thus similar to a current dipole in which the tube represents the current "sink" and the grain the current "source". The absolute current that flows through a growing lily pollen is about 100 to $300 \, pA$ and reaches the considerable density of approximately $60 \mu A \, cm^{-2}$ in the tube where it is concentrated by the reduced transversal section. Driving this current through the cell costs a pollen grain about 1% of its energy originating from oxidative phosphorylation.

The ions that participate in carrying the current through the lily pollen are mainly K^+ ions that flow passively into the tube and H^+ ions that are actively pumped out of the grain. This is concluded from the observations that the inward current decreases greatly when the K^+ concentration of the medium is suddenly reduced and increases when the K^+ concentration is raised. A drastic reduction of the K^+ concentration in the medium leads to a standstill of the transcellular current and the growth. A rise in the extracellular pH value increases the current, a fall leads to a reduction of the current (WEISENSEEL and JAFFE 1976).

In *barley roots*, current leaves the surface of the root hair zone and enters the region of root growth. This main current is superimposed by local currents that flow into the tips of the growing root hairs. The density of these currents is about 1 to $2 \mu A \, cm^{-2}$ on the surface of the root hair tips. Raising the pH in the medium, in the range of 5.6 to 7.1, leads to a decrease; lowering the pH leads to an increase in the influx within a few minutes. Changes in the Na^+, K^+, or Cl^- concentration in the medium have no

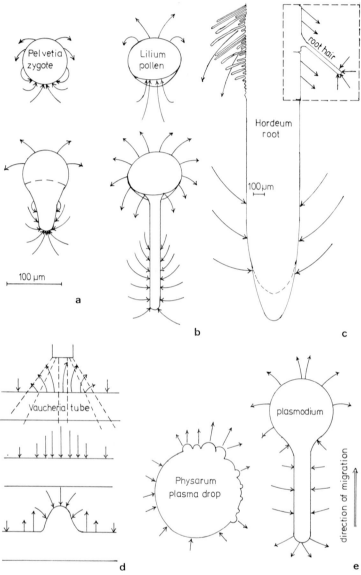

Fig. 4. Schematic representation of examples of the natural ionic currents traversing plant cells and tissues: *a* Zygotes of *Pelvetia* before and after the growth of a rhizoid (according to Nuccitelli and Jaffe 1974, Nuccitelli 1978). *b* Pollen grains of *Lilium* before and after germination (according to Weisenseel et al. 1975). *c* Root and root hairs of *Hordeum* (according to Weisenseel et al. 1979). *d* Three stages during the induction of an outgrowth by local irradiation with continuous blue light in *Vaucheria* tubes (Kicherer and Weisenseel, unpublished, Blatt, Weisenseel, and Haupt, submitted to Planta). *e* Drops of endoplasm and young plasmodium in *Physarum* (according to Aschenbach and Weisenseel 1981). Arrows indicate the direction of the current.

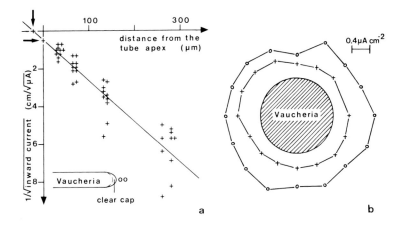

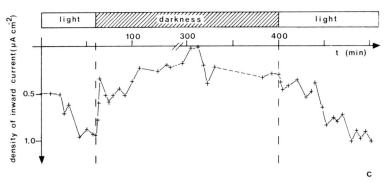

Fig. 5. Self-generated ionic currents in *Vaucheria* tubes: *a* Densities of the inward current at the growing tip as a function of the distance from the tube apex are shown here. The results from ten different tubes are plotted reciprocally. The intercept of the regression line with the x-axis (arrow) yields the distance of the current sink beneath the surface of the tube; the intersection with the y-axis (arrow) gives the average current density at the surface of a tube (inset: arrangement of tube and vibrating electrode, symbolized by two circles). *b* A representative example is shown here of the measurements of the current densities behind the tip of a vertically downward oriented *Vaucheria* tube. Little *x*s are current densities at measurement positions around the tube 75 μm behind the tip and 35 μm from the surface; little circles are current densities at measurement positions 100 μm behind the tip. In all positions, an inward current was measured. *c* Current densities at the growing tip in light and in darkness are shown here.

significant influence on the current (WEISENSEEL *et al.* 1979) (Fig. 4 *c*). Therefore, the main component of the current in the NO_3^- and $H_2PO_4^-$ free medium used seems to be H^+ ions. Indications that the current into the root hairs is correlated to their growth were obtained by observations that no current entered those root hairs that did not grow. The same current pattern found in barley roots was recently found in *Lepidium roots*, growing vertically downward in an aqueous medium (BEHRENS, WEISENSEEL and SIEVERS, unpublished). When a *Lepidium* root was turned from the vertical to the horizontal position, *i.e.*, placed in a geotropically stimulating position, a current was induced within a minute which left the top of the root cap and entered at the lower side of the root cap.

Endogenous ionic currents also traverse the coenocytic alga *Vaucheria sessilis.* Current enters the tips of the growing tubes and leaves at discrete areas behind the tips. With the help of linear regression analyses, we obtained a distance of the current sink of about 22 µm underneath the surface of the tip and an average surface density of the inward current of about 4 µA cm^{-2}. The density of the influx depends on light since it is about 40 to 60% stronger in white light-illuminated tubes than in tubes kept in the dark (Fig. 5 c). Measuring current densities behind but still close to the tips of vertically hanging tubes, we found an inward current of almost constant density over the whole perimeter (Fig. 5 b). At discrete sites somewhat further behind the tip we found efflux currents whose densities were about 25% less than those of the influx currents in the neighbouring regions. The direction and the density of the currents behind the tip may also change at each position in the course of a few hours.

In *Vaucheria* tubes, additional growth areas can be induced behind the tip by local irradiation (KATAOKA 1975, ÅBERG 1978, KICHERER and WEISENSEEL, unpublished) (Fig. 2). If tubes are locally irradiated with blue light with a maximum wavelength of about 450 nm, different current patterns appear in the irradiated area that can be grouped into three categories. Group 1: Seventeen of the 26 algae investigated responded within 1 to 2 minutes after the irradiation began with an out-flowing current that was restricted to the irradiated area and reached its highest value after 1 to 2 minutes more. After 10 to 130 minutes, a spontaneous change to an inflowing current took place. The influx was always stronger than the efflux. In 16 of these algae, branching occurred during the stage of the influx (Fig. 4 d; Table 1). Group 2: Eight other algae also reacted with an out-flowing current at first. This efflux changed to an influx, but then the current changed direction once or several times. Four of these algae showed branching anlagen that did not, however, grow further into new tubes within the 4 hours of observation. Group 3: One alga of the 26 investigated only showed an efflux after 70 minutes of irradiation. Moreover, this current reached its maximum only after another 70 minutes, and even about 300 minutes after the beginning of the irradiation, it was still present as an outward current. It was striking that the tube was rejuvenated for a length of about 70 µm in the irradiated area. During the observation time of about 5 hours, no visible branching occurred in this alga.

The growth anlagen induced by irradiation with blue light frequently appeared on the side facing the light source. The position and the shape of the subsequent growth anlage could already be recognized from the current pattern. The presence of a single, relatively narrow peak of inflowing current resulted in a single branching anlage at this point. Two peaks of inflowing current lead to two growth anlagen; however, only one of them subsequently grew further into a branch. A broad plateau of influx preceded a very broad growth anlage. In all cases, the irradiation with blue light induced an accumulation of chloroplasts in the exposed area (BLATT and BRIGGS 1980, BLATT *et al.* 1980) (see Fig. 2).

When *Vaucheria* tubes are irradiated with red light instead of blue light, the induced current pattern differs fundamentally. All six algae studied so far reacted immediately with a strong inward current. Irradiation with

Table 1. *The outward currents of 17 Vaucheria tubes responding to local blue light irradiation with outward current followed by inward current*

Maximal outward current [1] in nA cm^{-2}	Total outward current [2]		Duration of outward current in minutes	Beginning of visible growth in minutes after onset of irradiation
	in µC cm^{-2}	in nmol cm^{-2}		
67	133	1.33	96	126
206	419	4.19	70	73
50	35	0.35	22	84
36	48	0.48	36	86
41	96	0.96	72	92
170	78	0.78	25	90
230	439	4.39	50	120
397	1,071	10.71	81	101
160	255	2.55	47	114
245	449	4.49	71	112
205	28	0.28	45	166
225	1,349	13.49	130	160
67	18	0.18	11	—
137	105	1.05	33	38
196	57	0.57	22	106
132	163	1.63	34	96
88	20	0.20	8	107
$\bar{x} = 156$	280	2.80	50	106

[1] Measured 50 µm from the surface of the *Vaucheria* tubes.
[2] Calculated by integration of the outward current over its duration, assuming univalent ions as charge carriers.

red light for up to 4 hours produced no outward current, no branching, and no accumulation of chloroplasts in the irradiated area. In contrast to our results, ÅBERG (1978) reports branching induced by red light. From our observations, it can be concluded that an early efflux and a subsequent influx are necessary for the induction of a growth zone behind the tip of *Vaucheria* tubes. According to present results (BLATT, WEISENSEEL and HAUPT, submitted to Planta), the outflowing current seems to consist mainly of H$^+$ ions. This explains the observations that a rise in the pH value of the medium stimulates the outward current, a decrease in the pH value inhibits the current or stops it completely, and the H$^+$ ionophore CCCP inhibits the current even before the cell's ATP level declines. We speculate that the efflux could perhaps be related to a low cAMP concentration and the influx to a high cAMP concentration (cf., Kataoka 1977, Cohen 1974, Janistyn and Drumm 1972).

Between pH 5 and 8, in continuous white light, the density of the influx in the growing tips of the *Vaucheria* tubes is also, dependent on the pH value. The strongest influx in a medium containing only 0.1 mM $CaCl_2$ and 1 mM MES (or Tris) buffer occurred at a pH value of 5; the weakest, at a pH value of 8. These results support the view that the current in the growing tips consists at least in part of a passive H^+ influx (or a OH^- efflux). In this medium, the highest growth rate of the tubes (approx. 1 $\mu m\ min^{-1}$) was observed at a pH value of 7; and at a pH value of 5, the least growth rate was measured (approx. 0.01 $\mu m\ min^{-1}$). This does not support the idea of a direct proportionality between current density and growth rate. It rather indicates a relationship and dependence of the growth rate on specific ions comprising the inward current, perhaps Ca^{2+} ions.

If a strand of the slime mould *Physarum polycephalum* is pierced with a fine needle, a drop of endoplasm flows out at the injury point. This drop surrounds itself with a plasma membrane within a few seconds, begins to pulsate after about half an hour, and develops within several hours to a small plasmodium that begins to creep on the substratum (Wohlfarth-Bottermann and Stockem 1970). Ionic currents with densities up to 15 $\mu A\ cm^{-2}$ always flow into these drops at protrusions of the surface and flow out in areas that show a smooth surface (Fig. 4 e). During the development of a plasma drop into a small plasmodium, distinct displacements of membrane areas or redistribution of membrane functions probably occur, as current only enters the strand and flows out at the front pole (and back pole) in the young plasmodium (Achenbach and Weisenseel 1981).

In the water fungus *Blastocladiella,* the germination of a spore, induced by a depolarization of the cell membrane, begins with the formation of a rhizoid (Brunt and Harold 1980). This rhizoid grows probably via tip growth, whereas the thallus cell grows isotropically. After some time the plasma of the thallus cell divides into spores that are set free through a papilla that lies opposite the rhizoid pole. The development of this papilla is preceded by a considerable accumulation of vesicles at the future papilla site. Simultaneously, the rhizoid stops growing (Lovett 1975). Measurements with the vibrating electrode show that a current with a surface density of about 1 $\mu A\ cm^{-2}$ flows into the growing tip of the rhizoid (Stump et al., from Jaffe 1980). Conditions that lead to the formation of a papilla and subsequent sporulation, such as transferring the germlings into an inorganic medium, lead to a reversal of the current. The current then flows into the future papilla region. This reversal of the current occurs about 3 to 5 hours before any changes in the papilla area can be recognized in the light microscope.

Stalk segments of the marine alga *Acetabularia* regenerate a new tip when placed in sea water. If both ends of such a segment are electrically isolated from one another and the electrical potential between both ends is measured, voltage pulses are observed about 40 hours before the visible regeneration of a new tip begins. These pulses are about 100 s long, have an amplitude of 100 to 140 mV, and indicate that pulses of current of 10 to 100 μA cm^{-2} flow into the future tip (Novák and Bentrup 1972). These current pulses superimpose a very weak constant current that exits from the future tip. It has not been clarified, so far, which of the two currents, the pulse current or the constant current, is more important for the regeneration of the new tip. Other currents may flow in addition to the currents already observed, perhaps locally limited currents in the immediate vicinity of the regenerating tip. Such currents could not be detected by the methods used till now, but could play a decisive role in local regeneration (Goodwin, personal communication).

III. The Effect of Forced Transcellular Currents by Applied Electric Fields

Constant electric fields can polarize plant cells and control their direction of growth (cf., Jaffe and Nuccitelli 1977, Weisenseel 1979). For instance, the zygotes of *Fucus, Pelvetia,* and *Ulva;* the spores of *Equisetum* and *Funaria;* and the pollen grains of *Vinca* and *Lilium* react by germinating parallel to the applied field. The percentage of polarized cells in a population rises in proportion to the strength of the field, but the position of the germination point with respect to the anode and cathode varies between objects and even from population to population (Fig. 6).

Lund (1923) found that the zygotes of *Fucus inflatus* germinate toward the positive electrode in a constant electric field. The reaction began at about 20 mV per cell and reached saturation with almost 100% orientation of the population at approximately 30 mV per cell. Bentrup (1968) found that the zygotes of *Fucus serratus* formed rhizoids toward the negative electrode when placed in seawater of normal (i.e., 10 mM) or decreased K$^+$ concentration. Peng and Jaffe (1976) observed in *Pelvetia fastigiata* that 11 of the 16 investigated batches germinated toward the positive electrode, two batches grew toward the negative electrode, and three batches showed a mixed reaction—they germinated toward the negative electrode in weak fields and toward the positive electrode in strong fields. The zygotes of *Ulva mutabilis* all germinated toward the positive electrode in the investigations carried out by Sand (1973).

The spores of *Equisetum limosum* and *E. variegatum* germinated with a rhizoid toward the positive electrode (Bentrup 1968). The spores of the moss *Funaria hygrometrica,* which can be cultivated in a simple inorganic medium containing only the salt Ca(NO$_3$)$_2$, tend to form their rhizoids toward the positive electrode (Chen and Jaffe 1979). A reaction of 10% was already achieved with 2 mV per spore, a half maximal response with 4 to 8 mV. Chen and Jaffe found, furthermore, that the germination of the chloronema could not be influenced by electric fields up to 45 mV per cell. Electric fields that were applied perpendicularly to short growing rhizoids, however, induced the rhizoids to grow in the direction of the negative electrode.

About 50% reaction was achieved with 3 to 5 mV per cell, and almost
100% alignment took place at 12 mV across a rhizoid.

In *Vinca rosea* pollen, Marsh and Beams (1945) found that the pollen
tubes were preferentially formed at those pores that pointed toward the
negative electrode. In contrast to this, Sperber and Weisenseel (unpublished)
observed that the pollen of *Vinca rosea* and the pollen of *Lilium longiflorum*
formed tubes mainly toward the positive electrode. This was found in
90% of all cells; in the other 10%, the site of germination was directed
toward the negative electrode. The growing pollen tubes of *Lilium* reacted

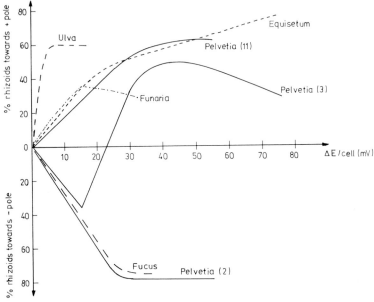

Fig. 6. Polarization of zygotes of *Ulva*, of *Fucus* and *Pelvetia*, and of spores of *Equisetum*
and *Funaria* by imposed electric fields. Data from Sand (1973), Bentrup (1968), Peng
and Jaffe (1976), Chen and Jaffe (1979). Ordinate: percent rhizoids toward positive or
negative electrode exceeding random distribution. (For details about the calculation of the
percentage see Peng and Jaffe 1976.). Abscissa: field strength expressed as the actual
voltage drop across each cell, *i.e.* applied voltage per cell diameter multiplied by a factor
of 1.5 to account for the distortion of the field near the cell. For *Pelvetia*, the three different
modes of response of 16 batches of zygotes are represented; the number of batches belonging
to each mode are shown in brackets. (The diameter of an *Ulva* zygote was estimated as
20 μm in order to adapt Sand's data for the figure. From Weisenseel 1979.

similarly to the *Funaria* rhizoids, showing a clear tendency to direct their
growth toward the negative electrode. While the orientation of the site of
germination in *Lilium* reaches a value of about 100% only at 100 mV per
cell, the orientation of the pollen tubes shows a 90% reaction at only 3 mV
per tube. Recently Brower and McIntosh (1980) and Brower and Giddings
(1980) showed an effect of applied electric fields on the morphogenesis of
Micrasterias. The lobes of this alga show a galvanotropism toward the
cathode (see p. 176 f.).

IV. Transcellular Ionic Currents as Cause and Consequence of Localization

The endogenous ionic currents observed up to the present always begin to flow through cells before cytomorphogenesis becomes visible, and they always indicate the place of subsequent growth. Suppression of these currents leads to standstill of growth and differentiation. Growth of cells without self-induced currents has not been hitherto observed. The natural currents seem to be, therefore, a necessary component in the mechanism of differentiation and growth. What drives these currents and what role could they play in cytomorphogenesis?

A. The Driving Force for Self-Generated Currents

The cell membrane, almost exclusively, can be considered the site of the electromotive force (emf) or battery for these currents because the cell membrane possesses an electrical potential, the so called membrane potential, and a sufficiently high internal resistance (approx. 10^9 to 10^{11} ohm cm) to prevent a short circuit of ion fluxes. As long as the membrane potential is of equal magnitude everywhere along the membrane, no current flows through the cell. Only when there is a local decrease or increase in the membrane potential, e.g., through opening or closing of ion channels or through segregation of ion channels and ion pumps, can a potential gradient, and thus an emf, arise across the cell (cf., JAFFE 1969). This emf can then drive an ionic current through the cell and the surrounding medium, as long as cytoplasm and medium contain mobile ions as charge carriers.

B. Electric Fields and Concentration Gradients as Mechanisms for Localization

The currents generated by the cells, or driven by externally applied electric fields, induce electric fields on the one hand and local differences in the ion milieu on the other. Already differences of a few mV in the membrane potential within a cell can displace electrically charged components of the membrane laterally (Fig. 7; cf., JAFFE 1977). Such lateral displacement of membrane components has been demonstrated, for instance, in experiments with embryonic muscle cells: field intensities of only about 1 mV per cell, which lasted for about 24 hours, lead to displacement of the concanavalin A receptor within the membrane (POO and ROBINSON 1977).

While the electric field along the membrane only depends on the difference in membrane potential, the electric field caused by the current traversing the cytoplasm depends on the current density and on the kind of ions involved. The magnitude of the ohmic field is defined by the product of the current density and the specific resistivity of the cytoplasm, which is generally between 100 and 300 ohm cm. For the endogenous currents, the magnitude of this ohmic field lies in the range of a few mV cm^{-1}. For ions like Ca^{2+}, the resistivity of the cytoplasm appears to be about 10^3 times higher than for ions like K^+ and Na^+ (HODGKIN and KEYNES 1957, JAFFE and NUCCITELLI 1977). This apparently low mobility of Ca^{2+} ions is probably due to strong binding of Ca^{2+} to fixed negative charges. Such binding could lead to strong local differences in the density of fixed charges in the cytoplasm and, therefore, to strong electric potential gradients, so called Donnan potentials. These Donnan potentials may reach intensities of several Volt cm^{-1}.

Ohmic potentials and Donnan potentials may both generate gradients of mobile charged particles. These gradients are the result of an electrophoretic movement of the molecules in one direction and the spontaneous diffusion in the opposite direction and are given by the relationship: G = m E/D, whereby G is the gradient, m the electrophoretic mobility of the molecules, E the electric field, and D the diffusion coefficient of the molecules. It seems to be quite important for the generation of a concentra-

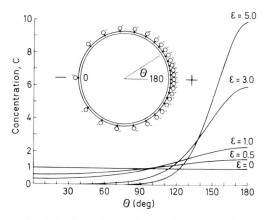

Fig. 7. Distribution of mobile charged particles on the surface of a sphere at electrophoretic equilibrium. Curves plotted according to the equation: $C = \bar{C} \ (\varepsilon \cdot \text{csch} \ \varepsilon) \ e^{\varepsilon \ \cos \Theta}$, where C is the equilibrium concentration on the surface; $\bar{C}$, the average concentration of the particles; $\varepsilon = (m/D) \cdot \bar{V}/2$ with m being the electrophoretic mobility; D, the diffusion coefficient; and $\bar{V}$, the voltage difference established across the sphere by the field. From Jaffe 1977.

tion gradient by this mechanism that the field lasts long enough. All measurements up to the present show that the endogenous currents that traverse cells during cytomorphogenesis flow for at least 10^4 to 10^6 seconds. The necessary time for molecules with diffusion coefficients of 10^{-8} to 10^{-11} cm^2 s^{-1}, probably present in the cytoplasm, to form a gradient of 10% is about 10^3 to 10^6 seconds, i.e., the natural currents flow long enough to build up considerable gradients.

Besides electric fields, the transcellular currents also seem to lead to local changes in the ion concentrations in cells and tissues. If, for instance, Ca^{2+} ions flow into the cell at the future rhizoid pole of *Fucus* zygotes (cf., Section II. C.) a local accumulation of calcium in the cytoplasm of the rhizoid probably takes place. Or when in *Lilium* pollen K$^+$ ions and Ca^{2+} ions enter at the tube and H$^+$ ions are pumped out of the pollen grain, differences probably result in the local ion milieu within the cell as well as in the immediate vicinity of the cell. These changes in the ion milieu probably have effects on the activation or inactivation of enzymes and hormones as well as on the development and function of cell structures. They could thus serve the cell directly as information for local differentiation and growth or, in the case of currents that flow through tissues, they could change the flow of information of other signal molecules.

C. Self-Generated and Forced Calcium-Currents as Important Effectors in Cytomorphogenesis

There are a series of recent observations that indicate that an influx of Ca^{2+} ions always precedes and accompanies the local growth of plant cells, and that Ca^{2+} ions always accumulate in growing cell regions (see also p. 289). For instance, when *Pelvetia* zygotes are put in small funnel-shaped holes of a thin nickel grid stretched between two chambers and illuminated from one side so that all future rhizoids are oriented toward one chamber and all thallus ends toward the other, the influx and efflux of ions at the rhizoid and thallus areas can be measured separately. Using this method, ROBINSON and JAFFE (1975) found a strong asymmetry in the influx of Ca^{2+} ions between the rhizoid and thallus poles during cell polarization and subsequent growth. Ca^{2+} ions are taken up more rapidly at the rhizoid pole than at the thallus pole, and at the thallus pole they were given off faster than at the rhizoid pole. The ionic current through *Pelvetia* zygotes consists, therefore, partly of Ca^{2+} ions. The efflux of Ca^{2+} ions must be active as it runs against the electrochemical gradient. It is especially noteworthy that the total transcellular flux of Ca^{2+} ions during the early development of the zygotes remains constant. Therefore, it seems that calcium channels and calcium pumps are not newly incorporated into the cell membrane but only redistributed.

Evidence for an intensified influx of Ca^{2+} ions into growing rhizoids, pollen tubes, and root hairs were obtained from experiments with radioactive Ca^{2+} and low temperature autoradiography. Brief immersion of embryos of brown algae and of germinated lily pollen in a medium containing $^{45}Ca^{2+}$ lead to an intense labelling of the growing tips of rhizoids and pollen tubes (ROBINSON and JAFFE 1975, L. A. JAFFE et al. 1975). Short labelling times and subsequent prolonged rinsing in an unlabelled solution made it almost certain that the heavy labelling of the tips was not just a matter of an intensified integration of $^{45}Ca^{2+}$ into the cell walls. The addition of $^{45}Ca^{2+}$ or chlorotetracycline lead to an increased labelling or fluorescence, respectively, in the tips of barley root hairs (WEISENSEEL and HOFER, unpublished).

Placing cells in electric fields is one method of investigating the role of transcellular ionic currents in cytomorphogenesis, as these fields drive current through the cells. The interpretation of the results thus obtained presents difficulties, however, as the ionic composition of the forced current is not known and has, perhaps, a rather nonphysiological composition. If one wants not only current but also particular ions to flow through cells, a way to achieve this is to expose the cells to gradients of ionophores.

Indications of a Ca^{2+} influx preceding and inducing local growth were obtained from experiments with the ionophore A 23187 in *Pelvetia* zygotes, *Blastocladiella,* and *Funaria* spores, and in hyphae of *Neurospora.* This ionophore induces, under physiological conditions, a flux of Ca^{2+} ions through the cell membrane in exchange for H^+ ions. This normally leads to a rise in the cytoplasmic Ca^{2+} concentration and to a fall in the H^+ion concentration. *Pelvetia* zygotes situated at a distance of less than 100 µm from a glass thread that was coated by A 23187, showed a strong tendency to form rhizoids in the direction of the glass thread (Robinson and Cone 1980). As the direction of the rhizoid anlage was independent of changes in the extracellular H^+ concentration between 6.5 and 8.1 and changes in the Mg^{2+} concentration between 5 and 50 mM (Mg^{2+} is also transported by A 23187), one can conclude that local growth is induced by a local increase in calcium concentration inside the cell. Also, the spores of *Funaria,* which are exposed to a local source of A 23187, show a tendency, weak but still significant, to form a rhizoid toward the source of the ionophore (Chen and Jaffe 1979). As the reaction proved to be independent of the external H^+ concentration at a pH of 5.5 and 8.0, the result can similarly be interpreted to indicate that a rhizoid is formed where the most Ca^{2+} ions flow into the cell. In control experiments, a purely random distribution of the rhizoid anlagen occured when the ionophore was previously inactivated by prolonged exposure to strong white light.

Flooding whole cells with the ionophore A 23187 leads to phenomena that can best be explained by an influx of Ca^{2+}at many sites of the cell. In the caulonema cells of *Funaria,* for instance, ionophore concentrations of 10^{-5} to 10^{-4} mol 1^{-1} in the medium induce irregular thickening in the cell wall, especially in the tip region (Schmiedel and Schnepf 1980). This points to an intensified exocytosis of Golgi vesicles at these sites. When growing tubes of lily pollen were transferred into a medium containing A 23187, an accumulation of Golgi vesicles at various places behind the tip could be demonstrated by electron microscopy, *i.e.,* vesicles appeared at sites where they are not normally found (Reiss and Herth 1979 a). Again an ionophore induced influx of Ca^{2+} at these sites seems to be responsible. In hyphae of *Neurospora,* A 23187 induces many lateral branches (Reissig 1977). About one hour after addition of the ionophore to a medium containing 10 to 100 µmol 1^{-1} of Ca^{2+}, an average of 1.7 tips per hypha was observed. In *Blastocladiella,* germination of a rhizoid also seems to be accompanied by a Ca^{2+} influx, as A 23187 induces the germination of the spores in the presence of Ca^{2+} ions in the medium (Brunt and Harold 1980). Moreover, substances like La^{3+} and D-600, which can block an influx of Ca^{2+} into the cell, prevent the formation of the papilla and the subsequent sporulation in *Blastocladiella.* On the other hand, Ca^{2+} ions in the medium proved to be necessary and sufficient to induce papilla formation and sporulation (Stump et al., *cf.* Jaffe 1980). The inflow measured by the vibrating electrode in the papilla region in *Blastocladiella* seems, therefore, at least partly to consist of Ca^{2+} ions.

Experiments with the fluorescent calcium indicator chlorotetracycline showed that a calcium gradient is present along tip growing cells. The calcium concentration is highest in the growing tip and falls quite rapidly behing it. One such gradient was demonstrated in pollen tubes of *Lilium*, in root hairs of *Lepidium*, in caulonema cells of *Funaria*, in hyphae of the fungus *Achlya*, and in the whorl cells of *Acetabularia* (REISS and HERTH 1978, 1979 b) (Fig. 8). Measurements with the proton microprobe confirmed the presence of a calcium longitudinal gradient in the pollen tubes of *Lilium* and showed, moreover, a transverse gradient with the highest calcium concentration at the periphery of the cell (BOSCH *et al.* 1980).

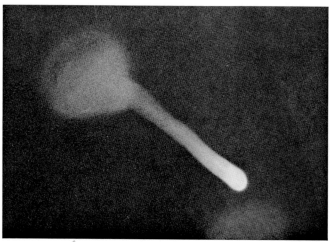

Fig. 8. Calcium gradient in the tube of a growing lily pollen. The growing pollen grain has been incubated in a medium containing the fluorescent dye chlorotetracycline. The intensity of fluorescence is maximal at the tip of the tube and decreases behind it. (According to REISS and HERTH 1978, microphotograph kindly provided by Dr. REISS.)

Application of A 23187 eliminated the longitudinal gradient by raising the intracellular concentration of calcium in the regions further behind the tip (SCHNEPF and REISS, personal communication).

In the light of all these observations, one is almost forced to draw the conclusion that local differences in calcium influx are part of the mechanism affecting localized growth. As it is known that local growth is associated with an exocytosis of vesicles, it can be assumed that an intracellular calcium gradient contributes to the accumulation of these vesicles at the growth pole and promotes their integration into the cell membrane.

V. Summary

Physical and chemical factors from the environment can influence and govern the cytomorphogenesis of plant cells. These factors are most likely perceived by receptors in the cell membrane and produce changes in it that lead to a transcellular ionic current that always enters the future or actual site of growth and leaves at nongrowing areas. In general, the current is more concentrated at the site of entry than at the site of exit. The current densities lie in the range of 1 to 10 μA cm^{-2} on the surface of the cells but might reach much higher densities within the cytoplasm due to constriction of the current paths. As far as is known, Ca^{2+} ions always are a part of the inward current.

The currents seem to be a cause of as well as a consequence of localization, leading to dominance of a region and inhibiting similar developments at other regions of the cell. The mechanism of localization probably involves the establishment of concentration gradients in the cell membrane and in the cytoplasm by electric fields and changes of the local ion milieu. The flux of Ca^{2+} ions seems of special importance as Ca^{2+} ions are strongly bound within the cytoplasm, leading to steep gradients of fixed negative charges and strong electric fields.

References

Åberg, H., 1978: Light and branch formation in the alga *Vaucheria dichotoma* (*Xanthophyceae*). Physiol. Plant. 44, 224—230.

Achenbach, F., Weisenseel, M. H., 1981: Ionic currents traverse the slime mould *Physarum*. Cell Biology International Reports 5, 375—379.

Bentrup, F.-W., 1968: Die Morphogenese pflanzlicher Zellen im elektrischen Feld. Z. Pflanzenphysiol. 59, 309—339.

— 1971: Räumliche Zelldifferenzierung. Umschau 10, 335—339.

— Sandan, T., Jaffe, L. F., 1967: Induction of polarity in *Fucus* eggs by potassium ion gradients. Protoplasma 64, 254—266.

Blatt, M. R., Briggs, W. R., 1980: Blue-light-induced cortical fiber reticulation concomitant with chloroplast aggregation in the alga *Vaucheria sessilis*. Planta 147, 355—362.

— Wessells, N. K., Briggs, W. R., 1980: Actin and cortical fiber reticulation in the siphonaceous alga *Vaucheria sessilis*. Planta 147, 363—375.

Bosch, F., El Goresy, A., Herth, W., Martin, B., Nobiling, R., Povh, B., Reiss, H.-D., Traxel, K., 1980: The Heidelberg proton microprobe. Nuclear Science Applications 1, 33—55.

Brower, D. L., McIntosh, J. R., 1980: The effects of applied electric fields on *Micrasterias*. I. Morphogenesis and the pattern of cell wall deposition. J. Cell Sci. 42, 261—277.

— Giddings, T. H., 1980: The effects of applied electric fields on *Micrasterias*. II. The distributions of cytoplasmic and plasma membrane components. J. Cell Sci. 42, 279—290.

Brunt, J. Van, Harold, F. M., 1980: Ionic control of germination of *Blastocladiella emersonii* zoospores. J. Bacteriol. 141, 735—744.

Chen, T.-H., Jaffe, L. F., 1979: Forced calcium entry and polarized growth of *Funaria* spores. Planta 144, 401—406.

Cohen, R. J., 1974: cAMP levels in *Phycomyces* during a response to light. Nature 251, 144—146.

Geddes, L. A., Hoff, H. E., 1971: The discovery of bioelectricity and current electricity. IEEE Spectrum 8, 38—46.

Haupt, W., 1957: Die Induktion der Polarität bei der Spore von *Equisetum*. Planta 49, 61—90.

Hodgkin, A. L., Keynes, R. D., 1957: Movements of labelled calcium in squid giant axons. J. Physiol. 138, 253—281.

Jaffe, L. A., Weisenseel, M. H., Jaffe, L. F., 1975: Calcium accumulations within the growing tips of pollen tubes. J. Cell Biol. 67, 488—492.

Jaffe, L. F., 1966: Electrical currents through the developing *Fucus* egg. Proc. Nat. Acad. Sci. U.S.A. 56, 1102—1109.

— 1968: Localization in the developing *Fucus* egg and the general role of localizing currents. Advan. Morphog. 7, 295—328.

— 1969: On the centripetal course of development, the *Fucus* egg, and self-electrophoresis. Develop. Biol. Suppl. 3, 83—111.

JAFFE, L. F., 1977: Electrophoresis along cell membranes. Nature 265, 600—602.

— 1980: Control of plant development by steady ionic currents. In: Plant membrane transport: current conceptual issues (SPANSWICK, R. M., LUCAS, W. J., DAINTY, J., eds.), pp. 381—388. Elsevier/North-Holland: Biomedical Press.

— NUCCITELLI, R., 1974: An ultrasensitive vibrating probe for measuring steady extracellular currents. J. Cell Biol. 63, 614—628.

— — 1977: Electrical controls of development. Ann. Rev. Biophys. Bioeng. 6, 445—476.

JANISTYN, B., DRUMM, H., 1972: Light-mediated changes of concentration of cAMP in mustard seedlings. Naturwissenschaften 59, 218.

KATAOKA, H., 1975: Phototropism in *Vaucheria geminata* II. The mechanism of bending and branching. Plant Cell Physiol. 16, 439—448.

— 1977: Phototropic sensitivity in *Vaucheria geminata* regulated by 3′,5′-cyclic AMP. Plant Cell Physiol. 18, 431—440.

LOVETT, J. S., 1975: Growth and differentiation of the water mold *Blastocladiella emersonii*: Cytodifferentiation and the role of ribonucleic acid and protein synthesis. Bacteriol. Reviews 39, 345—404.

LUND, E. J., 1923: Electrical control of organic polarity in the egg of *Fucus*. Botan. Gaz. 76, 288—301.

MARSH, G., BEAMS, H. W., 1945: The orientation of pollen tubes of *Vinca* in the electric current. J. Cell. Comp. Physiol. 25, 195—204.

NOVÁK, B., BENTRUP, F.-W., 1972: An electrophysiological study of regeneration in *Acetabularia mediterranea*. Planta 108, 227—244.

NUCCITELLI, R., 1978: Oöplasmic segregation and secretion in the *Pelvetia* egg is accompanied by a membrane-generated electrical current. Develop. Biol. 62, 13—33.

— JAFFE, L. F., 1974: Spontaneous current pulses through developing *fucoid* eggs. Proc. Nat. Acad. Sci. U.S.A. 71, 4855—4859.

PENG, H. B., JAFFE, L. F., 1976: Polarization of *fucoid* eggs by steady electrical fields. Develop. Biol. 53, 277—284.

POO, M.-M., ROBINSON, K. R., 1977: Electrophoresis of concanavalin A receptors along embryonic muscle cell membrane. Nature 265, 602—605.

REISS, H.-D., HERTH, W., 1978: Visualization of the Ca^{2+}-gradient in growing pollen tubes of *Lilium longiflorum* with chlorotetracycline fluorescence. Protoplasma 97, 373—377.

— — 1979 a: Calcium ionophore A 23187 affects localized wall secretion in the tip region of pollen tubes of *Lilium longiflorum*. Planta 145, 225—232.

— — 1979 b: Calcium gradients in tip growing plant cells visualized by chlorotetracycline fluorescence. Planta 146, 615—621.

REISSIG, J. L., 1977: The divalent cation ionophore A 23187 induces branching in *Neurospora*. J. Cell Biol. 75, 30 a.

ROBINSON, K. R., JAFFE, L. F., 1975: Polarizing *fucoid* eggs drive a calcium current through themselves. Science 187, 70—72.

— CONE, R., 1980: Polarization of *fucoid* eggs by a calcium ionophore gradient. Science 207, 77—78.

SAND, O., 1973: On orientation of rhizoid outgrowth of *Ulva mutabilis* by applied electric fields. Exp. Cell Res. 76, 444—446.

SCHMIEDEL, G., SCHNEPF, E., 1980: Polarity and growth of caulonema tip cells of the moss *Funaria hygrometrica*. Planta 147, 405—413.

WEISENSEEL, M. H., 1979: Induction of polarity. In: Encyclopedia of plant physiology, Vol. 7 (HAUPT, W., FEINLEIB, M. E., eds.), pp. 485—505. Berlin-Heidelberg-New York: Springer.

— NUCCITELLI, R., JAFFE, L. F., 1975: Large electrical currents traverse growing pollen tubes. J. Cell Biol. 66, 556—567.

— JAFFE, L. F., 1976: The major growth current through lily pollen tubes enters as K^+ and leaves as H^+. Planta 133, 1—7.

— DORN, A., JAFFE, L. F., 1979: Natural H^+ currents traverse growing roots and root hairs of barley (*Hordeum vulgare* L.). Plant Physiol. 64, 512—518.

WOHLFARTH-BOTTERMANN, K. E., STOCKEM, W., 1970: Die Regeneration des Plasmalemms von *Physarum polycephalum*. W. Roux' Archiv 164, 321—340.

Role of Lipid Self-Assembly in Subcellular Morphogenesis

P. SITTE

Institute for Biology II, University of Freiburg i. Br., Freiburg i. Br., Federal Republic of Germany

With 7 Figures

Contents

I. Introduction: Self-Assembly *versus* Instructed Morphogenesis

Every living organism can be characterized by its particular structure. At all levels of complexity, its structure is the basis for its specific functional activities. Since most of these structures change during development, one of the problems in biology is to understand how these structures are established. In dealing with morphogenesis at the subcellular level, it is advisable to differentiate between the processes of *synthesis* and *organization*. The molecular building blocks, as synthesized by the metabolic machinery, must become organized in order to form more complex functional entities such as organelles. Mitochondria or plastids are not just "synthesized". Whereas synthesis means the formation of covalent bonds by specific enzymes, any supramolecular organization comes about (and depends on) weak interactions among identical or different molecular components. As these interactions are not mediated by enzymes, the formation of supramolecular entities in living cells is not an immediate consequence of genetic activity. Most (eu-)cellular structures are formed outside the nucleus. Genes, although highly "organized" structurally and functionally, act as scalars in cellular life, not as vectors. Supramolecular structures, on the other hand, are invariably manifestations of vectors.

The supramolecular structures of living cells can come about following one of two different principles:

(1) In the case of *instructed morphogenesis,* the particular structure can be built from its components only by the help of additional information. In other words, the components themselves do not contain all the information necessary for their proper alignment. The obligate additional information is provided by a subsidiary structure that, in typical instances, consists of components other than the structure to be formed and will not take part in it. Thus, this "template" has something in common with a catalyst.

(2) In the case of *self-assembly,* or self-organization, the molecular components of the structure eventually formed bear all the information required to align themselves in the proper manner. No template is necessary. All that is needed is a sufficient concentration of building blocks and suitable conditions. Morphogenesis by self-assembly can be studied with relative ease in cell-free systems. One well-known paradigm of self-assembly is crystallization (*cf.,* KING 1969). The most spectacular aspect of self-assembly is that ordered structures will appear spontaneously in formerly amorphous and isotropic masses. In template-mediated morphogenesis, structural order comes about as a result of the action of another structure already present (*structura e structura*), whereas self-assembly can be envisaged as a *de novo* formation of specific structures.

There are many well-studied examples of self-assembly as a structure-generating principle, especially with protein involved (*cf., e.g.,* HOLMES 1975, COHEN 1977). Better known examples are virus capsids, nucleosomes, ribosomes and other RNP particles, polyenzyme complexes, microtubules, extracellular fibrous proteins such as silk or collagen, microfilaments and myofilaments, and two-dimensional lattices of connexin complexes in gap junctions. Even cell-cell recognition, depending on ligand-receptor interaction, is based on molecular self-assembly. However, in spite of all these (and many more) impressive examples, self-assembly is only of limited importance in subcellular morphogenesis. Apparently, there is a delicate balance of order and chaos in every living cell (SITTE 1976 a, 1980 a). Excessive structuralization must be fatal, as exemplified by metabolic disorders resulting in surplus production of, *e.g.,* structural proteins. Whereas product size is determined by the template in instructed morphogenesis, no corresponding limitations exist in the case of self-assembly. Accordingly, self-assembly often leads to structures of infinite dimensions in model experiments. The limitation of self-assembly processes in living cells often involves the participation of structures that may be regarded as templates. For example, the self-assembly of capsid protomers of viruses with helical symmetry (such as TMV) is known to be limited by the viral nucleic acid. In the case of the extracellular formation of collagen fibres, the premature fibre formation from intracellularly synthesized procollagen chains and tropocollagen particles if forestalled by a precise sequence of biochemical events during the exocytotic process (*cf.,* KÜHN 1974, MILLER and MATUKAS 1974, WEINSTOCK and LEBLOND 1974). Likewise, self-assembly of microtubules depends on nucleating sites (cf., DUSTIN 1978), and presumably there are some as yet unknown templates to stop microtubule growth.

I appears, then, that self-assembly, although extensively involved in basic morphogenetic processes, is normally subjected to regulative restrictions. The situation might have been different, however, in the earliest stages of evolution when effective templates had not yet been evolved. In those remote times, self-organization must have been of prime importance (cf., EIGEN 1971, SCHUSTER 1972, HAHN 1972, KUHN 1976, FOLSOME 1976, HAKEN 1978, EIGEN and SCHUSTER 1979).

When compared with self-assembly of proteins, polysaccharides, and nucleic acids, lipid self-assembly has been studied less intensely in recent years, possibly due to its relatively low specificity. This, however, appears inappropriate if one considers the general occurrence of lipid structures and their importance in so many fundamental processes of cellular life (SITTE 1979).

II. Lipid Dominated Subcellular Structures and Their Principal Functions

In organismal and cellular life, lipids are involved in quite different functions. Three of these functions are of general importance, e.g., compartmentation, storage, and pigmentation.

The separation of a cell from its surroundings, as well as separation of different compartments from each other within a cell, is secured by bio-membranes (cf., SITTE 1977, 1979). It is now generally agreed that all cellular membranes (not just the plasma membrane or the vacuolar membranes) separate plasmatic compartments from nonplasmatic ones. Correspondingly, every biomembrane, including the different intracellular membranes, faces a plasmatic bulk phase on one of its surfaces and an extra plasmatic bulk phase on the other (PS and ES, respectively, in contemporary freeze-etching nomenclature, cf., BRANTON et al. 1975). In functional terms, compartmentation invariably relies on the hindrance of free diffusion of polar compounds between aqueous domains, and it is the lipid moiety of biomembranes that ensures this very effect.

Biomembranes discriminate between solute and solvent, being readily penetrable for the latter (osmosis). Organisms not inhabiting aqueous environments need to restrict the loss of water by entirely impermeable surface layers. This life-securing function is again served by lipids (SITTE 1975, SCHÖNHERR 1976, HADLEY 1980). Outstanding examples of this kind are provided by the cuticles of higher plants (cf., MARTIN and JUNIPER 1970) and of insects (cf., BEAMENT 1968, ROCKSEIN 1974). In both cases, numerous, very thin films of apolar hydrocarbon and wax molecules are embedded in an extracellular macromolecular matrix of moderate lipophilia, the apolar lamellae being oriented parallel to the interface with the rod-like molecules perpendicular to it (SITTE 1962).

Among the different intracellular storage depots of metabolic fuel, lipids are particularly suitable with regard to their high proportion of stored energy to weight. The most abundant representative of storage lipids are the triacylglycerols. These neutral (apolar) lipids accumulate inside plasmatic compartments in the form of fat globules (oil droplets). Corresponding globular lipid accumulations are found extracellularly in the body fluid of many animals including mammals. These plasma lipoproteins serve transport functions (cf., Smellie 1971).

The mobilization of storage lipids apparently poses some problems for certain cell types, probably because of the insolubility of triacylglycerols in water. Lysosomes often possess only limited lipidolytic capabilities. Correspondingly, lipids prevail in residual bodies (telolysomes) that represent the final stages of compartmented intracellular digestion. Whenever an ageing cell is not able to extrude the residual bodies by exocytosis, they will eventually be transformed to irregularly shaped lipofuscin granules or "age pigment" (cf., Malkoff and Strehler 1963).

Among the different organismic pigments, hydrophilic and hydrophobic compounds are to be distinguished with entirely different intracellular localization (chymochromes and plasmochromes, respectively, sensu Seybold 1942). Hydrophobic pigments, such as the chlorophylls and the carotenoids, are found in close association with lipid-dominated subcellular structures such as, e.g., thylakoids or outer retinal segment membrane discs. How different carotenoid-accumulating structures actually can be is best exemplified by the chromoplasts of flower petals and fruits (cf., Sitte et al. 1980). In these yellow to red pigmented plastids, the carotenoids are associated with either lipid globules or filaments (chromoplast tubules, according to their appearance in cross section) or concentric membranes, and in some cases they are found in the form of carotene crystals.

III. Basic Principles of Lipid Self-Assembly

In aqueous phases, lipid molecules are forced together by the hydrophobic effect (Tanford 1973, 1978). Generally speaking, lipids are present, due to phase separation, either in high concentration or not at all. In spite of this, true lipid crystals are only rare instances in or on living cells. Among the exceptions are the carotene crystals of crystallous chromoplasts and the epicuticular wax crystallites (cf., Baker and Parsons 1971). More often, lipid agglomerations are in a fluid or semifluid condition. Yet, the structures formed by them are quite different because of the extent of interaction with water (Small 1970). With regard to this interaction, two principal groups of lipids can be distinguished. Truly apolar lipids (such as larger hydrocarbons, waxes, carotenes, and triacylglycerols) show practically no interaction with water, whereas amphipolar (amphiphilic) lipid molecules possess a hydrophilic region situated at one end of the rod-like molecule. Typical representatives of this latter group are phospholipids and glycolipids (as well as soaps and other tensides in technology). Lipids of this kind are surface-active, that is, they gather at water/air interfaces in the form of oriented monolayers and will reduce surface tension drastically so that, for example, foams can be formed. Within an aqueous bulk phase, amphiphils become arranged in different forms of *micelles* (*cf.*, MITTAL 1977, BOSCHKE 1980, TURRO *et al.* 1980) or bimolecular films (*bilayers*). Artificial lipid bilayers have been studied extensively by EM and X-ray techniques. (For a small selection of pioneering papers, the reader may refer to, *e.g.*, STOECKENIUS 1959, 1962, STOECKENIUS *et al.* 1960, LUZZATI and HUSSON 1962, BANGHAM and HORNE 1964, LUCY and GLAUERT 1964, REISS-HUSSON 1967, PAPAHADJO-POULOS and MILLER 1967.) These studies were done mainly in the hope of obtaining pertinent information on biomembranes (*cf.*, EISENBERG and McLAUGHLIN 1976). Artificial "membranes" of this kind ("liposomes"— BANGHAM 1972, and "black films"—HUANG *et al.* 1964) have been used widely in studies on membrane permeability and transport; it is hoped, moreover, that liposomes can be used therapeutically as a vector for administering drugs, enzymes, or hormones (*cf.*, COLLEY and RYMAN 1976, CELIS *et al.* 1980, GREGORIADIS and ALLISON 1980).

All the many different forms of lipid agglomerations, from crystals to droplets (globules) and finally membranes, come about by self-assembly, whereby the aqueous environment quite often plays a decisive role. In fact, it is largely due to the hydrophobic effect (entropic interaction) that lipid micromolecules (and not macromolecules as in the case of proteins, poly-saccharides, and nucleic acids) are endowed with the ability to build struc-tures of microscopic and even macroscopic dimensions—a unique situation indeed.

Fluid aggregates of apolar lipids tend to minimalize their contact area with water and correspondingly will assume a spherical shape. Since there is virtually no interaction between lipid molecules of this kind and water, the energy of cohesion in both phases is greater than the energy of adhesion between them. Such is the situation with storage lipid globules (see Section IV. D.). If, on the other hand, the lipid aggregate is built of amphiphils, adhesion prevails and the lipid/water contact area will be maximalized. The volume of such an aggregation correspondingly drops to a minimum, that is, monolayers or bilayers are being assembled. In fact, membrane-forming lipids such as phospholipids and glycolipids play an eminent role in cellular compartmentation. They have thus been collectively termed "structural lipids".

Many self-assembly aggregates of lipid molecules are both fluid and an-isotropic. Accordingly, they belong to the large class of liquid crystals (LC) or mesophases (*cf.*, *e.g.*, GRAY 1962, SAUPE 1968, JOHNSON and PORTER 1970, STEINSTRÄSSER and POHL 1973, DE GENNES 1974, KOBALE and KRÜGER 1975, FRIBERG 1976, BROWN and WOLKEN 1979, KELKER and HATZ 1980). More specifically, they represent lyotropic mesophases as they are formed under isothermic conditions, whereby they require the participation of a solvent (water).

Micelles and bilayers are *smectic* mesophases (Fig. 1). This type of an LC is essentially two-dimensional with the amphiphilic lipid molecules oriented perpendicularly to the plane of bilayer extension. In LC of the *nematic* type, the rod-shaped lipid molecules—apolar in this case—are again arranged in a parallel fashion. However, the whole aggregate is not two-dimensional but one-dimensional. The typical appearance of a nematic LC is threadlike. This is a consequence of the ease with which the lipid molecules can slide past each other. (In smectic mesophases, this kind of movement is prevented by the amphipolarity of the lipid molecules.) Unlike smectic LC, nematic LC often show high values of optical anisotropy since both intrinsic and structural birefringence are positive.

Until recently, smectic mesophases were the only lipid LC observed in living cells (see Section IV. A.). However, nematic mesophases can also be formed by cellular lipids (see Section IV. C.).

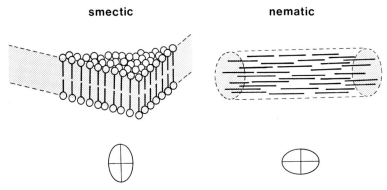

Fig. 1. Molecular build-up and optical character of smectic and nematic mesophases (original).

IV. Selected Examples of Lipid Self-Assembly

A. Biomembranes

According to literally all the evidence available, lipid bilayers are the basic structure in the molecular buildup of every biomembrane (for a few recent reviews, cf., FINEAN et al. 1978, BECK 1980, BITTAR 1980, see in particular ZWAAL et al. 1976). The lipid bilayer is assumed to be in a liquid-crystalline state of a smectic type. Rotation and lateral diffusion is therefore possible (and very fast indeed) not only among the lipid molecules but also among intrinsic membrane proteins inserted into the bilayer or even spanning it (cf., CHERRY 1979, PETERS 1980). On the other hand, the switching of lipid molecules from one side of the membrane to the other (flip-flop) is a thermodynamically unfavorable, and consequently rare, event. Because of this, and the different phases on both sides of every membrane, a pronounced lipid asymmetry has been observed among the two leaflets of the bilayer membrane (cf., BERGELSON and BARSUKOV 1977, OP DEN KAMP 1979). However, the basic bilayer structure is by no means perfect. It is disturbed to a varying extent not only by intrinsic membrane proteins (cf., LEE 1977, SANDERMANN 1978) but also by transient nonbilayer arrangements under physiological conditions (CULLIS and DE KRUIJFF 1979, DE KRUIJFF et al. 1980). Furthermore, the presence of sterol lipids in all eukaryotic membranes precludes abrupt thermal transitions and keeps the bilayer in a fluid form over a wide temperature range. All these features enable the cellular membranes to fuse (POSTE and NICOLSON 1978) or to separate in the course of membrane flow processes, to be extremely flexible, to exhibit a certain (if limited) amount of flip-flop, and to show a restricted permeability even for smaller polar particles.

The structural flexibility of biomembranes and their notorious ability to immediately seal up any artificial rupture (spectacularly exemplified by the formation of closed microsomal vesicles from disrupted ER cisternae in cell homogenates) are of paramount importance for the role biomembranes play as mediators of cellular compartmentation.

Nevertheless, it is possible that membranes could assume certain configurations due to forces generated by themselves. It is assumed that both the polar and the apolar lipid domains of a membrane bilayer are under considerable pressure (SCHUMACHER and SANDERMANN 1976, CONRAD and SINGER 1979). Due to the lipid asymmetry of biomembranes, it should be possible to intercalate suitable amphiphilic compounds either in the outer or the inner sheet of the bilayer, thus forcing it to form cup-like projections or indentations (bilayer couple hypotheses; SHEETZ and SINGER 1974). Although corresponding model experiments with cationic "cup formers" (e.g., chlorpromazine) and anionic "crenators" (e.g., oleates) have been partly successful, it remains to be shown whether processes of this kind are of any significance under in vivo conditions. In this connection it may be noted, nevertheless, that tubular transformations of membranes have been observed frequently (see, e.g., WILLIAMS 1977).

Accumulating evidence points to the amazing fact that biomembranes apparently cannot originate de novo (SITTE 1979, 1980 a), in spite of the fact that artificial lipid bilayers originate by simple self-assembly. According to the evidence presently available, biomembranes develop exclusively by enlargement and differentiation of pre-existing membranes, possibly as a consequence of their being barriers between plasmatic and nonplasmatic compartments. This finding shows that not only nucleic acid molecules but also more complicated, supramolecular structures of cells exhibit genetic continuity.

B. Prolamellar Bodies (PLB)

Etioplasts contain a remarkable agglomeration of vesicles and/or tubules known as the prolamellar body (PLB). In many instances, the tubules are regularly branched and arranged in paracrystalline arrays the structure of which has been elucidated by WEHRMEYER (1965). Recently it has been shown that crystalline PLB contain a large amount of saponins (KESSELMEIER and BUDZIKIEWICZ 1979). The saponins, which have been identified in the case of Avena as avenacoside A and B (Fig. 2), amount to about one-half the isolated PLB's dry weight. The complicated structure of crystalline PLB can be brought about by molecular self-assembly of their lipid components. When isolated PLB are entirely dissolved in SDS, tubules of the same diameter as in situ will appear after some time. These tubules are either unbranched or branched in a periodical manner (LÜTZ et al. 1977). Likewise, the "native" structure of PLB can be produced artificially *in vitro*, without any participation of proteins from an acetone extract of PLB (RUPPEL *et al.* 1978) and by self-assembly of pure saponins (Fig. 3). This latter experiment demonstrates that the complicated structure of crystalline PLB is primarily a result of lipid (saponin) self-assembly. A strict correlation seems to exist between the presence of amphipolar saponin glycosides in a given plant and its capability to form crystalline PLB in etioplasts (KESSELMEIER and RUPPEL 1979).

Primary thylakoids (PT, termed *prothylakoids* by LÜTZ) differ greatly from PLB tubules with regard to their chemical composition (LÜTZ 1978, LÜTZ and KLEIN 1979). They contain much more protein, their lipid-to-

Fig. 2. Avenacosid A (R = H) and B (R = Glc), according to TSCHESCHE after KESSELMEIER and BUDZIKIEWICZ 1979.

protein ratio being 1.9 as compared with 5.4 in the case of PLB tubules. Furthermore, the protein pattern as determined by SDS-PAGE is quite different. Among the proteins of the PT is also protochlorophyll(ide) reductase. The chlorophylloids seem to be located almost exclusively in the PT, whereas the PLB tubules contain less than 10% of them.

C. Chromoplast Tubules: Nematic Mesophases in Living Cells

Chromoplasts are carotenoid-bearing plastids devoid of photosynthetic activity due to the lack of chlorophyll and thylakoids. The carotenoids, often in the form of acylated xanthophylls (EGGER 1964), are associated with different internal substructures of the chromoplasts. Depending on the substructure, these organelles can be classified as globulous, tubulous, membraneous, and crystallous chromoplasts (SITTE 1974, SITTE *et al.* 1980). In the EM, chromoplast tubules appear as straight or slightly bent filaments with outer diameters of about 15 to 20 nm and different lengths of up to 10 µm in the elongated chromoplasts of ripe hips. In cross section these filaments show a cylindrical boundary heavily contrasted by OsO_4 or MnO_4^- and an electron-translucent core (Fig. 4). In most cases, many such tubules are aligned parallel to form bundles. Within these bundles, the

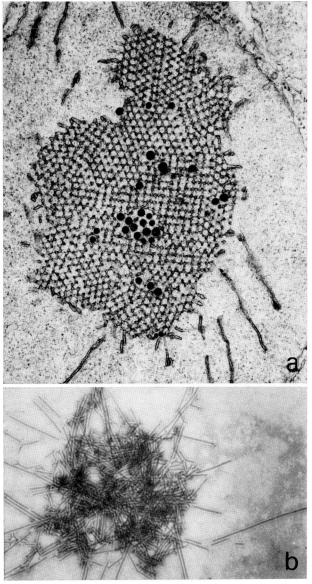

Fig. 3. *a* Paracrystalline prolamellar body of an etioplast of *Avena sativa*, 58,000 : 1.
b Re-aggregated avenacosid tubules, negative staining, 85,000 : 1. Courtesy of Dr. C. Lütz,
Cologne.

tubules are hexagonally arranged despite their varying diameters. In the polarizing microscope, the bundles exhibit an extraordinarily strong positive birefringence ($\bar{A} = 0.076$ in chromoplasts of the hip of *Rosa rugosa*, $\bar{A} = 0.046$ in petal chromoplasts of *Tropaeolum majus*; Sitte, unpublished). Chromoplast tubules of different sources have been isolated (Winkenbach et al. 1976, Wuttke 1976, Liedvogel et al. 1978). In negatively stained preparations, they appear as compact filaments (Fig. 5). According to chemical data, the tubules are lipid-dominated structures. The analytical data are also consistent with a structural model forwarded by Winkenbach et al. (1976) (*cf.*, Fig. 6) in which the tubules consist of an apolar central wick of pigment molecules and a less hydrophobic surface coat of polar lipids and proteins. This model is born out by EM studies on the distribution of contrast-mediating osmium or manganese compounds among lipid substructures of cells (*cf.*, *e.g.*, Bahr 1954, Sitte 1975, 1976 b, Mérida et al. 1981). According to these studies, the heaviest contrast is observed in regions of moderate lipophilia, whereas apolar domains remain uncontrasted.

Model experiments on the formation of such tubules by lipid self-assembly have been partly successful (Sitte 1980 b). If saturated solutions in chloroform of β-carotene or diacylluteine (which comprises more than 94% of the pigment fraction of flower chromoplasts of the nasturtium—Winkenbach et al., 1976) are allowed to lose the solvent by slow evaporation, flexible strands and clusters of small crystallites are formed. The strands, having different diameters, appear deep red in the intensely orange mother liquor. They exhibit positive phase contrast and show up very distinctively in Nomarski interference contrast (Fig. 7). Consequently, their refractive index is higher than that of the saturated solution. The strands exhibit positive birefringence, their optical anisotropy A being extremely high with values of over 0.24. These values surpass the corresponding ones of carotene crystals, which, in contrast to the strands, show oblique extinction in the polarizing microscope.

These observations demonstrate that pure carotenes and xanthophyll esters are able to form, under appropriate conditions, liquid crystals of the nematic type (Sitte 1980 b). The pigment molecules are oriented parallel to the longitudinal axis of the threadlike mesophases. The extraordinarily high optical anisotropy (also observed with other nematic mesophases) is attributable to the quasi-crystalline alignment of extended system of conjugated double bonds in axial direction, as contrasted to extremely weak bonds perpendicular to them. There is every reason to assume that the core of the chromoplast tubule is essentially a nematic, lyotropic mesophase of apolar pigment molecules. This appears to be the first observation of nematic liquid crystals in living cells. It might well be that the anisometric,

---

Fig. 4. Chromoplast tubules in sections. a, b: from hips of *Rosa rugosa*; cross section *a* fixation with glutaraldehyde and OsO₄; 55,000 : 1; electron micrograph by Dr. H.-G. Wuttke; longitudinal section *b* fixation with KMnO₄; 40,000 : 1, original. *c* Chromoplast from the petal epidermis of *Impatiens noli-tangere*, KMnO₄; 75,000 : 1, original.

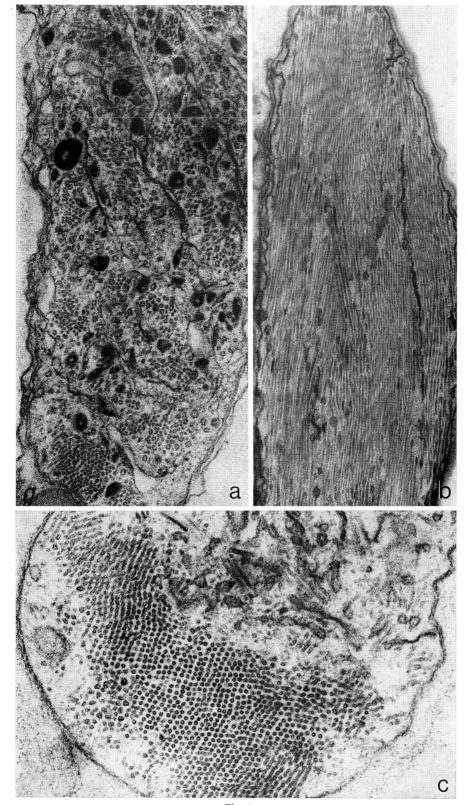

Fig. 4

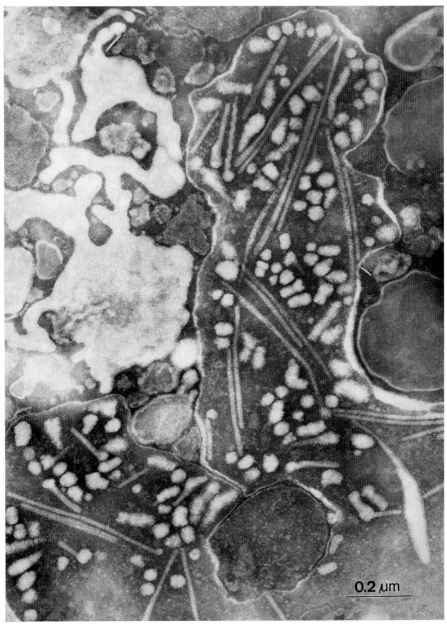

Fig. 5. Isolated chromoplast tubules, together with lipid bodies and membrane fragments from flower chromoplasts of *Tropaeolum majus*, negative staining. Courtesy of Dr. H. FALK.

and sometimes bizarre, shape of tubulous chromoplasts is dictated by the formation of tubule bundles and is to be regarded, therefore, as the consequence of a lipid self-assembly process.

According to this interpretation of the molecular structure of chromoplast tubules, the presence (in high concentration) of apolar carotene and/or carotenoid esters is a prerequisite for tubule formation. In fact, if the biosynthetic pathway leading to colored carotenoids is blocked in young petals of *Tropaeolum majus* or *Impatiens noli-tangere* by the herbizide SAN 9789 (FROSCH et al. 1979), unpigmented petals result whose plastids contain no filamentous or tubulous structures (EMTER and FALK, unpublished). This situation contrasts to the one encountered in chromoplasts of the globulous

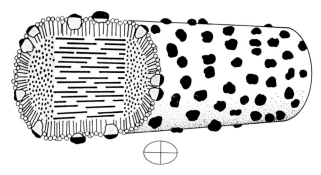

Fig. 6. Scheme of the molecular structure of chromoplast tubules (original).

(LJUBEŠIĆ 1973) or membraneous type (LIEDVOGEL and FALK 1980) in which the respective substructures are also formed in the absence of pigment.

D. Lipid Globules

Spherical lipid aggregations are common in practically all forms of eucytes. Prominent examples are found in the fat and oil producing and/or storing cells such as those of adipose tissue and fat bodies of mammals and insects and in the endosperm of oil seeds (cf., APPELQVIST 1975). Of particular interest is the general occurrence of "plastoglobules" in plastids (LICHTENTHALER 1968). The main function of these lipid droplets is the intracellular storage of reserve lipids, mainly in the form of triacylglycerols (cf., e.g., JACKS et al. 1967, YATSU et al. 1971, ALLEN et al. 1971, BANCHER et al. 1972, NORTON and HARRIS 1975, KLEINIG et al. 1978, HANSMANN 1980). In some cases they function additionally as accumulation sites of lipophilic pigments, as in the eyespots of many flagellates as well as in globulous chromoplasts and gerontoplasts.

The spherical shape of these lipid inclusions indicates their content to be in a fluid and nonordered state. Consequently, their shape seems dictated by the forces discussed on p. 405. The fluid state is apparently maintained by molecular heterogeneity and an appropriate level of desaturation of the fatty acid moiety, much as in biomembranes in which the corresponding

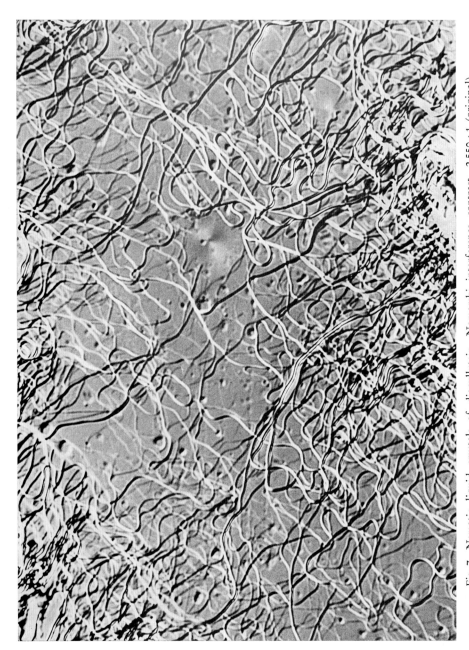

Fig. 7. Nematic liquid crystals of diacyllutein. Nomarski interference contrast, 2,550 : 1 (original).

phenomenon has been studied extensively (cf., e.g., TRÄUBLE 1971, WIESER 1973, KATES and KUKSIS 1980).

Nonspherical, polyhedral lipid inclusion bodies have been found only rarely in certain plastids (THOMSON and PLATT 1973, SITTE 1974). Presumably they contain crystallized lipids. After freeze-etching, lipid globules often appear to consist of concentric layers (cf. MOOR and MÜHLETHALER 1963, RUSKA and RUSKA 1969, LIU 1974, see also PRECHT 1979). However, these layers may be, in part, an artifact resulting from the cooling down of the samples during preparation.

The question of whether intracellular lipid globules possess a membrane is not definitely settled. That the surface of lipid globules is in a less fluid state than the interior is indicated by the observation that colliding globules do not fuse, even if their surfaces become flattened at the contact zone (SITTE 1963). However, from observations of this kind, no definite suggestions can be derived as to the real nature of the surface layer. Early EM investigations on the origin of oil droplets in plant cells have led to the assumption that the lipid globules are derived from the ER and thus become fitted with a surrounding unit membrane (FREY-WYSSLING et al. 1963, GRIESHABER 1964). Later studies indicated the accumulation of storage oil within the apolar interior of ER membranes, gradually forcing apart the two sheets of the bilayer (SCHWARZENBACH 1971, KWIATKOWSKA 1973, WANNER and THEIMER 1978, WANNER et al. 1981). Analytical data of lipid globules from different plant sources are consistent with findings that a half-unit membrane exists (cf., JACKS et al. 1967, YATSU et al. 1971, YATSU and JACKS 1972; see, however, KLEINIG et al. 1978; for plastoglobules, cf., HANSMANN 1980).

Similar conclusions have been reached for the heterogeneous class of lipoprotein particles circulating in the blood of mammals (for review, see SCANU 1972). Proteins, phospholipids, and free cholesterol form an outer coat corresponding roughly with a half-unit membrane, whereas the fluid core consists of apolar triacylglycerols and cholesterol esters (SHEN et al. 1977, SCANU 1978). Consequently, the diameter of the respective particle, corresponding to a certain surface-to-volume ratio, is closely related to chemical composition. This provides an impressive example of how shape and size of a structure can be determined by lipid self-assembly and chemical composition. In the course of thermal transitions, characteristic rearrangements of the lipid moiety have been observed (DECKELBAUM et al. 1977).

It seems, nevertheless, that observations and conclusions of this kind cannot be generalized. According to BERGFELD et al. (1978), the storage lipid bodies of cotyledon cells in Sinapsis alba originate in the ground plasm, usually in close proximity to plastids but separated from ER elements. At first they are devoid of any sort of membrane. Later on an osmiophilic coat, 3 nm thick, becomes detectable. After degradation of the storage lipids, the coat remains visible as a collapsed saccule. Its protein pattern is quite different from that of ER membranes.

In this connection it should be remembered that milk fat globules, which are wrapped in derivatives of the plasma membrane, are extruded from mammary gland cells by budding (*cf.*, PATTON and KEENAN 1975). This means that inside the cell those globules are devoid of a membrane (*cf.*, SITTE 1979).

Supposedly, these conflicting data are, in part, a consequence of the vast heterogeneity of intracellular lipid globules. Even if one confines one-self to plant cells, up to now no firm statement is possible as to whether all those lipid globules are homologous. In fact, there is no accord as to whether the small spherosomes (described as microsomes by v. HANSTEIN in 1880, *cf.*, PERNER 1958) found in nearly every plant cell are related to the larger oleosomes of lipid storing cells (YATSU *et al.* 1971). Whereas some authors assume both categories of lipid globules to be different mani-festations of the same thing (*cf.*, *e.g.*, YATSU *et al.* 1971), others presented evidence to the contrary be showing lipid bodies of different structural and/or cytochemical properties present in the very same cell (see, *e.g.*, SOROKIN 1967, SAUTER 1967, 1968, MOLLENHAUER and TOTTEN 1971, BANCHER *et al.* 1972, SMITH 1974).

References

ALLEN, C. F., GOOD, P., MOLLENHAUER, H. H., TOTTEN, C., 1971: Studies on seeds IV. Lipid composition of bean cotyledons. J. Cell Biol. 48, 542—546.

APPELQVIST, L. Ã., 1975: Biochemical and structural aspects of storage and membrane lipids in developing oil seeds. In: Recent advances in the chemistry and biochemistry of plant lipids (GAILLARD, T., MERCER, E. I., eds.), pp. 247—286. London-New York-San Francisco: Academic Press.

BAHR, G. F., 1954: Osmium tetroxide and ruthenium tetroxide and their reactions with biologically important substances. Exper. Cell Res. 7, 457—479.

BAKER, E. A., PARSONS, E., 1971: Scanning electron microscopy of plant cuticles. J. Microscopie 94, 39—49.

BANCHER, E., WASHÜTTL, J., GOLLER, H.-J., 1972: Untersuchungen der Lipide in den Sphaerosomen- und Mitochondrienfraktionen γ-bestrahlter Samen von Erdnuß (*Arachis hypogaea*) und Walnuß (*Juglans regia*). Z. Pflanzenphysiol. 67, 399—403.

BANGHAM, A. D., 1972: Lipid bilayers and biomembranes. Annu. Rev. Biochem. 41, 753—776.

— HORNE, R. W., 1964: Negative staining of phospholipids and their structural modification by surface-active agents as observed in the electron microscope. J. Mol. Biol. 8, 660—668.

BEAMENT, J. W. L., 1968: The insect cuticle and membrane structure. Brit. med. Bull. 24/2, 130—134.

BECK, J. S., 1980: Biomembranes. Fundamentals in relation to human biology. Washington-New York-London: Hemisphere Publ. Corp.

BERGELSON, L. D., BARSUKOV, L. I., 1977: Topological asymmetry of phospholipids in membranes. Science 197, 224—230.

BERGFELD, R., HONG, Y.-N., KÜHNL, T., SCHOPFER, P., 1978: Formation of oleosomes (storage lipid bodies) during embryogenesis and their breakdown during seedling develop-ment in cotyledons of *Sinapis alba* L. Planta 143, 297—307.

BITTAR, E. E. (ed.), 1980: Membrane structure and function. New York: J. Wiley.

BOSCHKE, F. L. (ed.), 1980: Micelles. (Topics in Curr. Chem., Vol. 87.) Berlin-Heidelberg-New York: Springer.

BRANTON, D., BULLIVANT, S., GULILA, N. B., KARNOVSKY, M. J., MOOR, H., MÜHLETHALER, K., NORTHCOTE, D. H., PACKER, L., SATIR, B., SATIR, P., SPETH, V., STAEHELIN. L. A., STEERE, R. L., WEINSTEIN, R. S., 1975: Freeze-etching nomenclature. Science 190, 54—56.

BROWN, G. H., WOLKEN, J. J., 1979: Liquid crystals and biological structures. New York-San Francisco-London: Academic Press.

CELIS, J. E., GRESSMANN, A., LOYTER, A. (eds.), 1980: Transfer of cell constituents into eukaryotic cells. New York-London: Plenum Press.

CHERRY, R. J., 1979: Rotational and lateral diffusion of membrane proteins. Biochim. Biophys. Acta **559**, 289—320.

COHEN, C., 1977: Protein assemblies and cell form. TIBS **2**, 51—55.

COLLEY, C. M., RYMAN, B. E., 1976: The liposome: from membrane model to therapeutic agent. TIBS **1**, 203—205.

CONRAD, M. J., SINGER, S. J., 1979: Evidence for a large internal pressure in biological membranes. Proc. Natl. Acad. Sci. U.S.A. **76**, 5202—5206.

CULLIS, P. R., DE KRUIJFF, B., 1979: Lipid polymorphism and the functional roles of lipids in biological membranes. Biochim. Biophys. Acta **559**, 399—420.

DECKELBAUM, R. J., SHIPLEY, G. G., SMALL, D. M., 1977: Structure and interactions of lipids in human plasma low density lipoproteins. J. Biol. Chem. **252**, 744—754.

DE GENNES, P. G., 1974: The physics of liquid crystals. Oxford: Clarendon Press.

DE KRUIJFF, B., CULLIS, P. R., VERKLEIJ, A. J., 1980: Non-bilayer lipid structures in model and biological membranes. TIBS **5**, 79—81.

DUSTIN, P., 1978: Microtubules. Berlin-Heidelberg-New York: Springer.

EGGER, K., 1964: Vergleichende Untersuchung der Xanthophyllveresterung in Blüten, Früchten und Herbstlaub. Ber. dtsch. bot. Ges. **77**, (145)—(150).

EIGEN, M., 1971: Selforganization of matter and the evolution of biological macromolecules. Naturwiss. **58**, 465—523.

— SCHUSTER, P., 1979: The hypercycle. Berlin-Heidelberg-New York: Springer.

EISENBERG, M., McLAUGHLIN, S., 1976: Lipid bilayers as models of biological membranes. BioScience **26**, 436—443.

FINEAN, J. B., COLEMAN, R., MICHELL, R. H., 1978: Membranes and their cellular functions, 2nd ed. Oxford-London-Edinburgh-Melbourne: Blackwell.

FOLSOME, C. E., 1976: Synthetic organic microstructures and the origins of cellular life. Naturwiss. **63**, 303—306.

FREY-WYSSLING, A., GRIESHABER, E., MÜHLETHALER, K., 1963: Origin of spherosomes in plant cells. J. Ultrastruct. Res. **8**, 506—516.

FRIBERG, S. (ed.), 1976: Lyotropic liquid crystals (Adv. Chem. Ser. No. 152). Washington, D.C.: Amer. Chem. Soc.

FROSCH, S., JABBEN, M., BERGFELD, R., KLEINIG, H., MOHR, H., 1979: Inhibition of carotenoid biosynthesis by the herbizide SAN 9789 and its consequences for the action of phytochrome on plastogenesis. Planta **145**, 497—505.

GRAY, G. W., 1962: Molecular structure and the properties of liquid crystals. London-New York: Academic Press.

GREGORIADIS, G., ALLISON, A. C. (eds.), 1980: Liposomes in biological systems. Chichester: J. Wiley.

GRIESHABER, E., 1964: Entwicklung und Feinbau der Sphärosomen in Pflanzenzellen. Vierteljahresschrift naturforsch. Ges. Zürich **109**, 1—23.

HADLEY, N. F., 1980: Surface waxes and integumentary permeability. Amer. Scientist **68**, 546—553.

HAHN, C. W., 1972: Composite protostructures: an exercise in model building. J. Polymer Sci. C **39**, 331—335.

HAKEN, H., 1978: Synergetics. An introduction. Nonequilibrium phase transitions and self-organization in physics, chemistry and biology, 2nd ed. Berlin-Heidelberg-New York: Springer.

HANSMANN, P., 1980: Isolierung und Charakterisierung der Globuli aus Chromoplasten von *Viola tricolor*. Dipl.-Work, Univ. Freiburg/Br.

HANSTEIN, J. v., 1880: Biologie des Protoplasmas. Bot. Abhdlg. Morph. Physiol. 4/2, Bonn.

HOLMES, K. C., 1975: Selbstorganisation biologischer Strukturen. Klin. Wochenschr. 53, 997—1005.

HUANG, C., WHEELDON, L., THOMPSON, T. E., 1964: The properties of lipid bilayer membranes separating two aqueous phases: formation of a membrane of simple composition. J. Mol. Biol. 8, 148—160.

JACKS, T. J., YATSU, L. Y., ALTSCHUL, A. M., 1967: Isolation and characterization of peanut spherosomes. Plant Physiol. 42, 585—597.

JOHNSON, J. F., PORTER, R. S., 1970: Liquid crystals and ordered fluids. New York-London: Plenum Press.

KATES, M., KUKSIS, A. (eds.), 1980: Membrane fluidity. Biophysical techniques and cellular regulation. Clifton, N.J.: Humana Press.

KELKER, H., HATZ, R., 1980: Handbook of liquid crystals. Weinheim: Verlag Chemie.

KESSELMEIER, J., BUDZIKIEWICZ, H., 1979: Identification of saponins as structural building units in isolated prolamellar bodies from etioplasts of *Avena sativa* L. Z. Pflanzenphysiol. 91, 333—344.

— RUPPEL, H. G., 1979: Relations between saponin concentration and prolamellar body structure in etioplasts of *Avena sativa* during greening and re-etiolating and in etioplasts of *Hordeum vulgare* and *Pisum sativum*. Z. Pflanzenphysiol. 93, 171—184.

KING, L. J., 1969: Biocrystallography—an interdisciplinary challenge. BioScience 19, 505—518.

KLEINIG, H., STEINKI, C., KOPP, C., ZAAR, K., 1978: Oleosomes (spherosomes) from *Daucus carota* suspension culture cells. Planta 140, 233—237.

KOBALE, M., KRÜGER, H., 1975: Flüssige Kristalle. Physik in uns. Zeit 6, 66—77.

KUHN, H., 1976: Model consideration for the origin of life. Naturwiss. 63, 68—80.

KÜHN, K., 1974: Struktur und Biochemie des Kollagens. Chemie in uns. Zeit 8, 97—103.

KWIATKOWSKA, M., 1973: Half unit membranes surrounding osmiophilic granules (lipid droplets) of the so-called lipotubuloid in *Ornithogalum*. Protoplasma 77, 473—476.

LEE, A. G., 1977: Annular events: lipid-protein interactions. TIBS 2, 231—233.

LICHTENTHALER, H. K., 1968: Plastoglobuli und die Feinstruktur der Plastiden. Endeavour 27, 144—149.

LIEDVOGEL, B., FALK, H., 1980: Leucoplasts mimicking membraneous chromoplasts. Z. Pflanzenphysiol. 98, 371—375.

— KLEINIG, H., THOMPSON, J. A., FALK, H., 1978: Chromoplasts of *Tropaeolum majus* L.: Lipid synthesis in whole organelles and subfractions. Planta 141, 303—309.

LIU, T.-P., 1974: Ultrastructure of the lipid inclusions of the yolk in the freeze-etched oocyte of an insect. Cytobiol. 8, 412—420.

LJUBEŠIĆ, N., 1973: Transformations of plastids in white pumpkin fruits. Acta Bot. Croat. 32, 59—62.

LUCY, J. A., GLAUERT, A. M., 1964: Structure and assembly of macromolecular lipid complexes composed of globular micelles. J. Mol. Biol. 8, 727—748.

LÜTZ, C., 1978: Separation and composition of prolamellar bodies and prothylakoids of etioplasts from *Avena sativa* L. In: Chloroplast development (AKOYUNOGLOU, G., et al., eds.), pp. 481—488. Amsterdam: Elsevier/North-Holland Biomed. Press.

— KLEIN, S., 1979: Biochemical and cytological observations on chloroplast development VI. Chlorophylls and saponins in prolamellar bodies and prothylakoids from etioplasts of etiolated *Avena sativa* L. leaves. Z. Pflanzenphysiol. 95, 227—237.

— KESSELMEIER, J., RUPPEL, H. G., 1977: Idem IV. Reaggregations of solubilized prolamellar bodies from etioplasts of *Avena sativa* L. Z. Pflanzenphysiol. 85, 327—340.

LUZZATI, V., HUSSON, F., 1962: The structure of the liquid-crystalline phases of lipid-water systems. J. Cell Biol. 12, 207—219.

MALKOFF, D., STREHLER, B., 1963: The ultrastructure of isolated and *in situ* human cardiac age pigment. J. Cell Biol. **16**, 611—616.

MARTIN, J. T., JUNIPER, B. E., 1970: The cuticles of plants. London: Edward Arnold.

MÉRIDA, T., SCHÖNHERR, J., SCHMIDT, H. W., 1981: Fine structure of plant cuticles in relation to water permeability: the fine structure of the cuticle of *Clivia miniata* Reg. leaves. Planta (in press).

MILLER, E. J., MATUKAS, V. J., 1974: Biosynthesis of collagen. Fed. Proc. **33**, 1197—1204.

MITTAL, K. L. (ed.), 1977: Micellization, solubilization and microemulsions, Vols. 1 and 2. New York: Plenum Press.

MOLLENHAUER, H. H., TOTTEN, C., 1971: Studies on seeds III. Isolation and structure of lipid-containing vesicles. J. Cell Biol. **48**, 533—541.

MOOR, H., MÜHLETHALER, K., 1963: Fine structure in frozen-etched yeast cells. J. Cell Biol. **17**, 609—628.

NORTON, G., HARRIS, J. F., 1975: Compositional changes in developing rape seed (*Brassica napus* L.). Planta **123**, 163—174.

OP DEN KAMP, J. A. F., 1979: Lipid asymmetry in membranes. Annu. Rev. Biochem. **48**, 47—71.

PAPAHADJOPOULOS, D., MILLER, N., 1967: Phospholipid model membranes I. Structural characteristics of hydrated liquid crystals. Biochim. Biophys. Acta **135**, 624—638.

PATTON, S., KEENAN, T. W., 1975: The milk fat globule membrane. Biochim. Biophys. Acta **415**, 273—309.

PERNER, E. S., 1958: Die Sphärosomen der Pflanzenzelle. (Protoplasmatologia III/A/2.) Wien: Springer.

PETERS, R., 1980: Translational diffusion in the plasma membrane of sea urchin eggs. Hoppe-Seyler's Z. physiol. Chem. **361**, 1605.

POSTE, G., NICOLSON, G. L. (eds.), 1978: Membrane fusion. (Cell Surface Rev., Vol. 5.) Amsterdam: North-Holland.

PRECHT, D., 1979: Die Mikrostruktur der Butter. Naturwiss. Rundschau **32**, 315—321.

REISS-HUSSON, F., 1967: Structure des phases liquide-cristallines de différents phospholipides, monoglycérides, sphingolipides, anhydres ou en présence d'eau. J. Mol. Biol. **25**, 363—382.

ROCKSTEIN, M. (ed.), 1974: The physiology of insecta, Vol. VI. (See, in particular, chapters 2—4, pp. 123—343.) New York-London: Academic Press.

RUPPEL, H. G., KESSELMEIER, J., LÜTZ, C., 1978: Biochemical and cytological observations on chloroplast development V. Reaggregations of prolamellar body tubules without proteins participation. Z. Pflanzenphysiol. **90**, 101—110.

RUSKA, C., RUSKA, H., 1969: Molekulare Schichtung in Tropfen von Speicherfett. Naturwiss. **56**, 332—333.

SANDERMANN jr., H., 1978: Regulation of membrane enzymes by lipids. Biochim. Biophys. Acta **515**, 209—237.

SAUPE, A., 1968: Neuere Ergebnisse auf dem Gebiet der flüssigen Kristalle. Angew. Chem. **80**, 99—106.

SAUTER, J. J., 1967: Untersuchungen zur zytochemischen Oxydase-Lokalisation in Sphärosomen verschiedener Gewebe von *Populus*. Z. Pflanzenphysiol. **57**, 352—367.

— 1968: Cytochemischer Nachweis von Redoxfermenten in Fetttropfen-assoziierten „Sphärosomen". Naturwiss. **55**, 351.

SCANU, A. M., 1972: Structural studies in serum lipoproteins. Biochim. Biophys. Acta **265**, 471—508.

— 1978: Plasma lipoproteins: structure, function, and regulation. TIBS **3**, 202—205.

SCHÖNHERR, J., 1976: Water permeability of isolated cuticular membranes: the effect of cuticular waxes on diffusion of water. Planta **131**, 159—164.

SCHUMACHER, G., SANDERMANN, jr., H., 1976: Solubility of phospholipid polar group model compounds in water. Biochim. Biophys. Acta **448**, 642—644.

SCHUSTER, P., 1972: Vom Makromolekül zur primitiven Zelle — die Entstehung biologischer Funktion. Chemie in uns. Zeit 6, 1—16.

SCHWARZENBACH, A. M., 1971: Observations on spherosomal membranes. Cytobiol. 4, 145—147.

SEYBOLD, A., 1942: Pflanzenpigmente und Lichtfeld als physiologisches, geographisches und landwirtschaftliches Problem. Ber. dtsch. bot. Ges. 60, (64)—(85).

SHEETZ, M. P., SINGER, S. J., 1974: Biological membranes as bilayer couples. A molecular mechanism of drug-erythrocyte interactions. Proc. Natl. Acad. Sci. U.S.A. 71, 4457—4461.

SHEN, B. W., SCANU, A. M., KÉZDY, F. J., 1977: Structure of human serum lipoproteins inferred from compositional analysis. Proc. Natl. Acad. Sci. U.S.A. 74, 837—841.

SITTE, P., 1962: Zum Feinbau der Suberinschichten im Flaschenkork. Protoplasma 54, 555—559.

— 1963: Hexagonale Anordnung der Globuli in Moos-Chloroplasten. Protoplasma 56, 197—201.

— 1974: Plastiden-Metamorphose und Chromoplasten bei Chrysosplenium. Z. Pflanzenphysiol. 73, 243—265.

— 1975: Die Bedeutung der molekularen Lamellen-Bauweise von Korkzellwänden. Biochem. Physiol. Pflanzen 168, 287—297.

— 1976 a: Elektronenmikroskopie und Biologie — Schicksal einer Symbiose. Mikroskopie 32, 145—190.

— 1976 b: Zur chemischen Fixierung von Lipidstrukturen. Mikroskopie 32, 208—209.

— 1977: Functional organization of biomembranes. In: Lipids and lipid polymers in higher plants (TEVINI, M., LICHTENTHALER, H. K., eds.), pp. 1—28. Berlin-Heidelberg-New York: Springer.

— 1979: General principles of cellular compartmentation. In: Cell compartmentation and metabolic channeling (NOVER, L., LYNEN, F., MOTHES, K., eds.), pp. 17—32. Jena: VEB G. Fischer Verlag and Amsterdam: Elsevier/North-Holland Biomed. Press.

— 1980 a: Electron microscopy and the understanding of life. In: Electron microscopy, Vol. 2 (BREDEROO, P., DE PRIESTER, W., eds.), pp. 818—825. Leiden.

— 1980 b: Nematic liquid crystals of lipid pigments in chromoplasts. Eur. J. Cell Biol. 22, 280.

— FALK, H., LIEDVOGEL, B., 1980: Chromoplasts. In: Pigments in plants, 2nd ed. (CZYGAN, F.-C, ed.), pp. 117—148. Stuttgart-New York: G. Fischer.

SMALL, D. M., 1970: Surface and bulk interactions of lipids and water with a classification of biologically active lipids based on these interactions. Fed. Proc. 29, 1320—1326.

SMELLIE, R. M. S. (ed.), 1971: Plasma lipoproteins. London-New York: Academic Press.

SMITH, C. G., 1974: The ultrastructural development of spherosomes and oil bodies in the developing embryo of Crambe abyssinica. Planta 119, 125—142.

SOROKIN, H. P., 1967: The spherosomes and the reserve fat in plant cells. Amer. J. Bot. 54, 1008—1016.

STEINSTRÄSSER, R., POHL, L., 1973: Chemie und Verwendung flüssiger Kristalle. Angew. Chem. 85, 706—720.

STOECKENIUS, W., 1959: An electron microscopic study of myelin figures. J. Biophys. Biochem. Cytol. 5, 491—500.

— 1962: Some electron microscopical observations on liquid-crystalline phases in lipid-water systems. J. Cell Biol. 12, 221—229.

— SCHULMAN, J. H., PRINCE, L. M., 1960: The structure of myelin figures and microemulsions as observed with the electron microscope. Kolloid-Z. 169, 170—180.

TANFORD, C., 1973: The hydrophobic effect: formation of micelles and biological membranes. New York: J. Wiley.

— 1978: The hydrophobic effect and the organization of living matter. Science 200, 1012—1018.

THOMSON, W. W., PLATT, K., 1973: Plastid ultrastructure in the barrel cactus, *Echinocactus acanthodes*. New Phytol. **72**, 791—797.

TRÄUBLE, H., 1971: Phasenumwandlungen in Lipiden. Mögliche Schaltprozesse in biologischen Membranen. Naturwiss. **58**, 277—284.

TURRO, N. J., GRÄTZEL, M., BRAUN, A. M., 1980: Photophysikalische und photochemische Prozesse in micellaren Systemen. Angew. Chem. **92**, 712—734.

WANNER, G., THEIMER, R. R., 1978: Membraneous appendices of spherosomes (oleosomes). Planta **140**, 163—169.

— FORMANEK, H., THEIMER, R. R., 1981: The ontogeny of lipid bodies (spherosomes) in plant cells. Planta **151**, 109—123.

WEHRMEYER, W., 1965: Zur Kristallgitterstruktur der sogenannten Prolamellarkörper in Plastiden etiolierter Bohnen, I—III. Z. Naturforsch. **20 b**, 1270—1296.

WEINSTOCK, M., LEBLOND, C. P., 1974: Formation of collagen. Fed. Proc. **33**, 1205—1218.

WIESER, W. (ed.), 1973: Effects of temperature on ectothermic organisms. Berlin-Heidelberg-New York: Springer.

WILLIAMS, M. C., 1977: Conversion of lamellar body membranes into tubular myelin in alveoli of fetal rat lungs. J. Cell Biol. **72**, 260—277.

WINKENBACH, F., FALK, H., LIEDVOGEL, B., SITTE, P., 1976: Chromoplasts of *Tropaeolum majus* L.: Isolation and characterization of liproprotein elements. Planta **128**, 23—28.

WUTTKE, H.-G., 1976: Chromoplasts in *Rosa rugosa*: Development and chemical characterization of tubular elements. Z. Naturforsch. **31 c**, 456—460.

YATSU, L. Y., JACKS, T. J., 1972: Spherosome membranes. Half unit-membranes. Plant Physiol. **49**, 937—943.

— — HENSARLING, T. P., 1971: Isolation of spherosomes (oleosomes) from onion, cabbage, and cottonseed tissues. Plant Physiol. **48**, 675—682.

ZWAAL, R. F. A., DEMEL, R. A., ROELOFSEN, B., VAN DEENEN, L. L. M., 1976: The lipid bilayer concept of cell membranes. TIBS **1**, 112—114.

Subject Index